# Lecture Notes in Computer Science 2768

Edited by G. Goos, J. Hartmanis, and J. van Leeuwen

Springer

*Berlin*
*Heidelberg*
*New York*
*Hong Kong*
*London*
*Milan*
*Paris*
*Tokyo*

Michael J. Wilson   Ralph R. Martin (Eds.)

# Mathematics of Surfaces

10th IMA International Conference
Leeds, UK, September 15-17, 2003
Proceedings

 Springer

Series Editors

Gerhard Goos, Karlsruhe University, Germany
Juris Hartmanis, Cornell University, NY, USA
Jan van Leeuwen, Utrecht University, The Netherlands

Volume Editors

Michael J. Wilson
The University of Leeds
School of Mathematics
Leeds LS2 9JT, UK
E-mail: mike@maths.leeds.ac.uk

Ralph R. Martin
Cardiff University
Department of Computer Science
5, The Parade, P.O. Box 916, Cardiff, Wales, CF24 3XF, UK
E-mail: ralph@cs.cf.ac.uk

Cataloging-in-Publication Data applied for

A catalog record for this book is available from the Library of Congress.

Bibliographic information published by Die Deutsche Bibliothek
Die Deutsche Bibliothek lists this publication in the Deutsche Nationalbibliografie;
detailed bibliographic data is available in the Internet at <http://dnb.ddb.de>.

CR Subject Classification (1998): I.3.5, I.3, I.1, G.2, G.1, F.2, I.4

ISSN 0302-9743
ISBN 3-540-20053-3 Springer-Verlag Berlin Heidelberg New York

Springer-Verlag Berlin Heidelberg New York
a member of BertelsmannSpringer Science+Business Media GmbH

http://www.springer.de

© Springer-Verlag Berlin Heidelberg 2003
Printed in Germany

Typesetting: Camera-ready by author, data conversion by Olgun Computergrafik
Printed on acid-free paper      SPIN: 10930991      06/3142      5 4 3 2 1 0

# Preface

This volume collects the papers accepted for presentation at the 10th IMA Conference on the Mathematics of Surfaces, held at the University of Leeds, UK, September 15–17, 2003. As with all earlier conferences in the series, contributors to this volume come from a wide variety of countries in Asia, Europe and North America. The papers presented here reflect the continued relevance of the subject, and, by comparison with the contents of earlier volumes in the series, the changing nature of application areas and mathematical techniques. To give the casual browser of these proceedings an idea of the diversity of the subject area, the papers in the present volume cover such topics and techniques as digital geometry processing, computer graphics, surface topology, medical applications, subdivision surfaces, surface reconstruction, surface triangulation, watermarking, data compression, data smoothing, human-computer interaction, extracting shape form shading, surface height recovery, reverse engineering, box-splines, the Plateau Problem, splines (a variety of papers), transfinite blending, and affine arithmetic. There is also a paper in memory of the late Prof. Josef Hoschek of the Technische Universität Darmstadt, co-authored by a former research student, Prof. Bert Jüttler, on the subject of using line congruences for parameterizing special algebraic surfaces.

We would like to thank all those who attended the conference and helped to make it a success. In particular we thank those who contributed to these proceedings. We are particularly grateful to Lucy Nye (at the Institute of Mathematics and Its Applications) for help in making arrangements, and to Alfred Hofmann and Christine Günther of Springer-Verlag for their help in publishing this volume. Following this preface is a list of distinguished researchers who expended much effort in refereeing the papers for this proceedings. Due to their work, many of the papers were considerably improved. Our thanks go to all of them and particularly to our fellow member of the Organizing Committee, Malcolm Sabin.

June 2003

Mike Wilson
Ralph Martin
Editors of these Proceedings

# Organization

The Mathematics of Surfaces X was organized by the Insitute of Mathematics and Its Applications (Catherine Richards House, 16 Nelson St., Southend-on-Sea, Essex, SS1 1EF, UK), and the University of Leeds, UK.

## Organizing Committee

Ralph Martin (Cardiff University, UK)
Malcolm Sabin (Numerical Geometry Ltd., UK)
Michael Wilson (University of Leeds, UK)

## Program Committee

C. Armstrong (Queen's University Belfast, UK)
N. Dyn (Tel Aviv University, Israel)
R. Farouki (University of California, Davis, USA)
A. Fitzgibbon (University of Oxford, UK)
M. Floater (Sintef, Norway)
D. Gossard (MIT, USA)
H. Hagen (University of Kaiserslautern, Germany)
M.-S. Kim (Seoul National University, Korea)
F. Kimura (University of Tokyo, Japan)
D. Kriegman (University of Illinois, USA)
T. Lyche (University of Oslo, Norway)
D. Manocha (University of North Carolina, USA)
R. Martin (Cardiff University, UK)
N. Patrikalakis (MIT, USA)
H. Pottmann (Technical University of Vienna, Austria)
M. Pratt (LMR Systems, UK)
M. Sabin (Numerical Geometry, UK)
K. Sugihara (University of Tokyo, Japan)
C. Taylor (Manchester University, UK)
M. Wilson (University of Leeds, UK)
W. Wang (Hong Kong University, China)

# Table of Contents

# Skeleton-Based Seam Computation for Triangulated Surface Parameterization

Xu-Ping Zhu[1], Shi-Min Hu[1], and Ralph Martin[2]

[1] Department of Computer Science and Technology, Tsinghua University
Beijing 100084, P. R. China
zhuxp@cg.cs.tsinghua.edu.cn, shimin@tsinghua.edu.cn
[2] Department of Computer Science, Cardiff University, Cardiff, CF24 3XF, UK
ralph@cs.cf.ac.uk

**Abstract.** Mesh parameterization is a key problem in digital geometry processing. By cutting a surface along a set of edges (a seam), one can map an arbitrary topology surface mesh to a single chart. Unfortunately, high distortion occurs when protrusions of the surface (such as fingers of a hand and horses' legs) are flattened into a plane. This paper presents a novel skeleton-based algorithm for computing a seam on a triangulated surface. The seam produced is a full component Steiner tree in a graph constructed from the original mesh. By generating the seam so that all extremal vertices are leaves of the seam, we can obtain good parametrization with low distortion.

## 1 Introduction

Due to their flexibility and efficiency, triangle meshes have been widely used in the entertainment industry to represent arbitrary surfaces for 3D games and movies during the last few years. Many new mesh techniques have been developed for different applications. In these techniques, parameterization is a key issue—surface parameterization is always an important problem in computer graphics. In this paper, we focus on the parameterization of triangle meshes. In particular, we wish to establish a mapping between a triangulated surface and a given domain.

Parameterization methods can be grouped into three categories, depending on whether the domain is a polyhedron, sphere or plane. Eck et al [6] first used a polyhedron as the parameter domain for a mesh. They called the domain polyhedron a *base-mesh*. A mesh parameterized using the base-mesh is called a semi-regular mesh; such an approach has been widely applied in multi-resolution modelling [14], remeshing [2] and morphing [13]. Recently, Khodakovsky et al [12] presented a globally smooth parameterization algorithm for semi-regular meshes. In contrast with polyhedral parameterization, there is little work [9,17] on spherical parameterization because of its limited application. Finally, we focus on planar parameterization. As is done for parametric surfaces, we want to map the mesh to a bounded region of the plane. This is trivial if the mesh surface is homeomorphic to a disc. Unfortunately, if the topology of the surface is complex, with

M.J. Wilson and R.R. Martin (Eds.): Mathematics of Surfaces 2003, LNCS 2768, pp. 1–13, 2003.

high genus, the surface is usually separated into several disc-like patches, and each patch mapped to a chart. This multi-chart method is often used in texture mapping [15]. Obviously, the more charts into which a surface is separated, the lower the distortion is, and in the extreme, if each triangle of the mesh has its own chart, the distortion is zero. Conversely, using fewer charts causes higher distortion. Hence, there exists a trade-off between the number of charts and the amount of distortion.

Our goal is to map an arbitrary topology surface mesh to a single chart with low distortion. Theoretically, any closed surface can be opened into a topological disc using a set of cut edges making up a *seam*. Cutting along the seam, we can get a disc-like patch. Unless the surface is developable, parameterization using a single chart inevitably creates distortion, which is especially great where protrusions of the surface (such as fingers of a hand and horses' legs) are flattened into the plane. Sheffer [20] and Gu et al [10] have found that to reduce the distortion, it is important for the seam to pass through the various so-called "extrema". Although extrema can be found accurately with their methods, it is still difficult to guide the seam through these extrema. In fact, it is a tough job even for human beings to choose an appropriate seam. Piponi et al [16] describe a system that allows the user to define a seam interactively. In this paper, we consider this problem as a minimal Steiner tree problem and present a novel skeleton-based method to approximate the minimal Steiner tree.

## 2   Related Work

Our goal is to map an arbitrary 2-manifold surface to a planar domain, so high genus surfaces must be cut by a set of edges or a seam. The problem of minimizing the length of the set of cut edges is known to be NP-hard [7]. Gu et al [10] present a method that approximates such a seam in $O(n \log n)$ time. Once a surface has been cut into a disc, additional cuts are usually necessary to reduce distortion. Choosing them includes two processes: firstly, all extremal vertices on the mesh must be detected; secondly, a path (always a tree) connecting all extremal vertices and the surface boundary must be found. Sheffer [20] detects extrema by searching for vertices with high curvature. Unfortunately, this method only solves problems associated with local protrusions. Gu et al give a more suitable algorithm for finding extrema which utilizes the shape-preserving feature of Floater's parameterization [8]. For each protrusion, he identifies the triangle with maximum stretch in the parametric domain, and picks one of its vertices as a new extremal vertex to augment the seam.

In this paper, we assume all extremal vertices are already given and concentrate on determining the layout of the seam. We pose the problem as a network (weighted graph) *Steiner tree* problem. A Steiner tree for a set of vertices (terminals) in a graph is a connected sub-graph spanning all terminals. Here terminals include all extremal vertices, and the boundary (the whole boundary is regarded as a terminal, but only one point on the boundary is required to be connected to the seam). Naturally, we want to minimize the length of the seam because it will eventually be added to the boundary of the parameterization. The problem of finding the minimal Steiner tree in a graph is NP-complete [11]. Sheffer [20] approximates it by the minimum spanning tree (MST). The Euclidean MST is never more than $2/\sqrt{3}$ times the length of the Euclidean Steiner minimal tree; but in a graph the corresponding worst-case approximation ratio is 2, not $2/\sqrt{3}$ [11]. Gu et al use a greedy approximation method: when a new extremal vertex is detected, the shortest path between the current seam and the new extremal vertex is added to the seam. This incremental method depends heavily on the sequence of adding extremal vertices; choice of the sequence is not discussed in his paper. Consider the model of a horse in Fig. 1. The most obvious extrema are located at the ends of the four legs and the head, and can be detected by Gu's method. The seam computed using the MST method is poor (See Fig. 1(a)), as the MST passes through most extrema more than once. However, extremal vertices are always at the ends of protrusions, and so it is reasonable to require that the seam should pass through each extremal vertex as few times as possible. It will be best if all extremal vertices are leaves on the seam.

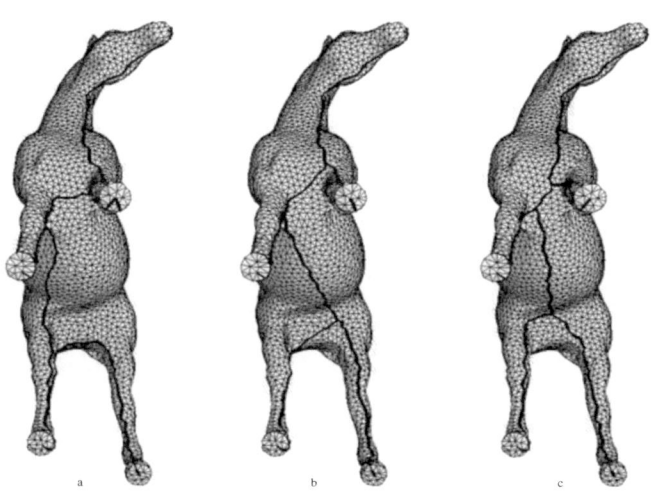

**Fig. 1.** Seam computed by: (a) Sheffer's method (seam length = 7.99), (b) Gu's method (seam length = 7.07), (c) our method (seam length = 5.97).

For this reason, we constrain the Steiner tree to be a *full component* of the mesh. (A full component is a subtree in which each terminal is a leaf). Robins [18] gives a method for constructing an approximation to a minimum Steiner tree in a weighted graph, with length approaching $1 + (\ln 3)/2 \approx 1.55$ times the minimal length. However, his method is not efficient if the Steiner tree is constrained to be a full component. Thus, in this paper we suggest a new method to compute an approximation to the minimal full component Steiner tree, deriving it from the *straight skeleton*. It produces good results in practice: comparisons with previous methods are given in Fig. 1.

The rest of this paper is organized as follows. In Section 3, we introduce the definition of straight skeleton. Our novel seam computation algorithm is presented in Section 4. Section 5 shows some experimental results of applying our new method to different models. Conclusions and future work are given in Section 6.

## 3    The Straight Skeleton

The *straight skeleton* of a planar straight-edged graph $G$ is obtained as the interference pattern of certain wavefronts propagated from the edges of $G$ [1]. If $G$ is a simple polygon, the part of the straight skeleton interior to $G$ is a tree whose leaves are the vertices of $G$ (see Fig. 2). If all vertices of the polygon are terminals, the interior straight skeleton must be a full component Steiner tree. Thus, the main idea of our method is to extend the concept of straight skeleton from the plane to a 3D triangle mesh. To do this, we must solve two problems. First, a shortest tour that visits all terminals is found. (Of course, one can also use other heuristics—see the discussion in Section 6). Secondly, the tour is shrunk to produce the skeleton which is a full component Steiner tree of terminals.

In the plane, the straight skeleton is found by shrinking a polygon via inward, uniform speed parallel movement of all edges, and finding where they intersect. On a triangle mesh, we analogously treat the part of the tour between two neighbouring terminals as an "edge" and shrink all "edges" inward in a uniform-speed way, as explained in Section 4.3.

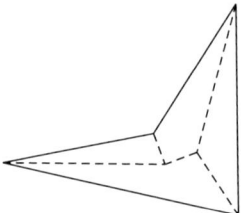

**Fig. 2.** The straight skeleton interior to a simple polygon.

## 4    Seam Computation

Before going into details of our algorithm, we now restate the problem. Given an arbitrary 2-manifold triangulated surface (closed or open) and several extremal vertices on it, we seek a minimum length seam connecting all extremal vertices which opens the surface into a disc. In practice, better results are obtained if all extremal vertices are leaves on the seam, so in principle we wish to compute a minimal full component Steiner tree. Because this is expensive, we approximate it using our new skeleton-based algorithm, which is explained in the following section.

### 4.1   Overview of the Algorithm

We carry out the following preprocessing before computing the seam:

- For surfaces of genus greater than zero, we use Gu's algorithm to compute an initial seam. Cutting along the initial seam, the surface is opened into a disc. After this process, the surface is either a closed genus-0 surface or a disc-like patch with one boundary loop.
- We find the shortest path between each pair of terminals, i.e. the shortest path between each pair of extremal vertices and the shortest paths between each extremal vertex and the boundary loop (if any). To compute the shortest paths, a modified Dijkstra's algorithm is used, as in [20].
- Our algorithm assumes that there are at least two terminals. If there are only two terminals, we just use the shortest path between them as the seam.

The Algorithm for computing the seam can be divided into three steps. Fig. 3 shows the results of each step.

- Find the shortest tour that visits all terminals.
- Shrink the tour to a skeleton.
- Straighten the skeleton.

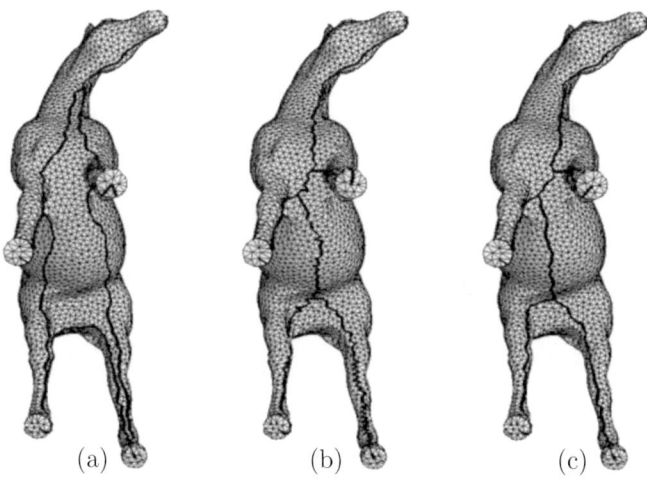

**Fig. 3.** Results after each step of seam computation.

### 4.2   Constructing a Tour with MST

For convenience, we use a half-edge data structure to describe the triangle mesh, i.e. we replace each edge with two opposed half-edges; the three half-edges inside a triangle are ordered anticlockwise (see Fig. 4(a)). With this data structure, the mesh can be represented by a directed weighted graph $G$ where the weight is the length of the path between terminals (see Fig. 4(b)). A directed tour must be found in $G$ which satisfies two requirements.

- The tour must visit all terminals exactly once.
- The tour must not self-intersect, i.e. the tour must separate the graph into two regions.

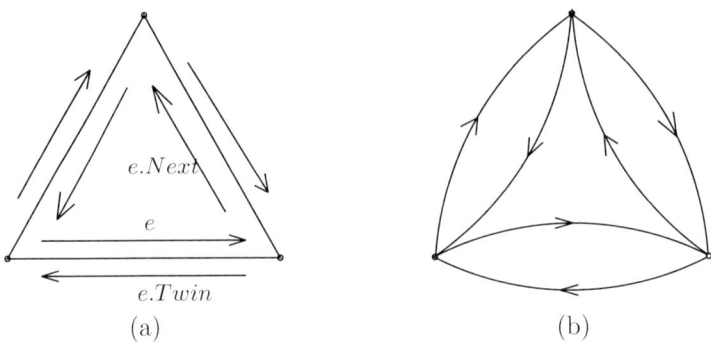

(a)                                                   (b)

**Fig. 4.** Half-edge data structure and the corresponding directed weighted graph.

Since the tour is directed, we define the region on the right hand side of the tour to be the interior of the tour. The first requirement is straightforward. Consider the second requirement. Fig. 5 shows two directed tours. They both satisfy the first requirement. We say that the tour on the left-hand side of the Figure is self-intersecting while the tour on the right-hand side is not. In the Figure, nodes marked "○" are terminal nodes, while nodes marked "●" are non-terminals. We may decide whether a tour is self-intersecting by the following algorithm.

**Algorithm 1 (Tour Self-intersection Test).**

```
For each half-edge e_i on a directed tour T
{
    Let e_j be the next adjacent edge to e_i in T
    e_k = e_i.Next; // Next half-edge in the data structure (see Fig. 4)
    While (e_k ≠ e_j ) // If e_k is not the next edge of the tour
    {
        If e_k is a half-edge in T
            return; // T is self-intersecting
        Else
        {
            e_k = e_k.Twin;
            e_k = e_k.Next;
            // Move to the next half-edge around the vertex
            // at the end of e_i (see Fig. 6)
        }
    }
}
```

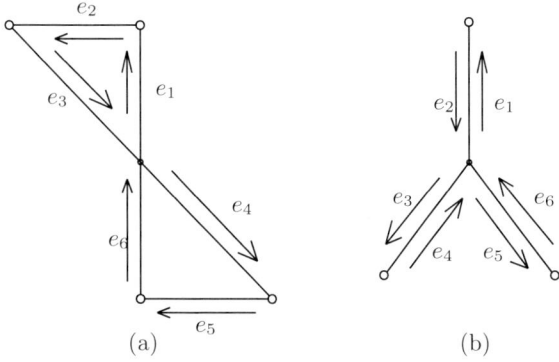

**Fig. 5.** Self-intersecting and non-self-intersecting tours.

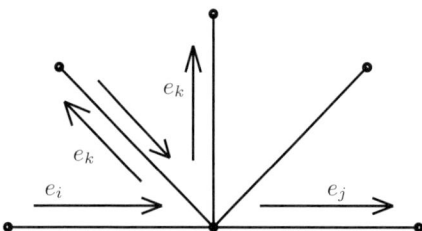

**Fig. 6.** Tour self-intersection testing.

We wish to find the shortest tour which satisfies the two requirements above. This is similar to the classic traveling salesman problem (TSP) which asks for a shortest tour that visits all nodes of a graph. Here we only restrict the tour to visiting all terminals. Since the part of the shortest tour between two terminals must be the shortest path between them, we can construct a complete graph $G_T$ based on all terminals. The weight of each edge in $G_T$ is the length of the shortest path between each pair of corresponding terminals. Thus we can reduce the problem to a traveling salesman problem in $G_T$. Unfortunately, $TSP$ is also NP-Hard [11]. Thus we use an approximate method, to be able to handle a large number of terminals. Christofides's algorithm [3] for $TSP$ approximation starts by computing a minimum spanning tree. Using the $MST$ alone gives a tour with length no more than twice the minimal answer. To improve the approximation, Christofides [3] uses minimum-length matching on the odd-degree vertices of the $MST$. Although this method improves the approximation ratio to 3/2, it can not guarantee the tour is not self-intersecting.

Thus, we use the basic $MST$ method together with a heuristic which not only approximately minimizes the tour, but also avoids self-intersections of the tour. The idea is as follows. Firstly we compute the minimal spanning tree of $G_T$. If we simply walk around the tree in half-edge order, we can obtain a directed tour which is not self-intersecting in $G$ (See Fig. 6(b)). However, this tour does not satisfy the requirement of visiting each terminal only once, and repeated visits to terminals must be eliminated from the tour. Starting from this initial tour, we remove repeat visits to terminals one-by-one, until all terminals appear in the tour only once. Let $t$ be a terminal visited more than once in the current tour, $t \rightarrow Pre$ be the terminal before $t$ in the tour and $t \rightarrow Next$ be the following terminal in the tour. When a visit to $t$ is removed, the part of the tour from $t \rightarrow Pre$ to $t \rightarrow Next$ is replaced by the shortest path from $t \rightarrow Pre$ to $t \rightarrow Next$. This operation may produce a self-intersection in the tour. Thus, each time we remove a visit to a terminal, we must check whether the resulting tour is self-intersecting. To decide which terminal to update, a heuristic rule is used. We choose the terminal $t$, and its $Pre$ and $Next$, with minimal shortest path between $t \rightarrow Pre$ and $t \rightarrow Next$. We now give the pseudocode of the algorithm as follows; Fig. 3(a) shows the resulting tour produced for the horse model.

**Algorithm 2 (Computing a Tour Using MST).**

   Use Kruskal's algorithm [4] to compute the $MST$ of $G_T$.
   Walk around the tree starting at any terminal.
   Record the sequence of terminals in a circular list $TourList$.

   While (size of $TourList$ > number of terminals)
   //Repeated terminals remain in the tour
   {
         For each repeated terminal $t$ in $TourList$
         {
               If the tour is not self-intersecting when the appropriate visit
               to $t$ is removed
                     $t.priority$ = Length of the shortest path between
                     $t \rightarrow Pre$ and $t \rightarrow Next$
               Else
                     $t.priority = \infty$
         }
         Remove the visit with minimum priority from $TourList$
   }

### 4.3    Shrinking the Tour to a Skeleton

As outlined in Section 3, we treat the part of the tour between two neighbouring terminals as an "edge" and shrink all "edges" inward in a uniform-speed way across the triangulation. The skeleton is the interference pattern of wavefronts propagated from these "edges". To carry out this idea, we mark each "edge" with a different number and propagate these numbers stepwise to the triangulation of the interior of the tour. Places where two different "edge" numbers meet are parts of the skeleton. Pseudocode for the algorithm for doing this follows; $q$ is a queue and all variables are explained in Fig. 7. The principle is to process all half edges adjacent to the tour first by placing them in a queue. Then, as their neighbours are processed, they are also added to the end of the queue, then their neighbours are processed in turn, and so on. Eventually, edges both of whose

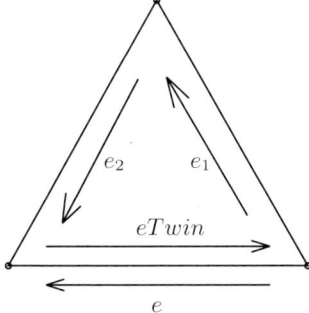

**Fig. 7.** Mark propagation in skeleton computation.

half-edges have already been marked with different numbers are encountered:
these are parts of the skeleton (see Fig. 3(b)).

**Algorithm 3 (Skeleton Computation).**

Order all terminals according to their sequence in the tour.

For each terminal $t_i$,

    For each triangulation half-edge $e$ on the shortest path
    between $t_i$ and $t_{i+1}$,

        $e.mark = i$;  $q.append(e)$;

//Shrink the tour

While ($q$ is not empty)

{

    $e = q.takefirst()$;

    If ($e$ is a boundary edge)

        continue; // Ignore it. Otherwise

    $eTwin = e.Twin$;

    If ($eTwin$ is not marked)

    // Then propagate the mark to the twin half-edge

    {

        $e_1 = eTwin.Next$;

        $e_2 = e_1.Next$;

        $eTwin.mark = e_1.mark = e_2.mark = e.mark$;

        $q.append(e_1)$;

        $q.append(e_2)$;

    }

    Else If ($eTwin.mark \neq e.mark$)

        Export $e$; // This edge is part of the skeleton

}

### 4.4   Straightening the Skeleton

The skeleton as produced above is always jagged and should be smoothed to
produce a better result. As intended, the skeleton is a tree spanning all terminals.

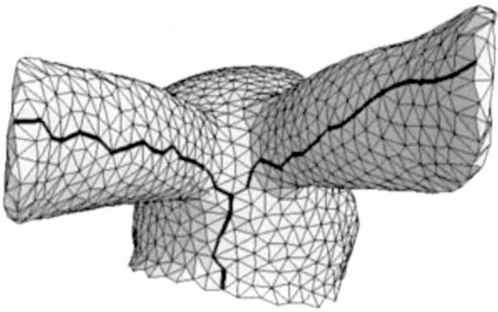

**Fig. 8.** Computed seam on an open model (a bunny head).

While all terminals are leaves of the tree, some interior vertices have valence greater than two. Such vertices are called *Steiner points*. We fix the positions of all Steiner vertices and all extremal vertices, and replace the skeleton path between each pair of such vertices by the shortest path connecting them in the triangulation (see Fig. 3(c)).

Finally, the set of smoothed paths is output as the seam.

## 5   Results

The following examples show seams computed by our method. Fig. 8 shows a bunny head model (with boundary). In this example, the three terminals include two extremal vertices at the tops of both ears, and a boundary loop. The seam computed by our method is almost the shortest path.

In Fig. 9, we demonstrate seams on a complex model produced by manually adding increasing numbers of extremal vertices. The model has 10062 facets. It only takes 1.6 seconds to compute the final seam (Fig. 9(f)). The bottleneck in our algorithm is computing the shortest path between any two terminals, which takes $O(E \log N)$ time [20], where $E$ is the number of edges and $N$ is the number of vertices in the mesh.

In Fig. 10(a), we can see the result is good even if the selected extremal vertices are not on protrusions. For such a head model, it is not appropriate for the seam to pass through the face. By using other heuristic rules when we construct the tour, the user can interactively guide seam generation (see the discussion in Section 6).

## 6   Conclusion and Future Work

The motivation for this paper was to reduce distortion when a 2-manifold tri-angulated surface is mapped to a single chart in the plane. Most previous ap-proaches reduce distortion by optimizing an energy function [5,19]. Unlike these

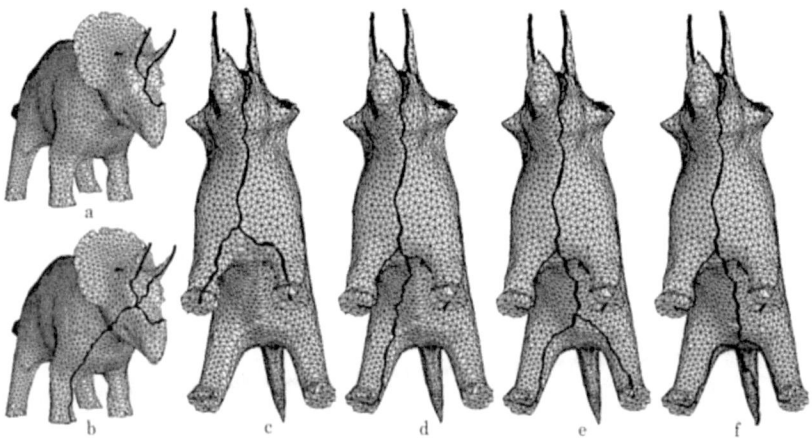

**Fig. 9.** Seams found by interactively adding new extremal vertices.

methods, we achieve the goal by finding an additional seam. Through experience, we have found that to get a good mapping, it is important for the seam to pass through all extrema just once. Such a seam is a constrained Steiner tree in which all terminals are leaves. Inspired by the straight skeleton of a simple planar polygon, we have presented a novel method for computing the seam. From Fig. 1, we can see that our method leads to shorter seams than previous methods. Furthermore, seams computed with our method treat extrema in a more-or-less symmetrical way, and produce seams similar to those produced by human intuition when mapping textures to a surface.

In Section 4.1, a heuristic rule was used to determine the seam: minimize the tour length. Now, the seam definitely lies on the part of the triangulation surrounded by the tour, and so we can use other heuristics based on this to find the seam. For example, we could minimize the total energy of all triangles interior to the tour, where the triangle energy may be defined in terms of area of triangle, or in terms of "visibility" [21]. Fig. 10 is the result when we define triangle energy as the dot product between the normal vector and the viewing direction. Using such a rule, the user can control the position of the seam according to the desired viewing direction.

Our method relies on correctly knowing all extremal vertices. If extremal vertices are found corresponding to all *large* protrusions, we can obtain very good results, as illustrated in Fig. 9. It is less important whether further extremal vertices corresponding to *small* protrusions are also provided. If not, the results are usually still acceptable (see Fig. 10). However, if an extremal vertex for some large protrusion is not provied, the seam may pass through it, resulting in a bad seam. Thus, in future work, we hope to develop a multi-resolution algorithm for selecting extremal vertices more carefully. We hope to first find all big protrusions at low resolution and then locate extremal vertices precisely at high resolution.

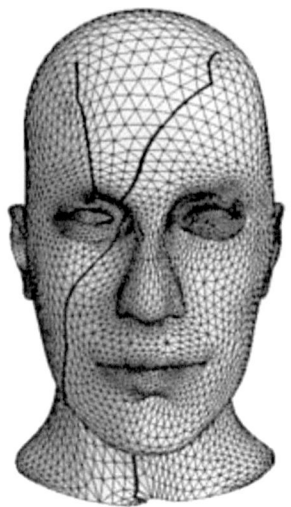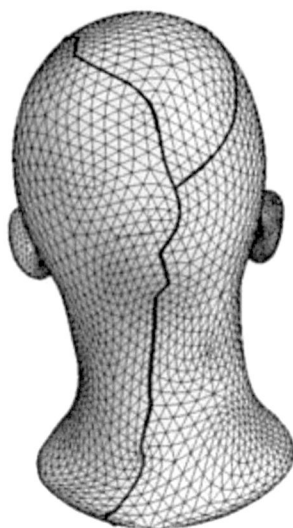

**Fig. 10.** Seams computed for different view points.

## Acknowledgement

This work was partly supported by the Natural Science Foundation of China (Grant 60225016), the Specialized Research Fund for the Doctoral Program of Higher Education (Grant 20020003051) and the National Basic Research Project of China (Grant 2002CB312102). We especially thank the reviewers for their helpful comments.

## References

1. Aichholzer O., and Aurenhammer F.: Straight skeletons for general polygonal figures in the plane. In: Proceedings of 2nd International Conference on Computing and Combinatorics, Hong Kong, (1996) 117–126.
2. Allize P., Meyer M. and Desbrun M.: Interactive geometry remeshing. In: Proceedings of SIGGRAPH 2002. ACM Transaction on Graphics, Vol. 21–3 (2002) 347–354.
3. Christofides N.: Worst-case analysis of a new heuristic for the traveling salesman problem. Technical Report 388, Graduate School of Industrial Administration, Carnegie Mellon University, Pittsburgh, 1976.
4. Cormen T. H., Lieseron C. E., and Rivest R. L.: Introduction to Algorithms. MIT Press, Cambridge, 2000.
5. Desbrun, M., Meyer, M., Alliez, P.: Intrinsic parameterizations of surface meshes. In: Proceedings of Eurographics 2002. Computer Graphics Forum 21–3 (2002) 209–218.
6. Eck, M., Derose, T., Duchamp, T., Hoppe, H., Lounsbery, M., and Stuetzle W.: Multiresolution Analysis of Arbitrary Meshes. In: Proceedings of SIGGRAPH 95. Computer Graphics 29 (1995) 173–182.

7. Erickson, J., and Har-peled, S.: Cutting a surface into a disk. In: Proceedings of the 18th ACM Symposium on Computational Geometry (2002), 244–253, ACM Press.
8. Floater, M.: Parameterization and smooth approximation of surface triangulations. Computer Aided Geometric Design 14–3 (1997) 231–250.
9. Gotsman, C., Gu, X., and Sheffer, A., Fundamentals of spherical parameterization for 3D meshes. In: Proceedings of SIGGRAPH 2003. ACM Transaction on Graphics, Vol. 22–3 (2003).
10. Gu, X., Gortler, S. and Hoppe, H.: Geometry images. In: Proceedings of SIGGRAPH 2002. ACM Transaction on Graphics, Vol. 21–3 (2002) 355–361.
11. Hochbaum S.: Approximation Algorithms for NP-hard Problems, PWS Publishing Company, 1997.
12. Khodakovsky, A., Litke, N. and Schröder, P.: Globally smooth parameterizations with low distortion, In: Proceedings of SIGGRAPH 2003. ACM Transaction on Graphics, Vol. 22-3 (2003).
13. Lee, A., Dobkin, D., Sweldens, W. and Schröder, P.: Multiresolution mesh morphing. In: Proceedings of SIGGRAPH 99, Computer Graphics 33(1999) 343–350.
14. Lee, A., Sweldens, W., Schröder, P., Cowsar, L., and Dobkin, D.: MAPS: Multiresolution adaptive parameterization of surfaces. In: Proceedings of SIGGRAPH 1998. Computer Graphics 32 (1998) 95–104.
15. Lévy, B., Petitjean, S., Ray, N., and Maillot, J.: Least squares conformal maps for automatic texture atlas generation. In: Proceedings of SIGGRAPH 2002. ACM Transaction on Graphics, Vol. 21-3 (2002) 362–371.
16. Piponi, D., and Borshukov, G. D.: Seamless texture mapping of subdivision surfaces by model pelting and texture blending. In: Proceedings of SIGGRAPH 2000. Computer Graphics 34 (2000) 471–478.
17. Praum, E. and Hoppe, H.: Spherical parameterization and remeshing. In: Proceedings of SIGGRAPH 2003. ACM Transaction on Graphics, Vol. 22-3 (2003).
18. Robins, G. and Zelikovsky, A.: Improved Steiner tree approximation in graphs. Proceedings of the 11th annual ACM-SIAM Symposium on Discrete Algorithms, San Francisco, California, (2000) 770–779.
19. Sander, S., Snyder, J., P., Gortler and Hoppe, H.: Texture mapping progressive meshes. In: Proceedings of SIGGRAPH 2001. Computer Graphics 35 (2001) 409–416.
20. Sheffer, A.: Spanning tree seams for reducing parameterization distortion of triangulated surfaces. In Geoff Wyvill(eds): Proceedings of Shape Modelling International 2002, IEEE CS Press (2002) 61–68.
21. Sheffer, A. and Hart, J. Seamster: Inconspicuous low-distortion texture seam layout. Proceedings IEEE Visualization 2002, IEEE CS Press, (2002) 291–298.

# Parameterizing $N$-Holed Tori

Cindy Grimm[1] and John Hughes[2]

[1] Washington University in St. Louis, Box 1045, St. Louis, MO
cmg@cs.wustl.edu
http://www.cs.wustl.edu/~cmg
[2] Brown University, Box 1910, Providence, RI
jfh@cs.brown.edu
http://www.cs.brown.edu/~jfh

**Abstract.** We define a parameterization for an $n$-holed tori based on the hyperbolic polygon. We model the domain using a manifold with $2n + 2$ charts, and linear fractional transformations for transition functions. We embed the manifold using standard spline techniques to produce a surface.

**CR Categories:** I.3.5 [Computer Graphics]: Computational Geometry and Object Modeling, Curve, Surface, Solid, and Object Representations, Splines, n-holed tori, hyperbolic octagon, linear fractional transformation

## 1   Introduction

We present a method for constructing $n$-holed tori for use in computer graphics. The requirements are that the construction be *explicit*: it is not sufficient to know that we have some description of a manifold known to be homeomorphic to an $n$-holed tori. Instead, we need explicit charts in $\mathbb{R}^2$ and transition functions that are easily computable.

We take a two step approach: first, we take a classical description of an $n$-holed torus and make that description explicit. Next, we use traditional planar embedding techniques, such as splines, to create an embedding of the $n$-holed torus. We differ from previous approaches [RP99,WP97] in that we cover the domain with $2n + 2$ charts, instead of modeling the domain as a single entity. The embedding is then built by blending between $2n+2$ individual, planar chart embeddings, rather than making a single, multiperiodic embedding function.

Our construction method is of interest from both a mathematical standpoint and an implementation one. Modeling arbitrary topology surfaces, especially ones with higher-order continuity, continues to be a challenging problem. The solution presented here is a computationally tractable natural parameterization of $n$-holed tori, and has many of the desirable features of traditional modeling techniques such as splines.

We begin with a global parameterization technique from topology [Lef49] which takes a $4n$-sided polygon and produces an $n$-holed torus by associating edges of the polygon. By recognizing (following classical complex analysis) that this polygon can be embedded in the unit disk so that the associations on the

M.J. Wilson and R.R. Martin (Eds.): Mathematics of Surfaces 2003, LNCS 2768, pp. 14–29, 2003.

edges are those induced by a group action on the disk, we see immediately that the object produced is indeed a complex manifold: our manifold structure, with $2n + 2$ charts, has transition functions that are all linear fractional transformations (see Section 4).

Once we have the explicit description of the manifold, we embed it into $\mathbb{R}^3$ using standard, planar embedding techniques (see Section 5).

# 2   Related Work

The modeling of arbitrary topology surfaces has received a great deal of attention. The approaches range from parameterized ones such as hole filling with $n$-sided patches [Ld89,Pet96], subdivision surfaces [CC78], and manifolds [GH95,Gri02,NG00], to implicit methods [Blo97] and other volumetric approaches [CBC+01,FPRJ00]. $n$-holed tori are easy to build using implicit methods, but they lack a parameterization.

Research in the area of hole filling and networks of patches is extensive, and beyond the scope of this paper to cover in detail. The basic approach is to apply constraints to the geometry of neighboring patches in order to ensure their continuity. These approaches are general, in that the approach works for any valid network, and hence any topology that can be modeled by that network. We differ from these approaches in that we specify the construction for each genus and do not rely on constraints to maintain continuity.

Manifolds and subdivision surfaces have both been used to build smooth, arbitrary topology surfaces without the use of constraints. Our approach differs from these in that our parameterization has few charts and reflects the standard structure for the $n$-holed surface, with no embedding-induced bias. Subdivision surfaces have proven to be very useful to the modeling community, but they do not easily extend to higher order continuities, nor are they parameterized, although the characteristic map can be used to derive a local parameterization [DKT98].

The mathematical foundations for hyperbolic-plane constructions of surfaces go back to the 19th century and the work of Poincaré and Klein; the topological basics are well-described in many elementary topology books [Mas77]. These ideas were first used by Ferguson and Rockwood [FR93] to build smooth surfaces with handles. Their basic approach begins with a function $f$ over the hyperbolic disk which does not agree at the boundaries. By applying a *multiperiodic basis* to $f$, they create a function $f'$ which does agree. The original paper created the function $f$ using a very specific construction process involving Coons patches arranged symmetrically in the domain; this process was replaced by a more general approach using radial basis functions in a later paper [RP99]. In this second approach, the function $f$ is built from a set of scattered points in the domain; the corresponding point locations in $\mathbb{R}^3$ define the function $f$. An embedded Delauny triangulation of the points provides a control polygon.

Spline orbifolds [WP97] are also built on the hyperbolic plane and use a spline-like embedding function. To create a tessellation they begin with a "sketch" polyhedron and use it to create a group-invariant triangulation of the domain. This sketch polyhedron is also used in a least-squares fitting approach to set the control points of the spline function.

We differ from these previous approaches in how we model the domain, our embedding function, and tessellation.

# 3   Manifold Representation, Group Actions, and a Fundamental Theorem

The standard definition of a manifold begins "Let $X$ be a Hausdorff space, and suppose that $\{V_i\}$ is an open cover of $X$, and that $\phi_i : V_i \longrightarrow U_i \subset R^k$ is a homeomorphism...." From this definition, one builds "transition functions" and eventually completes the structure of a manifold.

Evidently the disjoint union $Y$ of all of the $U_i$ in this standard definition have an equivalence relation on them: $p_i \in U_i$ is equivalent to $p_j \in U_j$ if and only if $\phi^{-1}(p_i) = \phi_j^{-1}(p_j)$. Indeed, we typically define "transition functions" $\psi_{ij} : U_{ij} \longrightarrow U_{ji}$, where $U_{ij}$ is defined as $\phi_i(V_j)$, and $\psi_{ij}$ is just $\phi_j \circ \phi_i^{-1}$. The equivalence relation then becomes $p_i \sim \psi_{ij}(p_i)$. The manifold $X$ is evidently in one-to-one correspondence with the quotient of $Y$ by this equivalence relation: the point $p \in X$ corresponds to the equivalence class of $\phi_i(p)$, where $p \in V_i$.

In the event that one does not have the space $X$ to start with, one can build a collection of open sets $U_i$ and a collection of functions $\psi_{ij}$ on appropriate subsets of them and then consider quotient of their disjoint union by the induced equivalence relation. Under suitable conditions, the quotient will be a manifold. This is the approach taken by Grimm and Hughes [GH95]; its advantage is that the description provided is particularly amenable to implementation. We will continue to use their notation, but in this particular case, we will know *a priori* that the space being constructed is a manifold. We will begin with the general notation and some mathematical preliminaries, and then, in Section 4.1, describe the $n$-holed torus manifold explicitly.

The components of the Grimm and Hughes description of a (two-dimensional) manifold are:

- A finite set, $A$, of $n$ nonempty subsets $\{c_1, \ldots, c_n\}$ of $\mathbb{R}^2$. $A$ is called an *atlas*. Each element $c_i \in A$ is called a *chart*.
- A set of subsets, $U_{ij} \subset c_i$ ($j = 1, \ldots, n$); the subset $U_{ii}$ must be all of $c_i$. These regions act as the "overlap regions" for the manifold: we will see that points of $U_{ij}$ and $U_{ji}$ get identified with one another. Note that $U_{ij}$ need not be a connected set; nor need it be nonempty, except in the case $i = j$.
- A set $\Psi = \{\psi_{ij} | i, j = 1, \ldots, n\}$ of $n^2$ functions called *transition functions*. For each $(i, j)$, the map $\psi_{ij} : U_{ij} \to U_{ji}$, where $U_{ij} \subset c_i$ and $U_{ji} \subset c_j$, must be smooth of some order $k$, and satisfy the requirements listed below. Note that because $U_{ij}$ may not be connected, $\psi_{ij}$ may be described by a set of functions, one for each connected component of $U_{ij}$.

As noted above, there is a relation $\sim$ defined on $Y = \sqcup_{c \in A} c$ (where $\sqcup$ denotes disjoint union) such that if $x \in c_i$, $y \in c_j$, then $x \sim y$ iff $\psi_{ij}(x) = y$. We require that the transition functions be symmetric ($\psi_{ij} = \psi_{ji}^{-1}$), that $\psi_{ii}$ is the identity for all $i$, and that they satisfy the *cocycle condition*, i.e., that $\psi_{ij} \circ \psi_{ki} = \psi_{kj}$ wherever this makes sense. These requirements ensure that the relation $\sim$ is an equivalence relation [Gri96]. The quotient of $Y$ by $\sim$ is then (under certain technical conditions[1]) guaranteed to be a manifold of class $C^k$.

Note that a "point" in this quotient manifold consists of a list of all of the chart points that are equivalent under $\sim$, *i.e.*, given a point $p_i$ in a chart $c_i$, the corresponding manifold point is all of the tuples $(c_j, p_j)$ such that $p_i \sim p_j$.

## 3.1   Group Actions on Manifolds

A group $G$ is said to act on a manifold $M$ if for each $g \in G$, there's an associated homeomorphism $f_g : M \longrightarrow M$ with the following properties:

- If $e$ is the identity element of $G$, then $f_e$ is the identity on $M$.
- If $g$ and $h$ are two elements of $G$, and $k = gh$, then $f_k = f_g \circ f_h$.

If the maps $f_g$ are diffeomorphism, we say that "$G$ acts by diffeomorphisms on $M$." If $M$ has a metric structure and the maps are isometries, we say that "$G$ acts by isometries on $M$."

Such a group action defines an equivalence relation: two points $m_1$ and $m_2$ of $M$ are equivalent if there's an element $g$ of $G$ such that $f_g(m_1) = m_2$.

As an example, the group of integers acts on the real line by translation: $f_n : \mathbb{R} \longrightarrow \mathbb{R} : x \mapsto x + n$. The set of equivalence classes for this action is in one-to-one correspondence with the interval $[0, 1]$, except that the equivalence classes for 0 and 1 are the same; in other words, the quotient space is a circle. (The topology on the quotient is that of the circle as well, but this requires careful working out of details beyond the scope of this paper.) The interval $[0, 1]$ is called a *fundamental domain* for the group action: it is a subset $K$ of the manifold with the property that $f_g(K) \cap K$ is entirely within the boundary of $K$ (for every $g$ in $G$), and such that $M = \cup_{g \in G} f_g(K)$.

The situation exemplified in the previous paragraph is not uncommon: frequently the quotient of a manifold by a group action is another manifold.

One general result [CDGM00] is all that we will need in this paper:

**Theorem 1.** *Suppose $G$ is a discrete group acting by isometries on a smooth manifold $M$, and for every compact subset $K$ of $M$, the set*

$$G(K; K) = \{g \in G | gK \cap K \neq \phi\}$$

---

[1] The technical conditions have to do with being Hausdorff: consider two copies of the real line, $U_1$ and $U_2$, and the maps $\psi_{12} : U_1 - \{0\} \longrightarrow U_2 - \{0\} : x \mapsto x$, $\psi_{21} = \psi_{12}^{-1}$, $\psi_{11} = id_R$ and $\psi_{22} = id_R$. These satisfy all the requirements of the description above, but the quotient space, "the line with two origins," is not actually a manifold, since the two copies of the point "0" cannot be separated by open sets.

*is finite. Further suppose that the only element of $G$ that fixes a point of $M$ is the identity. Then $M/G$ is a smooth manifold as well, and the map $M \to M/G$ is a local diffeomorphism.*

We will apply this to the case where $M$ is the open unit disk in the complex plane, and $G$ is a group of fractional linear transformations on the disk. A fractional linear transformation can be represented by a matrix:

$$\begin{bmatrix} a & b \\ c & d \end{bmatrix},$$

where the numbers $a, b, c,$ and $d$ are complex. The associated transformation takes the complex number $z$ to $\frac{az+b}{cz+d}$. One can easily check that the matrix:

$$Q = \begin{bmatrix} -1 & i \\ 1 & i \end{bmatrix}$$

takes the upper-half plane (the set of complex numbers with positive imaginary part) to the interior of the unit disk in the complex plane.

Furthermore, the matrices with *real* entries and determinant one (called $SL_2(\mathbb{R})$) send the complex plane one-to-one onto itself. That means that the matrices $QMQ^{-1}$ (for $M \in SL_2(\mathbb{R})$) send the unit disk one-to-one onto itself, so we can say that $SL_2(\mathbb{R})$ acts on the unit disk as well.

Serre [Ser90] describes this action and its properties in detail. In this paper, we will build a discrete subgroup of $SL_2(\mathbb{R})$; the quotient of the disk by this action will be an $n$-holed torus. This will actually be evident by the construction: a fundamental domain for the action will be a regular $4n$-gon in the hyperbolic disk, and the group action will be one that identifies edges with edges in the way described in elementary topology books. The fact that $G(K; K)$ is finite and that the only transformation fixing a point is the identity will be obvious from the construction. Hence the theorem above tells us that the resulting space is, in fact, a manifold as required.

## 4   Building a Manifold

From topology [Lef49], we know that an $n$-holed torus can be built from a $4n$-sided polygon by associating the edges of the polygon as shown in Figure 1. We use this structure to determine how many charts to use and how they overlap. We also use it to define the transition functions, by first mapping from chart $c_i$ to the polygon, then from the polygon to chart $c_j$. This two-step process is used to illustrate how the transition functions are built; the actual transition functions map directly from one chart to the other.

We first recall the description of how an $n$-holed torus is constructed from a $4n$-sided polygon and how we cover the polygon with $2n+2$ charts (Section 4.1). In Section 4.2 we describe the hyperbolic $4n$-gon. The charts and maps to and from the polygon are defined in Section 4.3.

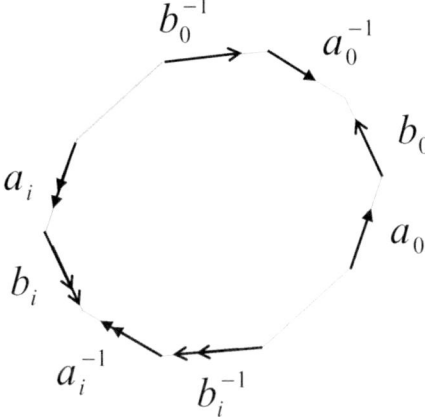

**Fig. 1.** An abstract $4n$-sided polygon with the associated edges labeled.

## 4.1   Building the *N*-Holed Tori

An $n$-holed torus can be built by associating the edges of a $4n$-sided polygon; the two-holed torus case is shown in Figure 2. Some observations about this construction:

- Each hole or handle is represented by a group of four consecutive edges.
- The first pair of associated edges $(a, a^{-1})$ correspond to a loop that goes around the hole. The second pair of associated edges $(b, b^{-1})$ corresponds to a loop that goes through the hole (once the torus is embedded in 3-space by our particular embedding; some other embedding might swap the roles).
- All of the vertices of the polygon correspond to a single point on the final surface. Each loop begins and ends at this point, in the order shown in Figure 2. Note that this is **not** the same order as the corners in the polygon.

We use $2n + 2$ charts to cover the polygon (see Figure 4). The first chart, termed the "inside" chart, is a disk that covers much of the interior of the polygon. The second chart, termed the "vertex" chart, is also a disk. It is centered on the polygon vertices and covers the corners of the polygon. The remaining $n$ charts are termed "edge" charts and are unit squares placed so that the midline of the chart covers most of the edge loop.

The vertex chart will cover the $4n$ corners of the polygon; this implies that each $2\pi/(4n)$ wedge of the vertex chart should map to its corresponding corner of the polygon. This mapping is greatly simplified if the corners of the polygon each have an angle of $2\pi/(4n)$, which is clearly not possible with Euclidean geometry. We therefore use a hyperbolic polygon.

## 4.2   Hyperbolic Polygon

In Figure 3 we see $4n$ circles arranged evenly around a unit disk. Each circle meets the boundary of the unit disk orthogonally. Neighboring pairs of circles

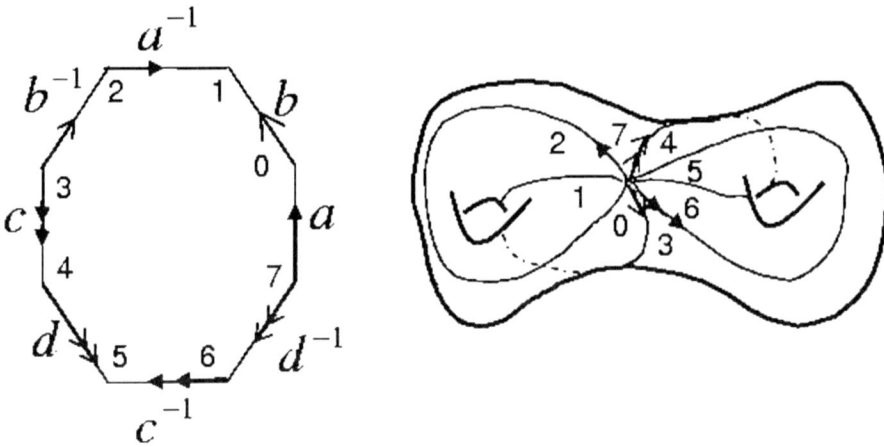

**Fig. 2.** Left: An 8-sided polygon with edges and vertex corners labeled. Right: A sketch of a 2-holed torus with the loops and vertex corners labeled.

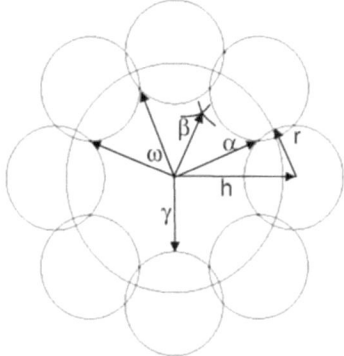

**Fig. 3.** The hyperbolic polygon for the two-holed torus, labeled with $\alpha$ (distance to the corner), $\omega = 2\pi/(4n)$, $\beta$ (edge-chart interior corner), $h$, $r$, and $\gamma$ (length to polygon boundary).

intersect at an angle of $2\pi/(4n)$. The part of the unit disk that is not inside any of the circles is therefore a regular $4n$-gon (with curved edges) and with vertex angles $2\pi/(4n)$.

A brief digression about this polygon is in order: the unit disk is one of several models for hyperbolic geometry. In this model, the geodesics ("straight lines") are (a) the diameters of the disk, and (b) circle-arcs that meet the boundary orthogonally; the isometries in this geometry are fractional linear transformations that map the disk to the disk. Therefore, what we have constructed is a regular $4n$-gon in the hyperbolic plane; the edges of this $4n$-gon, although curved from a Euclidean perspective, are straight lines in the hyperbolic perspective. Just as one can take an equilateral triangle in the Euclidean plane and flip it over one edge, then over another, and so on, and thus "fill out" the entire plane with a tiling by equilateral triangles, one can take this hyperbolic $4n$-gon and "flip" it over its edges repeatedly to fill out the entire hyperbolic disk. The "flips" are not reflections in the Euclidean sense, but rather are fractional linear transformations that map the disk to itself, leaving the "flip edge" fixed.

To return to the construction: the small circles are defined by their distance from the origin ($h$) and their radius ($r$). Their angular spacing is:

$$\omega = \frac{2\pi}{4n} \tag{1}$$

The constraint that the circles meet the disk-boundary orthogonally introduces the following relationship between $h$ and $r$:

$$r = \sqrt{h^2 - 1} \tag{2}$$

By changing $r$ (and consequently $h$) we can create nearly any desired angle between the intersecting circles. Given a desired angle $\omega$ we choose $h$ as follows:

$$h = \sqrt{\frac{\cos(\omega) + 1}{\cos(\omega) + 1 - 2\sin^2(\omega/2)}} \tag{3}$$

We will also need the distance to the vertex ($\alpha$) and to the middle of the edge ($\gamma$).

$$\alpha = h\cos(\omega/2) - \sqrt{(h\cos(\omega/2))^2 - 1} \tag{4}$$

$$\gamma = h - r \tag{5}$$

Finally, we need the point $\beta$, which will be used both to set the vertex chart size and in the tessellation (Section 5.2).

The derivations of these equations and of $\beta$ are given in Appendix A.

## 4.3   Mapping to the Polygon

As mentioned above, the isometries of the hyperbolic plane model are all linear fractional transformations (LFTs). We will be describing our transition functions in terms of these.

If $T$ and $S$ are LFTs with matrices $K$ and $L$, then the transformation with matrix $KL$ is $T \circ S$. If $K$ is the matrix $\begin{bmatrix} a & b \\ c & d \end{bmatrix}$ then the matrix

$$\begin{bmatrix} d & -b \\ -c & a \end{bmatrix} \tag{6}$$

represents the transformation $T^{-1}$.

We use two types of LFT. The first rotates and scales, the second translates:

$$S(p) = \begin{bmatrix} p & 0 \\ 0 & 1 \end{bmatrix} \qquad T(p) = \begin{bmatrix} 1 & p \\ p & 1 \end{bmatrix} \tag{7}$$

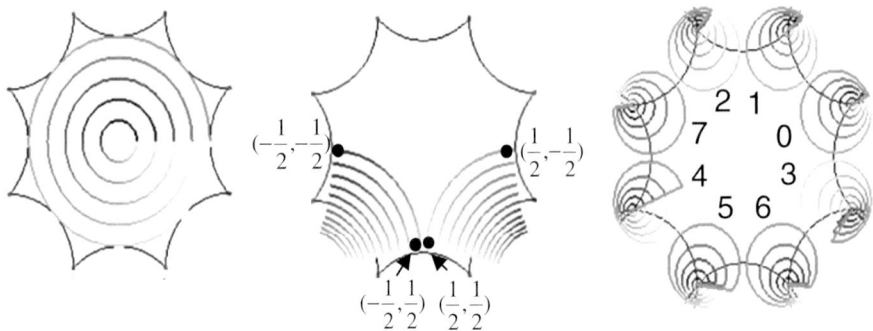

**Fig. 4.** From left to right: The inside chart, an example edge chart, and the vertex chart. The edge chart is split in two, with the left half mapped to the lower left edge, the right half mapped to the lower right edge. Each wedge of the vertex chart is mapped to a different corner.

The domain for the inside and vertex charts is the disk given in polar coordinates by $0 \leq r \leq 0.5$, i.e., the set of complex numbers $\{r \cos \theta + \mathbf{i}r \sin \theta | 0 \leq r < \frac{1}{2}, \theta \in [0, 2\pi]\}$. The domain for the edge charts is a square centered at the origin, i.e., $\{x + \mathbf{i}y | -\frac{1}{2} < x, y < \frac{1}{2}\}$.

The inside chart is mapped to the polygon by scaling the disk until it almost touches the edges of the polygon:

$$S(1.8\gamma) \tag{8}$$

The edge charts are mapped so that the chart is split vertically, with the left half mapping to the part of the disk near edge $a$, and the right half mapping to edge $a^{-1}$ (see Figure 4). Each piece of the chart is rotated, then translated out to the matching edge. The right half chart is rotated by $\pi$ before translating, so as to orient the mid-line in the opposite direction. The edge charts are first scaled by $7/8$ so they cover most, but not all, of the edge (the remainder will be covered by the vertex chart). Let $j$ be the chart number (going counter-clockwise around the circle). Recall that the edges come in groups of four, $aba^{-1}b^{-1}$, so that the $a$ charts overlap edges $4j$ and $4j + 2$, while the $b$ charts overlap edges $4j + 1$ and $4j + 3$.

$$p_l = \cos((4j + j \bmod 2)\omega) + \mathbf{i}\sin((4j + j \bmod 2)\omega) \tag{9}$$

$$p_r = \cos((4j + 2 + j \bmod 2)\omega) + \mathbf{i}\sin((4j + 2 + j \bmod 2)\omega) \tag{10}$$

$$p_f = \cos \pi + \mathbf{i}\sin \pi \tag{11}$$

$$S(p_l)T(h - r)S(7/8)(c + d\mathbf{i}) \qquad c < 0.5 \tag{12}$$

$$S(p_r)T(h - r)S(7/8)S(p_f)(c + d\mathbf{i}) \qquad c \geq 0.5 \tag{13}$$

The vertex chart is mapped to the polygon with $4n$ maps, one for each vertex of the polygon. Note that the order of the wedges $w_j$ in the chart is **not** the same as the order of the polygon corners. The order can be determined by following the associated edges around the polygon (see Figures 2 and 4). The mapping first rotates the chart by $(R_j)$, then translates it to the appropriate corner $C_j$ so that the correct chart wedge is facing inwards. The vertex chart is scaled by $\delta$ so that the boundary passes through $\beta$ (see Appendix A). The general form of the mapping is:

$$p_c = \cos((C_j + 1/2)\omega) + \mathbf{i}\sin((C_j + 1/2)\omega) \tag{14}$$

$$p_r = \cos(-R_j\omega) + \mathbf{i}\sin(-R_j\omega) \tag{15}$$

$$S(p_c)T(\alpha)S(p_r)S(\delta)(c + d\mathbf{i}), \qquad (c + d\mathbf{i}) \in w_j \tag{16}$$

The wedges in the vertex chart are indexed clockwise, with the first wedge centered on the negative real axis. The polygon corners are indexed counterclockwise, with the first corner at $\omega/2$ (or $C_0 = 0$). As we walk around the wedges in the vertex chart (incrementing $R$ by 1) then $C$ is updated by:

$$C_{j+1} = (C_j - 3) \bmod 4n \qquad (C_j \bmod 4) = 1, 2$$
$$C_{j+1} = (C_j + 1) \bmod 4n \qquad (C_j \bmod 4) = 0, 3$$

In other words, we either move to the next polygon corner, or back three.

### 4.4   Alternative Description

To describe the charts differently, we begin by describing a group of isometries of the disk. For each identified edge-pair (e.g., $a$ and $a^{-1}$) there is an orientation-preserving isometry of the disk that carries one edge to the other. In the case of the octagon, this map is a rotation by $\pi/2$ about the origin, followed by a (hyperbolic!) rotation of $\pi/2$ about the midpoint of the edge labelled $a^{-1}$. Similar maps take edge $b$ to edge $b^{-1}$, and so on. If one writes the maps out explicitly, it is clear that their only fixed points are on the unit circle, hence outside the domain we are considering. By taking all possible repeated compositions of these isometries, we get a tiling of the hyperbolic disk by hyperbolic octagons.

Each point of the disk is in exactly one of these polygons (or perhaps on the boundary of two or more of them). Thus we can define a chart that sends some domain into the disk, partly inside and partly outside of the fundamental octagon; the part that is outside the fundamental octagon corresponds (under one of the isometries) to some set of points that's *inside* another part of the octagon. In Figure 4 we have drawn the charts this way, showing, for example, an edge chart having an image that lies partly outside the fundamental octagon. We have also drawn the edge-chart followed by an isometry, and the part that lay outside is now within, and vice-versa.

### 4.5   The Transition Functions

The transition functions are the composition of chart $i$'s map to the polygon with chart $j$'s inverse map. Some care must be taken to ensure that the correct

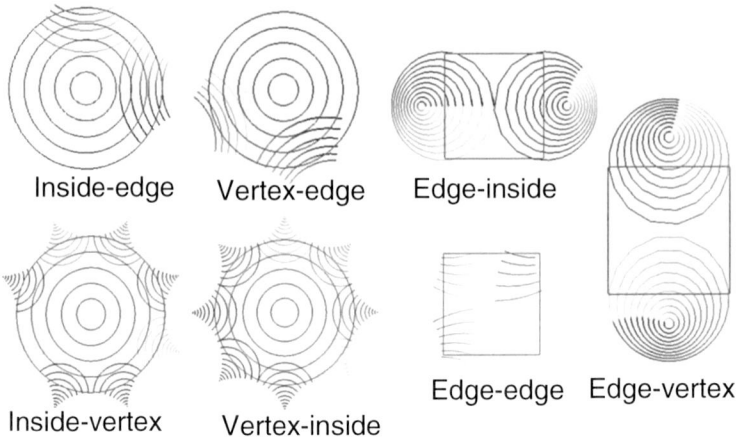

Fig. 5. The overlap regions $U_{ij}$ for the different cases.

pairs are combined, since each connected overlap region will have a different LFT. Refer to Figure 5:

- The inside chart overlaps with each edge chart in two regions, one in the left half and one in the right half.
- Each edge chart overlaps with two different edge charts. The overlap region has one, two, or three connected components each covering one of the corners.
- The vertex chart overlaps the inside chart in $4n$ places.
- The vertex chart overlaps each edge chart in two disconnected regions. Note that there are two possible constructions for these transition functions, but both yield the same LFT.

With these charts and transition functions, we have all the components necessary for the Grimm-Hughes description of a manifold. The only question that remains is whether the resulting quotient object is, in fact, a manifold. In general, that requires proving the quotient is Hausdorff, but in this case the answer is simple: the group of isometries described above evidently acts without fixed points (except for the identity transformation), and because the regular $4n$-gon is a fundamental region for the group action, it is clear that for any compact subset $K$ of the disk, the set $G(K; K)$ of all transformations $T$ such that $T(K)$ intersects $K$ is finite. Hence the theorem from Section 3 applies, and the quotient space is actually a manifold, indeed, a complex one-dimensional manifold.

## 5 Embedding

In this section we describe how to embed our manifold in 3-space and how to tessellate the domain in order to produce a mesh for rendering.

## 5.1   Defining an Embedding

To embed the manifold we first define a spline patch [BBB87] $S_c : c \to \mathbb{R}^3$ and a blend function $B_c : c \to \mathbb{R}$ for each chart $c$. We require that the blend function and its derivatives be zero by the boundary of the chart, and that the blend functions form a partition of unity. The surface embedding is then:

$$S(p) = \sum_{c \in A} B_c(\alpha_c(p)) S_c(\alpha_c(p)) \tag{17}$$

where $\alpha_c(p)$ is the point corresponding to $p$ in chart $c$, if there is one. If $p$ does not correspond to a point for chart $c$ then $B_c$ is defined to be zero. (Note that "corresponding to" in this case is simple: a point $p$ in the manifold is an equivalence class of points in the charts, i.e., a list of $2n+2$ or fewer chart-points.)

To build $C^k$ blend functions that form a partition of unity [GH95] we first build a proto-blend function in each chart using spline basis functions. We then "upgrade" these proto-blend functions to blend functions on the manifold by defining them to be zero for every manifold point not in the chart. This is why the proto-blend function, and its derivatives, must be zero by the boundary of the chart. To create a partition of unity we divide by the sum of all of the blend functions at that point; to ensure that this sum is non-zero we define the proto-blend functions so that their support covers the entire chart.

For the edge-chart blend functions we use the tensor product of two spline basis functions, each of which has a support of $[-1/2, 1/2]$. For the vertex and inside charts we use the distance from the chart center and a single spline basis function whose support is the diameter of the chart. This produces a radially symmetric blend function.

If the blend and embedding functions are $C^k$ continuous then the resulting surface is $C^k$. To produce a visually pleasing surface it is best if the spline patches agree where they overlap, i.e., $S_i(\alpha_i(p)) = S_j(\alpha_j(p))$; this ensures that the blending above is purely a formality.

To create our initial embedding we first build an $n$-holed mesh by taking a sphere and punching $n$ holes in it. We then split this mesh apart along boundary lines such as the ones shown in Figure 2. We flatten this mesh into the hyperbolic polygon using Floater's algorithm [Flo97] to produce a bijection between the mesh vertices and the points on the manifold[2]. We then solve for control points for each patch in a least-squares [FB91] fashion:

$$\sum_{c \in A} B_c(\alpha_c(p_i)) S_c(\alpha_c(p_i)) = P_i \tag{18}$$

The $p_i$ are the locations in the hyperbolic polygon (and hence on the manifold) found by embedding the split-open mesh into the hyperbolic polygon. The $P_i$ are the corresponding mesh locations in $\mathbb{R}^3$. For a given point $p_i$ we can evaluate all components of $B_c$ and $E_c$, creating a linear equation in the locations of the

---

[2] Because the boundary of the hyperbolic polygon is not convex this procedure is not guaranteed to produce a bijection; however, in this case it does.

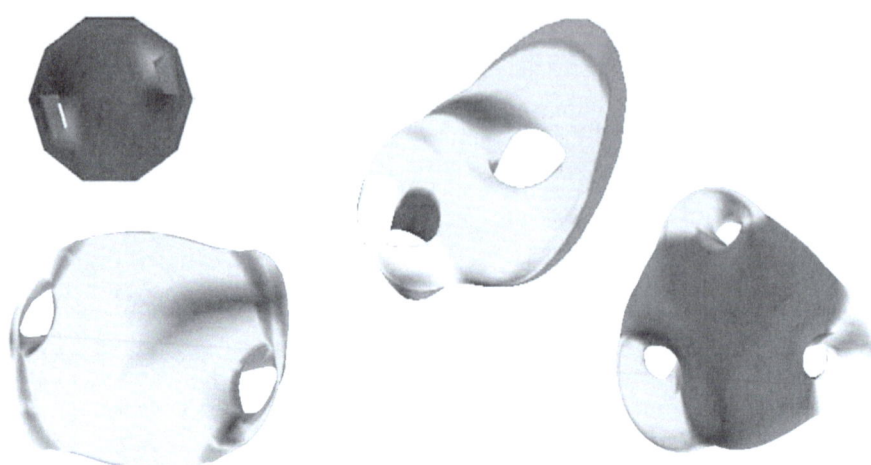

**Fig. 6.** Example embeddings. The mesh in the upper left hand corner is the $n$-holed mesh we fit to initially, to produce the surface in the lower left hand corner. The right two surfaces were created by editing this initial surface.

control points. Each mesh vertex will produce one linear constraint on some subset of the spline control points for some number of charts.

Once the initial surface is created we let the user edit it by direct manipulation. The user clicks on a point of the surface, which selects a point on the manifold. When the user moves the mouse we solve for the smallest change in the control points that moves the surface point to the new location (again using equation 18). Figure 6 shows the starting mesh and initial surface for the two-holed case, and two edited surfaces.

## 5.2  Tessellation

We tessellate the domain in order to produce a mesh to render. The tessellation has two parameters, $t_e$ and $t_r$, which control the resolution of the mesh. The coarsest level tessellation ($t_e = 1$, $t_r = 1$) partitions the domain into $2n + 2$ regions, one for each chart (see Figure 7).

The lowest resolution tessellation places a square in each edge chart. The size of this square is chosen so that the corners of the squares of neighboring edge charts map to the same point in the polygon, which is the point $\beta$ in Figure 3. The edges of these squares form $4n$ consecutive arcs in the vertex and inside charts, enclosing a wheel-shaped region in the center of the chart.

The first tessellation parameter, $t_e$, specifies the number of divisions in the square. This produces a grid inside each edge chart, and a wheel in the inside and vertex charts. The second parameter, $t_r$, specifies the number of divisions along the spokes of the wheel.

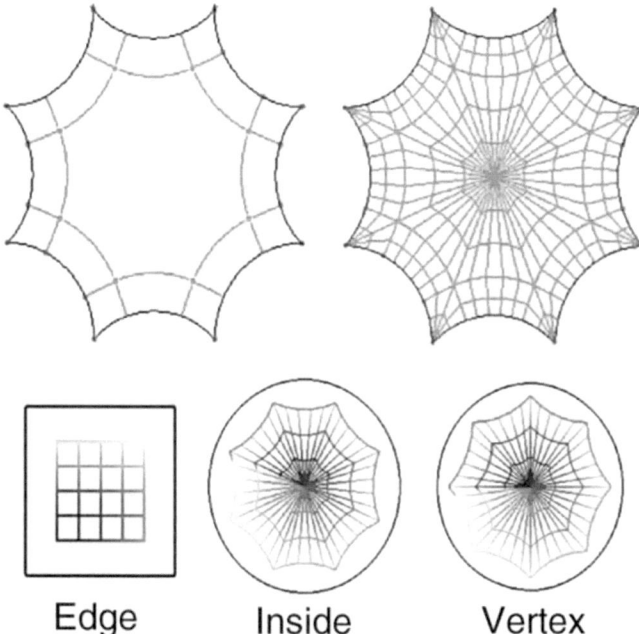

**Fig. 7.** Upper left: The coarsest tessellation. Upper right: Subdividing the tessellation. Bottom: The tessellation in the interior of each chart type.

## 6    Conclusion

We have presented a technique for modeling $n$-holed tori, based upon a natural parameterization for $n$-holed tori, suitable for implementation in computer graphics.

Several areas remain open, however. The group of fractional linear transformations that we have used can probably be adjusted to become a subgroup of the modular group ($2 \times 2$ matrices with integer entries and determinant one); this would result in a description of the $n$-holed torus as a natural branched cover of the unit sphere $S^2$. Meromorphic functions on these $n$-holed tori might well be used to define maps into the $k$-fold product $(S^2)^k \subset (\mathbb{R}^3)^k$, which might then be projected to give particularly nice maps from the $n$-holed torus to $\mathbb{R}^3$, rather than the somewhat ad-hoc spline-based embeddings we have described. We hope to pursue this in the future.

## A    Derivations

To derive $r = \sqrt{h^2 - 1}$ we note that the triangle formed by the center of the two circles and their intersection point is a right-angled one (by construction). Therefore $r^2 + 1^2 = h^2$.

Let $\boldsymbol{\omega} = \cos(\omega) + \mathbf{i}\sin(\omega)$ be the unit vector from the origin through the point where the first two circles intersect. Let $\alpha\boldsymbol{\omega}$ be the intersection point. Then

$$\cos(\omega/2) = \boldsymbol{\omega} \cdot \frac{(\alpha\boldsymbol{\omega} - h)^{\perp}}{r} \tag{19}$$

If $z$ and $w$ are unit vectors, then their dot product is $\mathrm{Re}(z\bar{w})$. Applying this, and $z^{\perp} = \mathbf{i}z$, we have:

$$\cos(\omega/2) = \mathrm{Re}(\boldsymbol{\omega}\mathbf{i}\frac{(\alpha\boldsymbol{\omega} - h)^{\perp}}{r}) \tag{20}$$

$$= \pm\frac{h}{r}(\sin\omega) \tag{21}$$

Applying $\cos(2x) = 2\cos^2 x - 1$:

$$\cos(\omega) = \frac{2h^2 + \sin^2\omega + 1}{h^2 - 1} \tag{22}$$

Solving for $h$ we get:

$$h = \sqrt{\frac{\cos(\omega) + 1}{\cos(\omega) + 1 - 2\sin^2(\omega/2)}} \tag{23}$$

To solve for $\beta$ we note that we are looking for the number $s$ such that:

$$m_0(-s + \mathbf{i}s) = m_1(-s - \mathbf{i}s) \tag{24}$$

where $m_0$ and $m_1$ are the LFT that take the upper half of chart 0 and chart 1 to the polygon. This reduces to:

$$\frac{m_0(0,0)(-s + \mathbf{i}s) + m_0(0,1)}{m_0(1,0)(-s + \mathbf{i}s) + m_0(1,1)} = \frac{m_1(0,0)(-s - \mathbf{i}s) + m_1(0,1)}{m_1(1,0)(-s - \mathbf{i}s) + m_1(1,1)} \tag{25}$$

which reduces to a quadratic equation in $s$. Once we have $s$ we can derive $\delta$ in a similar manner.

# References

BBB87.    R. Bartels, J. Beatty, and B. Barsky. *An Introduction to Splines for Use in Computer Graphics and Geometric Modeling*. Morgan Kaufmann, 1987.

Blo97.    Jules Bloomenthal, editor. *Introduction to Implicit Surfaces*. Morgan Kaufmann, 1997.

CBC+01.    Jonathan C. Carr, Richard K. Beatson, Jon B. Cherrie, Tim J. Mitchell, W. Richard Fright, Bruce C. McCallum, and Tim R. Evans. Reconstruction and Representation of 3D Objects With Radial Basis Functions. *Proceedings of ACM SIGGRAPH 2001*, pages 67–76, 2001.

CC78.    Ed Catmull and J. Clark. Reursively generated B-spline surfaces on arbitrary topological meshes. *Computer Aided Design*, 10:350–355, September 1978.

CDGM00.    Virginie Charette, Todd Drumm, William Goldman, and Maria Morrill. Complete Flat Affine and Lorentzian Manifolds. In *Proceedings of the Workshop "Crystallographic Groups and their Generalizations II," Contemporary Mathematics 262*, pages 135—146. American Mathematical Society, 2000.

DKT98.    Tony D. DeRose, Michael Kass, and Tien Truong. Subdivision Surfaces in Character Animation. *SIGGRAPH 98*, pages 85–94, 1998.

FB91.    B. Fowler and R. H. Bartels. Constraint based curve manipulation. *Siggraph course notes 25*, July 1991.

Flo97.    Michael S. Floater. Parametrization and smooth approximation of surface triangulations. *Computer Aided Geometric Design*, 14(3):231–250, 1997. ISSN 0167-8396.

FPRJ00.    Sarah F. Frisken, Ronald N. Perry, Alyn P. Rockwood, and Thouis R. Jones. Adaptively Sampled Distance Fields: A General Representation of Shape for Computer Graphics. In *Proceedings of ACM SIGGRAPH 2000*, Computer Graphics Proceedings, Annual Conference Series, pages 249–254. ACM Press / ACM SIGGRAPH / Addison Wesley Longman, July 2000. ISBN 1-58113-208-5.

FR93.    H. Ferguson and A. Rockwood. Multiperiodic functions for surface design. *Computer Aided Geometric Design*, 10(3):315–328, August 1993.

GH95.    Cindy Grimm and John Hughes. Modeling Surfaces of Arbitrary Topology using Manifolds. *Computer Graphics*, 29(2), July 1995. Proceedings of SIGGRAPH '95.

Gri96.    C. Grimm. *Surfaces of Arbitrary Topology using Manifolds*. PhD thesis, Brown University (in progress), 1996.

Gri02.    Cindy Grimm. Simple Manifolds for Surface Modeling and Parameterization. *Shape Modelling International*, May 2002.

Ld89.    Charles Loop and Tony deRose. A multisided generalization of bezier surfaces. *ACM Transactions on Graphics*, 8(3):204–234, July 1989.

Lef49.    S. Lefschetz. *Introduction to Topology*. Princeton University Press, Princeton, New Jersey, 1949.

Mas77.    William S. Massey. *Algebraic Topology: An Introduction*. Springer Verlag, New York, 1977.

NG00.    J. Cotrina Navau and N. Pla Garcia. Modeling surfaces from meshes of arbitrary topology. *Computer Aided Geometric Design*, 17(7):643–671, August 2000. ISSN 0167-8396.

Pet96.    Jorg Peters. Curvature continuous spline surfaces over irregular meshes. *Computer-Aided Geometric Design*, 13(2):101–131, February 1996.

RP99.    A. Rockwood and H. Park. Interactive design of smooth genus $N$ objects using multiperiodic functions and applications. *International Journal of Shape Modeling*, 5:135–157, 1999.

Ser90.    Jean Pierre Serre. *A Course in Arithmetic*. Springer Verlag, New York, 1990.

WP97.    J. Wallner and H. Pottmann. Spline orbifolds. *Curves and Surfaces with Applications in CAGD*, pages 445–464, 1997.

# Generic Parameterization
# of Bifurcating Structures

Malcolm I.G. Bloor and Michael J. Wilson

Department of Applied Mathematics, The University of Leeds, UK
{malcolm,mike}@maths.leeds.ac.uk
http://www.maths.leeds.ac.uk/Applied/staff.dir/wilson/wilson.html

**Abstract.** This paper shows how difficulties with singularities encountered in the geometry of bifurcating surfaces can be overcome and the shape parameterized efficiently enabling a designer to create and manipulate such geometries within an interactive environment.

## 1 Introduction

Nowadays there exist many methods for the design of complex objects, which use a variety of geometry representation schemes, e.g. B-Rep, CSG, feature based modelling and variational methods. Examples of existing commercial CAD systems include CATIA from Dessault Systemme, Pro/Engineer from Parametric Technology and PowerSHAPE from Delcam International. Using standard commercial CAD packages, complex mechanical parts can be built up from an intersecting series of geometric solids which form 'primary' surfaces and rolling ball blend surfaces between the primary surfaces. Thus, existing commercial CAD systems can create and manipulate the geometry of complex mechanical parts, although the design process may not always be straightforward.

However, as Imam [3] notes, when considering the design of engineering surfaces, it is essential that CAD systems, used to create the design, be able to parameterize the geometry of the design efficiently. In parametric design, the basic approach is to develop a generic description of an object or class of objects, in which the shape is controlled by values of a set of design variables or parameters. A new design, created for a particular application is obtained from this generic template by selecting particular values for design parameters so that the item has properties suited to that application. As noted above, it is often the case that commercial systems often fail to generate a parametric model of the complex part in question. This is often due to the lack efficient tools which enable one to define complex geometries in the form of a generic parametric model.

Bifurcations are common features of many objects that we wish to design or describe, and despite their ubiquitous nature, they present certain difficulties, if not for the designer directly, certainly for the designer of the software upon which the CAD system is based. For example, when designing bifurcating structures, it is almost inevitable that singularities will arise in the parameterization of

M.J. Wilson and R.R. Martin (Eds.): Mathematics of Surfaces 2003, LNCS 2768, pp. 30–39, 2003.

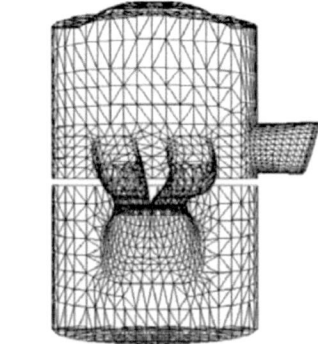

**Fig. 1.** Example of bifurcating transfer port on a 2-stroke internal combustion engine

the geometry, and it is unreasonable to expect a designer to have anything other than an intuitive appreciation of the problem. Hence when creating a system that allows for the interactive design of objects with bifurcating shapes, it is necessary that the system has a generic parameterization of the bifurcating geometry that implicitly incorporates the basic mathematical rules for describing the shape, and which frees the designer from having to know the techinical details of how the geometry is constructed.

In particular, the physical system with which we are concerned in this paper is that of tube like structures which have been formed by the amalgamation of a number of branches or formed from the bifurcation, or higher order splitting, of a tubular conduit. The geometry at the 'intersection' of the branches is such that there are no sharp corners and all junctions represent smooth transitions from one branch to another. The physical basis of these systems ranges from engineering applications to biological systems. For example, in the automotive industry, manifolds, either inlet or exhaust, are critical to the design and require smooth transitions at the tube junctions to avoid energy losses and to reduce noise, Mimis *et al.* [4]. Similarly in the arterial system, where bifurcation of the blood vessels is the mechanism by which blood is dispersed to all parts of the body, the geometry of the artery bifurcation is extremely important and smoothness of the splitting is a natural feature. A typical geometry of such physical system is shown in Fig.(1) which illustrates a bifurcated tube structure in an engineering application, while Fig. (2) shows an example that arises in a biophysical application [2].

In this paper we shall outline a method for parameterizing bifurcating surfaces using the PDE Method introduced by Bloor and Wilson, e.g. [1]; an example of an alternative approach to the problem may be found in the work of Ye *et al.* [6].

## 2    Generation and Parameterization of Geometry

The object of the present exercise is to use the PDE method to determine a mechanism for the generation of such shapes avoiding the introduction of the

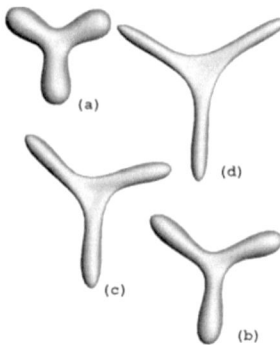

**Fig. 2.** Fluid vesicles showing bifurcating geometry

singularities which are an inherent part of bifurcated geometry. Thus it is necessary to obtain a suitable parametric form of the surface shape that has sufficient scope to cover a broad family of physically realistic branching configurations. In the discussion that follows, for the sake of illustration, we will consider a bifurcation consisting of the junction of three or more tube-like 'arms' or branches. The principles for generalising to the case of arms with more complicated cross-sections should be clear.

In a typical 'simple' bifurcation of one branch into two, one can think of three points, **P1**, **P2** and **P3** say, that denote the 'centres' of the constituent branches of the bifurcation and the unit normals **n1**, **n2** and **n3** give the 'directions' of the separate tubes at those locations. The nature of the bifurcation is restricted in the sense that **n1**, **n2** and **n3** are coplanar , or nearly so, in order that a 'perpendicular' to the bifurcation might be identified so as to lead to an acceptable geometry. Although there are not necessarily simple geometric symmetries in the system, there are strong similarities between various branches and it is consequently only necessary to consider one branch of the basic structure. This consideration would also apply in higher order splitting into three or more branches. In the absence of strict symmetries, in order to proceed in this way it is necessary to choose what might be called a centre of the bifurcation and in many circumstances this could be taken to be the mean of the points **P1**, **P2** and **P3**.

In Fig.(3) a single branch geometry is illustrated with the associated coordinate system set up and the centre of the bifurcation at the origin. The orientation of the Cartesian coordinate system is taken to be such that the y-axis is 'perpendicular' to the bifurcation i.e. parallel to the mean of $\mathbf{n1} \times \mathbf{n2}$, $\mathbf{n2} \times \mathbf{n3}$ and $\mathbf{n3} \times \mathbf{n1}$. It is clear that this constituent object has a tube like structure and consequently the surface shape $\mathbf{X}(u, v)$ may be taken to be a periodic function of $v$, say, where $u$ and $v$ are the usual parametric coordinates with $0 < u < 1$, and $0 < v < 2\pi$. Now $\mathbf{X}(u, v)$ is a mapping from the rectangular $(u, v)$ space, the domain $\Omega$ bounded by $\partial\Omega$, say, to Euclidean 3-space determined by the PDE

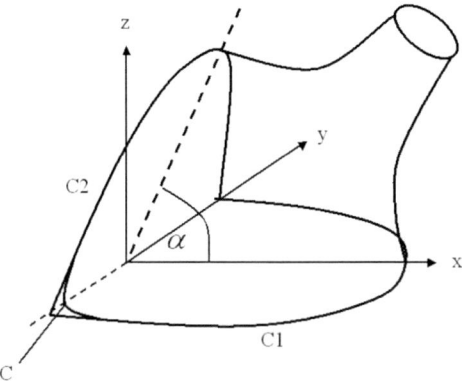

**Fig. 3.** Diagram showing coordinate axes and boundary conditions

$$\left(\frac{\partial^2}{\partial u^2} + a^2 \frac{\partial^2}{\partial v^2}\right)^2 \mathbf{X} = 0 \tag{1}$$

where $a$ is the smoothing parameter and the solution is subject to appropriate boundary conditions on $\mathbf{X}$ and its parametric derivatives $\mathbf{X}_u$ and/or $\mathbf{X}_v$ chosen from the geometry in Fig.(3). Consequently it is possible, in principle, to express $\mathbf{X}(u, v)$ as an infinite Fourier series in $v$ with the coefficients functions of u determined by the boundary conditions. However, an approximate solution, satisfying the boundary conditions exactly, can be found using a truncated series and a correction term [1] and this makes the problem tractable.

## 2.1  Boundary Conditions

The boundary conditions to be imposed on the problem are determined from continuity conditions at $u = 0$ and $u = 1$, the boundaries of the branch in Fig.(3). The curve $\mathbf{C1}(v)$ in the plane $z = 0$, referred to as $\pi1$, at $u = 0$, say, is combined with the curve $\mathbf{C2}(v)$ in the plane $z = x \tan(\alpha)$, where $\alpha$ is a constant, referred to as $\pi2$, to give the complete closed curve at $u = 0$. The boundary conditions at $u = 1$ arise from conditions on a closed curve which is the plane cross section of the tube. It can be seen that while the boundary conditions at $u = 1$ may be perfectly well behaved, those at $u = 0$ necessarily are singular owing to the discontinuities in the derivative with respect to $v$ of $\mathbf{X}(u, v)$ at $x = 0$, $z = 0$. In order to overcome the problems arising from this singularity, the surface in the neighbourhood of $x = 0$, $z = 0$ is given some asymptotic form, and a new boundary curve $u = 0$ is determined from the original boundary and a curve drawn on some centrally located surface to delineate the asymptotic surface which forms a part of it.

Across $\mathbf{C1}$ and $\mathbf{C2}$ continuity requirements with the adjacent branches must be enforced and, bearing in mind the choice of the direction of the $y$-axis, it is reasonable to impose the condition that the tangent plane to the surface at any point, $v$, on $\mathbf{C1}(v)$ and $\mathbf{C2}(v)$ is normal to the planes $\pi 1$ and $\pi 2$ respectively. Clearly, this condition has been chosen in order to keep the algebra as simple as possible and would prove suitable under most circumstances. However, other conditions could equally well be enforced providing the continuity conditions are not violated. It is a matter of finding the simplest way forward to achieve an acceptable design. These considerations not only apply to the PDE surface but also the asymptotic form of the surface near $x = 0$, $z = 0$.

A central surface, $\mathbf{S}(\xi, \eta)$, from which the asymptotic surface may be cut, can now be obtained in the following way. Parametric coordinates $\xi$ and $\eta$ are introduced such that boundary curves identical to $\mathbf{C1}(v)$ and $\mathbf{C2}(v)$ are defined at $\eta = 0$ and $\eta = 1$ respectively, with $v$ replaced by $\xi$. A surface $\mathbf{S}(\xi, \eta)$ is then formed by using a cubic interpolation for each coordinate between the two curves $\mathbf{C1}(\xi)$ and $\mathbf{C2}(\xi)$ in such a way that the boundary conditions mentioned above are satisfied.

Now consider the intersection of the surface $\mathbf{S}(\xi, \eta)$ with a circular cylinder of sufficiently small radius $r$, with its generators parallel to the $y$-axis, centred at $x = xc$, $z = r$. The intersection curve on $\mathbf{S}$ for $x < xc$, denoted by $\mathbf{C}$, is tangent to the boundary curves $\mathbf{C1}$ and $\mathbf{C2}$ given by Equations (1) and (3) (see below). The points of intersection are

$$x = xc, \ z = 0, \text{ corresponding to } \xi = v = v_1, \text{ say } \eta = 0$$

and also

$$x = xc - r\sin(\alpha) \ , \ z = r(1 + \cos(\alpha)), \text{ corresponding to } \xi = v = v_2, \text{ say } \eta = 1.$$

For the purposes of solving Equation (1), the boundary curve u=0 is now taken to be:

for $0 < v < v_1$ the curve $\mathbf{C1}$,
for $v_1 < v < v_2$ the curve $\mathbf{C}$,
for $v_2 < v < \pi$ the curve $\mathbf{C2}$

with equivalent conditions for $\pi < v < 2\pi$.

Thus, there need be no singularities in the boundary conditions at $u = 0$.

At $u = 1$, the positional conditions are straightforward, being defined along some cross-sectional curve on the connecting branch specified in terms of $v$.

In order to solve Equation (1), in addition to the positional boundary conditions discussed above, there needs to be normal derivative data supplied on $\partial\Omega$. Owing to the periodicity condition, it is sufficient to specify the derivatives $\mathbf{X}_u(u, v)$ on $u = 0$ and $u = 1$. First of all, at $u = 1$, the tangent plane continuity requirements demand that the branch of the bifurcation blends with the tube which would need to be specified.

On that part of the boundary $u = 0$ formed by the curve $\mathbf{C1}$, arguments mentioned above, indicate that for any $v$, the $y$ component of $\mathbf{X}_u(u, v)$ must be zero while the $x$ and $z$ components are shape parameters in the problem. Similar considerations apply on the curve $\mathbf{C2}$. On the curve $C$, the normal derivatives are taken in the direction towards the axis of the cylinder subject to the requirement that tangent plane continuity with the surface $\mathbf{S}$ is ensured.

With all of the conditions discussed above specified, an approximate solution to Equation (1) can be obtained in the form of a truncated Fourier series supplemented by a correction term to ensure that the boundary conditions are satisfied exactly [1].

## 3   Example

By way of an example, and in order to illustrate this approach, a particular case in which the geometry is kept as simple as possible is considered, and the boundary conditions determined explicitly in what follows. With this in mind, the boundary curve $\mathbf{C1}(v)$ in the plane $\pi1$ is given by,

$$
\begin{aligned}
x &= c_1 \cos(v) \\
y &= b \sin(v) \\
z &= 0
\end{aligned}
\tag{2}
$$

On $\pi2$, the boundary curve $\mathbf{C2}(v)$ is given by,

$$
\begin{aligned}
x &= -c_2 \cos(\alpha) \cos(v) \\
y &= b \sin(v) \\
z &= -c_2 \sin(\alpha) \cos(v)
\end{aligned}
\tag{3}
$$

In this simple example, the form of the surface $\mathbf{S}(\xi, \eta)$ is given by

$$
\begin{aligned}
x &= \cos(\xi)(c_1 + (3(c_2 \cos(\alpha) - c_1) - ss_1 \sin(\alpha))\eta^2 \\
&\quad -(2(c_2 \cos(\alpha) - c_1) + ss1 \sin(\alpha))\eta^3) \\
y &= b \sin(\xi) \\
z &= \cos(\xi)(ss_0 \eta + (3c_2 \sin(\alpha) \\
&\quad -2ss_0 - ss_1 \cos(\alpha))\eta^2 \\
&\quad +(ss_1 \cos(\alpha) + ss_0 - 2c_2 \sin(\alpha))\eta^3),
\end{aligned}
\tag{4}
$$

and its normal $\mathbf{n}_s$ is easily found from

$$
\mathbf{n}_s = \mathbf{S}_\xi \times \mathbf{S}_\eta
\tag{5}
$$

It is important to notice that the constants ss0 and ss1 are shape parameters for the surface $\mathbf{S}$. As such they influence the boundary conditions at $u = 0$, not only in terms of position but also the derivatives, and consequently are shape parameters for the final PDE surface. In certain applications it might be appropriate to determine these parameters automatically, by minimising mean curvature for example, whereas in others they could certainly be used as a supplementary design tool.

By using Equations (2) and (3) the values of $v1$ and $v2$ are determined by

$$
c_1 \cos(v_1) = xc
\tag{6}
$$

and

$$- c_2 \cos(\alpha) \cos(v_2) = xc - r \sin(\alpha) \tag{7}$$

Denoting any point on that part of the surface of the circular cylinder which intersects $\mathbf{S}$ by $x = xc - r \sin(\phi)$ and $z = r(1 - \cos(\phi))$, $0 < \phi < \pi - \alpha$, the corresponding values of $\xi$ and $\eta$ are found from Equation (4) for any particular value of $\phi$ in the range.

Hence, on $u = 0$, the positional data is given by

(i) for $0 \leq v \leq v_1$, and $2\pi - v_1 \leq v < 2\pi$, Equation (2)
(ii) for $v_1 < v \leq v_2$, and $2\pi - v_2 \leq v < 2\pi - v_1$, Equation (4) with $\xi$ replaced by $v$ and $\eta$ determined by the intersection condition
(iii) for $v_2 < v \leq 2\pi - v_2$, Equation (3)

and the derivative conditions are taken as,

(i) for $0 \leq v \leq v_1$, and $2\pi - v_1 \leq v < 2\pi$,

$$\mathbf{X}_u = s_0(0, 0, 1). \tag{8}$$

(ii) for $v_1 < v \leq v_2$, and $2\pi - v_2 \leq v < 2\pi - v_1$

$$\mathbf{X}_u = s_0 \hat{\mathbf{n}} \tag{9}$$

where $s_0$ is a scaling parameter and $\hat{\mathbf{n}}$ is the unit vector in the $\mathbf{n}$ direction defined by

$$\mathbf{n} = (xc - x, yn, r - z) \tag{10}$$

and $yn$ is such that $\mathbf{n}$ is normal to $\mathbf{ns}$,
(iii) and for $v_2 < v \leq 2\pi - v_2$

$$\mathbf{X}_u = s_0(\sin(\alpha), 0, \cos(\alpha)). \tag{11}$$

At $u = 1$, the positional conditions are taken as,

$$\begin{aligned} x &= h \cos(\alpha/2) + ra \sin(\alpha/2) \cos(v) \\ y &= ra \sin(v) \\ z &= h \sin(\alpha/2) - ra \cos(\alpha/2) \cos(v) \end{aligned} \tag{12}$$

and the derivative conditions,

$$\mathbf{X}_u = s_1(\cos(\alpha), 0, \sin(\alpha)). \tag{13}$$

where $s_1$ is another scaling parameter. Thus at $u = 1$ the above conditions represent a cylinder of circular cross section, radius $ra$, directed along the bisector of $\pi 1$ and $\pi 2$ symmetrical about the plane $y = 0$.

Equation (1) is solved subject to the conditions (4-8) and the resulting surface, is combined with two others, one the mirror image in the plane $x = 0$, to create a symmetric Y-shaped branch. The complementary asymptotic surfaces in the neighbourhood of the centre of the branch are also evaluated. The resulting complete surface shape is shown in Fig. (4) where the values of the

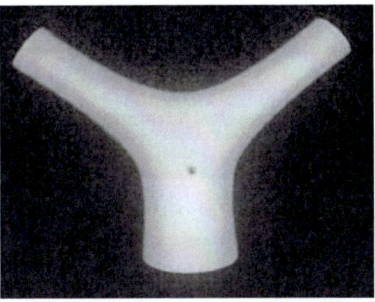

**Fig. 4.** Example Bifurcation

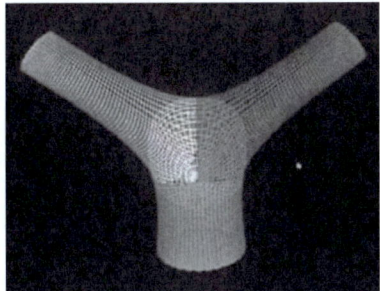

**Fig. 5.** Example bifurcation in wireframe

parameters are $c_1 = 1$, $c_2 = 1.5$, $b = 1$, $h = 4$, $ra = 0.5$, $\alpha = \pi/1.8$, $xc = 0.8$, $s0 = 2.5$, $ss_0 = 3$, $ss_1 = 3$, $a = 1.3$ for each symmetric branch, while for the third branch,$c_1 = 1$, $c_2 = 1$, $b = 1$, $h = 2$, $ra = 1$, $\alpha = 8\pi/9$, $xc = 0.8$, $s_0 = 2.5$, $ss_0 = 3$, $ss_1 = 3$, $a = 0.3$. In Fig. (4), isoparametric mesh of the constituent three branches of the surface are shown together with the underlying three central surfaces generated using Equation (4) and its equivalents for the two supplementary central surfaces.

The number of shape parameters used in this example is very small but the scope of accessible shapes is nevertheless quite extensive. The way in which the shape regime may be extended is quite straightforward through the use of more Fourier modes in the definitions of the boundary curves and derivative specifications. The approach outlined above for the removal of the singularities is still appropriate and the solution method would not require modification.

By way of illustration, another example is presented in Fig.(6), where the object consists of six branches smoothly joined at the 'centre' but terminating with rounded ends some distance from the centre. The conditions at $u = 0$ are very similar to those taken above but at $u = 1$, $ra = 0$ and the derivative conditions are such that $\mathbf{X}_u$ lies in the plane normal to $\mathbf{n}_i$, $i = 1...6$. It can be seen that the resulting shape is starfish like and of the type adopted by certain cells existing in a minimum surface energy state.

**Fig. 6.** Example structure showing higher-order branching

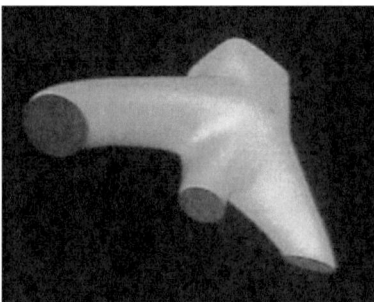

**Fig. 7.** Example bifurcation with non-coplanar branches

In Fig.(7), an example of an exhaust manifold for an internal combustion engine is shown. The geometry here is somewhat more complicated involving more Fourier modes in the cross-sectional definitions and also having the ni non-coplanar. Neverthless, the approach outlined above carries through and produces a realistic geometry for this functional part.

## 4   Discussion

In this paper we have described some preliminary results showing how a model parameterization of a bifurcating geometry can facilitate the design and manipulation of such structures. In earlier work, we have described how such generic models can be used as the basis of the interactive design of complex mechanical parts [5]. In particular, we have shown how the PDE parametric model allows complex designs to be created and manipulated in an interactive environment. Since the present work is a natural extension of this to cover the parameterization of bifurcating structures. the boundary conditions play a vital role in determining the shape of the surface and the design parameters have been introduced on the boundary curves which define the shape. These parameters are chosen in such a way that simple transformations of the boundary curves can be carried out. The reason for choosing such a model is that the design parameters introduced in this manner allows a designer to easily create and manipulate complex geometries without having to know the mathematical details of the solutions of PDEs.

# References

1. Malcolm I.G. Bloor and Michael J. Wilson. Spectral approximations to pde surfaces. *CAD*, 28(2):145–152, 1996.
2. Malcolm I.G. Bloor and M.J. Wilson. Method for efficient shape parameterization of fluid membranes and vesicles. *Physical Review E*, 61(4):4218–4229, 2000.
3. M.H. Imam. Three-dimensional shape optimisation. *International Journal for Numerical Methods in Engineering*, 18:661–673, 1982.
4. Angelos P. Mimis, Malcolm I.G. Bloor, and Michael J. Wilson. Shape parameterization and optimization of a two-stroke engine. *Journal of Propulsion and Power*, 17(3):492–498, 2001.
5. Hassan Ugail, Michael Robinson, Malcolm I.G. Bloor, and Michael J. Wilson. Interactive design of complex mechanical parts using a parametric representation. In Roberto Cipolla and Ralph Martin, editors, *The Mathematics of Surfaces IX: Proceedings of the Ninth IMA Conference on the Mathematics of Surfaces*, volume 4, pages 169–179. IMA, Springer, 2000.
6. X. Ye, Y-Y Cai, C. Chui, and J. Anderson. Constructive modeling of $g^1$. *Computer Aided Geometric Design*, 19:513–531, 2002.

# Direct Computation of a Control Vertex Position on any Subdivision Level

Loïc Barthe and Leif Kobbelt

Computer Graphics Group, RWTH Aachen
Ahornstrasse 55, 52074 Aachen, Germany
{barthe,kobbelt}@cs.rwth-aachen.de
http://www.rwth-graphics.de/

**Abstract.** In this paper, we present general closed form equations for directly computing the position of a vertex at different subdivision levels for both triangular and quadrilateral meshes. These results are obtained using simple computations and they lead to very useful applications, especially for adaptive subdivision. We illustrate our method on Loop's and Catmull-Clark's subdivision schemes.

## 1 Introduction

Since their introduction to Computer Graphics [1,2,3], subdivision surfaces have become a widely used surface representation in both animation and freeform shape modeling. A subdivision operator globally subdivides a coarse mesh into a finer one and after several subdivision iterations the meshes quickly converge to a smooth surface. Since the subdivision operator refines all faces of the mesh, the number of faces increases exponentially: For a classical diadic subdivision (like Loop's or Catmull-Clark's subdivision [4,1]) the number of polygons is multiplied by 4 after *one* subdivision step, and hence by $4^k$ after $k$ steps. A few steps are enough to generate a massive mesh that even the fastest hardware cannot handle interactively.

In order to overcome this disadvantage, adaptive subdivision [5,6,3] allows us to subdivide the mesh locally, only where some geometric detail is present. An accurate approximation of the surface is then provided while avoiding the generation of huge meshes (Fig. 1). Nowadays, adaptive subdivision is a standard technique used in applications like mesh editing, mesh simplification, and view-dependent rendering [7,8,9,10,11,12].

Since more steps of subdivision are performed in some parts of the mesh, vertices in the direct neighborhood of a central vertex can lie at different subdivision levels. The drawback is that when approximation schemes are used, vertices are displaced through subdivision, i.e. the same vertex has different positions when it is used to compute new vertices at different steps of subdivision. This requires an easy access to the position of a vertex at different subdivision levels and the limit position of every vertex has to be known in order to evaluate the current error. The application of *one* step of subdivision on a vertex requires

M.J. Wilson and R.R. Martin (Eds.): Mathematics of Surfaces 2003, LNCS 2768, pp. 40–47, 2003.

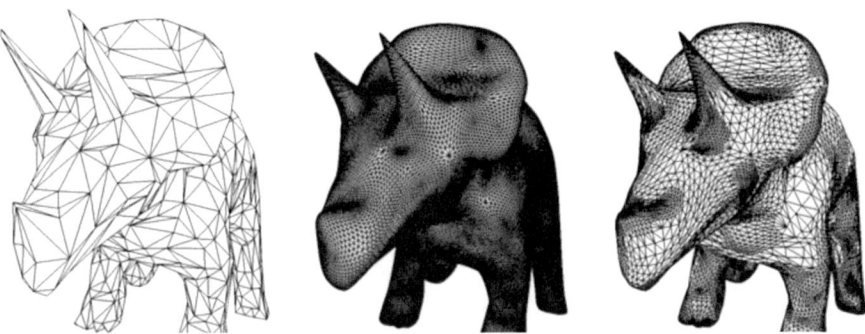

**Fig. 1.** *Three* steps of uniform Loop's subdivision performed on a $1.4K$ triangles triceratops model (left) yields a mesh composed of $90K$ triangles (center). An equivalent approximation of the limit surface is obtained with only $29K$ triangles when we use adaptive subdivision (right).

the knowledge of the position of its neighbors, and since vertices have always to be evaluated at different subdivision level, either expensive computations or large memory consumption is to be expected.

A simple method to avoid this phenomenon is rather to interpolate the control mesh so that the position of the vertices remains unchanged through subdivision. However standard stationary interpolatory subdivision schemes (like butterfly subdivision [13,14]) are known to provide limit surfaces with artifacts and oscillations, and small support approximation schemes providing better quality limit surfaces are still preferable. Therefore, the fundamental requirements that we have to address are the computation of the position of the mesh vertices at the limit and at an arbitrary subdivision step.

The computation of the limit position of a vertex has already been studied [15,16,17]. Closed form solutions are derived using the eigendecomposition of the subdivision matrix $S$, where $S$ is the operator which maps the vicinity of a central vertex into the same vicinity on the next refinement level. The limit position is then expressed as an affine combination of vertices of the control mesh, having its coefficients given by the dominant left eigenvector. Hence, our main contribution is the inexpensive computation of the vertex position at any intermediate step of subdivision. Indeed, if this evaluation is too expensive, it becomes preferable to simply store the different positions of the vertices after each subdivision step.

The size of the subdivision matrix $S$ grows linearly with the valency of the central vertex, and different valencies yield different eigendecompositions. Therefore, complications are to be expected if we try to derive closed form equations for the different vertex positions from the standard formulation of the subdivision matrix (as done in [16]). To overcome this disadvantage, a known solution is to use the Fourier transform of the matrix $S$ [15,17]. However as shown by Loop [4] for triangular meshes when he studied the convergence of his scheme, for all valencies, the subdivision matrix $S$ can be represented by a $2 \times 2$ matrix by exploiting its block circulant structure. This procedure avoids the computation of the Fourier transform and it has been later exploited by Kobbelt [18] to provide his $\sqrt{3}$ scheme with simple closed form equations to evaluate the positions of a vertex at both the limit and after $m$ subdivision steps.

In this paper, we present more general closed form equations for computing the position of a vertex at any intermediate subdivision level. After initialization, the evaluation of different vertex positions do not require any access to the neighbors, which allows us to avoid memory consumption while providing computationally inexpensive solutions.

The first part is the extension of this approach to its general form for triangular meshes. It illustrates how this simple computation process leads us to elegant closed form equations where the position of a vertex after $m$ subdivision steps is defined as the interpolation between its initial position and its position in the limit. The general solution is then applied to Loop's scheme [4] for illustration purposes.

The second part represents a more important contribution. We apply the same procedure on quadrilateral meshes and we show how the different structure of the subdivision matrix leads us to slightly more complicated equations that have to be handled with a mathematical software. Also in this case, after initialization, no neighbor information is required and we illustrate on *two* versions of Catmull-Clark's subdivision [1,2,19] that it still provides practical useful solutions.

## 2   Triangular Lattice

We begin with the study of triangular lattices. Since we only consider small support subdivision schemes, the vicinity of the central vertex is restricted to its "1-ring" neighborhood (as illustrated in Figure 2a). Each row of the subdivision matrix is a smoothing (or subdivision) rule which computes a new vertex as an affine combination of vertices of the original mesh, and once all the rotational symmetries are exploited, the subdivision of this set of vertices is written in the well known form:

$$
\begin{bmatrix} q^{n+1} \\ p_1^{n+1} \\ p_2^{n+1} \\ \vdots \\ p_v^{n+1} \end{bmatrix} = \left[ \begin{array}{c|cccc} a_0 & a_1 \cdots & \cdots & a_v \\ \hline b_0 & b_1 & b_2 & \cdots & b_v \\ b_0 & b_v & b_1 & b_2 & \cdots \\ \vdots & & & \ddots & \\ b_0 & b_2 & \cdots & b_v & b_1 \end{array} \right] \begin{bmatrix} q^n \\ p_1^n \\ p_2^n \\ \vdots \\ p_v^n \end{bmatrix},
\tag{1}
$$

with

$$
\sum_{i=0}^{v} a_i = 1, \sum_{i=0}^{v} b_i = 1, \text{ and hence } \sum_{i=1}^{v} a_i = 1 - a_0, \sum_{i=1}^{v} b_i = 1 - b_0.
\tag{2}
$$

In equation (1), $q^n$ is the central vertex at the $n^{th}$ step of subdivision, $v$ is its valency and vertices $p_j^n$ ($j = \{1, \ldots, v\}$) are its direct neighbors. Vertex $q^{n+1}$ is

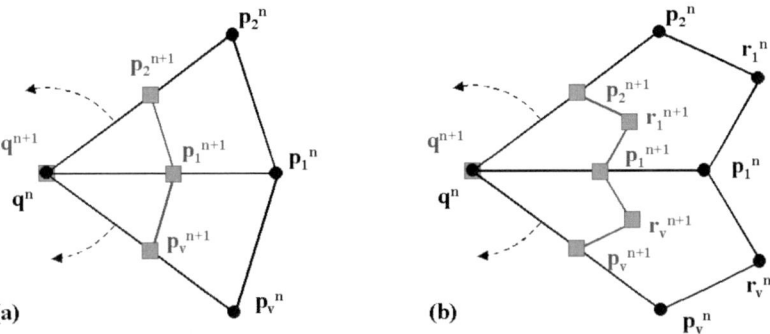

**Fig. 2.** One step of standard diadic subdivision performed on the central vertex $q^n$ and its 1-ring neighborhood: (a) on a triangular mesh, and (b) on a rectangular mesh. The superscript index indicates the subdivision level while the lowerscript index is the vertex number.

the new position of vertex $q^n$ after *one* subdivision step, vertices $p_j^{n+1}$ are new vertices which are the direct neighbors of vertex $q^{n+1}$. The subdivision matrix $S$ is the square matrix in Equation (1).

Let us denote $P^n$ as:

$$P^n = \frac{1}{v} \sum_{j=1}^{v} p_i^n.$$

Equations (1) and (2) allow us to express the sum of the new 1-ring vertices in terms of the old vertices as:

$$P^{n+1} = \frac{1}{v} \sum_{j=1}^{v} p_j^{n+1} = \frac{1}{v} \sum_{j=1}^{v} \left[ b_0 q^n + \sum_{i=1}^{v} b_{i-j+1} p_i^n \right]$$

$$= b_0 q^n + (1 - b_0) P^n.$$

Hence, relation (1) can be simplified in a form which simply involves a $2 \times 2$ matrix in the computation (independent from valencies $v$):

$$\begin{bmatrix} q^{n+1} \\ P^{n+1} \end{bmatrix} = \begin{bmatrix} a_0 & 1 - a_0 \\ b_0 & 1 - b_0 \end{bmatrix} \begin{bmatrix} q^n \\ P^n \end{bmatrix}.$$

When $m$ steps of subdivision are performed, the subdivision matrix $S$ is applied to the power $m$. This multi-step rule can also be written as an affine combination of the vertex $q^n$ and the average of its neighbors $P^n$, and once the $2 \times 2$ matrix is expressed in terms of its eigendecomposition we rewrite the subdivision rules as:

$$\begin{bmatrix} q^{n+m} \\ P^{n+m} \end{bmatrix} = \frac{b_0}{1 - a_0 + b_0} \begin{bmatrix} 1 & \frac{1-a_0}{b_0} \\ 1 & -1 \end{bmatrix} \begin{bmatrix} 1 & 0 \\ 0 & (a_0 - b_0)^m \end{bmatrix} \begin{bmatrix} 1 & \frac{1-a_0}{b_0} \\ 1 & -1 \end{bmatrix} \begin{bmatrix} q^n \\ P^n \end{bmatrix}. \quad (3)$$

In Equation (3), when $m$ tends to infinity, $(a_0 - b_0)^m$ tends to 0, because $0 < b_0 < a_0 < 1$ (variational diminishing property), hence, the limit position $q^\infty$ of vertex $q^n$ is directly computed using the following closed form equation:

$$q^\infty = \beta_\infty q^n + (1 - \beta_\infty)P^n \quad \text{with } \beta_\infty = \frac{b_0}{1 - a_0 + b_0} \tag{4}$$

From equations (3) and (4), we remove the neighbor information and we derive the expression of the position of vertex $q^n$ after $m$ subdivision steps in terms of its initial position $q^n$ and its limit position $q^\infty$. This leads us to the expected computation based only on the own vertex information:

$$\boxed{q^{n+m} = \mu(m)q^n + (1 - \mu(m))\,q^\infty \quad \text{with } \mu(m) = (a_0 - b_0)^m} \tag{5}$$

The closed form equations (4) and (5) applied to Kobbelt's $\sqrt{3}$ subdivision scheme are already given in [18] hence we present as an example their application in the case of **Loop's subdivision** [4]:

$$a_0 = 1 - \alpha_v \quad \text{with } \alpha_v = \tfrac{5}{8} - \left(\tfrac{3}{8} + \tfrac{1}{4}\cos\left(\tfrac{2\pi}{v}\right)\right)^2 \quad \text{and} \quad b_0 = \tfrac{3}{8}$$

$$\boxed{\beta_\infty = \frac{1}{1 + \tfrac{8}{3}\alpha_v} \quad \text{and} \quad \mu(m) = \left(\tfrac{5}{8} - \alpha_v\right)^m}$$

## 3  Quadrilateral Lattice

The case of a quadrilateral lattice is more complicated because the 1-ring neighborhood is composed of *two* different rotationally symmetric sets of vertices ($\{p_j^n\}$ and $\{r_j^n\}$, $j = \{1, \ldots, v\}$)(see Figure 2b). Therefore, the subdivision of a vertex $q^n$ and its 1-ring neigbors is defined by the following equation:

$$
\begin{bmatrix} q^{n+1} \\ p_1^{n+1} \\ \vdots \\ p_v^{n+1} \\ r_1^{n+1} \\ \vdots \\ r_v^{n+1} \end{bmatrix}
=
\left[\begin{array}{c|ccc|ccc}
a_0 & a_1 & \cdots & a_v & a_{v+1} & \cdots & a_{2v} \\
b_0 & b_1 & \cdots & b_v & b_{v+1} & \cdots & b_{2v} \\
\vdots & & \ddots & & & \ddots & \\
b_0 & b_2 & \cdots & b_1 & b_{v+2} & \cdots & b_{v+1} \\
c_0 & c_1 & \cdots & c_v & c_{v+1} & \cdots & c_{2v} \\
\vdots & & \ddots & & & \ddots & \\
c_0 & c_2 & \cdots & c_1 & c_{v+2} & \cdots & c_{v+1}
\end{array}\right]
\begin{bmatrix} q^n \\ p_1^n \\ \vdots \\ p_v^n \\ r_1^n \\ \vdots \\ r_v^n \end{bmatrix} .
$$

The subdivision matrix $S$ has *four* circulant blocks and in a similar manner than in Section 2 we express $S$ as a $3 \times 3$ matrix (independant from valencies $v$):

$$
\begin{bmatrix} q^{n+1} \\ P^{n+1} \\ R^{n+1} \end{bmatrix}
=
\begin{bmatrix}
a_0 & a_s & 1 - a_0 - a_s \\
b_0 & b_s & 1 - b_0 - b_s \\
c_0 & c_s & 1 - c_0 - c_s
\end{bmatrix}
\begin{bmatrix} q^n \\ P^n \\ R^n \end{bmatrix} , \tag{6}
$$

where $R^n = \dfrac{1}{v} \sum\limits_{j=1}^{v} r_j^n$ and $x_s = \sum\limits_{i=1}^{v} x_i$ for $x \in \{a, b, c\}$.

Following the procedure presented in Section 2 we use Matlab code to compute the different closed form equations for $\beta_\infty$, $\gamma_\infty$, $\mu(m)$ and $\nu(m)$ in order to evaluate the different positions of a vertex $q^{n+m}$.

From Equation (6) and its eigendecomposition we obtain the expressions:

$$q^{n+1} = a_0 q^n + a_s P^n + (1 - a_0 - a_s) R^n \tag{7}$$

$$q^{m+n} = \beta(m) q^n + \gamma(m) P^n + (1 - \beta(m) - \gamma(m)) R^n \tag{8}$$

$$q^\infty = \beta_\infty q^n + \gamma_\infty P^n + (1 - \beta_\infty - \gamma_\infty) R^n \tag{9}$$

Using Equations (7) and (9), we remove the neighborhood information from Equation (8) as follows:
Let

$$
\begin{bmatrix} q^n \\ q^{n+1} \\ q^\infty \end{bmatrix}
=
\begin{bmatrix} 1 & 0 & 0 \\ a_0 & a_s & 1 - a_0 - a_s \\ \beta_\infty & \gamma_\infty & 1 - \beta_\infty - \gamma_\infty \end{bmatrix}
\begin{bmatrix} q^n \\ P^n \\ R^n \end{bmatrix}
= T \begin{bmatrix} q^n \\ P^n \\ R^n \end{bmatrix},
$$

then $q^{n+m}$ is expressed as

$$
q^{m+n} = \begin{bmatrix} \beta(m) & \gamma(m) & 1 - \beta(m) - \gamma(m) \end{bmatrix} T^{-1}
\begin{bmatrix} q^n \\ q^{n+1} \\ q^\infty \end{bmatrix},
$$

which leads us to the final closed form equation:

$$\boxed{q^{n+m} = \mu(m) q^n + \nu(m) q^{n+1} + (1 - \mu(m) - \nu(m)) q^\infty}$$

In actual applications, the equations are greatly simplified and the formulas provided by the code lead us to practical solutions. These is illustrated for *two* versions of Catmull-Clark's subdivision.

**Standard Catmull-Clark's Subdivision [19]:** (Table 1)

$$a_0 = 1 - \frac{7}{4v}, \quad a_s = \frac{6}{4v}, \quad b_0 = \frac{3}{8}, \quad b_s = \frac{1}{2}, \quad c_0 = \frac{1}{4}, \quad c_s = \frac{1}{2}$$

**Bilinear Subdivision Plus Averaging Catmull-Clark's Subdivision [19]:**
In this case, the subdivision rules are independant of the vertex valency.

$$a_0 = \frac{9}{16}, \quad a_s = \frac{3}{8}, \quad b_0 = \frac{3}{8}, \quad b_s = \frac{1}{2}, \quad c_0 = \frac{1}{4}, \quad c_s = \frac{1}{2}$$

$$\forall v : \quad \beta_\infty = \gamma_\infty = \frac{4}{9}$$

$$\mu(m) = \frac{-1}{3} \left(\frac{1}{4}\right)^m + \frac{4}{3} \left(\frac{1}{16}\right)^m, \quad \nu(m) = \frac{16}{3} \left(\left(\frac{1}{4}\right)^m - \left(\frac{1}{16}\right)^m\right).$$

**Table 1.** Coefficients to compute the positions $q^\infty$ and $q^{n+m}$ of a vertex $q^n$ of the mesh when it is subdivided using the standard Catmull-Clark's subdivision.

| Valency | $\beta_\infty$ | $\gamma_\infty$ | $\mu(m)$ | $\nu(m)$ |
|---------|----------------|-----------------|----------|----------|
| 3 | $\frac{3}{8}$ | $\frac{1}{2}$ | $0$ | $6\left(\frac{1}{6}\right)^m$ |
| 4 | $\frac{4}{9}$ | $\frac{4}{9}$ | $\frac{-1}{3}\left(\frac{1}{4}\right)^m + \frac{4}{3}\left(\frac{1}{16}\right)^m$ | $\frac{16}{3}\left(\left(\frac{1}{4}\right)^m - \left(\frac{1}{16}\right)^m\right)$ |
| 5 | $\frac{1}{2}$ | $\frac{2}{5}$ | $-0.3165(0.3225)^m + 1.3165(0.0775)^m$ | $4.0825((0.3225)^m - (0.0775)^m)$ |
| 6 | $\frac{6}{11}$ | $\frac{4}{11}$ | $\frac{-2}{7}\left(\frac{3}{8}\right)^m + \frac{9}{7}\left(\frac{1}{12}\right)^m$ | $\frac{24}{7}\left(\left(\frac{3}{8}\right)^m - \left(\frac{1}{12}\right)^m\right)$ |

## 4   Conclusion

We have presented a general method for directly computing the position of a vertex on any subdivision level without accessing neighbor information, and this, in the case of triangular and quadrilateral lattices. These results can be directly applied to any subdivision scheme whose mask does not exceed the 1-ring neighborhood, as illustrated on Loop's and Catmull-Clark's subdivision. If the scheme's mask exceed *one* ring, the procedure for quad-meshes can be applied to derive closed form equations with more terms.

As explained in Section 1, the direct application to adaptive subdivision makes these closed form equations very useful for practical applications and they can be used to provide different approximation schemes with adaptive subdivision as a standard operator for meshes [20].

## Acknowledgments

The first author has partially been funded by the E.U. through the MINGLE project. Contract HPRN-CT-1999-00117

## References

1. Catmull, E, Clark, J.: Recursively generated B-spline surfaces on arbitrary topological meshes. Computer Aided Design **10**,6, (1978) 350–355
2. Doo, D., Sabin, M.A.: Analysis of the behaviour of recursive subdivision surfaces near extraordinary points. Computer Aided Design **10**,6, (1978) 356–360
3. Zorin, D., Schröder, P.: Subdivision for modeling and animation. SIGGRAPH 2000 course notes, (2000)
4. Loop, C.: Smooth subdivision surfaces based on triangles. Master's thesis, University of Utah, (1987)
5. Vasilescu, M., Terzopoulos, D.: Adaptive meshes and shells: Irregular triangulation, discontinuities and hierarchical subdivision. Proceedings of Computer Vision and Pattern Recognition, (1992) 829–832
6. Verfürth, R.: A review of a posteriori error estimation and adaptive mesh refinement techniques. Wiley-Teubner, (1996)

7.  Zorin, D., Schröder, P., Sweldens, W.: Interactive multiresolution mesh editing. Proceedings of SIGGRAPH 1997, ACM, (1997) 259–268

8.  Xu, Z., Kondo, K.: Local subdivision process with Doo-Sabin subdivision surfaces. Proceedings of Shape Modeling International, (2002) 7–12

9.  Lee, A., Moreton, H., Hoppe, H.: Displaced subdivision surfaces. Proceedings of SIGGRAPH 2000, ACM, (2000) 85–94

10. Hoppe, H.: View-dependent refinement of progressive meshes. Proceedings of SIG-GRAPH 1997, ACM, (1997) 189–198

11. Kamen, Y., Shirman, L.: Triangle Rendering Using Adaptive Subdivision. IEEE Computer Graphics and Applications **18**,2, (1998) 356–360

12. Hoppe, H.: Smooth view-dependent level-of-detail control and its application in terrain rendering. IEEE Visualization, (1998) 35–42

13. Dyn, N., Levin, D., Gregory, J.: A butterfly subdivision scheme for surface interpolation with tension control. ACM Transaction on Graphics, **9**,2, (1990) 160–169

14. Zorin, D., Schröder, P., Sweldens, W.: Interpolating subdivision for meshes with arbitrary topology. Proceedings of SIGGRAPH 1997, ACM, (1996) 189–192

15. Halstead, M., Kass, M., DeRose, T.: Efficient, fair interpolation using Catmull-Clark surfaces. Proceedings of SIGGRAPH 1993, ACM, (1993) 35–43

16. Hoppe, H., DeRose, T., Duchamp, T., Halstead, M.: Piecewise smooth surfaces reconstruction. Proceedings of SIGGRAPH 1994, ACM, (1994) 295–302

17. Stam, J.: Exact evaluation of Catmull-Clark subdivision surfaces at arbitrary parameter values. Proceedings of SIGGRAPH 1998, ACM, (1998) 395–404

18. Kobbelt, L.: $\sqrt{3}$-subdivision. Proceedings of SIGGRAPH 2000, ACM, (2000) 103–112

19. Warren, J., Weimer, H.: Subdivision methods for geometric design: a constructive approach. San Fransisco: Morgan Kaufman, (2002) 209–212

20. OpenMesh subdivision tool: http://www.openmesh.org/

# Optimising Triangulated Polyhedral Surfaces
# with Self–intersections

Lyuba Alboul

Sheffield Hallam University, Sheffield S1 1WB, UK
L.Alboul@shu.ac.uk
http://www.shu.ac.uk/scis/artificial_intelligence/L.Alboul.html

**Abstract.** We discuss an optimisation procedure for triangulated polyhedral surfaces (referred to as $(2-3)D$ triangulations) which allows us to
process self–intersecting surfaces. As an optimality criterion we use minimisation of total absolute extrinsic curvature (**MTAEC**) and as a local
transformation – a diagonal flip, defined in a proper way for $(2-3)D$ triangulations. This diagonal flip is a natural generalisation of the diagonal
flip operation in $2D$, known as Lawson's procedure. The difference is that
the diagonal flip operation in $(2-3)D$ triangulations may produce self-
intersections. We analyze the optimisation procedure for $(2-3)D$ closed
triangulations, taking into account possible self–intersections. This analysis provides a general insight on the structure of triangulations, allows
to characterise the types of self–intersections, as well as the conditions
for global convergence of the algorithm. It provides also a new view on
the concept of optimisation on the whole and is useful in the analysis of
global and local convergence for other optimisation algorithms. At the
end we present an efficient implementation of the optimality procedure
for $(2-3)D$ triangulations of the data, situated in the convex position,
and conjecture possible results of this procedure for non–convex data.

## 1    Introduction

Triangulating a set of points (a data set) is a key technique in solving problems in various applications, where one deals with reconstructing or modelling
the surface of a three-dimensional object; for instance, in computer graphics,
cartography, geology, reverse engineering, architecture, visual perception, and
medicine.

The first step in reconstructing a surface is to obtain a triangulation of the
data sites. A 'good' triangulation can help to solve many problems. It is the
quickest way to obtain an initial look at the data before referring to higher-
order interpolation or approximation methods. If the data are taken from an
irregular (non–smooth) surface, a triangulation (a polyhedral surface) might be
the only 'reliable' approximation of this surface.

The main goal of a triangulation is to give an initial representation of the
surface, which is to be reconstructed. The problem of finding a suitable triangulation of the given data comprises, in general, two steps: first, constructing

M.J. Wilson and R.R. Martin (Eds.): Mathematics of Surfaces 2003, LNCS 2768, pp. 48–72, 2003.
© Springer-Verlag Berlin Heidelberg 2003

some initial triangulation, and second, its optimisation in order to meet certain requirements. For example, one can require that a triangulation be as smooth as possible (i.e. without unnecessary sharp edges), or reflect certain features of the shape of the object [9,13]. These requirements determine the choice of one or another optimality criterion. Optimisation usually consists of transforming an initial triangulation via a sequence of transformations to some final triangulation, which is better with respect to the given criterion. The operation of transformation should preferably be simple as well as general enough in order to be able to reach the optimal triangulation from any initial one.

We concentrate on the optimisation problem for triangulations of closed surfaces in $3D$. We designate such triangulations as $(2-3)D$ triangulations (a justification for this notation is given in Section 2). In [2,26] the optimisation procedure was introduced, in which **minimisation** of **total absolute extrinsic curvature**, or, shortly, **TAEC**, was used as the optimisation criterion. This criterion was initially called the **tightness criterion** and the triangulations that are obtained on the basis of this criterion were called **tight triangulations**. Some of its properties are investigated in [2], where it is also shown that tight triangulations are evidently better than Delaunay triangulations, as, for example, the tight triangulation automatically preserves convexity.

In this paper we refer to the criterion based on minimising TAEC, as **MTAEC**. Triangulations, on which TAEC reaches its minimum, are referred to as **triangulations of minimal total absolute extrinsic curvature (MTAEC triangulations)**. We replace tight triangulations with MTAEC triangulations in order to avoid any confusion with the concept of tight polyhedral surfaces, known in the theory of polyhedral immersions [19].

TAEC belongs to the class of so-called discrete curvatures. Concepts of discrete curvatures are of growing interest for geometric modelling [22,13,10,18,17]. After its introduction in 1994 MTAEC has been reflected in a few research publications [13,22,25,28,4,21], and even in commercial software [16]. However, the initial optimisation procedure, based on minimisation of TAEC, has several shortcomings, which we correct in this paper. One of its shortcomings is that it uses a *diagonal flip*, or swap, as a local transformation procedure, under the assumption that its application may not produce self–intersections. This assumption influences the algorithm, designed to obtain the MTAEC triangulation, as well as the conclusion with respect to the convergence of the algorithm.

In this paper a generalisation of the optimisation procedure is given, which crucially affects the general view on the concept of optimisation procedure on the whole. First of all we skip the conventional agreement on the diagonal flip operation, that this operation can be applied only if it doesn't produce self–intersections in the triangulation [1]. However, doing so demands analyzing the structure of self-intersections. We show that they might be of two types: **local self–intersections**, when the triangulated polyhedral surface is not even immersed in the space, but possess singularities; and **global self–intersection**, when the surface is immersed in the space. Each type of self–intersection has a different impact on the performance of the optimisation procedure and influence the convergence of the algorithm. Studying the optimisation procedure gives also a more detailed view on the structure of $(2-3)D$ triangulations, constructed from scattered three–dimensional data, and might be useful in other optimisation procedures.

The results, shown in this paper, require understanding of concepts and notions, originated from different mathematical fields. In order to make our presentation self–contained we include, whenever it is appropriate, brief expositions of the theoretical background. The paper is organised as follows. In Section 2 we discuss the diagonal flip operation and outline the main differences between Lawson's swap and the swap in $(2-3)D$. We introduce a $(2-3)D$–geometric flip, which might produce self-intersecting triangulations in the optimisation process. We demonstrate its soundness and show, that $(2-3)D$ triangulations of the same data set $V$ (situated in the general position), that topologically equivalent to the $2D$ sphere, are connected under the $(2-3)D$–geometric flip. In Section 3 we give a short overview of discrete curvatures and provide all necessary definitions and concepts, concerning TAEC, in a revised form. In Section 4 self–intersections are studied in detail. A new insight on an arbitrary triangulation of convex data is provided. In Section 5 the principal differences between the current procedure and the one, presented earlier in [3,26,2], are thoroughly outlined. In the same section a new condition, concerning the convergence of our procedure, is given. At the end of Section 5 we test the procedure on $(2-3)D$ triangulations of the data, situated in convex position. We present an example of its implementation, which shows how the procedure works with self–intersections. This numerical example is based on apparently simple data, but of fundamental nature. We conclude the paper by conjecturing the outcome of the procedure in the case of non–convex data.

## 2    Diagonal Flip as a Local Transformation in $(2-3)D$ Triangulations

### 2.1    Theoretical Background

The diagonal flip as a local transformation in triangulations has a long history. Lawson introduced this operation for triangulations in the plane [20]. Lawson's operation consists of swapping (flipping) a diagonal $bd$ of the convex quadrilateral formed by two adjacent triangles $abd$ and $dbc$ to the other diagonal $ac$, thus replacing one edge by a new one and obtaining another triangulation of the given data (with triangles $abc$ and $acd$) (see Fig. 1).

Flipping is not allowed in a non-convex planar quadrilateral, because it does not produce a triangulation. It has also been shown that for any two triangulations $T_1$ and $T_2$ of a given planar point set $V$, there is always a sequence of Lawson's operations transforming $T_1$ into $T_2$. Later this result was extended to general polygonal domains (see, for example, [12]).

The diagonal flip operation has also been extensively studied in *topological graph theory* (see [24] for a survey). The difference with the flip in the geometrical

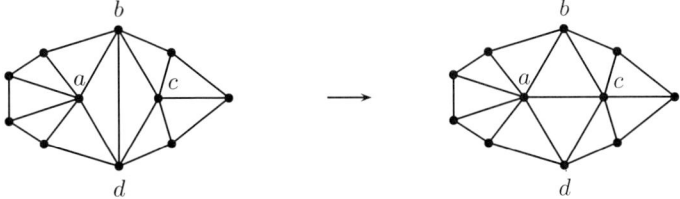

**Fig. 1.** A diagonal flip in a triangulation

setting (as in computational geometry, geometric modelling, surface reconstruction), is that in topological graph theory the positions of vertices can be changed freely and the edges may be bent. In what follows we refer to the flip in geometric setting as a *geometric flip*.

We differentiate several types of triangulations. We refer to triangulations of points in the plane as **2D triangulations**, as usual. The situation in the space is more complex. If three-dimensional point data are given, solid–based and surface–based triangulations can be constructed. Solid–based triangulations represent collections of tetrahedra, and surface–based triangulations – collections of triangles. In applications solid–based triangulations are often referred to as *tetrahedralisations*, and surface–based closed triangulations - as *3D triangulations*. However, a solid–based triangulation is a three-dimensional object, and a surface–based triangulation is a *two–dimensional* polyhedral surface. Therefore, it seems more accurate to refer to solid–based triangulations in space as **3D** triangulations. In order to distinguish surface–based triangulations from their planar counterparts, as well as from solid–based triangulations, we designate them as **(2-3)D triangulations**. For surface–based triangulations for point data with elevation, *i.e.*, in the *functional setting* of the triangulation problem, the conventional notation is **2.5D triangulations**. In general, $(2-3)D$ triangulations can be divided into three types: closed $(2-3)D$ triangulations, $(2-3)D$–triangulations with boundary, that are not $2.5D$ triangulations, and $2.5D$ triangulations. We prefer the above–mentioned notations, as they are devoid of ambiguities. Moreover, a $(2-3)D$ triangulation is not always a bounding surface of a three–dimensional solid object, as self–intersections might occur.

An arbitrary $(2-3)D$ triangulation is denoted by $\triangle$, and symbol $T$ refers to a $2D$ triangulation. In this paper we always mean that $\triangle$ is closed (i.e. a triangulation of the data taken by some measurements from the surface of a closed object). The *star* $\mathbf{Str}(v)$ of a vertex $v$ is by definition the union of all the faces and edges that contain the vertex, and the *link* $\mathbf{Lnk}(v)$ of the star (the boundary of the star) is the union of all those edges of the faces of the star $\mathrm{Str}(v)$ that are not incident to $v$.

As a $(2-3)D$ triangulation represents a collection of triangles, a very appealing idea is to use the geometric flip as a transformation operation also for these triangulations [9,26,23,7,13,6]. In a $(2-3)D$ triangulation the diagonal (geometric) flip also exchanges two adjacent triangles $abc$ and $cbd$ in a triangulation $\Delta$ to the other two possible ones, namely to $abd$ and $adc$, however it does not make sense to speak about convex or non-convex quadrilaterals anymore. The flip operation in $(2-3)D$ triangulations has not yet been thoroughly studied, though some remarks are given in [3,23]. In [2,26,3,23] swapping edges in a non–convex quadrilateral was allowed, with the only restriction that swapping does not produce the already existing edge.

In [3] the problem of possible generation of self–intersections was noted. However, it was presumed that the *conventional agreement* on the diagonal flip operation had to be valid: the *diagonal flip in $(2-3)D$ triangulations may not produce self-intersections* [1], despite the fact that a $(2-3)D$ triangulation is situated in space, and not in a plane.

Indeed, in general, one assumes that a $(2-3)D$ triangulation, as a polyhedral triangulated surface, is embedded in $3D$, and therefore, has no self-intersections. This condition is supposed to be valid also for intermediate triangulations that one encounters in the optimisation procedure. We refer to the geometric flip operation which does not allow self–intersections as a *conventional geometric flip*.

In the same paper [3] the following problem was stated as open:

*Can any initial $(2-3)D$ triangulation of the same data set (and the same topological type) be transformed to any other by swapping appropriate edges in such a way that no self–intersections will occur (a direct analogue of Lawson's procedure)?*

However, it has been recently shown that self-intersections inevitably occur [1]. The authors demonstrated that the set of all triangulations, homeomorphic to the sphere that span the same data $V$, might be separated into at least two disjoint subsets with respect to the diagonal flip operation, namely, one cannot move from one subset to another by flipping an edge without producing a self-intersecting triangulation. A corresponding example has been also provided. The example is based on the following idea. Suppose we take a triangular prism and make a hole in it in the form of the prism as in Fig. 2.

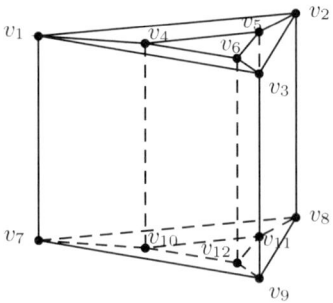

**Fig. 2.** Prism with a hole

Suppose now that we make a vertical cut in the obtained body (through $v_3v_6$ and $v_9v_{12}$) and then make a small opening along this cut in the body. Each of vertices $v_6$, $v_3$, $v_{12}$ and $v_9$ is split into two vertices. We denote these new vertices by $v_l'$ and $v_l''$, $l = 3, 6, 9, 12$. By 'closing' the holes in the obtained figure, *i.e.*, quadrilaterals $v_6', v_3', v_9', v_{12}'$ and $v_6'', v_3'', v_9'', v_{12}''$, we obtain a simple polyhedron (see the left picture in Fig 3).

 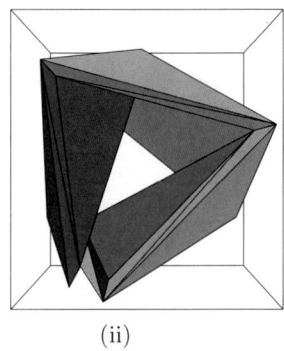

(i)                                      (ii)

**Fig. 3.** Left: 'Double prism', right: 'twisted double prism'

For simplicity, we call the left body in Fig. 3 a 'double prism'. Its vertices lie in two parallel planes (eight in the upper plane, and eight in the lower plane), therefore they are in convex position. The parts of the polyhedron that lie in these parallel planes are simple polygons. If we twist the upper polygon with respect to the lower one, for example in the counter–clock direction, we get a second polyhedron ('twisted double prism'). The vertices remain in the same planes, therefore are still in the convex position. Some of its edges of the 'internal' part of its surface are concave and some – convex (see the right picture in Fig. 3).

If we now want to swap the 'internal' concave edges, the new edges will penetrate the body of the polyhedron and we get the object as in Fig. 4.

We still can hope that if we first swap convex edges to concave ones in the 'internal' part of the surface of the object, and just make all internal edges concave, then, maybe, we can proceed with the procedure of swapping and avoid self–intersections. However, this is not the case, as demonstrated in [1].

Therefore, it seems logical to reconsider the restriction on selfintersections for the diagonal flip operation in $(2-3)D$ triangulations. In contrast to planar triangulations, a valid triangulation can still be determined even if a self-intersection has occurred after applying the flip operation. Moreover, the restriction impoverishes also the class of polyhedral surfaces: many surfaces in nature as well as in applications may be self-intersecting.

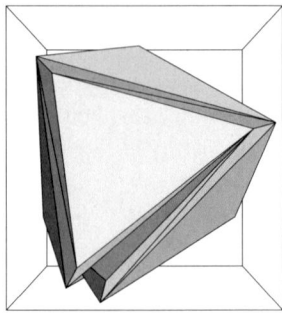

**Fig. 4.** Polyhedron with self–intersections

## 2.2   Notations and Definitions

A new strategy was announced in [5]. The strategy is to allow self–intersections in a triangulation and to define properly a local swapping procedure for a non-flat quadrilateral. This operation is called an extended diagonal flip, or simply, an **EDF**. Besides self-intersections, the **EDF** might even produce the 'gluing' of adjacent triangles. For this reason *generalised (2-3)D triangulations* have been defined, which admit multiple edges. It has also been proved that all generalised $(2-3)D$ triangulations of the data, which are in general position, and that are topologically equivalent to the 2D sphere are connected under the EDF.

Let us recall the main notions related to the EDF, such as *spatial quadrilaterals* and *orientation*. The figure, built of two adjacent triangles in $(2-3)D$ triangulations, is called a **spatial quadrilateral**. The edges of its two triangles except the common edge form a closed polygonal line. This line is called the **boundary of the quadrilateral**. If a triangulation is given, the boundary of a quadrilateral is fixed, and then there are only two possible spatial quadrilaterals with the same boundary. The common side of two adjacent triangles in a spatial quadrilateral is called a **spatial diagonal**. The orientation of the triangles forming the spatial quadrilateral must be coherent, as one deals with orientable surfaces (triangulations) (see Fig. 5).

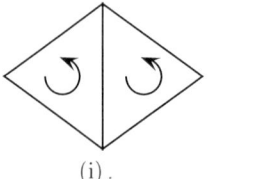
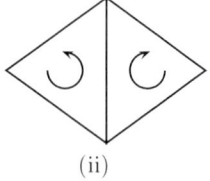

(i).                                                    (ii)

**Fig. 5.** Two triangles with a common edge: (i) coherently oriented,(ii) - not coherently oriented

For an oriented surface we can determine two directions of normals to a surface (and, therefore, to each triangle of the surface). Usually, the direction of outwards pointing normals is said to be positive. Then an edge (diagonal in the spatial quadrilateral) is said to be **concave** (sometimes, it is called also *reflex*), if two lines, determined by the unit normals to the adjacent triangles, sharing this edge, are intersecting in the positive direction, otherwise, the edge is called **convex**. The edge is **flat** if the normals are parallel.

If the data are situated in the general position, then flipping in space means exchanging the convex edge to the reflex edge and vice versa. Flipping then is well defined, even if self–intersections might occur, because flipping switches from one well-defined spatial quadrilateral to the other one, also well defined (see Fig. 6).

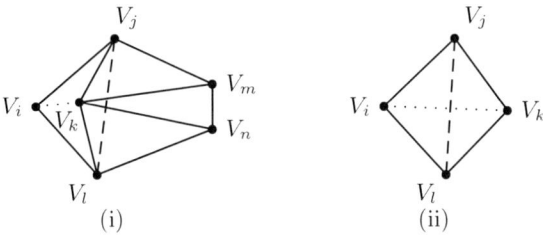

**Fig. 6.** Swapping an edge in the space: (i) Flipping an edge in a non–convex quadrilateral; (ii) Development of the obtained quadrilateral in the plane. In both cases the dotted edge swapped to the dashed one

The spatial quadrilateral $V_i V_j V_k V_l$ in Fig. 6 can always be developed in the plane, as presented in the right figure. Flipping involves only two adjacent triangles, and therefore always can be performed. One can, however, encounter local self-intersections in the star of a vertex (see Fig. 7).

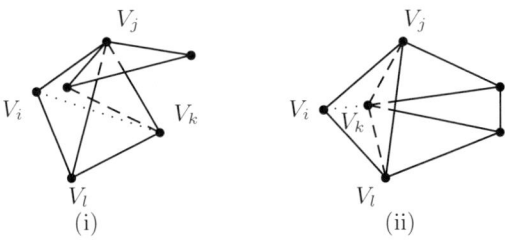

**Fig. 7.** Swapping diagonal $V_i V_k$ in a spatial quadrilateral: (i) With self–intersections. (ii) Without self–intersections. Vertex $V_j$ in the left picture is a so–called pinch point

A vertex with such a star, as in the left picture in Fig. 7, is known as a **pinch point** [15]. A pinch point is a singular point such that every neighbourhood of the point intersects itself. Pinch points are also called Whitney singularities or branch points. In the case of triangulations (polyhedral triangulated surfaces) a pinch point can be only a vertex and we would have an intersection in the star.

## 2.3   $(2-3)D$–Geometric Flip

We refer to $(2-3)D$ triangulations that are topologically equivalent to the $2D$ sphere as $(2-3)D_0$ triangulations. The diagonal flip operation, which we use, corresponds to the notions from subsection 2.2. Unlike to [5], we don't allow performing this operation if it produces multiple edges, and therefore 'gluing' of triangles doesn't occur. But we do not forbid self–intersections. We call this operation the *(2-3)D*–**geometric flip** (swap). We can show that the following result is valid :

**Theorem 1.** $(2-3)D_0$ *triangulations of the same data set* $V$ *(situated in the general position) are connected under the* $(2-3)D$*–geometric flip.*

Outline of the proof. The proof is based on putting a triangulation into one-to-one correspondence with its planar graph, and then on using the results from topological graph theory.

We outline here the main steps of the proof without going in details.

Suppose we have some initial $(2-3)D_0$ triangulation $\triangle$ with or without self–intersections. It can be viewed as an image of a sphere under some mapping $f$.

*Remark 1.* When we refer to the $2D$ *sphere* we mean the standard topological form, and when to *a sphere* we mean its usual realisation in the 3-space (as the boundary of a ball).

The mapping $f$ might be not one-to-one for all points on the sphere (if our $\triangle$ is not embedded in the space). However, if we consider both triangulations as graphs: one as a graph with $n$ vertices embedded on the sphere and the second as a graph on the given $\triangle$, we can put these graphs (triangulations) in one–to–one correspondence, in the sense that they are homeomorphic (there exists a graph isomorphism that induces also a bijection between correspondent faces).

The proof requires three main steps.

**1.** We can single out some triangle in the given $\triangle$ triangulation. Without loss of generality, let it be $v_1v_2v_3$. We can always renumber the vertices. To our triangulation corresponds a triangulation embedded on the sphere. This triangulation represents a planar graph and can be depicted in a schematic form in the plane as in Fig. 8. (Some internal vertices and edges are omitted).

Suppose, we perform a sequence of $(2-3)D$–geometric flips, starting from $\triangle$ and without affecting the edges in triangle $v_1v_2v_3$, thus obtaining a sequence

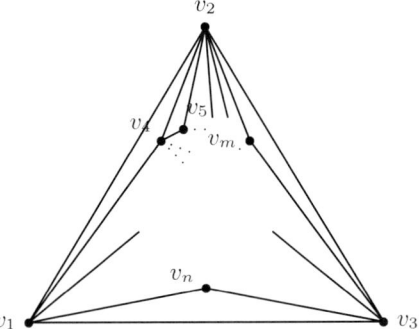

**Fig. 8.** A planar graph of some $(2-3)D_0$ closed triangulation

of $(2-3)D_0$ triangulations, each containing triangle $v_1v_2v_3$. As our transformations involve only edges we can always execute the corresponding topological diagonal flip operations in the graph on the sphere. One of the basic theorems in topological graph theory is Wagner's theorem [27]:

**Theorem 2 (Wagner).** *Any two triangulations on the sphere with the same number of vertices are equivalent to each other under diagonal flips, up to homeomorphism.*

According to the theorem we can reduce any triangulation on the sphere to the standard form $\Delta_n$ as in Fig. 9.

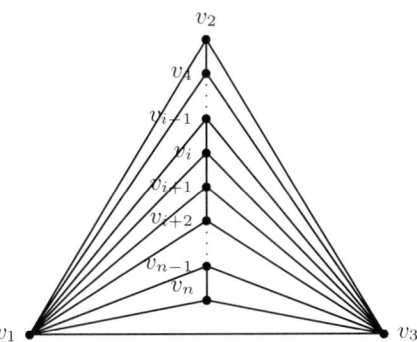

**Fig. 9.** The standard form of spherical triangulations $\Delta_n$

The proof of Wagner's theorem can be found in [24]. The transformation from the graph in Fig. 8 to the graph in Fig. 9 can be performed by flipping edges without producing multiple edges (but the edges might be bent in order to draw the graph without self–intersections). In $(2-3)D_0$ triangulations the corresponding transformation will be performed by swapping edges without producing multiple edges as well, but with possible self–intersections. Because the data are situated in the general position, no overlapping of triangles occur.

**2.** The next step is to show that we can exchange internal vertices in the graph, presented in Fig. 9. This can be done in four sub–steps. First we swap edge $v_1 v_i$ to edge $v_{i-1} v_{i+1}$, and then edge $v_3 v_{i+1}$ to edge $v_i v_{i+2}$. The third sub–step is to swap edge $v_{i-1} v_i$ to edge $v_{i+1} v_3$, and the last sub–step – to swap edge $v_{i+1} v_{i+2}$ to edge $v_i v_1$. The graph, obtained after the second sub–step, is depicted in Fig. 10. One can also see that in order to draw the resulting graph without self–intersections the edges must be bent, that might produce a self–intersection in the corresponding $(2-3)D_0$ triangulation.

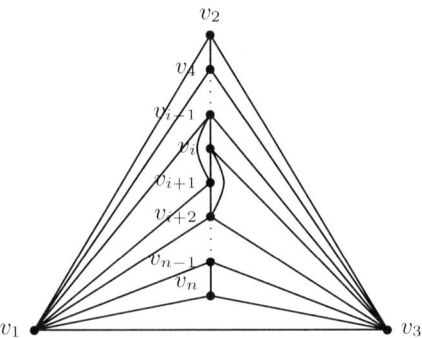

**Fig. 10.** The second step in exchanging two internal vertices

**3**. The last step is to show that any vertex in the outer triangle $v_1 v_2 v_3$ can be exchanged with an internal vertex. This can also be performed without producing multiple edges by exchanging appropriate vertices in a similar way as it was demonstrated for internal vertices in Step 2.

This demonstrates that any standard form can be transformed to any other standard form, and hence, completes the proof.

## 3  Minimisation of Total Absolute Extrinsic Curvature. Revision

### 3.1  Discrete Curvatures

Curvatures are among the fundamental concepts of the geometry of curves and surfaces. Their estimations provide important information about the object under study. For two–dimensional surfaces two main curvatures are the Gaussian and Mean Curvatures, and their combination describes sufficiently well the shape of a surface [11]. Calculation of the values of Gaussian and Mean curvatures is relatively easy if evaluating derivatives is possible. But in geometric modelling, dealing with three-dimensional discrete data, the evaluation of derivatives must be performed in a discretised form, and this yields a large number of operations and therefore a high computational cost.

Therefore, one attempts to use other techniques to estimate curvatures directly from the given data without referring to high-order formulas of differential geometry. Such curvatures are called discrete curvatures in order to distinguish them from their counterparts in the continuous setting.

Concepts of discrete curvatures are of growing interest for geometric modelling [22,10]. We can single out several reasons for this. Besides the attempts to reduce high computational expenses, one of the reasons is that many applications deal with three-dimensional discrete point data, and therefore, only a discrete approach makes sense. Another reason is that a polyhedral model of the underlying object, which is represented by discrete data, is the simplest way to obtain a preliminary sketch of the given object. In general, such a model is given in the form of a triangulated polyhedral surface, or simply a triangulation. Polyhedral surfaces belong to so-called non–regular surfaces, and for non-regular surfaces concepts of curvatures similar to classical curvatures are well determined [2]. One can define for polyhedral surfaces analogues of Gaussian and Mean curvature. The basic background idea is to use integral forms of Gaussian and Mean curvatures, defined for smooth surfaces. Indeed, the notion of curvature is strictly related to the notion of angle, and this relation manifests in the integral form of curvatures. For example, the Gaussian integral curvature is determined by the excess of a geodesic triangle, i.e. by the expression $\alpha + \beta + \gamma - \pi$, where $\alpha$, $\beta$, $\gamma$ are the angles of the triangle. However, one can determine more types of curvatures for polyhedral surfaces than for smooth ones. This is due to the fact that curvatures in polyhedral surfaces are concentrated exactly around the vertices and along the edges. Informally speaking, curvatures of a domain are 'glued' together [8,2].

One of the first works in which discrete curvatures were explored regarding approximation of surfaces was [8]. The authors investigated the problem of approximating by a smooth surface a given polyhedral surface so that the (discrete) curvatures of the polyhedral surface would be approximated sufficiently well by the curvatures of the approximating smooth surface. In applications one deals essentially with just the opposite problem: to find a suitable polyhedral surface (triangulation) such that it will approximate sufficiently well the smooth surface of the object under investigation. Even if in some applications the object under study is unknown a priori, one often requires its sufficient smoothness. The authors of [14] used discrete Gaussian and Mean curvatures in order to characterise a given geographic region (presented as a triangulated polyhedral surface). In [26,2,4] discrete curvatures were used in the procedure of optimisation for irregularly located $3D$ data. In [2] a comparative analysis of the differences between curvatures determined for smooth (regular) surfaces and polyhedral ones was provided. In the recent 5–6 years several other researchers have explored optimising procedures, using cost functions based on measurements of various discrete curvatures [13,22,10].

In this paper we provide a new insight on the optimality criterion which is based on minimising the discrete analogue of the integral absolute Gaussian curvature, namely, **total absolute extrinsic curvature (TAEC)**. To make our presentation self–contained and to emphasise the differences between the optimisation procedure, presented in [3], and the procedure, presented in this article, we briefly provide an overview of notions related to MTAEC with some revisions.

### 3.2   Curvatures around Vertices in $\triangle$

On the basis of the notion of the angle, the following curvatures for stars of vertices in triangulation $\triangle$ can be determined:

**1. Curvature $\omega(v)$ around Vertex $v$.**

This curvature is an analogue of the integral Gaussian curvature of a domain around some point in a smooth surface. Sometimes one refers to $\omega$ simply as the Gaussian curvature around a vertex. $\omega$ is defined as follows:

$$\omega(v) = 2\pi - \theta(v) \tag{1}$$

$\theta(v_i)$ is the total angle around vertex $v$, $\theta(v_i) = \sum \alpha_i$, and $\alpha_i$ are those angles of the faces in the star of $v$ that are incident to $v$. Expression 1 is also valid for any point $x$ in $\triangle$, but only for vertices $\omega$ might be not equal to zero (see Fig. 11).

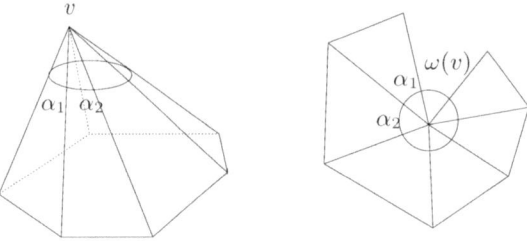

**Fig. 11.** Curvature $\omega$ around vertex $v$

**2. Positive (Extrinsic) Curvature $\omega^+(v)$.**

Curvatures that are analogues of the integral Gaussian curvature are concentrated in vertices, and positive and negative 'parts' of the curvature, if they both exist, are 'glued' together. But it is not difficult to separate them. If vertex $v$ belongs to the boundary of the convex hull of its star (i.e. the convex hull of $v$ and all vertices in its star), then we can single out another star $\mathrm{Str}^+(v)$ with $v$ as the vertex and those edges of $\mathrm{Str}(v)$ that belong to the boundary of the convex hull. The edges of $\mathrm{Str}^+(v)$ will determine the faces of $\mathrm{Str}^+(v)$. We refer to $\mathrm{Str}^+(v)$ as the convex cone of vertex v. Then

$$\omega^+(v) = 2\pi - \theta^+(v) \tag{2}$$

where $\theta^+(v)$ is the total angle around $v$ in $\mathrm{Str}^+(v)$. $\omega^+(v)$ is equal to zero, if the vertex and all the vertices in its star lie in the same plane. If the convex cone around $v$ doesn't exist, i.e. $v$ lies inside the convex hull of $\mathrm{Str}(v)$, then $\theta^+(v)$ is, by definition, equal to zero.

### 3. Negative (Extrinsic) Curvature $\omega^-(v)$.

We can now 'extract' the negative part of $\omega(v)$ as follows:

$$\omega^-(v) = \omega^+(v) - \omega(v) \tag{3}$$

### 4. Absolute (Extrinsic) Curvature $\omega(v)_{abs}$.

$$\omega(v)_{abs} = \omega^+(v) + \omega^-(v) \tag{4}$$

On the basis of the curvatures around a vertex one distinguishes three basic types of vertices: *Convex vertices* $(\omega^+(v) = \omega(v))$, *saddle vertices* $(\omega^-(v) = -\omega(v))$, and *mixed vertices* $(\omega^+(v) > 0, \omega^+(v) \neq \omega(v))$. In Fig. 12 examples of all three types of vertices are presented.

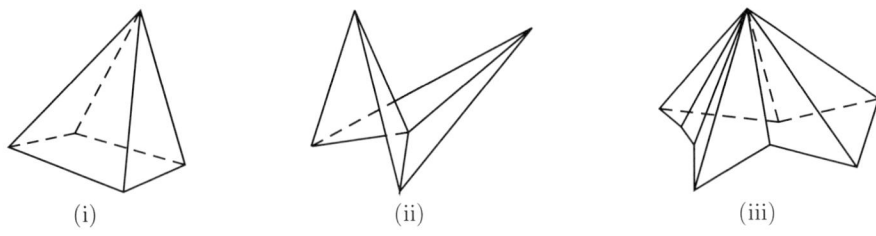

|            (i)            |            (ii)            |            (iii)            |

**Fig. 12.** Stars of vertices: (i) Convex (ii) Saddle (iii) Mixed

|            (i)            |            (ii)            |

**Fig. 13.** A mixed vertex (i) and its convex star (ii)

You can see a mixed vertex and its correspondent convex cone in Fig. 13.

Total absolute extrinsic curvature $\Omega_{abs}$ of $\triangle$ then defined as the sum of absolute extrinsic curvatures of all the vertices in the triangulation:

$$\Omega_{abs}\triangle = \sum_i \omega_{abs}(v_i) = \sum_i [\omega^+(v_i) + \omega^-(v_i)] \tag{5}$$

Convex, saddle and mixed vertices form three disjoint subsets of the triangulation $\triangle$. If we denote these subsets by {Vconv}, {Vsaddle} and {Vmix}, then expression 5 can be rewritten in the following form:

$$\Omega_{abs}(\triangle) = \sum_{V convex} \omega^+(v_{convex}) + \sum_{V saddle} \omega^-(v_{saddle}) + \sum_{V mix} [\omega^+ + \omega^-](v_{mix}) \quad (6)$$

### 3.3   MTAEC as Optimality Criterion

Let us note that we didn't make any assumption about the topological type of $\triangle$. Indeed, the above definitions are valid for an orientable polyhedral surface, which is determined by $\triangle$, of any genus. However, for a given closed three-dimensional point data set $V$, where under 'closed data' we mean the data originated or presumed to be originated from the surface of a closed object, we can, in general, construct triangulated surfaces of different genera with the given points as the vertices.

We are interested only in orientable polyhedral surfaces in this paper. All orientable triangulations $\triangle$ of the data set $V$ fall into disjoint subclasses of triangulations $\{\triangle(V)\}j$, $j = 1, ..., n$, in such a way that all triangulations of the same class have the same genus (i.e. topologically equivalent to the $2D$ sphere, the torus, and so on).

On the other hand, the data are situated, in general, on an existing surface. Even if we cannot view the surface, nevertheless, we can have the idea about its basic topological features as orientability and genus.

Therefore, it makes sense to apply an optimality criterion only to one subclass $\triangle(V)_l$ of the set $\{\triangle(V)\}$ of all triangulations, that span $V$. Then to each of this subclass we can put into correspondence the value $\Omega_{abs}(\triangle)$, and thus define a function $\{\Omega_{abs}(\triangle)\}_j$.

As we announced at the beginning, we will use minimisation of total absolute extrinsic curvature as the optimality criterion. A motivation for this is that in global differential geometry two–dimensional surfaces of minimal total absolute curvature possess some nice properties. For example, every local supporting plane is also global; surfaces of genus 0 are convex surfaces. We refer a reader for a survey to [2]. The precise definition of MTAEC triangulations is given below:

**Definition 1.** *A triangulation $\tilde{\triangle} \in \{\triangle(V)\}_l$ where $\{\triangle(V)\}_l$ is some subclass of possible triangulations that span data set $V$, is said to be a triangulation of minimal total absolute extrinsic curvature (MTAEC triangulation), if $\{\Omega_{abs}(\triangle)\}_l$ reaches its minimal value $\tilde{\Omega}_{abs}$ on $\tilde{\triangle}$.*

MTAEC triangulation may be not unique for general data, but for data in the convex and general position MTAEC triangulation is unique and coincides with the convex triangulation of the data (this fact follows from the classical theory of surfaces with minimal total absolute curvature). If the data are convex but not in the general position, then some flat domain consisting of several triangles might occur. These domains can be triangulated in different ways, so MTAEC triangulation is unique up to re-triangulation of flat domains in the triangulation. We use this nice property in Section 4.

# 4   Self–intersections

## 4.1   Mixed Vertices

The definitions, presented in this subchapter, are valid for any orientable $(2-3)D$ triangulations of any genus.

Note that convex vertices can be split into two categories: convex and concave vertices, depending how the normals to their faces are intersecting. If their intersection is in the negative direction (see subsection 2.2), then a vertex is called *convex*, if the intersection of the normals is in the positive direction, then the vertex is said to be *concave*.

Convex cones of mixed vertices fall also into these two categories: convex 'convex cones' and concave 'convex cones'. An easy way to determine whether a vertex possesses the convex cone, is to determine if its star has a local supporting plane that passes through the vertex and some vertices (at least two) in the its link. Each face of the convex cone determines a local supporting plane. Therefore, we can further decompose the types of vertices into two subsets: *convex–mixed* vertices if the convex cones of their stars are convex, and *concave–mixed* vertices, if their stars posses concave convex cones.

The structure of a mixed vertex and saddle vertex might be rather complicated. For example, in [22] several examples of complicated mixed vertices are given. Here some remarks are needed. In [22] the authors distinguish six types of vertices: convex, concave, saddle, convex-fan, concave-fan, and saddle-fan. In our terminology, a convex-fan vertex is convex–mixed, and a concave–fan vertex is concave–mixed respectively.

The authors distinguish a saddle-fan vertex in order to make some comparison with a mixed vertex. Indeed, both types of vertices have 'folds' in their stars, but a mixed vertex, however, possesses always the convex cone around its star, and a saddle–fan does not have this property. The name 'mixed vertex' reflects the fact, that its curvature has two parts: positive and negative, while the curvature of a saddle vertex, however complicated it might be, is always negative. The 'folds' in the star of a saddle–fan vertex is due to the fact that in its star several simple saddles are 'glued' together. An example of a saddle–fan vertex is the *monkey saddle* [2].

However, since we allow self–intersections, a new type of the mixed vertex must be introduced. This type is called a **pinch vertex** and has self–intersections in its star. For a mixed pinch vertex, however, the convex cone can be defined in the same way, as for 'normal' (embedded) mixed vertices. In Fig. 14 an example of a mixed pinch vertex and its correspondent convex cone is given.

*Remark 2.* A saddle vertex might also have self–intersections in its star, and in this case, is defined as a **saddle pinch vertex**. However, it doesn't possess positive curvature, and, therefore, not a mixed vertex.

## 4.2   Triangulations of Convex Data. Revisions of the Properties

In this section we discuss the optimisation procedure, based on MTAEC, with respect to the properties of $(2-3)D_0$ triangulations of convex data. As said

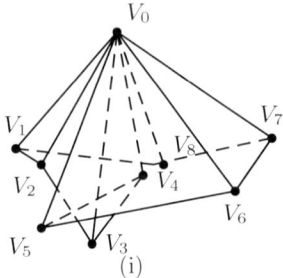 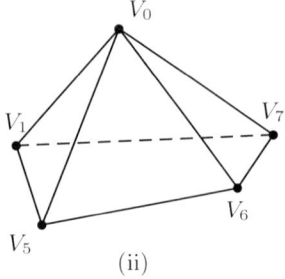

**Fig. 14.** Pinch vertex (i) and its convex star (ii)

above, the MTAEC triangulation of convex data coincides with the convex triangulation. Nevertheless, an arbitrary $(2-3)D_0$ triangulation of the given data points $V_i$, $i = 1, ..., n$, that are situated in the convex position, is not convex, and might even be very complicated, if the number of data points is large. But even for a small data set, a triangulation of convex data can strongly deviate from the convex triangulation, as we saw in Section 2. Therefore, testing our implementation procedure on triangulations of convex data is informative. Triangulations of convex data are partially studied in [3], where minimisation of TAEC was chosen as the optimality criterion, and the geometric flip as the transformation operation. In [3] it was also claimed that the algorithm was global, i.e. one can recover a convex triangulation starting from any initial one. However, there is a flaw in the reasoning. It assumes that no self–intersections might occur; we now know that this is not the case.

Actually in [3] has been shown that if an initial triangulation $\triangle_{init}$ belongs to the subset $\{D_{good}\}$ of all possible $(2-3)D_0$ triangulations of the given convex data, then by using MTAEC as the optimality criterion and the geometric flip as a transformation one obtains the convex triangulation without self-intersections in the process of optimisation. One of the properties of $\{D_{good}\}$ is that it contains the convex triangulation.

One of the open problems still is *whether this subset $\{D_{good}\}$ is unique for the given data?*

The answer to this question seems positive, and it is based on the following argumentation: Suppose we have two such subsets, $\{D_{good}\}_1$ and $\{D_{good}\}_2$, and $\triangle_1$ is in $\{D_{good}\}_1$ and $\triangle_2$ is in $\{D_{good}\}_2$. Both $\{D_{good}\}_1$ and $\{D_{good}\}_2$ contain the convex triangulation $\triangle_{conv}$. Therefore, starting from $\triangle_1$ we reach $\triangle_{conv}$, and by reversing the process of optimisation we get $\triangle_2$.

If the data are in the convex position, then all vertices lie in the boundary of the convex hull of the data, and, hence, in the convex triangulation all local supporting planes are global. One can therefore conclude, that any triangulation of convex data doesn't contain saddle vertices. See [3] for a detailed argumentation. Starting from this fact and the assumption that no self-intersections might occur, [3] concluded, that any triangulation of convex data does not contain concave vertices or mixed vertices with concave 'convex cone'. This conclusion is wrong. See, for example, Fig. 15.

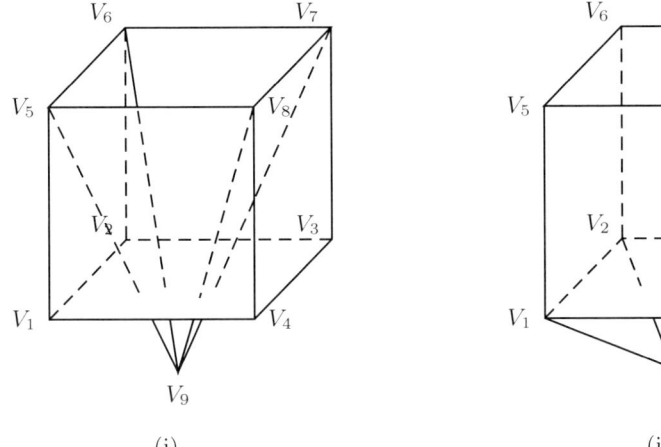

**Fig. 15.** Two triangulations of convex data: (i) with a global–self–intersections; (ii) convex triangulation

In this figure you can see two triangulations of the data set, consisting of nine vertices that are situated in the convex position. (The flat domains in the triangulations are not triangulated). The first triangulation has a self-intersection, and vertex $V_9$ is a concave vertex according to the definition. (One can see the negative side of the surface, protruding through the flat face $V_1V_2V_3V_4$). This self-intersection is of a global character and the first triangulation doesn't essentially differ from a triangulation of non-convex data. Indeed, compare this triangulation with the triangulation of non–convex data in Fig. 16.

It has been shown that the algorithm based on minimisation of TAEC might not reach the global minimum for triangulations of non-convex data [3]. Therefore, the algorithm based on minimisation of TAEC might not reach the global minimum even on triangulations of convex data. In the following section we discuss the procedure of optimisation based on minimisation of TAEC in the view of the remarks concerning self–intersections.

## 5   Recovering the Convex Triangulation. Revision

In this section we present a new algorithm to obtain MTAEC triangulations that allows to work with self–intersections. We present a comparative analysis of this algorithm with the previous algorithm, introduced in [2,26]. Then we make some adjustments to ensure convergence of our new algorithm in most cases. Finally, we give an example of recovering the convex triangulation from a very 'bad' initial triangulation.

### 5.1   Algorithm

Recall, that the notion of mixed vertex includes the 'pinch vertex'. The pinch vertex is characterised by local self-intersections in its star. Mixed vertices which

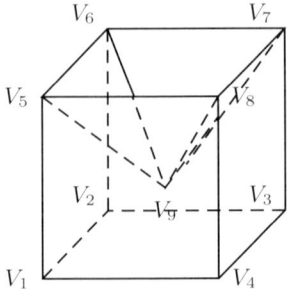

**Fig. 16.** A triangulation of non–convex data

are not pinch vertices, are called *simple mixed vertices*. As it was said in subsection 4.1, the convex cone of the star of a pinch vertex is defined in the same way as that of a simple mixed vertex.

Taking into consideration the existence of mixed pinch vertices, we make necessary adaptations to the algorithm, presented in [2,26].

This algorithm can schematically be described in the following way:

1. i=1.
2. Given a triangulation $\triangle_i$. Determine the convex cones for all its vertices.
3. Compute TAEC according to formula 5.
4. Minimise TAEC by applying the (conventional) geometric flip. Transform $\triangle_i$ to $\triangle_l$ by using the geometric flip, only if TAEC($\triangle_l$) is less than TAEC($\triangle_i$). If no transformation is possible, then Stop. $\triangle_i$ is the final triangulation. Otherwise, l=i, and go to Step 2.

The most serious flaw the algorithm of [26,2] is in the procedure, which determines the convex cone of a vertex (Step 2).

**Procedure 'Convex Cone' in [26,2] (Schematically).**

1. Determine a supporting plane for the star of a vertex. If the supporting plane does not exist, than write 'vertex is saddle' and go to the next vertex.
   To determine a supporting plane it is enough to detect one face of the convex cone of the star. This face is determined by the vertex itself and two neighbouring edges in the convex cone of the star of the vertex. See, for example, Fig. 17.
   Without loss of generality, suppose that the face of the convex cone which is first determined, is the face $v_0v_1v_4$.
2. Determine the coherent orientation of the face $v_0v_1v_4$. Determine the direction of the outwards normal to this face. This singles out the positive and negative sub–spaces in which the supporting plane, determined by $v_0v_1v_3$, divides the space.

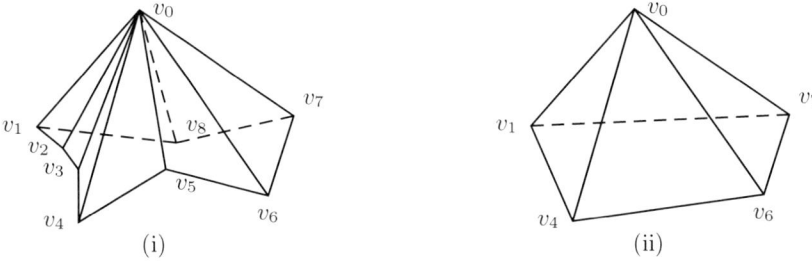

**Fig. 17.** A simple mixed vertex (i) and its convex star (ii)

3. Find out, whether the end–points of the remaining edges in the star of $v_0$ lie in the positive or negative sub-space. If they lie in the negative sub–space, than the convex cone is convex 'convex', otherwise, it is concave 'convex'.
4. Determine other edges of the convex cone.

The first three steps are simple computing exercises.

*Remark 3.* In [26] it was assumed, that no self–intersections occur in the star, so the last step is simplified. A simple observation is that the edges in a convex 'convex cone' must all be convex, and in a concave 'convex cone' they are concave. Without loss of generality, suppose that the star in Fig. 17 is convex–mixed, then the other edges of its convex cone can be determined as follows (the simplified version):

– Take the faces $v_0v_1v_4$ and $v_0v_4v_5$, determine if the edge $v_0v_4$ is convex. If yes, add the face $v_0v_4v_5$ to the convex cone of the star. If not, add the face $v_0v_4v_6$. Proceed in the similar way till all edges of the convex cone are determined.

In Fig. 14 you can see that the above simplified procedure does not work. By trying to apply the simplified procedure 'Convex cone' to a triangulations with pinch vertices, we detected that this procedure did not work with self-intersections. The corrected version of this procedure is presented below:

**Procedure 'Convex Cone' (Corrected).**

– Determine an initial *non–flat* convex cone, with vertex $v_0$ as the apex, and edges $v_0v_{i_1}$, $v_0v_{i_2}$ and $v_0v_{i_3}$. The cone is *flat* if $v_0$, $v_{i_1}$, $v_{i_2}$ and $v_{i_3}$ lie in one plane. If no non–flat cone is found, then $v_0$ and all vertices in its link lie in the same plane, and $\omega^+(v_0) = \omega(v_0) = 0$
– If a non–flat cone is found, then check if you can maximise it by adding new edges from remaining ones.

At the end of this procedure either the convex cone of the star of the vertex is determined, or the vertex is a saddle vertex. This procedure determines properly the convex cone of a pinch vertex.

We can make one more adjustment in the algorithm of [2,26], namely in Step 3.

Note that the expression 5 for $\Omega_{abs}(\triangle)$ can be rewritten as follows

$$\Omega_{abs}(\triangle) = \sum_i \omega_{abs}(v_i) = \sum_i [2\omega^+(v_i) - \omega(v_i)] \tag{7}$$

or

$$\Omega_{abs}(\triangle) = 2\sum_i [\omega^+(v_i)] - \sum_i [\omega(v_i)] \tag{8}$$

Then

$$\Omega_{abs}^+ = \sum_i \omega^+(v_i), \tag{9}$$

where $\Omega_{abs}^+$ is called **total positive curvature** (**TPC**) of triangulation $\triangle$. Expression $\Omega = \sum_i \omega(v_i)$ is the total curvature of $\triangle$, and is a direct analogue of total Gaussian curvature of a surface in a classical sense, and is constant. Its value depends on the genus of a surface, and for surface, topologically the $2D$ sphere, this value is $4\pi$.

Therefore, minimisation of TAEC is equivalent to minimisation of TPC. TPC reaches its minimum on convex triangulations, and this minimum value is also equal to $4\pi$. Computing only TPC reduces the computational cost.

Then the algorithm can be rewritten:

1. i=1.
2. Given a triangulation $\triangle_i$. Determine the convex cones for all its vertices (by using the corrected version of this procedure).
3. Compute TPC according to formula 9.
4. Minimise TPC by applying the $(2-3)D$–geometric flip. Transform $\triangle_i$ to $\triangle_l$ by using the geometric flip, only if TPC($\triangle_l$) is less than TPC($\triangle_i$). If no transformation is possible, then Stop. $\triangle_i$ is the final triangulation. Otherwise, l=i, and go to Step 2.


## 5.2   Remarks on the Convergence of the Algorithm

In [3] on the basis of the assumption, that no self–intersections may occur, was concluded that the algorithm, based on MTAEC, converged to the global minimum in the case of convex data, *i.e.*, the convex triangulation was always to be recovered. As we now can see, this is not always. However, here we claim, that if one starts with an initial triangulation of the given convex data, that doesn't have self-intersections or has only local self-intersections, then no global self-intersection might occur in the process of optimisation and the algorithm converges to the global minimum. We correct the result from [3] in the following proposition:

**Proposition 3** *The algorithm, based on minimising total absolute extrinsic curvature, or, equivalently, on minimising total positive curvature, converges to the global minimum for convex data, if the initial triangulation does not contain global self–intersections.*

Outline of the proof. We outline the proof by correcting some of the conclusions in the proof, presented in [3].

Let us recall some definitions and lemmas from [3].

1. Minimisation of the total absolute curvature of a triangulation of closed data (not necessarily convex) is equivalent to maximisation of the convex cones of the vertices of a triangulation.
   A convex cone of the star of vertex $v$ is called maximal if $\omega^+(v)$ is minimal with respect to all triangulations of the given data.
2. Any triangulation of convex data does not contain proper saddle vertices.
3. Any star in a triangulation of convex data has a convex cone.
4. In a convex triangulation all vertices are proper convex vertices.
5. Two convex cones in a triangulation are called *coherent*, if the normals to their triangles intersect in the same direction (positive or negative).
6. There are no concave vertices in any proper (i.e without self-intersections) triangulation of convex data.
7. All convex cones in a triangulation of convex data are coherent with corresponding maximal convex cones, or, in other words, all convex cones are positive.
8. If a triangulation of convex data is not convex, then there is always an edge that will be swapped.

The last proposition in [3] leads to the conclusion, that after swapping all edges that might be swapped, the convex triangulation of the data is obtained, or an internal loop has been generated, that does not contain the edges of convex 'convex cones' of vertices. Internal loop might be a degenerated one, when one of its boundary contains only one vertex. Actually, all the above–mentioned considerations are true. But on the base of these considerations the following conclusion is made in [3]:

– The existence of a loop implies non–convexity of the data.

This conclusion we rewrite in the following form:

**Proposition 4** *The existence of a loop implies global self–intersections in a triangulation of the data.*

But a global self–intersection means, that some convex cones are not coherent with the corresponding maximal convex cones (i.e. convex cones in the convex triangulation). If we have only local self–intersections, then all convex cones are coherent with the corresponding maximal cones, and by flipping an edge in the algorithm, that yields minimisation of TPC, we cannot change the property of a convex cone to be coherent. Thus we get proposition 3.

## 5.3   Example of the Implementation

We have tested our optimisation procedure on several convex data sets. As we expected, the algorithm, based on minimising TPC, doesn't converge to global minimum (convex triangulation) if the initial triangulation has global self–intersections. In the case of local self–intersections it results in the convex triangulation. The example, presented here, shows how the algorithm works if the initial triangulation is the 'twisted double prism' (presented in the left picture in Fig. 18). This polyhedron plays a key role in [1] in order to show that there is no sequence of geometric (conventional) flips from the triangulation, represented by this polyhedron, to the convex triangulation of the data. The data set is apparently very simple, but its choice doesn't influence the fundamental character of the obtained results. This example shows, that even for such simple data, local self–intersections are inevitable. Our optimisation procedure, with TPC as the minimisation criterion and $(2-3)D$–geometric flip as the local transformation, allows to eliminate local self–intersections. We have been able to recover the convex triangulation, starting from the twisted double prism. The twisted prism, after a sequence of transformations, is transformed to the final convex triangulation, as presented in Fig. 18. Intermediate triangulations contain self–intersections. Any of these intermediate triangulations can be taken as an initial one.

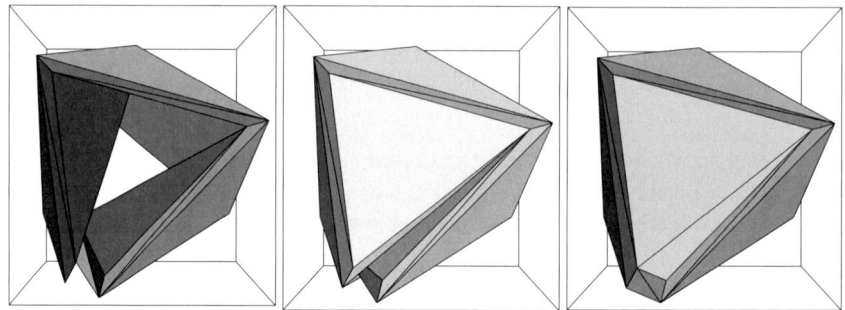

**Fig. 18.** Transformation of the 'twisted prism'

## 6   Conclusion

We introduce an optimisation procedure for triangulated polyhedral surfaces that allows to process self–intersecting surfaces. We present an analysis of the nature of self–intersections and on the basis of this analysis the limitations of the optimisation procedure are explained. Our results provide also a new view on the concept of surface–based triangulations of three–dimensional data sets, as well as on their optimisation, and might be useful in the analysis of global and local convergence for other optimisation algorithms.

An open problem is how to treat global self–intersections. Another open problem is to study the optimisation procedure for non–convex data. We conjecture, that the optimisation procedure for non–convex data, based on minimising TPC, will result in 'unfolding' all local self–intersections as well. Surprisingly, pinch points, that are considered to be cumbersome singularities in the continuous setting, seem to be easily dealt with in the case of triangulations.

The last open problem which we want to emphasise is complexity. The number of operations in the algorithm was linear with respect to the number of vertices. However, a rigorous analysis of complexity of the procedure is required.

The research on optimisation procedures that allow self-intersections is in its beginning, but it opens new perspectives for surface modelling from discrete scattered three-dimensional data.

## Acknowledgements

I wish to thank Prof. W. Kühnel for numerous discussions, and Prof. S. Negami for useful information. Also many thanks to my friend Carol Waddington for editorial help, and to my colleges Prof. M. Rodrigues, Dr. J. Penders and Alan Robinson for helpful suggestions. I would like also to thank the anonymous reviewers for useful comments.

## References

1. Aichholzer, O., Hurtado, F., and Alboul, L.: On flips in polyhedral surfaces. Int. J. of Found. of Comp. Sc., 13(2), 2002, pp. 303–311.
2. Alboul, L., and van Damme, R.: Polyhedral metrics in surface reconstruction. In: Mullineux, G. (ed.), The Mathematics of Surfaces VI, Clarendon Press, Oxford, 1996, pp. 171–200.
3. Alboul L., and van Damme, R.: Polyhedral metrics in surface reconstruction: Tight triangulations. In: Goodman, T. (ed.), The Mathematics of Surfaces VII, Clarendon Press, Oxford, 1997, pp. 309–336.
4. Alboul, L., Kloosterman, G., Traas, C. R., and van Damme R.,: Best data-dependent triangulations. J. of Comp. and Applied Math., 119, 2000, pp. 1–12.
5. Alboul, L., and van Damme, R.: On flips in polyhedral surfaces: a new development. In Proc. of EWCG 2002, April 10-12, 2002, Warsaw, pp. 80–84.
6. Alliez, P., Meyer, M., and Desbrun, M.: Interactive geometry remeshing. ACM Transactions on Graphics (TOG), 21(3), 2002, pp. 347–354.
7. Bischoff, S., and Kobbelt, L.: Towards robust broadcasting of geometry data. 2001. To appear in Computers and Graphics.
8. Brehm, U., and Kühnel, W.: Smooth approximation of polyhedral surfaces with respect to curvature measures. In: Global differential geometry, 1979, pp. 64-68.
9. Choi, B. K., Shin, H. Y., Yoon, Y. I., and Lee, J. W.: Triangulations of Scattered Data in $3D$ Space. Comp. Aided Design, 20(5), 1988, pp. 239–248.
10. Desbrun, M., Meyer, M., Schröder, P., and Barr, A. H.: Discrete Differential-Geometry Operators for Triangulated 2-Manifolds. Proc. of the Int. Workshop on Visualization and Mathematics (VisMath'02), 2002, Berlin, 27 pp.
11. do Carmo, M. P.: Differential geometry of curves and surfaces. Prentice-Hall, 1976.
12. Dyn, N., Goren, I., and Rippa, A.: Transforming Triangulations in Polygonal Domains. Computer Aided Geometric Design, 10, 1993, pp. 531–536.

13. Dyn N., Hormann, K., Kim S.-J., and Levin, D.: Optimising 3D triangulations using discrete curvature analysis. In: Lyche, T., and Schumaker (eds.), Mathematical Methods in CAGD, Vanderbilt University Press, 2001, pp. 1–12.

14. Falcidieno, B., and Spagnuolo, M.: A new method for the characterization of topographic surfaces. Int. J. Geographical information systems, 5(4), 1991, pp. 397–412.

15. Francis, G. K.: A Topological Picturebook. Springer–Verlag, Berlin Heidelberg New York, 1987.

16. Ignatiev, Y., Matveev, I., and Mikushin, V.: Generating consistent triangulation for b-rep models with parametric face representation in 3D–Vision system, Grid Generation: Theory and Applications, RAS, 2002, p. 241–266. ISBN 5-201-09784-7. 3D–Vision system, industrial software.
    (Internet: http://www.ccas.ru/gridgen/ggta02/papers/Matveev.pdf)

17. Kobbelt, L.: Fairing by finite difference methods. In: Mathematical Methods for Curves and Surfaces II, M. Daehlen, T. Lyche, and L. L. Schumaker (eds.), Vanderbilt University Press, 1998, pp. 279–286.

18. Krsek, P., Lucács, G., and Martin, R. R.,: Algorithms for computing curvatures from range data. In: R. Cripps (ed.), The Mathematics of Surfaces VIII 1998, pp. 1-16.

19. Kühnel, W.: Tight Polyhedral Submanifolds and Tight Triangulations. Lecture Notes in Mathematics, Vol. 1612. Springer–Verlag, Berlin Heidelberg New York, 1995.

20. Lawson, C., Transforming triangulations. Discrete mathematics, 3, 1972, pp. 365–372.

21. Ma, L., Wang, Q., and Kai Yun, T.C.: Recursive Algorithms for 3D Triangulation Decomposition and 2D Parameterization. 5th International Conference on Computer Integrated Manufacturing (ICCIM) 2000, Singapore, 28-30 March 2000, 12 pp.

22. Maltret, J.-L., and Daniel, M.: Discrete curvatures and applications: a survey. http://creatis-www.insa-lyon.fr/~frog/curvature

23. Morris, D. D., and Kanade, T.: Image-Consistent Surface Triangualtion. Proc. 19th Conf. Computer Vision and Pattern Recognition, IEEE, 1, 2000, pp. 332–338.

24. Negami, S.: Diagonal flips of triangulations on surfaces. Yokohama Math. J., 47, 1999, pp. 1–40.

25. Rocchini, C., Cignoni, P., Montani, C., Pingi, P., and Scopigno, R.: A low cost 3D scanner based on structured light. EUROGRAPHICS 2001, 20 (3), 2001, pp. 298–307.

26. van Damme, R., and Alboul, L.: Tight triangulations. In: Daehlen M. et al (eds.), Mathematical methods in CAGD III, 1995, pp.517-526.

27. Wagner, K.: Bemekungen zum Vierfarbenproblem. J. der Deut. Math., 46 (1), 1936, pp. 26–32.

28. Weinert, K., Mehnen, J., Albersmann, F., and Drerup, P.: New solutions for surface reconstruction from discrete point data by means of computational intelligence. In Int'l seminar on intelligent computation in manufacturing engineering (ICME '98), pp. 431–438.

# Combinatorial Properties of Subdivision Meshes

Ioannis Ivrissimtzis and Hans-Peter Seidel

Max-Planck Institute für Informatik
Stuhlsatzenhausweg 85, Saarbrücken, 66123, Germany
{ivrissimtzis,hpseidel}@mpi-sb.mpg.de
http://www.mpi-sb.mpg.de

**Abstract.** In this paper we study subdivision from a graph-theoretic point of view. In particular, we study the chromatic numbers of subdivision meshes, that is the number of distinct colors we need for a vertex, face or edge coloring of a subdivision mesh. We show that, unlike the size, the chromatic numbers of subdivision meshes are not larger than the corresponding chromatic numbers of the initial mesh and sometimes are even smaller.

## 1  Introduction

In this paper we are dealing with the combinatorial properties of subdivision meshes, that is, meshes constructed from other meshes by applying one step of a subdivision scheme. Our aim is to pose, and answer when possible, questions usually arising within the context of graph-theory, such as the existence of vertex, face or edge colorings. Although the study of the subdivision meshes from a graph-theoretic point of view has not attracted a lot of interest yet, except of giving new insights into subdivision processes, can also have more concrete applications. In [1], for example, it is shown that meshes with a particular combinatorial structure accept refinements much slower than the refinements that can be applied to general meshes.

The relation between meshes and graphs is straightforward. A mesh is an embedding of a graph into a surface such that every face is homeomorphic to an open disk. The basic primitives of a mesh are the vertices the edges and the faces. The vertices and the edges are the primitives of the abstract graph while the faces arise from its embedding in a surface. We call two vertices adjacent if they are connected with an edge, two edges are adjacent when they share a common vertex while two faces are adjacent if they have a common edge.

The basic combinatorial structures we will deal with are the colorings of these primitives. A vertex $n$-coloring of a mesh is a coloring of the vertices of the mesh with $n$ colors, or from another point view a function from the set of vertices of the mesh to the set of colors

$$V \to C = \{0, 1, 2, \ldots, n-1\} \tag{1}$$

such that no two adjacent vertices have the same color. Similarly a face $n$-coloring is a coloring of the set of faces

$$F \to C = \{0, 1, 2, \ldots, n-1\} \tag{2}$$

M.J. Wilson and R.R. Martin (Eds.): Mathematics of Surfaces 2003, LNCS 2768, pp. 73–84, 2003.

such that no two adjacent faces have the same color. An edge $n$-coloring is a coloring of the set of edges

$$E \rightarrow C = \{0, 1, 2, \ldots, n-1\} \tag{3}$$

such that no two adjacent edges have the same color.

In the section dealing with the edge-colorings we will also briefly deal with a combinatorial structure related to edge-colorings, the flows. A $k$ flow on a graph $G$ assigns to each edge a direction and a positive integer weight from the set $\{1, 2, \ldots, k-1\}$ such that for each vertex the sum of weights directed into it is equal to the sum of weights directed away from it.

Fig. 1 shows a vertex 2-coloring a face 3-coloring an edge 3-coloring and a flow on the mesh corresponding to the cube.

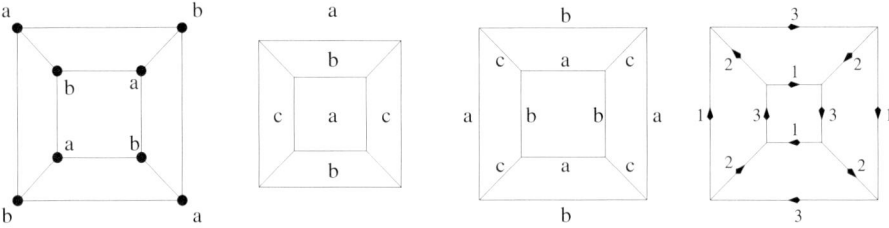

**Fig. 1.** Combinatorial structures on the mesh of a cube. Vertex coloring, face coloring, edge coloring and flow.

As far as possible we will try to work independently of the particular subdivision scheme. We will only assume the usual regular behavior of the subdivision, that is, if $M$ is the initial mesh and $M'$ the refined mesh we have

- All the vertices of $M$ retain their valence in $M'$.
- The new inserted vertices of $M'$ are regular in that they have valence 6 if $M$ is triangular and they have valence 4 if $M$ is quadrilateral.
- Any two vertices of $M$ connected by an edge are not connected in $M'$

We can easily verify that most of the known primal subdivision schemes satisfy these conditions, while the dual schemes like the Doo-Sabin [2] can be treated in an analogous way through the vertex-face duality. If we cannot have any satisfactory results under the above general assumption we will treat particular subdivision schemes, like the Loop [3] and the $\sqrt{3}$ [4] scheme for triangular meshes and the Catmull-Clark [5] for quadrilateral.

Our results will show that although the subdivision process increases the number of vertices, faces and edges of a mesh, nevertheless, it generally reduces its combinatorial complexity. An intuitive explanation is that the initial mesh is assumed to belong to the general class of meshes, while after one subdivision step the mesh belongs to a proper subclass of this class.

Throughout the paper we will keep a consistent notation denoting by $M$ an initial arbitrary mesh, $M'$ the mesh arising after one subdivision step, and $G, G'$ the corresponding underlying graphs. $M_d$, $M'_d$ will denote the duals of $M, M'$ corresponding, and $G_d$, $G'_d$ the underlying graphs.

## 2  Terminology and Some Basic Results

In this section we give some standard terminology and some basic graph-theoretic results, following mainly the two book references [6], [7].

The chromatic number of a graph $G$ is a positive integer $\chi(G)$, with the property that $G$ has a vertex coloring with $\chi(G)$ colors but not with $\chi(G) - 1$ colors.

The maximum valence of the vertices of $G$ is denoted by $\Delta(G)$. A graph with all its vertices having valence 3 is called cubic.

The complete graph $K_n$ consists of $n$ vertices and has the property that every vertex is connected with all the other vertices. Obviously $\chi(K_n) = n$.

A graph with $n$ vertices $x_0, x_1, \ldots, x_{n-1}$ and $n$ edges $x_0 x_1, x_1 x_2, \ldots x_{n-1} x_0$ is called a cycle of length $n$. A cycle is called even or odd according to its length. Usually we are interested in cycles as subgraphs of larger more general graphs.

A first basic theorem regarding the chromatic number of a graph is

**Proposition 2.1 (Brooks).** $\chi(G) \leq \Delta(G) + 1$ holds for every graph $G$. Moreover; $\chi(G) = \Delta(G) + 1$ if and only if either $\Delta(G) \neq 2$ and $G$ has a complete $(\Delta(G) + 1)$-graph $K_{\Delta(G)+1}$ as a connected component, or $\Delta(G) = 2$ and $G$ has an odd cycle as a connected component.

See [6], p.7. Usually we suppose that $G$ is connected and because its vertices have valences at most $\Delta(G)$ we can easily see that $G$ has $K_{\Delta(G)+1}$ as a connected component only if $G = K_{\Delta(G)+1}$, while, if $\Delta(G) = 2$ it has an odd cycle as a connected component only if it is an odd cycle.

Another useful result is

**Proposition 2.2 (König).** $\chi(G) = 2$ if and only if it does not contain an odd cycle.

See [6], p.9. Notice that this result is essentially different from Proposition 2.1 as does not assume anything about the maximum valence of the vertices of $G$. For planar graphs we can substitute the condition $G$ has not an odd cycle with the weaker assumption that $G$ has not a face with an odd number of edges.

We will also need the following simple lemma

**Proposition 2.3.** Let $G$ be a graph embedded in a surface as a triangulation. If $\chi(G) = 3$ then $G$ has a face coloring in two colors.

**Proof.**  Let $\chi(G) = 3$. The vertices of any triangle have all different colors, and so, going anticlockwise around the vertices of a triangle we get either the permutation $(123)$ or the inverse permutation $(123)^{-1} = (132)$. We notice that to any two adjacent faces correspond inverse permutations. Coloring the faces in two colors, according to the type of the permutation of the vertices we obtain a face 2-coloring. $\qquad\square$

We notice that the face 2-coloring and the vertex 3-coloring of $G$ are unique up to a permutation of the colors, and also, by Proposition 2.2 the face 2-coloring is equivalent to the non-existence of an odd cycle on the dual of $G$.

## 3   Face Colorings

A face coloring of a mesh is equivalent to the vertex coloring of its dual. Here we will study the face colorings of subdivision meshes mainly through this duality. We will find propositions regarding the minimum number of distinct colors we need for a face coloring of the mesh $M'$, a number which obviously is equal to the chromatic number of the dual of $M'$.

### 3.1   Triangle Meshes

If the mesh $M$ and thus the $M'$ are triangular meshes, then the graphs of their duals $G_d, G'_d$ are cubic graphs. Hence, we have

**Proposition 3.1.** The inequalities

$$\chi(G_d) \leq 4 \quad \text{and} \quad \chi(G'_d) \leq 3 \tag{4}$$

hold.

**Proof.** The first inequality is a direct consequence of Proposition 2.1 and the fact that $G_d$ is cubic. The second inequality holds because $G'_d$ cannot be the complete graph $K_4$, that is, the graph of the tetrahedron. Indeed, if $G'_d = K_4$ then the dual of $G'_d$, that is $G'$ is also the $K_4$ and so, all its vertices have valence 3. But we assumed that the new inserted have valence 6, giving a contradiction.   □

By the above proposition the only possible values of $\chi(G_d)$ are 4,3,2. We have,

**Corollary 3.2.** If $\chi(G_d) = 4$ then $\chi(G'_d) = 3$.
Proof. As we noticed in the proof of proposition 3.1, if $\chi(G_d) = 4$ then $G = K_4$. After the subdivision step the vertices of $G$ will retain their valence 3, and thus $G'_d$ has triangular faces, that is, it contains odd cycles. So, by proposition 2.2, $\chi(G'_d) \neq 2$ and thus $\chi(G'_d) = 3$.   □

Fig. 2 shows a tetrahedron with a 4-coloring of its faces. After one subdivision step with the Loop scheme, a 3-coloring of the faces is possible.
We will also prove that in the cases of Loop and $\sqrt{3}$-scheme if $\chi(G_d) = 2$ then also $\chi(G'_d) = 2$. The latter together with Proposition 3.1 will also imply that if $\chi(G_d) = 3$ then $\chi(G'_d) = 3$, completing the study of the face coloring numbers for these two schemes. We have

**Proposition 3.3.** In the Loop subdivision scheme $\chi(G_d) = 2$ if and only if $\chi(G') = 2$.

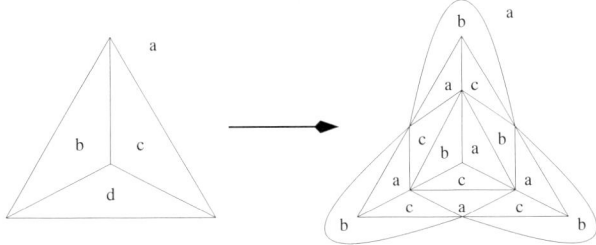

**Fig. 2.** The tetrahedron and the resulting mesh after a Loop subdivision step with a 4-coloring and a 3-coloring of the faces, respectively.

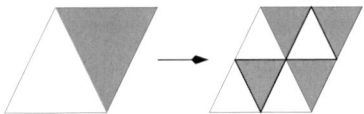

**Fig. 3.** Face 2-coloring for a Loop subdivision mesh.

**Proof.** In Loop subdivision each triangle of the initial mesh $M$ splits in 4. From a face 2-coloring of $M$ we can obtain a face 2-coloring of $M'$ by altering the of the interior triangle and keeping the color of the other 3 triangles. See Fig. 3. In the same manner, from a face 2-coloring of $M'$ we get a face 2-coloring of $M$ by altering the color of the central faces. □

In an alternative proof we can work on the dual meshes $M_d, M'_d$. The $M'_d$ is obtained from $M_d$ by inserting in the interior of each face $F$ a new face $F'$ with the same number of edges, assuming that the edges of $F'$ are parallel to the edges of $F$, and then connecting the corresponding vertices and removing the old edges of $M$. See Fig. 4.

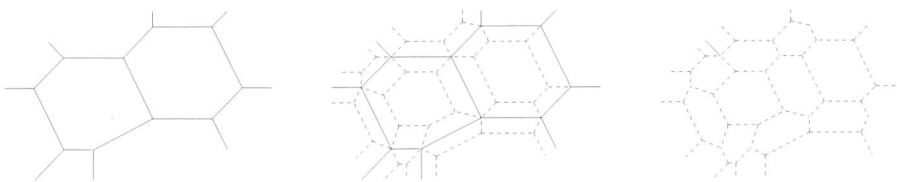

**Fig. 4.** The dual mesh after one Loop subdivision step.

It is straightforward to check that every cycle of $M'_d$ lifts to a cycle of $M_d$ with the same parity and so by proposition 2.2 the face 2-colorability of the one mesh implies the face 2-colorability of the other.

We also have

**Proposition 3.4.** In the $\sqrt{3}$ subdivision scheme $\chi(G_d) = 2$ if and only if $\chi(G') = 2$.

**Proof.** Suppose first $M$ has a face 2-coloring and so $\chi(G_d) = 2$. In [8] we saw that the $\sqrt{3}$ operator can be factorized as a duality operator followed by a stellation or capping operator where a vertex is inserted in the interior of each face and connected with edges with the vertices of that face. So, if $\chi(G_d) = 2$ then by coloring the vertices inserted by the stellation operator with a third color we get vertex 3-coloring of $M'$ which by Proposition 2.3 implies that $M'$ has a face 2-coloring.

To prove the converse we will first find a construction of the dual of an $\sqrt{3}$ subdivision mesh similar to the one for the Loop subdivision shown in Fig. 4. This time the dual mesh $M'_d$ is obtained from $M_d$ again by inserting in the interior of each face $F$ a new face $F'$ with the same number of edges, but rotating it so that its vertices correspond to the edges rather than the vertices of $F$. Then we connect the vertices of adjacent faces corresponding to the same edge of $M$ and finally remove the edges of $M$. See Fig. 5.

Assuming a fixed orientation, there is a natural 1-1 correspondence between the vertices of $M'_d$ and the directed edges of $M_d$. Under this correspondence every closed path of $M_d$, that is every union of cycles of $M_d$, lifts to a closed path of $M'_d$ with the same parity. If $\chi(M'_d) = 2$ then, by Proposition 2.2 all the cycles on it have even parity and so do all the closed paths. Therefore every closed path of $M_d$ has also even parity. In particular, every cycle in $M_d$ is even and by Proposition 2.2 $\chi(M_d) = 2$ as required.                    □

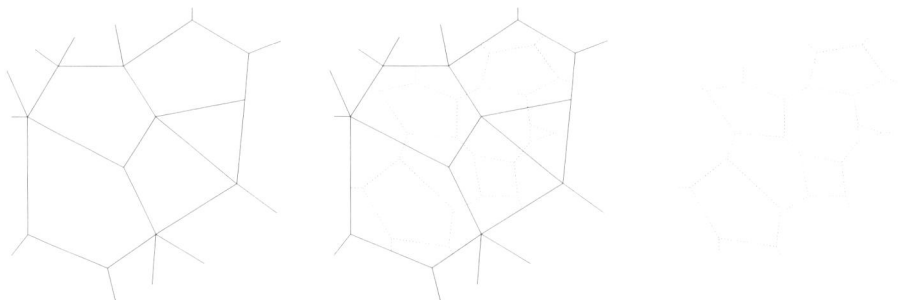

**Fig. 5.** The dual mesh after one $\sqrt{3}$ subdivision step.

## 4   Flows and Edge Colorings

In this section we will prove a proposition for the $n$ flows of the Loop subdivision meshes and we will see an immediate corollary on the edge colorings of these meshes.

**Proposition 4.1.** In Loop subdivision, if $G_d$ has an $n$-flow then $G'_d$ has also an $n$-flow.

**Proof.** Fig. 4 shows the effect of one Loop refinement step on the dual graph of the mesh. Based on an $n$-flow on $G_d$ we can draw an $n$-flow on $G'_d$ as Fig. 6

shows. The edges of $G_d$ that are parallel to the edges of $G'_d$ in the figure have the same orientation with the corresponding edges of $G_d$. The three edges around the black vertex, which is a common vertex of $G_d$ and $G'_d$ are obtained from the edges of $G_d$ after a rotation by $\pi$ and a change of orientation.                    □

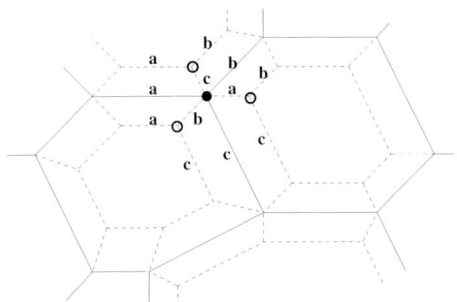

**Fig. 6.** By marking the edges of $G'_d$ as shown, and following the procedure described in Proposition 4.1, we obtain a flow on it.

An edge 3-coloring of a cubic graph is called Tait coloring. The existence of a Tait coloring is equivalent to the existence of a 4-flow on that graph, see [7] p. 241. As a the existence of a 4-flow for the graph $G_d$ implies the existence of a 4-flow for $G'_d$, we have that if $G_d$ has a Tait coloring then $G'_d$ has also a Tait coloring.

The Tait coloring of a cubic graph $G_d$ and the obvious 1-1 correspondence between the edges of a graph and its dual give a coloring of the edges of $M$ which is not a proper edge coloring as defined in the Introduction but has the characteristic property that every triangle has all the three edges in different colors.

The converse proposition is not true in general. The graph $G'_d$ can have an $n$-flow while $G_d$ has not. In particular a graph $G'_d$ can have a 4-flow while the $G_d$ has not. We can see the latter using the equivalence of the existence of a 4-flow with a Tait coloring. Fig. 7 shows the Petersen graph which is the simplest example of a graph with no Tait coloring, see [7] p. 242. The graph obtained after one Loop or $\sqrt{3}$ subdivision step on the Petersen graph has a Tait-coloring as Fig. 7 shows.

## 5   Vertex Colorings

### 5.1   Triangle Meshes

In contrast to the chromatic numbers of cubic graphs we studied in section 3, the chromatic number of a triangular mesh can be arbitrarily large. Indeed, if $n \equiv 0, 3, 4, 7 \mod 12$ then the complete graph $K_n$ has an embedding as a triangulation

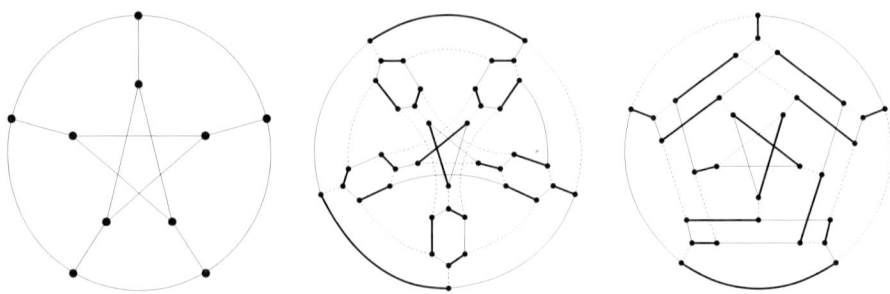

**Fig. 7. Left:** The Petersen graph. **Middle and Right:** A Tait coloring of the graph $G'_d$ where $G$ is the Petersen graph and the refinement rule is Loop and $\sqrt{3}$ subdivision, respectively, shown with thin, thick and dashed lines.

of a orientable surface, see for example [9]. Nevertheless, the chromatic number of a Loop or $\sqrt{3}$ subdivision mesh is always relatively small. We have

**Proposition 5.1.** In Loop subdivision

$$\chi(G') \leq 5 \tag{5}$$

**Proof.** Let $G_v$ be the graph obtained from $G'$ after the old vertices are removed. See Fig. 8 (Left). We notice that all the vertices of $G_v$ are 4-valent because they were regular 6-valent as vertices of $G'$ and being midedges of $G$ were connected with exactly two of the old vertices. By proposition 2.1 we have $\chi(G_v) \leq 5$. Also, $\chi(G_v) \neq 5$, giving $\chi(G_v) \leq 4$, because $G_v$ can not be the complete graph $K_5$. To see the latter we notice that if there is a vertex of $G$ with valence greater than 3, then there are two edge sharing a common vertex but not a common face. The midedges of these edges are non-adjacent vertices of $G_v$. If all the vertices of $G$ are trivalent then $M$ is a cubic triangular mesh, that is, it is the mesh of the tetrahedron and we immediately verify that $G_v \neq K_5$.

Finally, by coloring all the old vertices with a fifth color and joining them with the vertices of $G_v$ we get a 5-coloring of $G'$. □

Thus, there are three cases for the chromatic number of $G'$, that is, $\chi(G') = 3, 4, 5$. The case $\chi(G') = 3$ is easy to deal with because, as we noticed in section 2, a vertex 3-coloring is essentially unique. We have

**Proposition 5.2.** In Loop subdivision $\chi(G') = 3$ if and only if $\chi(G) = 3$.

**Proof.** A vertex 3-coloring of $G$ extends to a vertex 3-coloring of $G'$ by coloring the midedges of $G$ with the color not used on the two ends. Conversely, a 3-coloring of $G'$ colors the ends of an edge of $G$ with different colors. Fig. 8 (Right) shows that by coloring both ends of an edge by **a** we fall on a contradiction. □

In the cases $\chi(G') = 4, 5$ it is not so easy to have a conclusive answer, but still we can have some interesting results. We consider the mesh $M_v$ corresponding to the graph $G_v$ defined above. A triangle of $M_v$ can either be in the interior

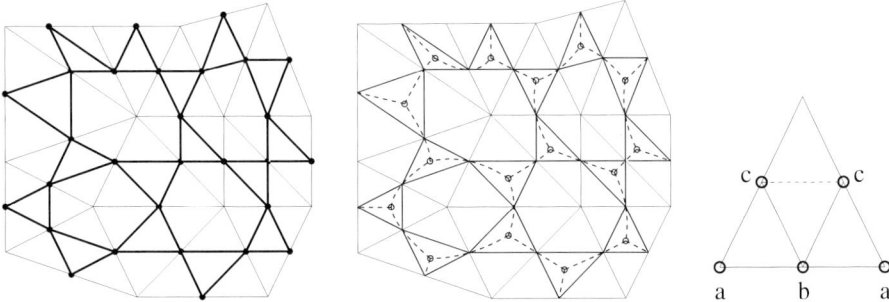

**Fig. 8.** Loop subdivision. **Left:** The thick lines and the large vertices show the edges and the vertices of $G_v$. **Middle:** If two triangles in thick solid line have a common vertex we join their centerfaces with a thick dashed line. **Right:** By coloring both ends of an edge of $G$ with the same color **a** we soon fall to a contradiction.

of a triangle of $M$ or correspond to a 3-valent vertex of $M$. We insert a vertex in the center of the triangles of $M_v$ of the first kind and we connect two such centerface vertices if the corresponding triangles have a common vertex.

This new graph shown in Fig. 8 (Middle) with thick dashed lines is the $G_d$. We notice that the edges of $G_d$ correspond to the vertices of $G_v$ and that a vertex coloring of $G_v$ corresponds to an edge coloring of $G_d$. If $G_d$ has an edge 3-coloring (Tait coloring), as it usually happens, then $\chi(G_v) = 3$. By using a fourth color for the old vertices of $G$ we get $\chi(G') = \chi(G_v) + 1 = 4$.

If $G_d$ does not have a Tait coloring, for example when $G_d$ is the Petersen graph, then it is more difficult to have a conclusive result for the chromatic number of $G'$. In this case we need 4 colors for an edge coloring of $G_d$, see [6] p.16, therefore $\chi(G_v) = 4$, and we need 5 colors for a vertex coloring of $G'$ such that all the old vertices have the same color. Nevertheless this does not mean that $\chi(G') = 5$. In particular, if there is an edge 4-coloring of $G_d$ such that every face has edges of only 3-colors, then we can get a vertex 4-coloring of $M_v$ such that every face is vertex 3-colored. Then by inserting at the center of each face a vertex with the fourth color and joining it with edges with the vertices of that face we get a 4-coloring of $G'$. So, the theoretical question for the existence of a Loop subdivision mesh such that $\chi(G') = 5$ is equivalent to the question if every cubic mesh can be edge 4-colored such that every face has edges of only 3 colors.

In the case of $\sqrt{3}$ the chromatic number of the subdivision mesh is usually even lower. We have

**Proposition 5.3.** In $\sqrt{3}$ subdivision we have $\chi(G') \leq 4$.

**Proof.** Similarly to the proof of Proposition 5.1 we consider the mesh $M_v$ with underlying graph $G_v$ obtained from $M'$ after removing the old vertices. See Fig. 9.

In this case $M_v$ is the dual of $M$ and $G_v = G_d$, see [8]. As $G_d$ is a cubic graph we have $G_d \leq 4$, with the equality holding when $G_d$ is the complete graph $K_4$. The latter happens exactly when $G$ is also the $K_4$ in which case $\chi(G') = 4$ see

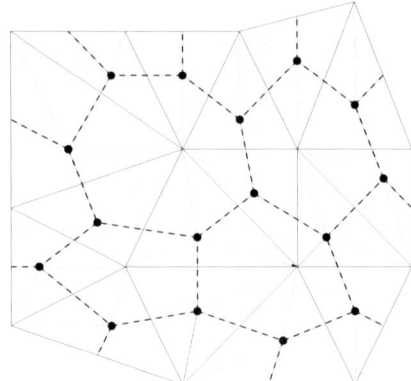

**Fig. 9.** $\sqrt{3}$ subdivision: The thin continuous lines are the edges of the original mesh $M$. The thin dotted lines are edges of the refined mesh $M'$ but not edges of $M_v$. The thick dashed lines and the large vertices are the edges and the vertices of $M_v$.

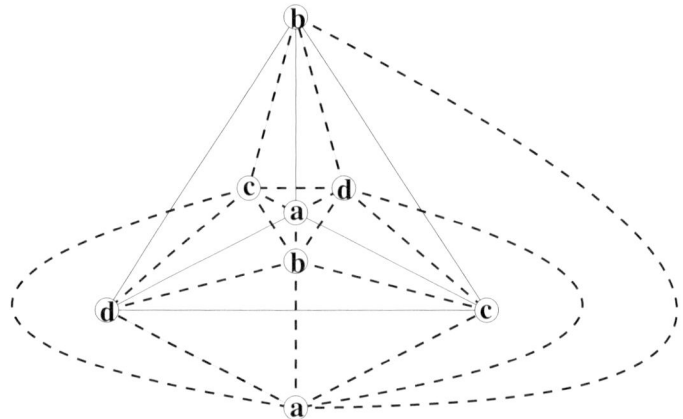

**Fig. 10.** $\sqrt{3}$ subdivision: Although we need 4 colors for a vertex coloring of $G_d$, each of its faces is vertex 3-colorable, being just a triangle. Using the fourth color for the centerfaces of $G_d$ we get a vertex 4-coloring of $G'_d$.

Fig. 10. Generally, when $G$ is not the $K_4$ we have $\chi(G_v) = 3$ and by coloring the old vertices of $G'$ with a fourth color we get a vertex 4-coloring of $G'$.      □

We also have

**Proposition 5.4.** In $\sqrt{3}$ subdivision if $\chi(G) = 3$ then $\chi(G') = 3$.

**Proof.**   We will use the factorization of $\sqrt{3}$ subdivision into a truncation or corner-cutting followed by a duality operator, see [8]. The truncation operator on the vertex 3-colored mesh $M$ will give a face 3-colored mesh $M_t$ and then the duality operator will give a vertex 3-colored mesh which will be $M'$.      □

In the proof of Proposition 3.4 we saw that $\chi(G') = 3$ when $G$ is face 2-colorable. As there exist face 2-colorable meshes which are not vertex 3-colorable we conclude that the converse of Proposition 5.4 does not hold in general. Fig. 11 (Left) shows such a mesh.

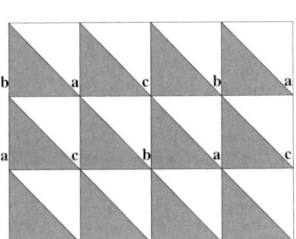
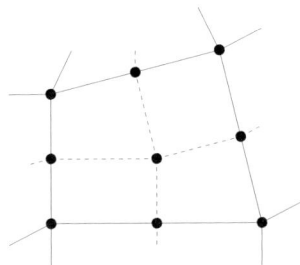

**Fig. 11. Left:** The opposite sides of the rectangle are identified to give a face 2-colored toroidal mesh. We can verify that the construction of a vertex 3-coloring on it will lead to a contradiction. **Right:** The effect of one Catmull-Clark step on a single face of the original mesh.

## 5.2   Quad Meshes

The situation is much simpler in the case of the Catmull-Clark scheme. In every step new vertices are inserted in the midedges and the centerfaces, and then joined with edges, splitting the original face in 4. See Fig. 11 (Right). We notice that there always exists a vertex 2-coloring of the refined mesh because every new edge has an endpoint at the middle of an old edge and the second endpoint is either an old vertex or an old centerface. So, we need one color for the vertices corresponding to midedges and one color for vertices corresponding to old vertices and centerfaces.

By Proposition 2.2 that means that after one Catmull-Clark subdivision step every cycle has even length.

## 6   Conclusion – Future Work

We showed that the chromatic numbers of subdivision meshes are generally smaller than the corresponding chromatic numbers of the initial meshes. That is, the combinatorial structures on subdivision meshes are generally simpler than the corresponding structures on the initial meshes, or in a sense, the subdivision processes reduce the combinatorial complexity of a mesh.

Our work can be expanded to more advanced topics of graph-theory. Almost every question asked in the context of graph-theory can also be examined in the more narrow class of subdivision meshes. In particular we conjecture that the Proposition 4.1 still holds when Loop subdivision is substituted with the $\sqrt{3}$ subdivision.

In another direction of generality we can study some simple combinatorial properties of the general subdivision schemes classified in [1] with the use of lattice transformations. In that paper, each subdivision scheme corresponds to a lattice transformation and there is evidence suggesting that the lattice transformation and the combinatorial properties of the subdivision mesh are related.

Finally, what is probably more important is to see how can we incorporate this combinatorial approach into the study of the other properties of subdivision.

## Acknowledgement

We would like to thank Malcolm Sabin for several suggestions that have improved the paper considerably.

## References

1. Ivrissimtzis, I., Dodgson, N., Sabin, M.: A generative classification of mesh refinement rules with lattice transformations. Research Report UCAM-CL-TR-542, Cambridge University, Cambridge, UK (2002)
2. Doo, D., Sabin, M.: Behaviour of recursive division surfaces near extraordinary points. Computer-Aided Design **10** (1978) 356–360
3. Loop, C.T.: Smooth subdivision surfaces based on triangles (1987)
4. Kobbelt, L.: sqrt(3) subdivision. In: Siggraph 00, Conference Proceedings. (2000) 103–112
5. Catmull, E., Clark, J.: Recursively generated B-spline surfaces on arbitrary topological meshes. Computer-Aided Design **10** (1978) 350–355
6. Jensen, T.R., Toft, B.: Graph coloring problems. New York, NY: John Wiley & Sons. (1995)
7. Tutte, W.: A census of planar triangulations. Can. J. Math. **14** (1962) 21–38
8. Ivrissimtzis, I., Seidel, H.P.: Polyhedra operators for mesh refinement. In: Proceedings of Geometric Modeling and Processing 2002, Wako, Saitama, Japan, IEEE (2002) 132–137
9. Grannell, M., Griggs, T., Sirn, J.: Face 2-colourable triangular embeddings of complete graphs. J. Comb. Theory, Ser. B **74** (1998) 8–19

# Watermarking 3D Polygonal Meshes Using the Singular Spectrum Analysis

Kohei Murotani and Kokichi Sugihara

Department of Mathematical Informatics Graduate School of Information Science and Technology University of Tokyo 7-3-1, Hongo, Bunkyo-ku, Tokyo, 113-0033, Japan
muro@simplex.t.u-tokyo.ac.jp

**Abstract.** Watermarking is to embed a structure called a watermark into the target data such as images. The watermark can be used, for example, in order to secure the copyright and detect tampering. This paper presents a new robust watermarking method that adds a watermark into a 3D polygonal mesh in the spectral domain. In this algorithm, a shape of a 3D polygonal model is regarded as a sequence of vertices called a vertex series. The spectrum of the vertex series is computed using the singular spectrum analysis (SSA) for the trajectory matrix derived from the vertex series. Watermarks embedded by this method are resistant to similarity transformations and random noises.

## 1 Introduction

Digital watermarking is a technique for adding secret information called a watermark to various target objects data. The watermark must not interfere with the intended purpose of the target object data (e.g., the watermark should not decrease the geniality of a 2D image), and the watermark should not be separable from the target object data. Embedded watermarks can be used to secure copyright, to add comments, to detect tampering and to identify authorized purchasers of the object.

In general, watermarks are classified into private watermarks and public ones. Private watermarks are retrieved by both of the original data and the watermarked data, while public watermarks are retrieved by only the watermarked data. A lot of papers on watermarking these data objects have been published [13]. However most of the previous research on watermarking have been concentrating on watermarking "classical" object data types, such as texts, 2D still images, 2D movies, and audio data. Recently, on the other hand, 3D objects data, such as 3D polygonal meshes and various 3D geometric CAD data, become more and more popular and important, and hence techniques to watermark 3D models also become more important [1] [4]-[12].

This paper presents an algorithm that embeds watermark data into 3D polygonal meshes. The watermark embedded by the algorithm is robust against similarity transformation (i.e., rotation, translation, and uniform scaling). It is also resistant against such random noises added to vertex coordinates.

M.J. Wilson and R.R. Martin (Eds.): Mathematics of Surfaces 2003, LNCS 2768, pp. 85–98, 2003.

In many techniques for watermarking, secret data (watermarks) are hidden in the spectral coefficients. Therefore, in those techniques of watermarking, some kind of spectrum decomposition is required. The eigenvalue decomposition of a Laplacian matrix derived only from connectivity of the mesh in [9] and the wavelet decomposition for only meshes of certain classes in [4] are computed. Since both methods depend on the connectivity of the mesh, the methods are not robust for altering the connectivity of the mesh. While our method is robust for altering the connectivity of the mesh, since our method do not depend on the connectivity of the mesh.

In section 2, we will review the singular spectrum analysis, which is a basic tool for our algorithm. In section 3, the algorithm of embedding and extracting watermarks will be described. In section 4, we will present experimental results, and in section 5 we conclude this paper with summary and future work.

## 2   Basic SSA

The singular-spectrum analysis (SSA) [2][3] is a novel technique for analyzing time series incorporating the elements of classical time series, multivariate statistics, multivariate geometry, dynamical systems and signal processing. Recently, SSA is one of the popular statistical methods for signal detection in climatology and meteorology.

In many techniques for watermarking, secret data (watermarks) are hidden in the spectral coefficients. Therefore, in those techniques of watermarking, some kind of spectrum decomposition is required. In this paper, SSA is applied to watermarking, since SSA performs a spectrum decomposition of time series.

In this section, we describe the basic algorithm of SSA. In the next section, a generalized version of SSA to multivariate series is applied to our purpose.

### 2.1   Algorithm of the Basic SSA

Let $N > 2$. Consider a real-value time series $F = (f_0, f_1, \ldots, f_{N-1})$ of length $N$. Assume that $F$ is a nonzero series; that is, there exist at least one $i$ such that $f_i \geq 0$. The basic SSA consists of two complementary stages: the decomposition stage and the reconstruction stage.

**Decomposition Stage.** The decomposition stage consists of the next two steps.

*1st Step: Embedding.* In the first step, the original time series is mapped to a sequence of lagged vectors in the following way. Let $L$ be an integer (window length) such that $1 < L < N$. We define $K = N - L + 1$ lagged vectors $X_i$ by

$$X_i = (f_{i-1}, \ldots, f_{i+L-2})^T, \qquad 1 \leq i \leq K. \tag{1}$$

We shall call $X_i$'s $L$-lagged vectors. The $L$-trajectory matrix (or simply trajectory matrix) of the series $F$ is defined by

$$X = [X_0 : \ldots : X_K], \tag{2}$$

whose columns are the $L$-lagged vectors. In other words, the trajectory matrix is

$$X = (x_{ij})_{i,j=1}^{L,M} = \begin{pmatrix} f_0 & f_1 & f_2 & \cdots & f_{K-1} \\ f_1 & f_2 & f_3 & \cdots & f_K \\ f_2 & f_3 & f_4 & \cdots & f_{K+1} \\ \vdots & \vdots & \vdots & \ddots & \vdots \\ f_{L-1} & f_L & f_{L+1} & \cdots & f_{N-1} \end{pmatrix} \tag{3}$$

Obviously $x_{ij} = f_{i+j-2}$ and the matrix $X$ has equal elements on the 'diagonal' $i + j = const$. Thus, the trajectory matrix is a Hankel matrix. Certainly if $N$ and $L$ are fixed, then there is a one-to-one correspondence between the trajectory matrix and the time series.

*2nd Step: Singular Value Decomposition.* In the second step, the singular value decomposition (SVD) is applied to the trajectory matrix. Let $S = XX^T$. Denote by $\lambda_1, \ldots, \lambda_L$ the eigenvalues of $S$ taken in the decreasing order of magnitude ($\lambda_1 \geq \ldots \geq \lambda_L \geq 0$), and by $U_1, \ldots, U_L$ the orthonormal system of the eigenvectors of the matrix $S$ corresponding to these eigenvalues. Let $d = Max\{i,$ such that $\lambda_i \geq 0\}$. We defines $V_i = X^T U_i / \sqrt{\lambda_i}$ and $X_i^T = \sqrt{\lambda_i} U_i V_i^T$ ($i = 1, \ldots, d$). Then the SVD of the trajectory matrix $X$ can be written as

$$X = X_1 + X_2 + \ldots + X_d. \tag{4}$$

The matrix $X_i$ have rank 1; therefore they are elementary-matrices. The collection $(\lambda_i, U_i, V_i)$ is called ith eigentriple of the SVD (4).

**Reconstruction Stage.**

*3rd Step: Diagonal Averaging.* In the last step of the basic SSA, each matrix in the decomposition (4) is transformed into a new series of length $N$; this step is called the diagonal averaging. Let The matrix $Y$ be an $L \times K$ matrix with elements $y_{ij}, 1 \leq i \leq L, 1 \leq j \leq K$.

Diagonal averaging transfers the matrix $Y$ to the series $(g_0, \ldots, g_{N-1})$ by the formula:

$$g_k = \begin{cases} \frac{1}{k+1} \sum_{m=1}^{k+1} y_{m,k-m+2} & for \quad 0 \leq k < L-1, \\ \frac{1}{L} \sum_{m=1}^{L} y_{m,k-m+2} & for \quad L-1 \leq k < K, \\ \frac{1}{N-k} \sum_{m=k-K+2}^{N-K+1} y_{m,k-m+2} & for \quad K \leq k < N. \end{cases} \tag{5}$$

The expression (5) corresponds to averaging of the matrix elements over the 'diagonal' $i + j = k + 2$: for $k = 0$ we have $g_0 = y_{11}$, for $k = 1$ we have $g_1 = (y_{12} + y_{12})/2$, and so on. Note that if $Y$ is the trajectory matrix of some series $(h_0, \ldots, h_{N-1})$ (in other word, if $Y$ is the Hankel matrix), then $g_i = h_i$ for all $i$. Diagonal averaging (5) applied to the decomposition matrix $X_k$ produces the series $\tilde{F}^{(k)} = (\tilde{f}_0^{(k)}, \ldots, \tilde{f}_{N-1}^{(k)})$ and therefore the initial series $F = (f_0, \ldots, f_{N-1})$ is obtained by the sum of $d$ series:

$$f_n = \sum_{k=1}^{d} \tilde{f}_n^{(k)}. \tag{6}$$

## 2.2   Optimality of SVD and the Hankel Matrix

Here, we describe two optimal its features in the process of SSA. The first optimality is related to SVD.

Proposition 2.1. Let $\boldsymbol{X} = [X_0 : \ldots : X_K]$ be the matrix define by equation (2), and let $\boldsymbol{X_i}$ be the matrices define by equation (4). Then the following two statements hold.

1. The vector $Q = U_1$ is the solution of the problem

$$\nu_1 := \sum_{i=1}^{K}(X_i, Q) = Max_P \sum_{i=1}^{K}(X_i, P), \tag{7}$$

where the maximum on the right hand side of (7) is taken over all $P \in \boldsymbol{R}^L$ with $||P|| = 1$, and also $\nu_1 = \lambda_1$ holds.

2. Let $Q$ be the solution of the following optimization problem

$$\nu_k := \sum_{i=1}^{K}(X_i, Q) = Max_P^{(k)} \sum_{i=1}^{K}(X_i, P), \tag{8}$$

where the maximum on the right hand side of (8) is taken over all $P \in \boldsymbol{R}^L$ such that $||P|| = 1$ and $(P, U_i) = 0$ for $1 \leq i < k$ . If $k \leq d$, then the $Q = U_k$ and $\nu_k = \lambda_k$. If $k > d$, then $\nu_k = 0$.

Proposition 2.1 enables us to call the vector $U_i$ the ith principal vector of collection $X_0, \ldots, X_K$. The second optimality is related to diagonal averaging. When a general matrix is transformed to the Hankel matrix, diagonal averaging have the optimality as stated in the following proposition. Proposition 2.2. Assume that $\boldsymbol{Z} = \psi(\boldsymbol{Y})$ is a Hankel matrix of the same dimension as $\boldsymbol{Y}$ such that the difference $\boldsymbol{Y} - \boldsymbol{Z}$ has the minimal Frobenius norm. Then the element $\tilde{y}_{ij}$ of the matrix $\psi(\boldsymbol{Y})$ is given by

$$\tilde{y}_{ij} = \begin{cases} \frac{1}{s-1}\sum_{l=1}^{s-1} y_{l,s-l} & for \quad 2 \leq k \leq L-1, \\ \frac{1}{L}\sum_{l=1}^{L} y_{l,s-l} & for \quad L \leq k \leq K+1, \\ \frac{1}{N-s+2}\sum_{l=s-K}^{L} y_{l,s-l} & for \ K+2 \leq k \leq N+1. \end{cases} \tag{9}$$

The linear operator $\psi$ is called the Hankelization operator.

## 3   Algorithm for Watermarking in the Spectral Domain

The watermarking algorithm in this paper inserts a watermark to a given 3D polygonal mesh. Previous methods using connectivity of the mesh are not robust against modifications of the connectivity of the mesh. In our method, on the other hand, connectivity of the mesh is not used. Instead, a 3D polygonal model is regarded as a sequence of vertices called a vertex series. Therefore, our method is robust against modifications of the connectivity of the mesh. Next, the spectra of the vertex series are computed using SSA for the trajectory matrix derived from the vertex series. The watermarks are added in the spectra domain in such a way that their singular values are modified. To recover the watermarks, the watermarked matrix is converted into the vertex series by the diagonal averaging. This method is for private watermarking, meaning that the watermark extraction requires both the watermarked vertex series and the original non-watermarked vertex series. The watermark can be extracted by comparing singular values of the watermarked data and the original data in the spectral domain.

## 3.1   Spectral Decomposition of the Vertex Series

Though a scalar-value time series $F = (f_0, f_1, \ldots, f_{N-1})$ is considered in the basic SSA, we expand the basic SSA into a multivariate version in order to use tri-value time series $\boldsymbol{F} = (F_0, \ldots, F_{N-1})$ with $F_i = (f_{i,x}, f_{i,y}, f_{i,z})^T$. The trajectory matrix (2) is expanded into a $3L \times K$ matrix:

$$\boldsymbol{X} = \begin{pmatrix} F_0 & F_1 & F_2 & \cdots & F_{K-1} \\ F_1 & F_2 & F_3 & \cdots & F_K \\ F_2 & F_3 & F_4 & \cdots & F_{K+1} \\ \vdots & \vdots & \vdots & \ddots & \vdots \\ F_{L-1} & F_L & F_{L+1} & \cdots & F_{N-1} \end{pmatrix} \tag{10}$$

and we perform the singular value decomposition for the trajectory matrix (10). The SVD produces a sequence of singular values and a corresponding sequence of elementary-matrices.

Approximately, large singular values correspond to lower spatial frequencies, and small singular values correspond to higher spatial frequencies. Elementary-matrices associated with higher singular values represent global shape features, while elementary-matrices associated with lower singular values represent local or detail shape features.

## 3.2   Embedding Watermark

Suppose that we want to embed an m-dimensional bit vector $\boldsymbol{a} = (a_1, a_2, \ldots, a_m)$ where each bit takes value 0 or 1. Each bit $a_i$ is duplicated by chip rate $c$ to produce a watermark symbol vector $\boldsymbol{b} = (b_1, b_2, \ldots, b_{mc})$, $b_i \in \{0, 1\}$ of length $m \times c$;

$$b_i = a_j, \qquad jc < i \le (j+1)c \tag{11}$$

Embedding the same bit $c$ time increases resistance of the watermark against additive random noises, because averaging the detected signal $c$ times reduces the effect of the additive random noises. Let $b' = (b'_1, b'_2, \ldots, b'_{mc})$, $b'_i \in \{-1, 1\}$, be the vector defined by the following simple mapping:

$$b'_k = \begin{cases} -1 & for \ b_i = 0, \\ 1 & for \ b_i = 1, \end{cases} \qquad (12)$$

For $i = 1, 2, \ldots, mc$, let us choose $p_i \in \{-1, 1\}$ in an appropriate way described later. Moreover let $\alpha$ be a positive constant, called the watermark embedding amplitude. The $i$-th singular value is converted by the following formula:

$$r_i = \sqrt{\lambda_i} + b'_i p_i \alpha. \qquad (13)$$

Using $r_i$, $i = 1, 2, \ldots, d$, we construct the trajectory matrix by

$$\boldsymbol{X'} = \sum_{i=1}^{d} r_i U_i V_i^T = \sum_{i=1}^{d} \sqrt{\lambda_i} U_i V_i^T + \sum_{i=1}^{d} b'_i p_i \alpha U_i V_i^T. \qquad (14)$$

From this matrix, the vertex coordinates $\boldsymbol{F'} = (F'_0, \ldots, F'_{N-1})$ with $F'_i = (f'_{i,x}, f'_{i,y}, f'_{i,z})^T$ are computed by using the formula (4). As a result, the vertices of the original polyhedral mesh are converted into watermarked vertices, which are slightly altered from the original positions.

Finally, if $p_i \in \{-1, 1\}$ is chosen randomly, the both centers of gravity $\phi(\boldsymbol{F})$ and $\phi(\boldsymbol{F'})$ are quite different where $\phi(\boldsymbol{F})$ is the center of gravity of vertex series $\boldsymbol{F}$. The center of gravity is a very important invariant for recovering the watermark, and hence it is desirable select $p_i$ so that $\phi(\boldsymbol{F}) = \phi(\boldsymbol{F'})$. Since we can not satisfy $\phi(\boldsymbol{F}) = \phi(\boldsymbol{F'})$ accurately, we try to minimize the absolute value of the second term in the right-hand side of equation (14), i.e., we obtain the optimal $p_i$ by solving the optimal problem:

$$Min\|\sum_{i=1}^{d} b'_i p_i \phi \circ \psi(U_i V_i^T)\|_2 \qquad s.t. \quad p_i \in \{-1, 1\}. \qquad (15)$$

## 3.3   Extracting Watermark

In this method, extraction of the watermark requires both of the original vertex series and a watermarked vertex series. The extraction stars with fitting the original vertex series and the watermarked vertex series by translation, rotation and scaling. First, the data are translated so that the center of gravity of the watermarked vertex series considers with the center of gravity of the original vertex series. Next, coarse approximations of their vertex series are reconstructed from the first (highest-frequency) 15 singular values and the corresponding elementary-matrices. (Though number 15 is reasonably determined by our experiences, this number need not always be 15.) Then, each set of eigenvectors (i. e., three principal axis for each vertex series) is computed from a $3 \times 3$ covariance matrix derived from each reconstructed shape. Then the data rotated so that the directions of two sets of eigenvectors coincide with each other.

Next, SVD is performed for the original vertex series $\boldsymbol{F}$ to produce the singular values $\sqrt{\lambda_i}$ and the associated elementary-matrices. For the watermarked

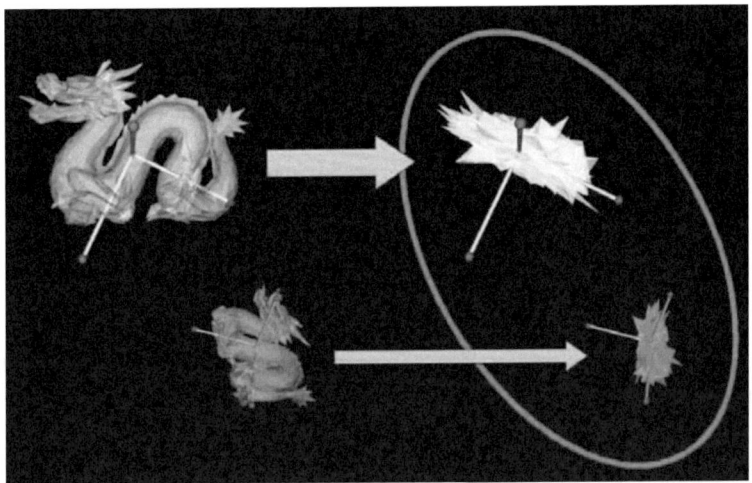

**Fig. 1.** The perception of similarity transformation.

vertex series $F'$, we do not apply SVD, but use the orthogonal matrices $U$ and $V$ for $F$ and compute $\sqrt{\lambda_i'}$ by the equations (7) and (8). Here, if we multiply the difference $(\sqrt{\lambda_i'} - \sqrt{\lambda_i})$ with $p_i$ and sum the result over $c$ times, then we obtain $q_j$:

$$q_j = \sum_{i \in I_j}(\sqrt{\lambda_i'} - \sqrt{\lambda_i})p_i \approx \sum_{i \in I_j} b_i' p_i^2 \alpha$$
$$I_j = \{j, j+m, j+2m, \ldots, j+(c-1)m\}. \tag{16}$$

As the $p_i$ for the embedding and the extraction are synchronize, and if disturbances applied to the vertex coordinates are negligible,

$$q_j = b_i' \alpha c. \tag{17}$$

where $q_j$ takes one of the two values $\{-ac, ac\}$. Since $\alpha$ and $c$ are always positive, simply testing the signs of $q_j$ recovers the original message bit sequence $a_j$:

$$a_j = \{sign(q_j) + 1\}/2. \tag{18}$$

As a result, we can extract the embedded watermark.

### 3.4 Vertex Series Partitioning

When we want to watermark a large mesh, we partition the vertex series into smaller vertex sub-series of a treatable size, as shown in Color Plate 2, so that the calculations times are decreased and the accuracies are increased. The embedding watermarks and extracting watermarks are performed for individual vertex sub-series separately. In the experiments in this paper, as seen in Color Plate 2 (c-1) (d-1), the vertex series is partitioned into 5 groups of the same size according to the Voronoi regions.

## 4    Experiments and Results

### 4.1    Parameters

**Chip Rate $c$.** In case of watermarking 2D still images and 2D movies, the chip rate $c$ can be quite large, since there are at least a few ten thousands of numbers to be manipulated for watermarking. But, in case of 3D polygonal meshes, the numbers to be manipulated for watermarking is often not as much as the 2D data. In this paper, we use two popular mesh models, the "bunny" model (1494 vertices, 2915 faces) and the "dragon" model (1257 vertices, 2730 faces), both of which have a little more then a thousand vertices. We chose the chip rates as 15, since we embed a 50 bits data. (In this case, the maximum chip rates are respectively 20 and 18.) If a mesh is fixed, a higher chip rate means a lower data capacity and more robustness.

**The Watermark Embedding Amplitude $\alpha$.** The watermark embedding amplitudes $\alpha$ is defined as a function of the largest length $l$ of edge of the axis-aligned bounding box of the target mesh. In this paper, we define as $\alpha = \beta \times l$, in the Color Plate, the appearances for $\beta = 0.01, 0.1, 1$ are presented. If the a is larger, the watermark withstand against much disturbances, (for example, adding uniform random noises and mesh smoothing) and the appearance is not preserved.

**Vertex Series.** In our method, we can construct the vertex series in various methods, and the appearance and the robustness of the watermarked meshes may be much different. In Color Plate 2, we show two kinds of the vertex series. The vertex series (c-2) (d-2) are obtained by solving TSP (Traveling Salesman Problems) for vertices and the vertex series (c-3) (d-3) are obtained by choosing vertices in random order. We compare the appearance in section 4.2 and the robustness in section 4.3 between the watermarked meshes obtained from two kinds of vertex series.

### 4.2    Appearances of the Watermarked Meshes

Color plate 1 shows appearances of the watermarked meshes. In color Plate 1, (a-1) and (b-1) show the original meshes, while (a-2)-(a-7) and (b-2)-(b-4) show the watermarked meshes for $\beta = 0.01, 0.1$ and 1, respectively. The appearances of (a-2), (a-3), (a-5), (a-6) or (b-2), (b-3), (b-5), (b-6) can hardly be distinguished from the appearances of (a-1) or (b-1); thus they are watermarked successfully. On the other hand, the appearances of the original meshes are not preserved in (a-4), (a-7) or (b-4), (b-7); thus the watermarks are too large in those cases.

In the appearances of the watermarked meshes, we can not see the much difference between the methods selecting vertex series.

### 4.3    Robustness

We experimentally evaluate the robustness of our watermarks against the similarity transformation and random noises.

**Similarity Transformation.** Watermarks embedded by our method are robust against similarity transformation, for we insert the watermarks in such a way that the center of gravity of all the vertices moves as small as possible, and consequently the transformation can be identified and inverted by the method described in section 3.3.

In case of the dragon model mesh, if the $p_i$ was randomly selected, the center of gravity moved by about $0.01 \times \alpha$ in the 2-norm. On the other hand, when the optimal $p_i$ is selected by (15), the center of gravity moved by about $10^{-5} \times \alpha$ in the 2-norm. We perform some kind of similarity transformations for the watermarked meshes, investigate the number of the values answered correctly from the mesh and repeated the experiment 100 times. As the result, we obtained average false values of the 18.91 bits data out of the 50 bits data if the $p_i$ was randomly selected and we reconstructed most of the bits correctly if the optimal $p_i$ is selected by (15). We can see the great effect selecting the optimal $p_i$ from the experiment.

**Added Uniform Random Noises.** Color Plate 3 shows the appearances of the mesh whose vertex coordinates are disturbed with uniform random noises with amplitude $\alpha \times \gamma$. In the case of $\beta = 0.1$ and $\gamma = 0.01, 0.1, 1$, we investigate the number of the values answered correctly out of 50 bits. We repeated the experiment 100 times, and shown in Color Plate 4.

Color Plate 4 shows the numbers of the value answered correctly. Each vertical line represents one experiment; the red bars show the number of the values answered correctly for $\gamma = 1$, the yellow bars show for $\gamma = 0.1$, the green bars show for $\gamma = 0.01$ and their bars are superposed in this order.

As shown in Color Plate 4, we investigated three kinds of methods for the bunny and dragon models. (g-1)-(g-3) are the results of experiments for bunny model and (f-1)-(f-3) are the results of the experiments for dragon model. (g-1) (f-1) are the results of experiments for the watermarked mesh using vertex series obtained by TSP, (g-2) (f-2) are the results of experiments for the watermarked mesh using vertex series obtained by choosing vertices in random order and (e-3)-(f-3) are the experiments for the watermarked mesh using Ohbuchi's method [9].

From this experiment, we can see that the watermark can withstand against uniform noises for about $\gamma \leq 0.1$. Moreover, we can not see much difference between our method and the Ohbuchi's method [9].

## 5    Summary and Future Work

We present a new watermarking algorithm that embeds data into 3D polygonal meshes. The algorithm employs the singular values of the trajectory matrix as the feature to be modified for watermarking. The trajectory matrix is produced by the vertex series of the 3D polygonal mesh. Since our method is a private watermark, we require both the original mesh and the watermarked mesh for extracting the watermark.

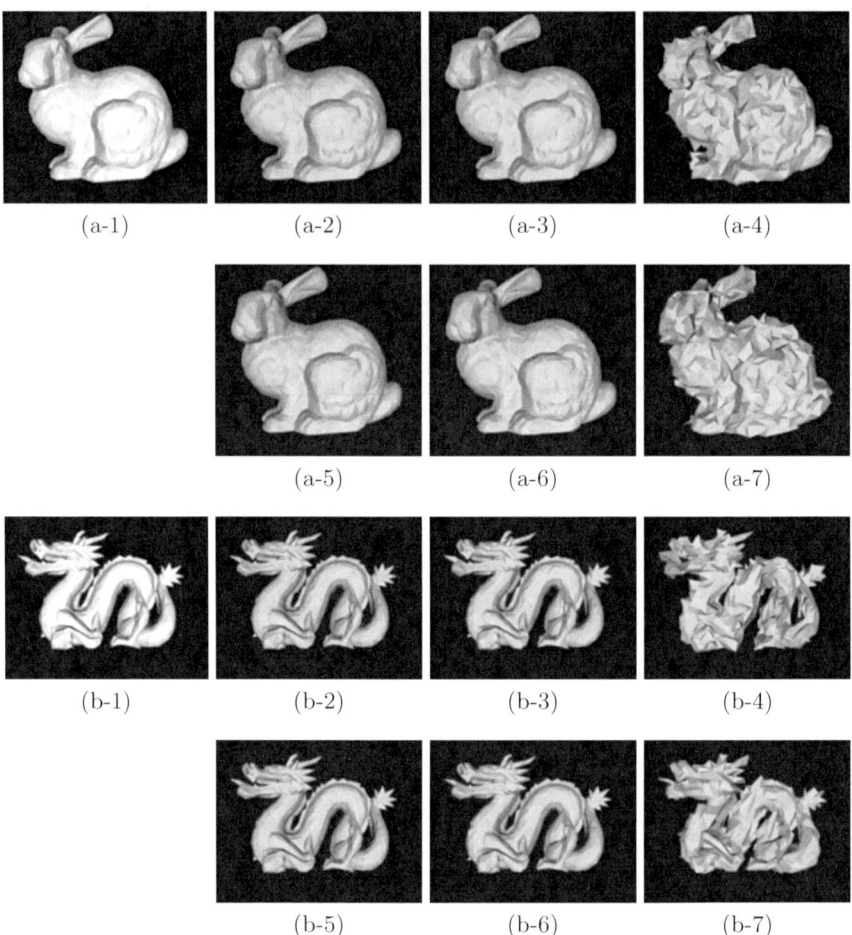

**Color Plate 1.** Appearance of the "bunny" model (1494 vertices, 2915 faces) and "dragon" model (1257 vertices, 2730 faces) watermarked with the watermark embedding amplitudes $\alpha$ and the chip rate $c = 15$. The embedded data is 50 bits.

(a-1)(b-1) The original 3D polygonal meshes of the bunny and dragon models.

(a-2)(b-2) The watermarked meshes with $\beta = 0.01$ using the vertex series obtained by TSP.

(a-3)(b-3) The watermarked meshes with $\beta = 0.1$ using the vertex series obtained by TSP.

(a-4)(b-4) The watermarked meshes with $\beta = 1$ using the vertex series obtained by TSP.

(a-5)(b-5) The watermarked meshes with $\beta = 0.01$ using the vertex series obtained by choosing the vertices in random order.

(b-6)(b-6) The watermarked meshes with $\beta = 0.1$ using the vertex series obtained by choosing the vertices in random order.

(b-7)(b-7) The watermarked meshes with $\beta = 1$ using the vertex series obtained by choosing the vertices in random order.

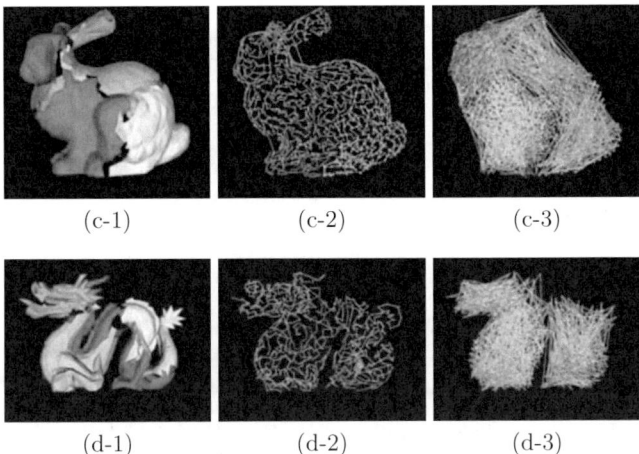

**Color Plate 2.** Partition of the vertices and two kind of the vertex series.

(c-1)(d-1) Meshes obtained by partitioning the vertices into 5 groups of the same size according to the Voronoi regions.

(c-2)(d-2) Vertex series obtained by solving TSP for the vertices in the respective regions.

(c-3)(d-3) Vertex series obtained by choosing the vertices in random order for the vertices in the respective regions.

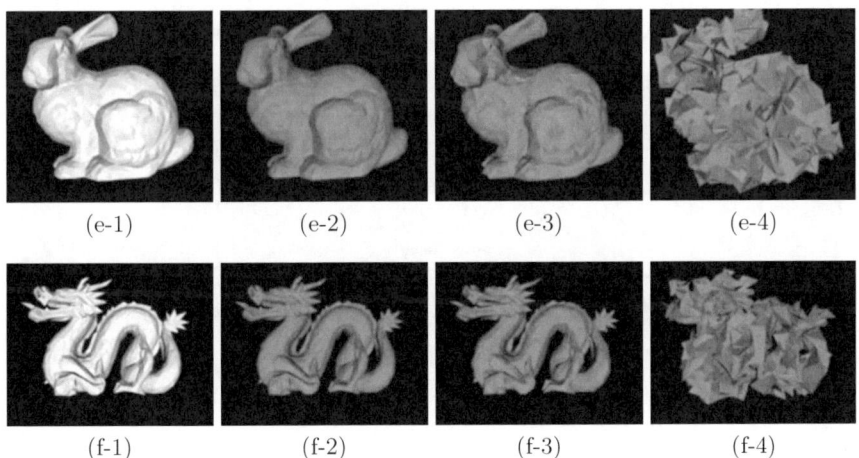

**Color Plate 3.** Appearance of the bunny and dragon modeles to which uniform random noises with amplitude $\alpha \times \gamma$ ( $\beta = 0.1$ ) are added.

(e-1)(f-1) The original 3D polygonal meshes of the models.

(e-2)(f-2) The models with uniform random noises with $\gamma = 0.01$.

(e-3)(f-3) The models with uniform random noises with $\gamma = 0.1$.

(e-4)(f-4) The models woth uniform random noises with $\gamma = 1$.

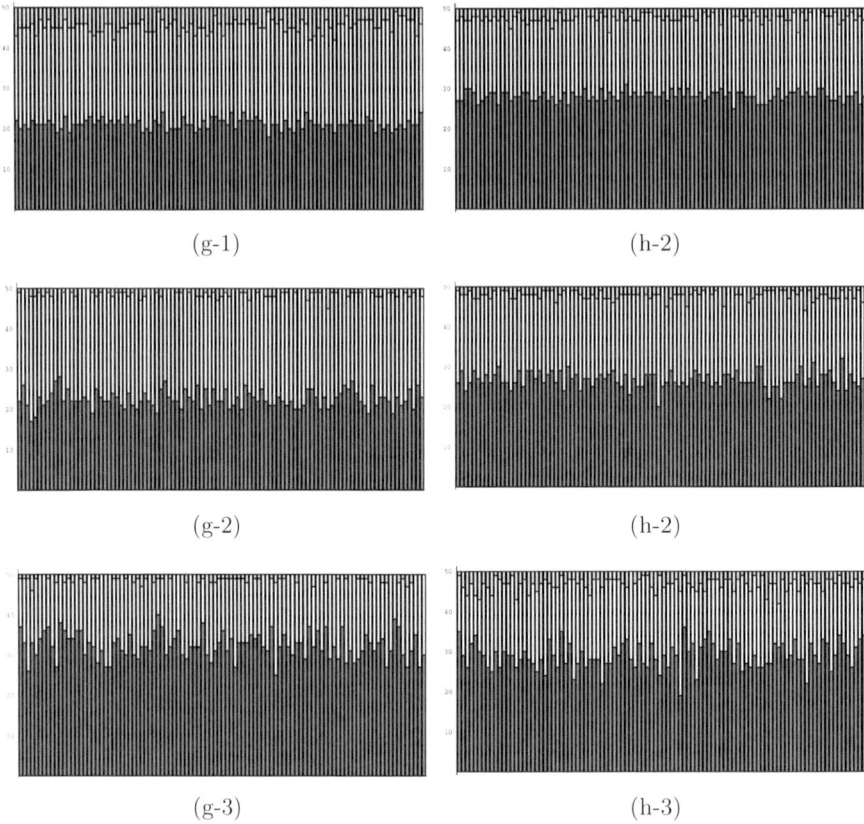

(g-1)          (h-2)

(g-2)          (h-2)

(g-3)          (h-3)

**Color Plate 4.** The number of the values answered correctly out of 50 bits in 100 times. Each vertical line represents one experiment; the red bar shows the number of the values answered correctly for $\gamma = 1$, the yellow bar shows for $\gamma = 0.1$, the green bar shows for $\gamma = 0.01$, and these bars are superposed in this order.

(g-1) Case where uniform random noises were added to (a-3). The average numbers of the values answered correctly are $\{21.14, 45.47, 50.00\}$ for $\gamma = \{1, 0.1, 0.01\}$.

(g-2) Case where uniform random noises were added to (a-6). The average numbers of the values answered correctly are $\{22.53, 48.97, 50.00\}$ for $\gamma = \{1, 0.1, 0.01\}$.

(g-3) Case where uniform random noises were added to the bunny model watermarked by Ohbuchi's method [9]. The average number of the values answered correctly are $\{32.31, 49.13, 50.00\}$ for $\gamma = \{1, 0.1, 0.01\}$.

(h-1) Case where uniform random noises were added to (b-3). The average numbers of the values answered correctly are $\{28.14, 47.77, 50.00\}$ for $\gamma = \{1, 0.1, 0.01\}$.

(h-2) Case where uniform random noises were added to (b-6). The average numbers of the values answered correctly are $\{26.73, 48.02, 50.00\}$ for $\gamma = \{1, 0.1, 0.01\}$.

(h-3) Case where uniform random noises were added to the dragon model watermarked by Ohbuchi's method [9]. The average number of the values answered correctly are $\{29.59, 47.13, 50.00\}$ for $\gamma = \{1, 0.1, 0.01\}$.

The watermark embedded by our method is robust against similarly transformation and moderate uniform noises added to vertex coordinates. This method has a relatively high information density; we require a small mesh having only the $\lfloor 3(m+1)/4 \rfloor$ vertices to embed $m$ bit data without duplicating the watermarks. If we want to watermark a large mesh, the vertex series is partitioned into smaller vertex sub-series with a treatable size, so that the calculations times are decreased and the accuracies are increased. In the future, we would like to investigate the relations between the geometric features of the polygonal meshes and the performance of the watermarking. First, we should modify the vertex series. In this paper, we use two kinds of sequences of vertices so that we can see much difference between two kinds of the vertex series. The vertex series may not be give great effects the appearance and the robustness of the watermarks. Second, the various parameters in our method should be adapted for several purposes. We would like to investigate the features of those parameters.

## Acknowledgement

This work is partly supported by the 21st Century COE Program on Information Science Strategic Core, and Grant-in-Aid for Scientific Research (S) of the Japanese Ministry of Education, Culture, Sports, Science and Technology.

## References

1. Benedens, O., Geometry-Based Watermarking of 3D Models, *IEEE CG&A*, pp. 46-55, January/February 1999.
2. Galka, A., *Topics in Nonlinear Time Series Analysis*, World Scientific, pp. 49-71, 2001.
3. Golyandina, N., Nekrutkin, V., and Zhigljavsky, A., *Analysis of Time Series Structure–SSA and Related Techniques*, Chapman & Hall/CRC, 2001.
4. Kanai, S., Date, H., and Kishinami,T., Digital Watermarking for 3D Polygons Using Multiresolution Wavelet Decomposition, *Proceedings of the Sixth IFIP WG 5.2 International Workshop on Geometric Modeling: Fundamentals and Applications (GEO-6)*, pp. 296-307, Tokyo, Japan, December 1998.
5. Ohbuchi, R., Masuda, H., and Aono, M., Watermarking Three-Dimensional Polygonal Models, *Proceedings of the ACM International Conference on Multimedia '97*, pp. 261-272, Seattle, USA., November, 1997.
6. Ohbuchi, R., Masuda, H., and Aono, M., Watermarking Three-Dimensional Polygonal Models Through Geometric and Topological Modifications, *IEEE Journal on Selected Areas in Communication*, Vol. 16, No. 4, pp. 551-560, May, 1998.
7. Ohbuchi, R., Masuda, H., and Aono, M., Targeting Geometrical and Non-Geometrical Components for Data Embedding in Three-Dimensional Polygonal Models, *Computer Communications*, Vol. 21, pp. 1344-1354, October, 1998.
8. Ohbuchi, R., Masuda, H., and Aono, M., A Shape-Preserving Data Embedding Algorithm for NURBS Curves and Surfaces, *Proceedings of the Computer Graphics International'99*, pp. 180-177, Canmore, Canada, June 7-11, 1999.
9. Ohbuchi, O., Takahashi, S., Miyazawa, T., and Mukaiyama, A., Watermarking 3D Polygonal Meshes in the Mesh Spectral Domain, *Proceedings of the Graphics Interface 2001*, pp. 9-17, Ontario, Canada, June 2001.

10. Praun, E., Hoppe, H., Finkelstein, A., Robust Mesh Watermarking, *ACM SIG-GRAPH 1999*, pp. 69-76, 1999.
11. Wagner, M. G., Robust Watermarking of Polygonal Meshes, *Proceedings of Geometric Modeling & Processing 2000*, pp. 201-208, Hong Kong, April 10-12, 2000.
12. Yeo, B-L. and Yeung, M. M., Watermarking 3D Objects for Verification, *IEEE CG&A*, pp. 36-45, January/February 1999.
13. Matsui, K., *Basic of watermarks* (in Japanese), Morikita Shuppan Publishers, Tokyo, 1998.

# Compression of Arbitrary Mesh Data Using Subdivision Surfaces

Hiroshi Kawaharada and Kokichi Sugihara

Department of Mathematical Informatics
Graduate School of Information Science and Technology
University of Tokyo
{kawarada,sugihara}@simplex.t.u-tokyo.ac.jp

**Abstract.** This paper proposes a new method for mesh data compression. In this method, the original mesh is fitted by a subdivision surface. Thus, our method approximates irregular meshes to semi-regular meshes. The volume bounded by the mesh data is first partitioned into star-shaped volumes, and then each star-shaped volume is approximated. For the approximation we establish a nearly one-to-one correspondence between the resulting vertices of the subdivision surface and the vertices of the original mesh using rays emanating at a kernel point, and fit the surface by the least square method. The resulting approximated data preserves rough shape of the original mesh. Our method can be considered as an approach to a multiresolution mesh. Moreover, our method has advantages for interactively deformed meshes.

## 1   Introduction

Although there are many methods to model the 3-D form by parametric surfaces in the field of CAD or CG, if animation is taken into consideration, it is hard to treat a complicated model since it is difficult to maintain the connection between patches by the conventional form expression. Therefore, higher attention is paid to the research of subdivision surfaces which can be adapted for arbitrary models [1,2,3,4].

Moreover, mesh data can be applied to the subdivision surfaces, and simplifying or compressing large-scale mesh data has been studied from the purposes such as the transmission of mesh data and 3D contents on web [5,6,7,8,9,10,11].

Hoppe et al. [12] proposed a method for the simplification of a mesh using the evaluation function. This method estimates the energy function between dense data and the simplified mesh. Hoppe applied it to a lossless compression method [13]. Furthermore, Sander et al. [14] developed texture mapping method for the compressed data.

Hoppe et al. [15] performed fitting of subdivision surfaces. Their method uses an evaluation function between the subdivision surface, and the data scanned densely.

Moreover, Lee et al. [16] proposed an application of fitting of subdivision surfaces. They took the error between the fitting mesh and the original mesh, and raised the accuracy and the rate of compression using displacement.

M.J. Wilson and R.R. Martin (Eds.): Mathematics of Surfaces 2003, LNCS 2768, pp. 99–110, 2003.
© Springer-Verlag Berlin Heidelberg 2003

Derose et al. [17] used the subdivision surface for making nonsmooth meshes. Their subdivision which can change the grade of the angle omission consists of the usual subdivision and the subdivision with the tag attachment [15]. They also described the texture mapping of the subdivided mesh.

Lounsbery et al. [18] used multiresolution analysis based on wavelet. Although the error between the subdivided mesh and the original mesh is described as a wavelet coefficient, their method is valid only for semi-regular meshes. Furthermore, Lee et al. [19] showed a technique for parameterization in multiresolution analysis.

Our method can be regarded as a technique for approximating an irregular mesh by a semi-regular mesh. There are some methods to approximate irregular meshes to semi-regular meshes, called "remesh" algorithm. Eck et al. [20] first proposed such method. They approximated irregular meshes with bounding error. Their method added vertices into an initial mesh, too. However, they added vertices, in order to bound the error. And their method is automatic too.

The method developed by Lee et al. [16] can be regarded as remesh algorithm, too. The fit residual of the method is expressed as a semi-regular "scalar" displacement map. Khodakovsky et al. [21] used the MAPS method [19] to partition meshes and create parameterizations. This method achieved tremendous compression rates using zero-tree coding of wavelet coefficients. Guskov et al. [22] used a MAPS-like scheme, in which they expressed most of wavelet terms as scalars.

The above three algorithms used scalar wavelet or zero-tree coding. These tools may be extended to other types of mesh approximation algorithms as far as they treat only static mesh data. However, they can not be used for our purpose, because we want to treat dynamic mesh data deformed interactively.

Guskov et al. [23] proposed a "hybrid" mesh. The hybrid mesh is a multiresolution mesh that is an expansion of a semi-regular mesh. That structure has irregular operations allowing a hybrid mesh to change topology throughout the hierarchy. So the initial mesh of hybrid hierarchy is very simple. However, hybrid meshes are unsuitable to wavelet boundary element method (wavelet BEM) [24] because of the topological changes. Moreover, the method which creates hybrid meshes is interactive. Thus, the method can not be used.

Moreover, Xianfeng et al. [25] proposed a remesh algorithm. They approximated irregular mesh to completely regular mesh. Therefore, meshes are transformed to 2-D images, called "geometry image".

In this paper, we propose a new method for approximation of an arbitrary triangular mesh to a semi-regular mesh in such a way that wavelet BEM can be used for animation. By this method, mesh data are divided into star-shaped blocks and a subdivision surface is considered for fitting. Therefore, this compression is also a simplification, which also means that this compression has a loss. However, we can modify our method into a lossless one, if we store the difference between the fitted data and the original data.

The main features of this technique are as follows. The first feature is that our method is applicable to coarse mesh data. Previous methods such as [15]

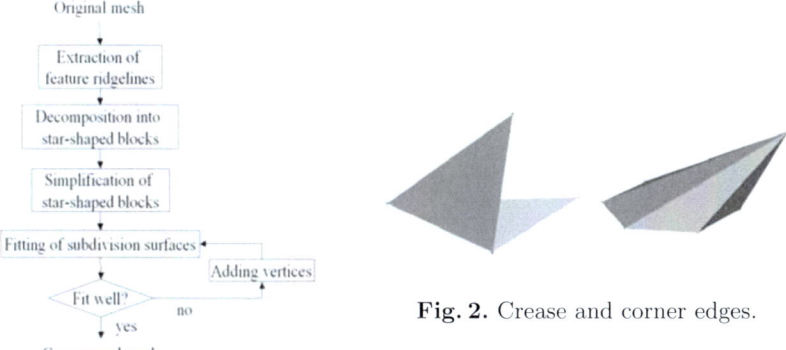

Original mesh

Extraction of
feature ridgelines

Decomposition into
star-shaped blocks

Simplification of
star-shaped blocks

Fitting of subdivision surfaces

Adding vertices

Fit well?   no

yes

Compressed mesh

**Fig. 1.** Flowchart.

**Fig. 2.** Crease and corner edges.

can be used only for the dense mesh, because a coarse data sometimes generates inconsistency. On the other hand, since our method uses the kernel of star-shape, there is no fear of the topology being confused. Therefore, it can be applied also to coarse data.

The second feature is that our method is applicable to arbitrary meshes. By the method of [18], the target mesh was restricted to the semi-regular mesh. On the other hand, our method has no such restriction.

The third feature is that our method has nearly one-to-one correspondence between original irregular meshes and approximated semi-regular meshes. We consider this a key of wavelet BEM on irregular meshes.

## 2    Outline of Our Method

Fig.1 is the flow chart of the approximation algorithm proposed in this paper. The ridgelines which express the features of the forms are first extracted from the given mesh as shown in this figure. Next, the mesh is divided into star-shaped blocks. Finally each star-shaped block is approximated. Below, each step is explained in details.

### 2.1    Extraction of the Feature Ridgelines

Let M be an original triangular mesh. $M$ consists of vertex data $V$ and complex $K$, where $K$ is a 2-dimensional complex consisting of the set of vertex numbers, and the set of edges represented by pairs of vertex numbers, and the set of faces represented by triples of vertex numbers.

Before applying our approximation method we want to extract edges that represents important feature of the surface, and treat them in a special manner so that the important features are not changed much by the approximation. For this purpose we give the following definitions. Let $\theta$ and $\theta'$ be two small fixed

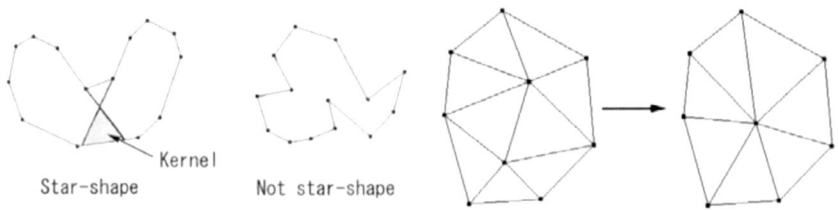

**Fig. 3.** Examples of two dimensional star-shapes.

**Fig. 4.** Examples of short cut.

angles, which are used as thresholds. An edge is called a crease edge if its two side faces form a dihedral angle smaller than a threshold $\theta$. A vertex is called a corner vertex if the maximum of the angles between two edges incident to the vertex is smaller than $\theta'$. And the edges incident to the corner vertex are called corner edges. Fig.2 shows the examples of the crease edge and the corner vertex. Let $C$ be the set of all crease edges and corner edges in $M$. An element of $C$ is called a ridgeline. Note that Hoppe et al. gave another definition of the ridgeline in [15], but we use the above definition because their definition generates too many ridgelines for our purpose.

## 2.2 Decomposition into Star-Shaped Blocks

Next, the given mesh $M$ is decomposed into star-shaped blocks. Set $S$ of points is called a star-shape if $S$ has a point from which one can see all the surfaces of $S$. More formally let us define $K(S) = \{x \in S \mid \forall y \in S, \, 0 \leq \forall \lambda \leq 1, \, \lambda x + (1-\lambda)y \in S\}$. If $K(S) \neq \phi$, $S$ is called a star-shape, and $K(S)$ is called the kernel of $S$. $S$ is convex if $K(S) = S$. But in general star-shape is not necessarily convex. Fig.3 is an example of the star-shape and an example of a non-star-shape. Polyhedron $S$ can be divided into star-shaped blocks by the following methods.

For face $f_i$ in $M$, let $g_i$ denote the center of gravity of $f_i$, and let $n_i$ denote the normal to $f_i$. Function $w_i$ is defined as $w_i = e^{n_i \cdot (r - g_i)}$, where $r$ is the position vector. When the point $r$ is in the side which $n_i$ points out, we have $w_i > 1$, and when it is in an opposite side, we have $w_i < 1$. An arbitrary face, say $f_1$, is chosen from $M$, and the face set $F$ is initialized by $F = \{f_1\}$. $r$ is chosen so that it satisfies $w_1 < 1$. The three edges of $f_1$ are put into $B$. Next, the face set $F$ which constitutes star-shape is augmented one by one, maintaining the condition that the point $r$ is in the kernel.

Therefore, face $f_k$ which has two edges in $B$ is taken. If there is no such face, the face which has one edge is taken. Then the gradient of $w = \sum_{i=1}^{k} w_i$ is computed. Suppose that $r$ was moved to $r'$ by the gradient descent method. If all the values of $w_i$ are smaller than 1 (this means that $r'$ is a kernel point of $F \cup \{f_k\}$), $f_k$ is added to $F$ and $r$ is updated to $r'$. At the same time $B$ is updated in such a way that the edge(s) of $f_i$ that is in $B$ is deleted from $B$ and the edge(s) of $f_i$ that is rut in $B$ is added to $B$. If a ridgeline is contained in $B$, we do not carrying out the addition of a face with this edge. $(F, B, r)$ is updated in this procedure.

If $r$ can not be updated on the other hand, the current $(F, B, r)$ is considered as one star-shaped block. Moreover, $B = \phi$ holds when the closed surface is obtained as one block.

Thus, mesh $M$ is decomposed into a set of star-shaped block $\bar{S} = (F, B, r)$, $r$ is a kernel point of this star-shaped block.

## 2.3   Simplification of a Star-Shaped Block

In order to obtain an initial surface for subdivision, each star-shaped block $\bar{S}$ is simplified. For this purpose the short cut operation is used. This operation removes one edge of a mesh and merges the two end vertices to one, as shown in Fig.4. This operation is performed only to an edge whose short cut does not change the topology of the mesh, called "legal move" [12].

Moreover, the short cut operation is not performed to feature ridgelines, and the position of a vertex incident to a ridgeline is not moved in short cut. In this simplification stage, only the number of vertices is taken into consideration, we do not care how the shape changes by the short cut.

We fix a subdivision scheme, and choose a positive number $t$ by which we apply the subdivision repeatedly (usually $t$ is chosen as 1, 2 or 3). The short cut operation are performed until the number of vertices becomes small enough so that the number of vertices obtained after $t$-time subdivision is almost the same as that of $\bar{S}$. In the simplification, $r$ is kept inside the kernel of the subdivision surface of simplified mesh. Let us denote by $M'$ the simplified mesh obtained by the above procedure.

## 2.4   Fitting of Subdivision Surface

The simplified mesh $M'$ is next changed so that the subdivision of $M'$, and original star-shaped block $\bar{S}$ become as close as possible.

Let $M'_s$ be the mesh obtained by applying the subdivision $t$ times to $M'$. In this paper we use the Loop subdivision [26], which divides each face into four. In this subdivision process, we do not move the edges in $C$ and in $B$. That is, new vertices generated on edges in $C$ and $B$ are placed at the middle point, and terminal vertices of edges in $C$ and $B$ are not be moved at all. The subdivision which generate such a piecewise smooth surface is introduced in [15].

Next, we try to move the vertices of $M'$ so that two meshes $\bar{M}$ and $M'_s$ come close in the same of the least-square. For this purpose we a half line from $r$ to ward each vertex of $M'_s$, and the nearest vertex on the face of $\bar{M}$ through which the half line passes is regard as the corresponding point of the vertex of $M'_s$. When the nearest vertex is already taken as a corresponding point of another vertex, the nearer one is taken among the other vertices of the face which are not yet taken as a corresponding point. If all three points are already taken, the nearest vertex will be taken again, in this case, the correspondence is not one-to-one. Fig.5 shows how to take the corresponding point.

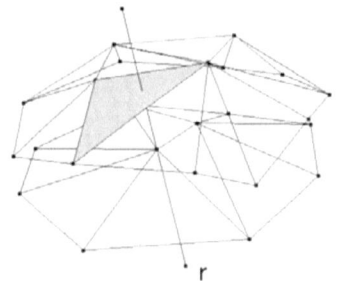

**Fig. 5.** Correspondence.

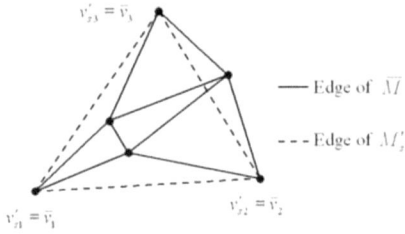

**Fig. 6.** Example of adding on vertex.

Now, since $r$ is a kernel point in star-shaped block $\bar{M}$, there is only one face of $\bar{M}$ through which a half line passes. However, $r$ is not necessarily a kernel point of $M'_s$. If $r$ is not a kernel point of $M'_s$, we cannot get a good correspondence and hence $r$ is updated by the same gradient descent method as 2.2, in order for $r$ to become inside the kernel of $M'_s$. The block is divided into two blocks if such $r$ can not be taken.

If a correspondence between $\bar{M}$ and $M'_s$ is taken, the fitting problem can be written as follows. Let $V_{\bar{M}}$ be the matrix each of whose rows consists of the $x$, $y$, and $z$ coordinates of a vertex of $\bar{M}$. We define the matrices $V_{M'}$ and $V_{M'_s}$ similarly. Let $P$ be the subdivision matrix, that is, if we multiply $V_{M'}$ from the right, then the vertices after the subdivision will be obtained:$V_{M'_s} = PV_{M'}$. Then we want to fined vertices in $M'$ that achieve

$$\min ||V_{\bar{M}} - PV_{M'}||^2 \, ,$$

where $|| \cdot ||$ is the square root of the sum of the squares of elements of the matrix.

Since we are given $V_{\bar{M}}$ and $P$, the above minimization problem is reduced to a system of linear equation, and hence the unknown matrix $V_{M'}$ (i. e., unknown coordinates of the vertices of $M'$) that achieves the minimum can be obtained by solving the system of linear equations.

Our method for establishes the correspondence might be combined with the algorithm of [15], However to apply the subdivision only in a finite number of times in this paper, the number of the corresponding points is regarded as a grade of fitting. Thus we propose a original termination condition.

## 2.5    Termination Conditions

Starting with the simplified mesh $M'$ obtained by the method of Section 2.3, we repeat fitting and addition of vertices to $M'$, until $PV_{M'}$ comes close enough to $V_{\bar{M}}$.

Whether a new vertex is added or not is determined depending on how the corresponding points on the mesh $\bar{M}$ is taken. Suppose that the three vertices $v'_{s1}, v'_{s2}, v'_{s3}$ on a face in the subdivided mesh $M'_s$ corresponds to vertices $\bar{v}_1, \bar{v}_2, \bar{v}_3$

on $\bar{M}$. We consider the three edges $(v'_{s1}, v'_{s2})$, $(v'_{s2}, v'_{s3})$, $(v'_{s3}, v'_{s1})$, and a new vertex. If (1) three or more edges in $\bar{M}$ correspond to any one edge of $M'_s$, or (2) two edges of $\bar{M}$ correspond to each of three edges in $M'_s$ and the three vertices between two correspond edges in $\bar{M}$ do not coincide at all three correspond edges, or (3) two of the three edges in $M'_s$ correspond two edges in $\bar{M}$ and two vertices between two correspond edges do not coincide, the face of $M'$ which generates $\{v'_{s1}, v'_{s2}, v'_{s3}\} \in M'_s$ is divided into three by adding a new vertex at the center of gravity.

We may continue the addition of vertices to $M'$ until all vertices of $\bar{M}$ have thier own corresponding points in $M'_s$. In that case, calculation cost decreases since what is necessary is just to check whether each edge in $\bar{M}$ corresponds to an individual edge in $M'_s$. In this paper, however, we use the above-mentioned method in order to add as few vertices as possible. Although there are some vertices of $\bar{M}$ which do not have the corresponding points, the fitting procedure is not violated, since those vertices are not on the feature ridgelines, and their neighbor vertices have the corresponding points.

The additions of vertices are performed simultaneously is one repetition for all necessary faces of $M'$. At this time, the self-similarity of the Loop subdivision shows that the vertices to which we must search for the corresponding points are the vertices on the faces of $M'_s$ generated by new three faces of $M'$ divided by adding the vertex.

When no face in $M'$ requires the addition of a vertex, the repetition terminates.

### 2.6 Merge of the Star-Shaped Blocks

The compressed data for all the star-shaped blocks are merged together.

Since the vertices on the boundary edges i. e., the vertices in $B$, each star-shaped block were not moved in the approximation operation, the blocks can be joined together.

The resulting compressed data consists of the simplified meshes $M'$ together with the boundary edge set $B$ and the feature ridgeline set $C$ (which must not be moved in the subdivision procedure).

## 3   Conclusions

In this paper, we have proposed a method for approximating mesh data using the subdivision surface. This method can be applied to an arbitrary mesh.

Moreover, this method works even if the distance of the surfaces of a mesh is smaller than a sampling interval, or even if there is self-intersection.

Examples of the mesh data approximated using our algorithm are shown.

Fig.7 is an original mesh data of a human body used for this experiment. First, we extract the feature ridgeline set $C$ according to our definition in section 2.1. Next, we augment $C$ by adding edges that do not move on the occasion of animation, i. e., deformation. Actually the edges on the line of the shoulder is

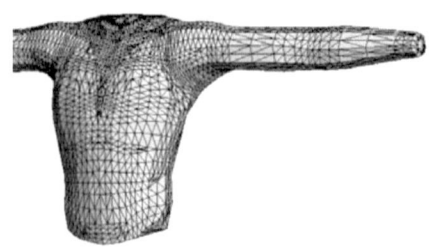

**Fig. 7.** Original mesh.

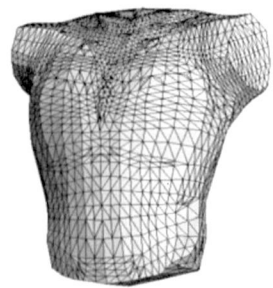

**Fig. 8.** A star-shaped block.

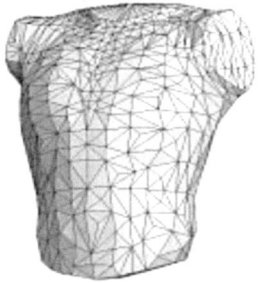

**Fig. 9.** Compressed star-shaped block.

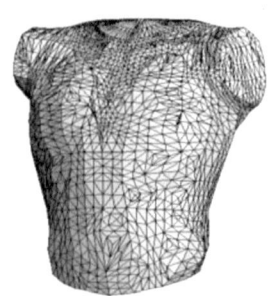

**Fig. 10.** Reconstructed star-shaped block.

added to $C$. The resulting main star-shaped block was obtain, as shown in Fig.8. We simplified the star-shaped block of Fig.8 and carried out fitting subdivision surface. The compressed data is as shown in Fig.9 and the reconstructed mesh is as shown in Fig.10. The number of original vertices was 3845 which was reduced to 1369 after compression. The compression rate was 37.1, where the data size is measured in bytes position of the vertices of the united mesh data, the face information, the boundary edge set $B$, and the feature ridgeline set $C$.

We next show compression of a self-intersected model. Fig. 11 is the original mesh of the self-intersected model. Fig. 12 is the reconstructed blocks of that. Fig. 13 shows the compressed blocks of that.

In this method, the number of times of the subdivision which carries out fitting determines a rough compression rate. Therefore, improvement in the rate can be expected if we increase the number of times of applying the subdivision, but the error also increases. Sufficient compression will be expected if fitting is practically carried out by applying the subdivision 2 or 3 times.

The structure of the hybrid mesh [23] has a very simple initial mesh. So the initial mesh can be considered as a compressed data, whose compression rate is high. Since our method requires the correspondeces between the vertices of $\bar{M}$ and those of $M'_s$, the compression rate of our method is inferior to that of the

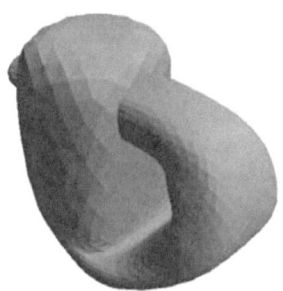

**Fig. 11.** A intersected model.

**Fig. 12.** Reconstructed star-shaped blocks of the intersected model.

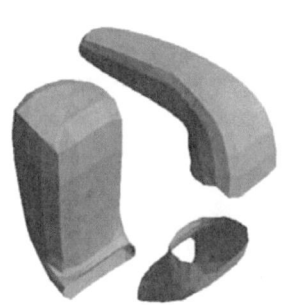

**Fig. 13.** Compressed blocks of the intersected model.

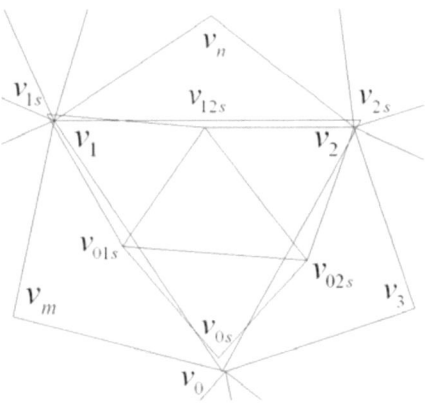

**Fig. 14.** Self-similar triangulation.

hybrid mesh. However, in the first place, previous remesh algorithms including the hybrid mesh algorithm only cares the shape of objects; the connectivity and number of vertices is ignored. On the other hand, we consider that user who makes the original mesh wants to try wavelet BEM on the original mesh, and hence we care the connectivity and the number of vertices of original mesh.

In this method, the computational complexity of each part is as follows. The decomposition into star-shaped blocks requires $O(n^2)$ time as long as a computation of the exponential is constant, where $n$ is the number of triangles of the star-shaped block. The simplification of a star-shaped block requires $O(n)$ time. The fitting of subdivision surface requires $O(n^3)$ time in general. However if the basis function of subdivision has full orthogonal wavelets, the time reduces to $O(n)$. If the basis function has bi-orthogonal wavelets, the time reduces also to $O(n)$ [3,18]. The checking the termination conditions requires $O(n^2)$ time, if the number of edges connecting a vertex is at most constant. Since the number of iteration of the fitting is constant, the time complexity of this algorithm is $O(n^2)$ or $O(n^3)$.

Since the mesh data compressed by this method represent a rough form of the original shape in a similar manner as the algorithm in [15], compressed data can be used like progressive JPG; we first show the rough form by displaying the compressed data directly, and then generate the precise form by decoding the data. That is, progressive refinement is possible and this property is useful for the distribution via web. And, the subdivision operation for decoding can be applied locally, and, therefore, this method can be used for selective refinement. In short, the resulting mesh of our method belongs to the multiresolution mesh.

Moreover, the correspondence is constructed using the rays starting at a kernel point of the star-shaped block and hence corresponding pair of vertices are near to each other. Therefore, animation (motion of the shape) defined by the motion of the vertices of the given original mesh can also be represented by the motion of the vertices of the corresponding subdivision surface, and can be further represent by the motion of the vertices in the compressed data. Hence, we specify only the motion of the vertices in the compressed data, and still the resulting motion of the decoded shape achieve almost the same grade of smoothness as the original animation. If we consider the animation more important, we may add the edges which do not move in animation to the feature ridgeline set $C$, and also we may augment the vertices so that all the vertices of given mesh have the corresponding vertices in $M'_s$.

There is a correspondence nearly one-to-one between $\bar{M}$ and $M'_s$ (since there is some duplicate vertices, it is not a strict one-to-one), and there is also a correspondence between $M'_s$ and $M'$. Hence, we can consider the correspondence between $M'$ and $\bar{M}$. This correspondence enables us to map the texture from $\bar{M}$ to $M'$. Therefore, this compression method is advantageous to animation.

## 4   Future Work

There are remaining work for future. First, the short cut operation for edges was used for simplifying a star-shape block, but other simplification operation can also be considered. Probably, a simplification which can suppress the addition of vertices should also be considered. Moreover, when a face is divided into three the new vertex is generated at the center of gravity in our method, but it is also useful to add in such a way that the correspondence between vertices becomes easier. Furthermore, although the nearest vertex was taken as the corresponding vertex, it may be better to choose other than the nearest in order to avoid one-to-many correspondence.

Secondly, we add a vertex by dividing a face into three, but this restricts the topology (in the sense of the vertex degree) of the compressed mesh greatly, and hence other operations such as the edge division in [12] etc. should also be considered. Moreover, if you want to add a regular vertex what is necessary is just to swap the edges of the three triangles incident to the new vertex. However, the number of times of applying the least-squares method can be decrease by adding two or more vertices simultaneously. For that purpose also we need to local operations to add new vertices.

Thirdly, we need a systematic method for changing the kernel point $r$ to keep it inside $M'_s$. By our method, if $r$ is outside $M'_s$, fitting can not be carried out. So it is necessary to consider a method which adjusts the step size of the renewal of $r$, or a method for keeping $r$ inside $M'_s$. Since $M'_s$ approaches $\bar{M}$ by fitting, the possibility that $r$ will come outside $M'_s$ in the repetition is small, probably, there are only few blocks which must be divided.

Fourthly, $r$ should be inside the kernel of $M'$. For example, when the mesh $M'$ with vertices $v_0, v_1, \cdots$ is subdivided into $M'_s$ with vertices $v_{0s}, v_{1s}, \cdots, v_{01s}, v_{02s}, v_{12s}, \cdots$ as shown in Fig.14, $r$ must be in the rear sides of triangles $\{v_{01s}, v_{02s}, v_{12s}\}$ and $\{v_{0s}, v_{01s}, v_{02s}\}$. However, $r$ is also in the kernel of original mesh. We have not get found an efficient method for checking all these conditions. Therefore, in our present method, we check whether $r$ is inside the kernel of the mesh every time the subdivision is applied to a triangle. this effort requires $O(n)$ time, where $n$ is the number of triangles of the star-shaped block. Since the time complexity of subdivision is also $O(n)$, the order of the time complexity as the whole does not change.

Fifthly, we want to adjust the difference between $\bar{M}$ and $M'_s$. One advantage of the method in [15] is that they can specify the upper bound of the error of $M'_s$ from $\bar{M}$, arbitrarily no matter whether the vertices are given on the same surface densely or sparsely. In our method, on the other hand, we may adjust the error by the number of times of subdivision. We need to find the best times of subdivision subject to the trade-off between the number of the additional vertices, and the error of the compression result.

## Acknowledgments

This work is supported by the 21st Century COE Program on Information Science Strategic Core of the Ministry of Education, Science, Sports and Culture of Japan.

## References

1. Capell, S., Green, S., Curless, B., Duchamp, T., Popovic, Z.: A multiresolution framework for dynamic deformations. In: SIGGRAPH '02 Proceedings, ACM (2002) 41–47
2. Grinspun, E., Krysl, P., Schroder, P.: Charms: A simple framework for adaptive simulation. In: SIGGRAPH '02 Proceedings, ACM (2002) 281–290
3. Stollnitz, E.J., DeRose, T.D., Salesin, D.H.: Wavelets for Computer Graphics: Theory and Applications. Morgan Kaufmann Publishers (1996)
4. Warren, J., Weimer, H.: Subdivision Methods for Geometric Design: A Constructive Approach. Morgan Kaufmann Publishers (1995)
5. Debunne, G., Desbrun, M., Cani, M.P., Barr, A.H.: Dynamic real-time deformations using space and time adaptive sampling. In: SIGGRAPH '01 Proceedings, ACM (2001) 31–36
6. Gandoin, P.M., Devillers, O.: Progressive lossless compression of arbitrary simplicial complexes. In: SIGGRAPH '02 Proceedings, ACM (2002) 372–379

7. Lindstrom, P.: Out-of-core simplification of large polygonal models. In: SIG-GRAPH '00 Proceedings, ACM (2000) 259–262

8. Szeliski, R., Terzopoulos, D.: From splines to fractals. In: SIGGRAPH '89 Proceedings, ACM (1989) 51–60

9. Terzopoulos, D., Metaxas, D.: Dynamic 3d models with local and global deformations: Deformable superquadrics. IEEE Transactions on Pattern Analysis and Machine Intelligence **13** (1991) 703–714

10. Vemuri, B.C., Mandal, C., Lai, S.H.: A fast gibbs sampler for synthesizing constrained fractals. IEEE Transactions on Visualization and Computer Graphics **3** (1997) 337–351

11. Vemuri, B.C., Radisavljevic, A.: Multiresolution stochastic hybrid space models with fractal priors. ACM Transactions on Graphics **13** (1994) 177–207

12. Hoppe, H., Derose, T., Duchamp, T., Mcdonald, J., Stuetzle, W.: Mesh optimization. In: SIGGRAPH '93 Proceedings, ACM (1993) 19–26

13. Hoppe, H.: Progressive meshes. In: SIGGRAPH '96 Proceedings, ACM (1996) 99–108

14. Sander, P.V., Snyder, J., Gortler, S.J., Hoppe, H.: Texture mapping progressive meshes. In: SIGGRAPH '01 Proceedings, ACM (2001) 409–416

15. Hoppe, H., Derose, T., Duchamp, T., Halsted, M.: Piecewise smooth surface reconstruction. In: SIGGRAPH '94 Proceedings, ACM (1994) 295–302

16. Lee, A., Moreton, H., Hoppe, H.: Displaced subdivision surface. In: SIGGRAPH '00 Proceedings, ACM (2000) 85–94

17. Derose, T., Kass, M., Truong, T.: Subdivision surfaces in character animation. In: SIGGRAPH '98 Proceedings, ACM (1998) 85–94

18. Lounsbery, M., Derose, T., Warren, J.: Multiresolution analysis for surfaces of arbitrary topological type. ACM Transactions on Graphics **16** (1997) 34–73

19. Lee, A.W.F., Sweldens, W., Schroder, P., Cowsar, L., Dobkin, D.: Maps: Multiresolution adaptive parameterization of surfaces. In: SIGGRAPH '98 Proceedings, ACM (1998) 95–104

20. Eck, M., DeRose, T., Duchamp, T.: Multiresolution analysis of arbitrary meshes. In: SIGGRAPH '95 Proceedings, ACM (1995) 173–182

21. Khodakovsky, A., Schroder, P., Sweldens, W.: Progressive geometry compression. In: Proceedings of the 27th annual conference on Computer graphics and interactive techniques, ACM (2000) 271–278

22. Guskov, I., Vidimce, K., Sweldens, W., Schroder, P.: Normal meshes. In: Proceedings of the 27th annual conference on Computer graphics and interactive techniques, ACM (1995) 95–102

23. Guskov, I., Khodakovsky, A., Schroder, P., Sweldens, W.: Hybrid meshes: multiresolution using regular and irregular refinement. In: Proceedings of the eighteenth annual symposium on Computational geometry, ACM Press (2002) 264–272

24. James, D.L., Pai, D.K.: Multiresolution green's function methods for interactive simulation of large-scale elastostatic objects. ACM Transactions on Graphics **22** (2003) 47–82

25. Gu, X., Gortler, S.J., Hoppe, H.: Geometry images. In: Proceedings of the 29th annual conference on Computer graphics and interactive techniques, ACM (2002) 355–361

26. Loop, C.T.: Smooth subdivision surfaces based on triangles. Master's thesis, University of Utah, Department of Mathematics (1987)

# Triangle Mesh Duality:
# Reconstruction and Smoothing

Giuseppe Patanè and Michela Spagnuolo

Istituto di Matematica Applicata e Tecnologie Informatiche
Consiglio Nazionale delle Ricerche
Via De Marini 6, 16149 Genova, Italia
{patane,spagnuolo}@ge.imati.cnr.it
http://www.ge.imati.cnr.it

**Abstract.** Current scan technologies provide huge data sets which have to be processed considering several application constraints. The different steps required to achieve this purpose use a structured approach where fundamental tasks, e.g. surface reconstruction, multi-resolution simplification, smoothing and editing, interact using both the input mesh geometry and topology. This paper is twofold; firstly, we focus our attention on duality considering basic relationships between a 2-manifold triangle mesh $\mathcal{M}$ and its dual representation $\mathcal{M}'$. The achieved combinatorial properties represent the starting point for the reconstruction algorithm which maps $\mathcal{M}'$ into its primal representation $\mathcal{M}$, thus defining their geometric and topological identification. This correspondence is further analyzed in order to study the influence of the information in $\mathcal{M}$ and $\mathcal{M}'$ for the reconstruction process. The second goal of the paper is the definition of the "*dual Laplacian smoothing*", which combines the application to the dual mesh $\mathcal{M}'$ of well-known smoothing algorithms with an inverse transformation for reconstructing the regularized triangle mesh. The use of $\mathcal{M}'$ instead of $\mathcal{M}$ exploits a topological mask different from the 1-neighborhood one, related to Laplacian-based algorithms, guaranteeing good results and optimizing storage and computational requirements.

**Keywords:** Computational geometry, mesh duality, Laplacian smoothing, filtering.

## 1   Introduction

Recent applications to compression [13], smoothing [18], subdivision [21] reveal an increasing attention to the correspondence between a mesh $\mathcal{M}$ and its dual representation $\mathcal{M}'$. The growing interest on primal-dual correspondence is due to a greater regularity of the dual mesh topology which corresponds to storage and computational optimization. The first part of the paper analyzes in detail the topological and geometric identification between $\mathcal{M}$ and $\mathcal{M}'$; more precisely, we provide a reconstruction algorithm of the geometry of $\mathcal{M}$ through that of $\mathcal{M}'$, also achieving basic combinatorial properties of the 1-neighborhood of each internal vertex in $\mathcal{M}$. This correspondence results in the definition of a discrete

M.J. Wilson and R.R. Martin (Eds.): Mathematics of Surfaces 2003, LNCS 2768, pp. 111–128, 2003.

homeomorphism between $\mathcal{M}$ and $\mathcal{M}'$ whose computational cost is linear in the number of faces in $\mathcal{M}$. The stability to noise on the mesh vertices is studied underlining the correlation between adjacent neighborhoods in $\mathcal{M}$. The developed framework is used to look at the signal processing theory of triangle meshes by considering the dual mesh as noised one, and successively defining the regularized mesh through a process different from the primal-dual identification which cannot be applied due to the violation of the derived combinatorial properties. Therefore, the "*dual Laplacian smoothing*" reveals the way the regularization process affects the input mesh geometry. This approach to smoothing enables to consider a new topological mask for the mesh regularization whose effectiveness is compared with that of the Taubin's $\lambda|\mu$ algorithm [17].

The paper is organized as follows: in Section 2 definitions and properties of 3D polygonal meshes and duality are given. Combinatorial relations and triangle mesh reconstruction through duality are discussed in Section 3, providing several considerations on the primal-dual correspondence. The dual Laplacian smoothing is analyzed in Section 4 underlining its relationship with analogous methods and triangle mesh duality previously analyzed. Conclusions and future work are discussed in the last section.

## 2   Polygonal Meshes and Duality

A *polygonal mesh* is defined by a pair $\mathcal{M} := (P, F)$ where $P$ is a set of vertices $P := \{p_i := (x_i, y_i, z_i) \in \mathbb{R}^3, \ i = 1, \ldots, n_V\}$, and $F$ an abstract simplicial complex which contains the connectivity information, i.e. the mesh topology. In particular, if we consider a *triangle mesh* each element in the complex $F$ comes into one of these elements: vertex $\{i\}$, edge $\{i, j\}$, face $\{i, j, k\}$. Traversing the mesh is achieved by using the relations [14]:

- vertex-vertex $VV(v) = (v_1, \ldots, v_k)$, face-face $FF(f) = (\tilde{f}_1, \ldots, \tilde{f}_m)$;
- face-vertex $VF(v) = (f_1, \ldots, f_k)$, vertex-face $FV(f) = (v_1, \ldots, v_q)$.

In the following of the paper we assume that the previous relations are consistently evaluated. A vertex $v$ is defined as *internal* if its *1-neighborhood* $VV(v)$ is closed, i.e. $v$ is not on the boundary of $\mathcal{M}$. Different authors have proposed optimized data structures [10,14] for efficiently representing and traversing a polygonal mesh; specializations of these techniques to triangle meshes are described in [6]. The duality of structures arises in different scientific contexts such as functional/numerical analysis (e.g. dual of Hilbert spaces) and computational geometry. In the last field one of the fundamental data structure is the Voronoi diagram [1,5,8,10,16] of a discrete set of points. Its study, which has influenced different application areas such as math, computer and natural science, is strictly related to the Delaunay triangulation and their duality relationships. In the sequel of the section we briefly review the duality in the plane and its extension to 3D meshes. If $P := \{p_i\}_{i=1}^{n_V}$ is a set of $n_V$ points in $\mathbb{R}^d$, its *Voronoi diagram* $V(P)$ is a cell complex which decomposes $\mathbb{R}^d$ into $n_V$ cells $\{V(p_i)\}_{i=1}^{n_V}$ where $V(p_i)$ is defined as

$$V(p_i) := \{x \in \mathbb{R}^d : \|x - p_i\|_2 < \|x - p_j\|_2, j \neq i\}$$

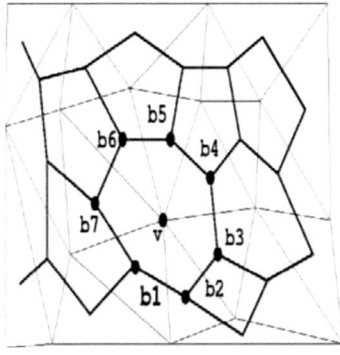

**Fig. 1.** Triangle mesh and dual graph (marked line).

and $\| \ \|_2$ denotes the Euclidean distance. This definition introduces a proximity relation among points in $\mathbb{R}^d$. In the planar case, i.e. $d = 2$, the *dual graph* $\mathcal{G}$ of $V(p)$ has a node for every cell and it has an arc between two nodes if the corresponding cells share an edge. The *Delaunay graph* of $P$ is defined as the straight-line embedding of $\mathcal{G}$ obtained by identifying the node corresponding to the cell $V(p_i)$ with $p_i$ and the arc connecting the nodes of $V(p_i)$ and $V(p_j)$ with the segment $\overline{p_i p_j}$. The Delaunay graph of a planar point set is a plane graph and, if $P$ is in *general position*, i.e. no four points lie on a circle, all vertices of the Voronoi diagram have degree three. This result guarantees that all bounded faces in the Delaunay graph are triangles, thus defining the *Delaunay triangulation* of $P$. The extension of this theory for triangulating a set of points in $\mathbb{R}^3$ is not trivial; as a result important properties of the two dimensional Delaunay triangulation, e.g. optimal storage requirement, computational cost, partially apply to the 3D setting. Previous considerations have required the definition of new algorithms [3,11], and 3D triangulation remains a challenging problem in Computer Graphics. From a geometric point of view, this has also brought a diminishing attention to duality mainly due to the use of other geometric structures.

The use of several algorithms for constructing a polygonal mesh of a 3D point cloud requires to define the dual graph in a general way, taking out of consideration the method that has been used for the mesh construction. Given a polygonal mesh $\mathcal{M}$, its *dual graph* $\mathcal{G}$ has a node $v^\star$ for each face $f(v^\star)$ in $\mathcal{M}$ and it has an arc between two nodes $v^\star$ and $w^\star$ if and only if $f(v^\star)$ and $f(w^\star)$ share an edge (see Figure 1). In analogy with the previous definitions, the *barycenter dual graph* of $\mathcal{M}$ is defined as the straight-line embedding $\mathcal{M}'$ of $\mathcal{G}$ obtained by identifying each one of its nodes with the barycenter $b_{f(v^\star)}$ of the corresponding face $f(v^\star)$, and the arc connecting the nodes $v^\star$ and $w^\star$ with the segment $\overline{b_{f(v^\star)} b_{f(w^\star)}}$. Therefore, in the dual mesh $\mathcal{M}' := (B, G)$ each vertex $b_{f(v^\star)}$ corresponding to the face $f(v^\star) = (v_1, \ldots, v_l)$ is computed as

$$b_{f(v^\star)} := \frac{1}{l} \sum_{i=1}^{l} p_{v_i}$$

and the connectivity $G$ is completely defined by $F$. From the previous description it follows that each face of the dual mesh generally has a different number of vertices even if the input mesh is triangular, quadrilateral, etc. . Clearly, if $\mathcal{M}$ is a triangle mesh $l$ is three.

Finally, we note that the genus of the dual mesh is equal to that of the input mesh, thus preserving its topology. This simply follows observing that the Euler characteristic $\chi(\mathcal{M}) = n_V - n_E + n_F$ is the same of $\mathcal{M}'$ being $n'_V = n_F$, $n'_E = n_E$, $n'_F = n_V$, where $n_V$, $n_E$, $n_F$ and $n'_V$, $n'_E$, $n'_F$ are the number of vertices, edges and faces of $\mathcal{M}$ and $\mathcal{M}'$ respectively. Therefore, we can summarize this property as: *the genus of a polygonal mesh is invariant under the duality transformation.*

## 3   Combinatorial Properties of Triangle Meshes

In the previous section we have derived the invariance of the genus of $\mathcal{M}$ under the duality transformation. We are now concerned with the analysis of the geometry of $\mathcal{M}$ and $\mathcal{M}'$. The most general and strictly related questions which give a deeper understanding of the relationships and differences between $\mathcal{M}$ and $\mathcal{M}'$ can be summarized as follows.

- Is it possible to locally characterize the 1-neighborhood structure of each vertex of $\mathcal{M}$?
- Is it possible to reconstruct the input mesh by using only the dual mesh? Which is the minimal number of information, if any, required for this purpose?

Answering these questions is not trivial and in the next section we take into consideration the case where $\mathcal{M}$ is a *2-manifold triangle mesh*; the analysis of the general problem is discussed in Section 3.3.

### 3.1   1-Neighborhood Analysis

In this section we are going to derive two basic combinatorial properties of the 1-neighborhood structure of each internal vertex $v$ in $\mathcal{M}$. These relationships are used in the sequel for describing the linear reconstruction algorithm of $\mathcal{M}$ from $\mathcal{M}'$ thus defining a complete topological and geometric identification between a mesh and its dual representation.

**Theorem 1.** *Let $\mathcal{M}$ be a 2-manifold triangle mesh with two adjacent vertices $v$ and $w$, $VF(v) = (f_1, \dots, f_k)$ the faces incident in $v$, and $b_i$ the barycenter of the face $f_i$, $\forall i = 1, \dots, k$ (see Figure 2). If $v$ is an internal vertex, the following conditions hold:*

- *if $k$ is even,*

$$\sum_{i=1}^{k} (-1)^i b_i = 0 \qquad (1)$$

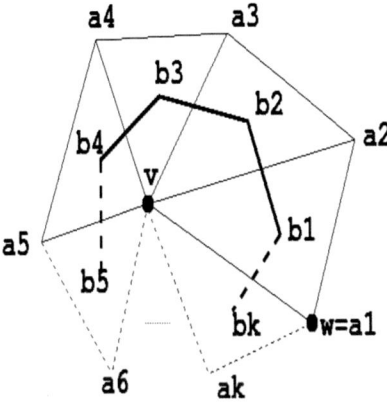

**Fig. 2.** 1-neighborhood of the vertex $v$.

– *if k is odd,*

$$3\sum_{i=1}^{k}(-1)^{i+1}b_i = v + 2w \tag{2}$$

*Proof. Considering $VV(v) = (v_1, \ldots, v_k)$, with*

$$a_1 := p_{v_1} = w, \ a_l := p_{v_l}, \ l = 2, \ldots, k$$

*the idea is to express each vertex $a_l$ as a linear combination of $\{b_i\}_{i=1}^{l-1}$, $2 \leq l \leq k$, $v$ and $w$. For each triangle $f_l$, we have that*

$$b_l = \frac{1}{3}(a_{l+1} + a_l + v) \iff a_{l+1} = 3b_l - a_l - v.$$

*Substituting in the last equality the expression of $a_l$ in terms of $b_{l-1}$, $a_{l-1}$, $v$, and recursively applying this process we achieve that*

$$a_{l+1} = \begin{cases} 3\sum_{i=1}^{l}(-1)^i b_i + a_1 & \text{if } l \text{ is even} \\ \\ 3\sum_{i=1}^{l}(-1)^{i+1}b_i - a_1 - v & \text{if } l \text{ is odd.} \end{cases}$$

*The condition $a_{k+1} = a_1$ implies (1) if $k$ is even, and (2) if $k$ is odd.*

The interesting element in (1), (2) is that the coefficients which appear in the linear combination of the barycenters are constant and not related to their positions. Furthermore, if $k$ is even the manifold structure on $\mathcal{M}$ ensures that $k \geq 4$; therefore, identifying $b_i$ with the vector $(b_i - v)$ (2) gives their linear dependency relationship in the vector space $\mathbb{R}^3$ only using constant coefficients.

### 3.2   Triangle Mesh Reconstruction Through Duality

The first step of the reconstruction algorithm (see Figure 3) chooses two internal vertices $v$, $w$ of an edge in $\mathcal{M}$ and associated to the neighborhoods

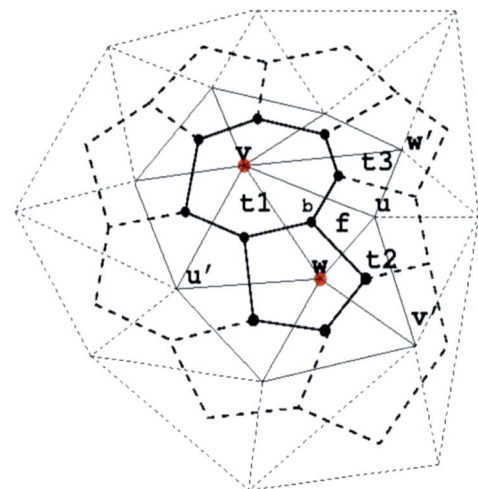

**Fig. 3.** 1-neighborhoods of $v$ and $w$ used for the reconstruction of the mesh geometry.

$$VV(v) = (v_1, \ldots, v_{n_1}), \quad VV(w) = (w_1, \ldots, w_{n_2}).$$

Selected the vertex $v$, each one of its incident triangles $VF(v) = (f_1, \ldots, f_{n_1})$, $n_1 \geq 3$, is represented by a vertex in the dual mesh $\mathcal{M}'$ which is the barycenter of the related triangle in $\mathcal{M}$. Supposed to have calculated the vertices $v$, $w$ and indicated with $f$ one of the two triangles which have the segment $\overline{vw}$ as its edge, the third vertex $u$ is evaluated as

$$u = 3b - v - w \tag{3}$$

where $b$ is the barycenter of $f$.

After this calculation, the triangle $f$ is marked as visited and its adjacent ones $(t_1, t_2, t_3) = FF(f)$ are considered. Using (3), the new vertices $u', v', w'$ are calculated marking these triangles as visited. Growing from the visited faces by using their adjacent triangles, and recursively applying this criterion to the non-marked faces of $\mathcal{M}$ enables to reconstruct the geometry of the input mesh with exactly $n_F$ steps.

It remains to describe the method for evaluating the two vertices $v$ and $w$ which have been used for reconstructing the input mesh geometry. Without loss of generality[1] we can suppose that $n_1$ and $n_2$ are odd; applying (2) to $v$ and $w$ leads to the symmetric linear system

$$\begin{cases} 3\sum_{i=1}^{n_1}(-1)^{i+1}b_i = 2w + v \\ \\ 3\sum_{i=1}^{n_2}(-1)^{i+1}b'_i = 2v + w \end{cases}$$

where $VF(w) = (f'_1, \ldots, f'_{n_2})$ and $b'_i$ is the barycenter of the face $f'_i$. Because these relations are linearly independent, its unique solution is

---

[1] For instance, if $n_1$ is even we can split $f_{n_1}$ into two new triangles; i.e. joining its vertex $v$ with the middle point of the edge opposite to $v$ in $f_{n_1}$.

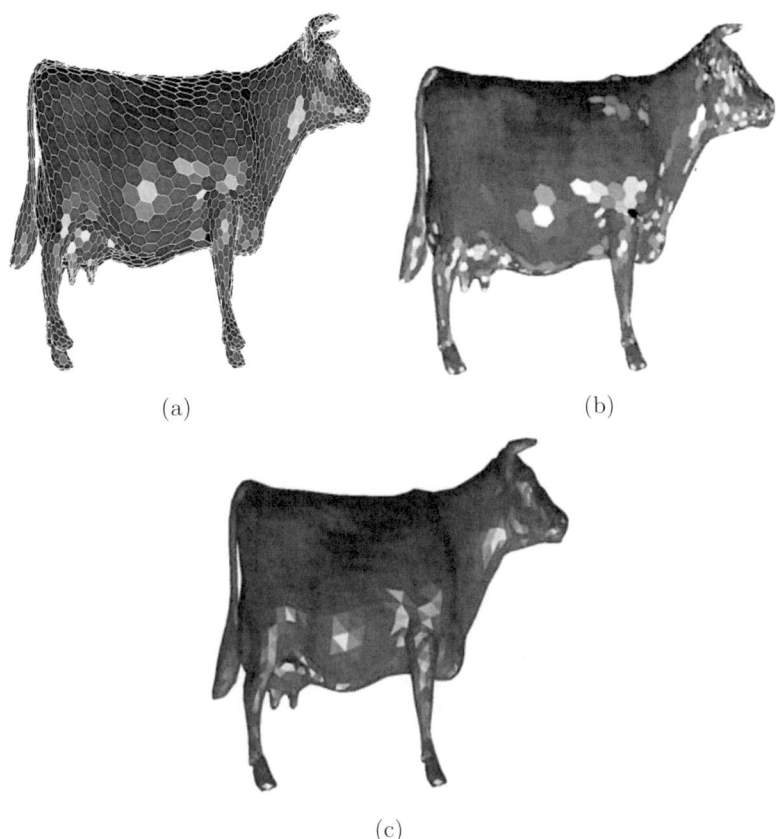

(a)                                                        (b)

(c)

**Fig. 4.** (a) Input dual mesh with 5.804 vertices and 2.904 faces, (b) dual graph colored with respect to the number of vertices in each face (see Table 1), (c) reconstructed triangle mesh.

$$\begin{cases} w = 2\alpha - \beta \\ v = 2\beta - \alpha \end{cases} \qquad (4)$$

with $\alpha = \sum_{i=1}^{n_1}(-1)^{i+1}b_i$ and $\beta = \sum_{i=1}^{n_2}(-1)^{i+1}b_i'$.

The relation (4) expresses these vertices as a linear combination of the barycenters of the triangles of their 1-neighborhoods; we also underline the symmetry in the expression of $v$ and $w$ with respect to $\alpha$, $\beta$. The computational cost of the proposed algorithm is optimal because it only requires to visit all the triangles of the input mesh, and the expression (3) is computationally stable minimizing the numerical instability of the algorithm. Therefore, the transformation which maps $\mathcal{M}$ to $\mathcal{M}'$ is linear in $n_F$ as its inverse. An example of dual mesh and of the reconstruction process is given in Figure 4.

**Table 1.** Face coloring in the dual graph.

| Color | Number of face vertices $k$ |
|-------|------------------------------|
| yellow | $1 \leq k < 2$ |
| cyan | $2 \leq k < 6$ |
| blue | $k = 6$ |
| red | $7 \leq k < 10$ |
| black | $k \geq 10$ |

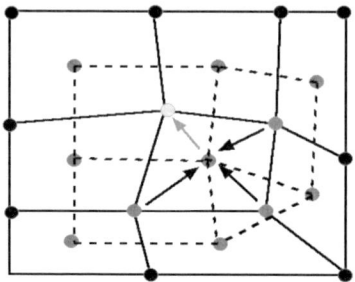

● **Face barycenter**
● **Starting vertices**
● **Non reconstructed vertices**

**Fig. 5.** Quadrilateral mesh: dual representation (dotted line) and non-reconstructed geometry.

## 3.3 Considerations on the Primal-Dual Correspondence

We present in this section several considerations on the primal-dual correspondence which is related to the dual Laplacian smoothing described in the following of the paper. The extension of the reconstruction process from the dual mesh of a $q$-mesh, $q \geq 4$, cannot be directly derived from the approach previously described. In fact, supposed that $(l - 1)$ vertices $(p_{v_1}, \ldots, p_{v_{l-1}})$ of a face $f = (v_1, \ldots, v_l)$ in $\mathcal{M}$ have been calculated, the last vertex $p_{v_l}$ is evaluated using the barycenter of $f$

$$b_f = \frac{1}{l} \sum_{i=1}^{l} p_{v_i}$$

as

$$p_{v_l} = l b_f - \sum_{i=1}^{l-1} p_{v_i};$$

however, this information is not sufficient for finding the position of all the vertices in the adjacent faces of $f$ (see Figure 5).

We answer questions given in Section 3 in a simple way as summarized by Theorem 2.

**Theorem 2.** *Given a 2-manifold triangle mesh $\mathcal{M}$ with or without boundary and with at least two internal vertices, the following conditions hold:*

- $\mathcal{M}$ *and its dual mesh* $\mathcal{M}'$ *are topologically equivalent, i.e.* $\chi(\mathcal{M}) = \chi(\mathcal{M}')$;
- $\mathcal{M}$ *and* $\mathcal{M}'$ *are geometrically equivalent, i.e.* $\mathcal{M}$ *(resp.* $\mathcal{M}'$*) is reconstructed in* $n_F$ *steps (resp.* $n_V$*) from its dual representation* $\mathcal{M}'$ *(resp.* $\mathcal{M}$*);*
- *the vertices of* $\mathcal{M}$ *and* $\mathcal{M}'$ *satisfy conditions (1), (2) given in Theorem 1.*

In Theorem 2, it has been pointed out that the dual mesh is sufficient for identifying the input triangle mesh geometry and topology without storing additional information. We want to analyze the influence of the information in $\mathcal{M}'$ for the reconstruction process; equivalently, we study how the geometry of $\mathcal{M}$ is affected by changing the position of $v$ and $w$. To this end, we add a noise $e$ to each one of them considering the new points

$$\tilde{v} := v + e, \qquad \tilde{w} := w + e.$$

Denoted with $\mathcal{M}$ the triangle mesh reconstructed from $\mathcal{M}'$, $v$, $w$ and with $\mathcal{M}_{noise}$ the one achieved with $\mathcal{M}'$, $\tilde{v}, \tilde{w}$, we want to estimate their deviation by using a norm for the error evaluation. The comparison of two triangle meshes with different geometry and connectivity has been studied in [4]. Because $\mathcal{M} := (P, F)$ and $\mathcal{M}_{noise} := (\tilde{P}, F)$ share the same topology, a simpler comparison between vertex positions and triangle normals is introduced using the following vectors:

$$d_v(\mathcal{M}, \mathcal{M}') := \left( \frac{\|p_i - \tilde{p}_i\|_2}{C_V} \right)_{i=1}^{n_V}, \quad C_V := \max_{i=1,\ldots,n_V} \{\|p_i - \tilde{p}_i\|_2\}$$

$$d_n(\mathcal{M}, \mathcal{M}') := \left( \frac{\|n_i - \tilde{n}_i\|_2}{C_N} \right)_{i=1}^{n_F}, \quad C_N := \max_{i=1,\ldots,n_F} \{\|n_i - \tilde{n}_i\|_2\}$$

with $n_i$ and $\tilde{n}_i$ unit normals to the faces $f_i$ and $f_i'$. For a better visualization, the increasing reorder of $d_v(\mathcal{M}, \mathcal{M}')$ and $d_n(\mathcal{M}, \mathcal{M}')$ is plotted without normalization (i.e. $C_V := C_N := 1$). As underlined in Figure 6(a), a small perturbation $e$ creates a wrong reconstruction of the input triangle mesh showed in Figure 4(c). This phenomena is mainly due to the high correlation between vertices in $\mathcal{M}$ and $\mathcal{M}'$ resulting in an error propagation which grows in parallel with the visiting triangle process. This aspect is a consequence of the fact that each new vertex is calculated starting from those ones previously evaluated; indeed, after $k$ steps (3) results affected by an error which is proportional to $e^k$. These considerations also apply if we add a noise to the vertices in $\mathcal{M}'$ as underlined Figure 6(b). Figure 7 shows all steps of the proposed framework.

We associate to a mesh three matrices which code in a compact form its topology. If $VF(v) = (f_1, \ldots, f_m)$ is the set of faces incident in a vertex $v$ of $\mathcal{M}$, we construct the corresponding face-vertex matrix $I_{VF} \in M_{n_V, n_F}(\mathbb{R})$

$$I_{VF}(i, j) := \begin{cases} 1 & \text{if } f_j \in VF(v_i), \\ 0 & \text{else} \end{cases} \tag{5}$$

and its normalization

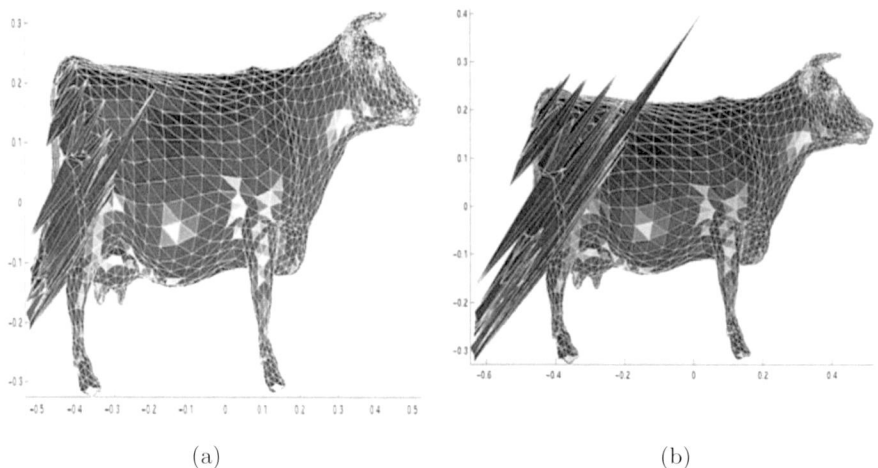

(a)                                   (b)

**Fig. 6.** Noise influence on the reconstruction process using: (a) noised vertices $\tilde{v}$, $\tilde{w}$ in $\mathcal{M}$, $e := 5.0 * 10^{-6}$, (b) a noised vertex in $\mathcal{M}'$, $e := 5.0 * 10^{-6}$.

$$N_{VF}(i,j) := \begin{cases} 1/m_i & \text{if } f_j \in VF(v_i), \ \#VF(v_i) = m_i, \\ 0 & \text{else.} \end{cases}$$

In a similar way, we define the normalized vertex-vertex matrix $N_{VV} \in M_{n_V, n_V}(\mathbb{R})$ as

$$N_{VV}(i,j) := \begin{cases} 1/m_i & \text{if } v_j \in VV(v_i), \ \#VV(v_i) = m_i, \\ 0 & \text{else} \end{cases}$$

and the normalized vertex-face matrix $N_{FV} \in M_{n_F, n_V}(\mathbb{R})$

$$N_{FV}(i,j) := \begin{cases} 1/m_i & \text{if } v_j \in FV(f_i), \ \#FV(f_i) = m_i, \\ 0 & \text{else.} \end{cases}$$

The properties of their spectrum have important connections with the topological characteristics (e.g. number of connected components) of the input mesh [15] and with numerical properties of the Laplacian smoothing and re-sampling operator studied in [18,17,20]. Using these matrices, we want to settle the numerical approach to the reconstruction process from $\mathcal{M}$ to $\mathcal{M}'$, which is expressed as

$$N_{FV}P = B, \tag{6}$$

that is, a linear system with $n_V$ unknowns $(p_i)_{i=1}^{n_V}$ and $n_F$ equations.

Because $n_F \approx 2n_V$, (6) is undetermined if $rank(N_{FV}) < n_V$, over-determined if $rank(N_{FV}) > n_V$ and it has a unique solution otherwise. In order to define a well-posed problem, (6) can be replaced by the *least-square problem* [9]

$$\|N_{FV}\bar{P}_i - B_i\|_2 = \min_{P_i}\{\|N_{FV}P_i - B_i\|_2\},$$

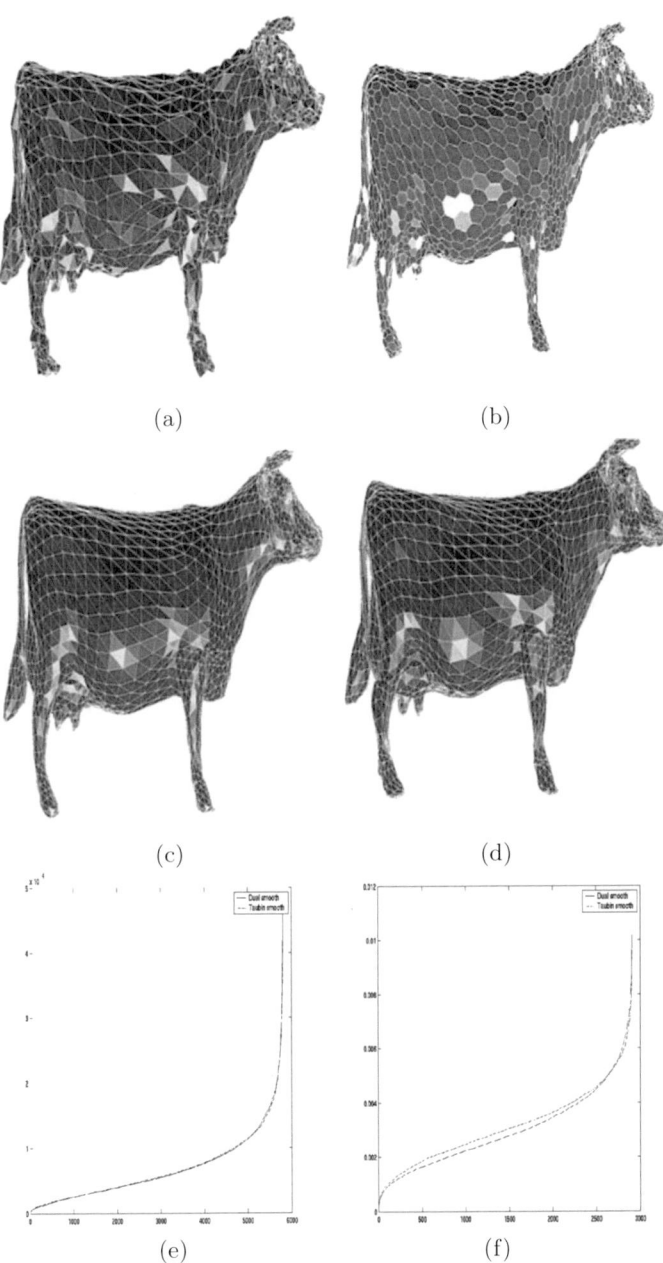

**Fig. 7.** (a) Input data set: num. vertices 2.904, num. triangles 5.804, (b) Laplacian smoothing applied to the dual mesh $\lambda = 0.6$, $\mu = -0.5640$, $k = 10$, (c) average reconstruction, (d) Taubin's smoothing with previous $\lambda$, $\mu$, $k$, (e) error evaluation on vertices, (f) error evaluation on normals.

or equivalently $N_{FV}^T N_{FV} \bar{P}_i = N_{FV}^T B_i$, $i = 1, 2, 3$ where $P_i$ and $B_i$ is the $i$-th column of $P$ and $B$ respectively. This choice produces a family of triangle meshes $\{\tilde{P}\}$

$$\tilde{P}_i := \bar{P}_i + E_i, \qquad E_i \in ker(N_{FV}), \qquad i = 1, 2, 3$$

each one represents an approximated solution of (6), and the computational cost is $O(n_V^3)$. With respect to the primal-dual correspondence described in Section 3.2, this strategy faces-up to its expensive computational cost providing a family of approximated triangle meshes $(\tilde{P}, F)$ instead of the initial mesh $\mathcal{M}$.

## 4    Dual Laplacian Smoothing

In the previous section we have focused our attention on the relationships between a triangle mesh $\mathcal{M}$ and its dual representation underlining their correlation. Here, we consider applications of the dual representation for smoothing noised data sets. The key observation is that, considering the 1-neighborhood structure related to each point, $\mathcal{M}$ has a little regularity while $\mathcal{M}'$ can be considered with more simplicity because each of its vertices has three links if the related triangle of $\mathcal{M}$ is internal, and one/two if it belongs to the mesh boundary. This observation is the base in [13] for the compression of triangle meshes; furthermore, primal-dual correspondence is partially exploited in [21] for primal-dual subdivision schemes, and in [18] for the definition of dual re-sampling and non-shrinking smoothing operators. Firstly, we review Laplacian and Taubin's smoothing algorithms [18,17,20] which are strictly related to our approach, and we refer the reader to [2] for a complete description and comparison of mesh regularization methods.

– **Laplacian Smoothing.** Each internal vertex $p_v$ of the input mesh is updated using its 1-neighborhood structure $VV(v) := (v_1, \ldots, v_n)$ as described by the following procedure:

$$p_v^{(1)} := (1 - \lambda)p_v + \frac{\lambda}{\sum_{i=1}^n w_i} \sum_{i=1}^n w_i p_{v_i}$$

where $\lambda \in [0, 1]$ is a positive parameter controlling the smoothing process. The weights $(w_i)_{i=1}^n$ can be chosen in different ways even if the following ones are commonly used:

• *constant weights:* $w_i = 1$, $i = 1, \ldots, n$, i.e.

$$p_v^{(1)} := (1 - \lambda)p_v + \frac{\lambda}{n} \sum_{i=1}^n p_{v_i}. \tag{7}$$

• *adaptive weights* [17]: $w_i$ is proportional to the inverse of the distance between $p_v$ and its neighbor $p_{v_i}$, i.e. $w_i := \|p_v - p_{v_i}\|_2^{-1}$. A general choice is given by $w_{ij} \geq 0$, $\sum_j w_{ij} = 1$ whose properties rely on the stochastic matrix theory [9,19].

We write (7) as

$$P^{(1)} = [(1 - \lambda)I_{n_V} + \lambda N_{VV}]P = f_\lambda(L)P$$

where $I_{n_V}$ is the identity matrix of order $n_V$, $f_\lambda(t) = (1-\lambda t)$, $L := I_{n_V} - N_{VV}$ and

$$P = \begin{pmatrix} p_1 \\ \vdots \\ p_{n_V} \end{pmatrix} \in M_{n_V,3}(\mathbb{R}).$$

The Laplacian smoothing reduces all non-zero frequencies of the signal corresponding to the mesh and tends to shrink its geometry. To partially solve this drawback, in [7] each smoothing iteration is combined with a mesh volume-restoring and re-scaling step.

- **Taubin's Smoothing.** The solution to shrinkage proposed in [17] is based on the alternation of two scale factors of opposite signs $\lambda$, $\mu$ in the Laplacian smoothing, i.e.

$$P^{(1)}_{Taubin} := f_\lambda(L)f_\mu(L)P$$

where $-\mu > \lambda > 0$. Using this filter enables to suppress high frequencies while preserving the low ones. Good results are achieved by choosing the input parameters which satisfy the condition

$$\frac{1}{\lambda} + \frac{1}{\mu} = 0.1.$$

The application of $k$ iteration steps gives

$$P^{(k)}_{Taubin} = f^{(k)}_{\lambda,\mu}(L)P$$

with $f^{(k)}_{\lambda,\mu}(t) := [f_\lambda(t)f_\mu(t)]^k$.

## 4.1   Dual Approach to Triangle Mesh Smoothing

Considered a noised mesh $\mathcal{M}_{noise} := (P, F)$, the idea is to apply the Laplacian smoothing and its extensions proposed in [18,17,20] to the dual mesh $\mathcal{M}'_{noise} := (B, G)$ which is affected by noise as well as $\mathcal{M}_{noise}$. The merit of using the dual mesh instead of the input one is mainly due to the following considerations. Firstly, the normalized vertex-vertex matrix $N_{FF}$ of $\mathcal{M}'_{noise}$, which will be used for the smoothing process, has at most three non zero elements in each row. This implies the optimization of storage and computational requirements which grow with the complexity of the input mesh in terms of the number of vertices. Furthermore, the construction of the incident matrix of $\mathcal{M}'_{noise}$ is simply achieved with the constant relation $FF$ applied to $\mathcal{M}$. Secondly, the dual smoothing considers at each vertex of $\mathcal{M}$ a different topology for the regularization with respect to the 1-neighborhood structure used for the (primal) Laplacian smoothing (see Figure 8).

Denoted with with $L'$ the Laplacian matrix of $\mathcal{M}'_{noise}$, and with $\mathcal{M}'_{smooth}$ the smoothed dual mesh, the last step reconstructs the regularized mesh $\mathcal{M}_{smooth}$.

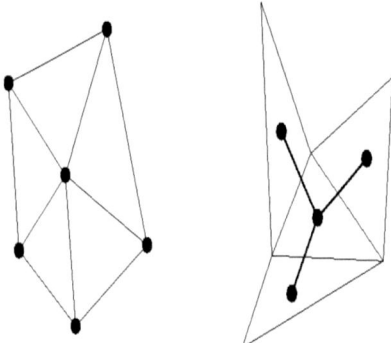

**Fig. 8.** Topological masks used for the primal and dual Laplacian smoothing.

Because of the vertices of $\mathcal{M}'_{smooth}$ do not satisfy (1), (2), the considerations about the high correlation between $\mathcal{M}'_{smooth}$ and $\mathcal{M}_{smooth}$ (see Section 3.3) highlight the impossibility of reconstructing $\mathcal{M}_{smooth}$ by using the primal-dual correspondence previously described. The solution to this problem is achieved by defining the new vertices of $\mathcal{M}_{smooth} := (P_{smooth}, F)$ as the barycenters of the faces in $\mathcal{M}'_{smooth}$ which are exactly $n_V$. This process can be summarized as

$$P^{(k)}_{smooth} = N_{VF} \underbrace{f^{(k)}_{\lambda,\mu}(L')B}_{\text{Dual smooth}} = N_{VF} f^{(k)}_{\lambda,\mu}(L')N_{FV}P \tag{8}$$

with $L' := I_{n_F} - N_{FF}$. The previous relation expresses the regularized mesh geometry only using the information on $\mathcal{M}_{noise}$.

We now compare (8) with the mesh achieved by applying the Taubin's smoothing using the same number of iterations $k$, and parameters $\bar{\lambda}$, $\bar{\mu}$

$$P^{(k)}_{Taubin} = f^{(k)}_{\bar{\lambda},\bar{\mu}}(L)P.$$

From the previous relation, it follows that[2]

$$\frac{\|P^{(k)}_{smooth} - P^{(k)}_{Taubin}\|_2}{\|P\|_2} \le \|N_{FV}\|_2\|N_{VF}\|_2 g^k(\lambda,\mu) + g^k(\bar{\lambda},\bar{\mu})$$

with $g(x,y) := \frac{(x-y)^2}{-4xy}$. Because $|\lambda| < 1$, $|\mu| < 1$ (resp. $|\bar{\lambda}| < 1$, $|\bar{\mu}| < 1$), we have $|g(\lambda,\mu)| < 1$ (resp. $|g(\bar{\lambda},\bar{\mu})| < 1$) thus guaranteeing that

$$\lim_{k \to +\infty} \|P^{(k)}_{smooth} - P^{(k)}_{Taubin}\|_2 = 0,$$

i.e. *the asymptotic behavior of $P^{(k)}_{smooth}$ resembles that of $P^{(k)}_{Taubin}$ and its computational cost is linear in the number of vertices $n_V$.* All previous considerations

---

[2] The inverted parabola $f_{\lambda,\mu}(t) := (1-\lambda t)(1-\mu t)$ has its minimum at $\bar{t} := \frac{1}{2}(\frac{1}{\lambda} + \frac{1}{\mu}) \in (0,1)$ and $f_{\lambda,\mu}(\bar{t}) = \frac{(\lambda-\mu)^2}{-4\lambda\mu}$.

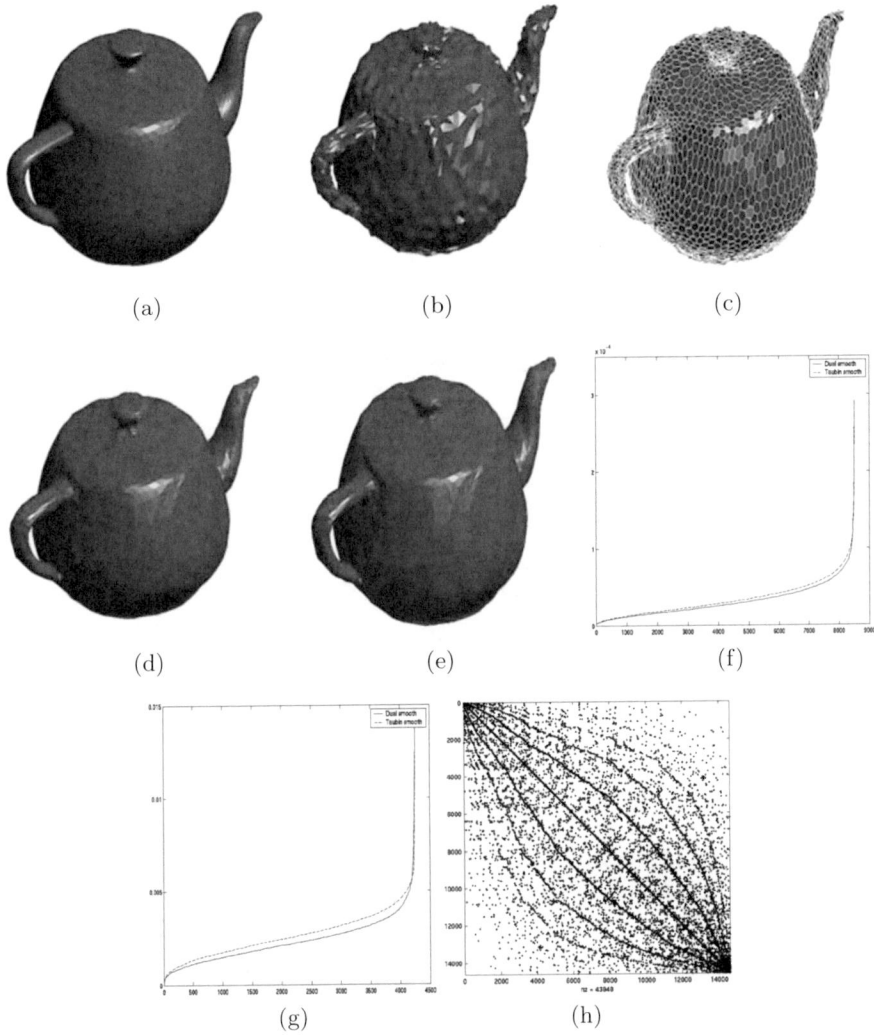

**Fig. 9.** (a) Input data set: num. vertices 7.308, num. triangles 14.616, (b) noised data set with normal error, (c) Laplacian smoothing applied to the dual mesh $\lambda = 0.6$, $\mu = -0.5640$, $k = 20$, (d) average reconstruction, (e) Taubin's smoothing with previous $\lambda$, $\mu$, $k$, (f) error evaluation on vertices, (g) error evaluation on normals, (h) Laplacian matrix sparsity.

also apply if we consider adaptive weights instead of constant ones. Finally, we observe that we have not applied the least square approach (6) for reconstructing $\mathcal{M}$ from $\mathcal{M}'$ because it is computationally expensive and without evident benefits with respect to the previous choice. Other examples of the proposed approach are given in Figure 9, 10.

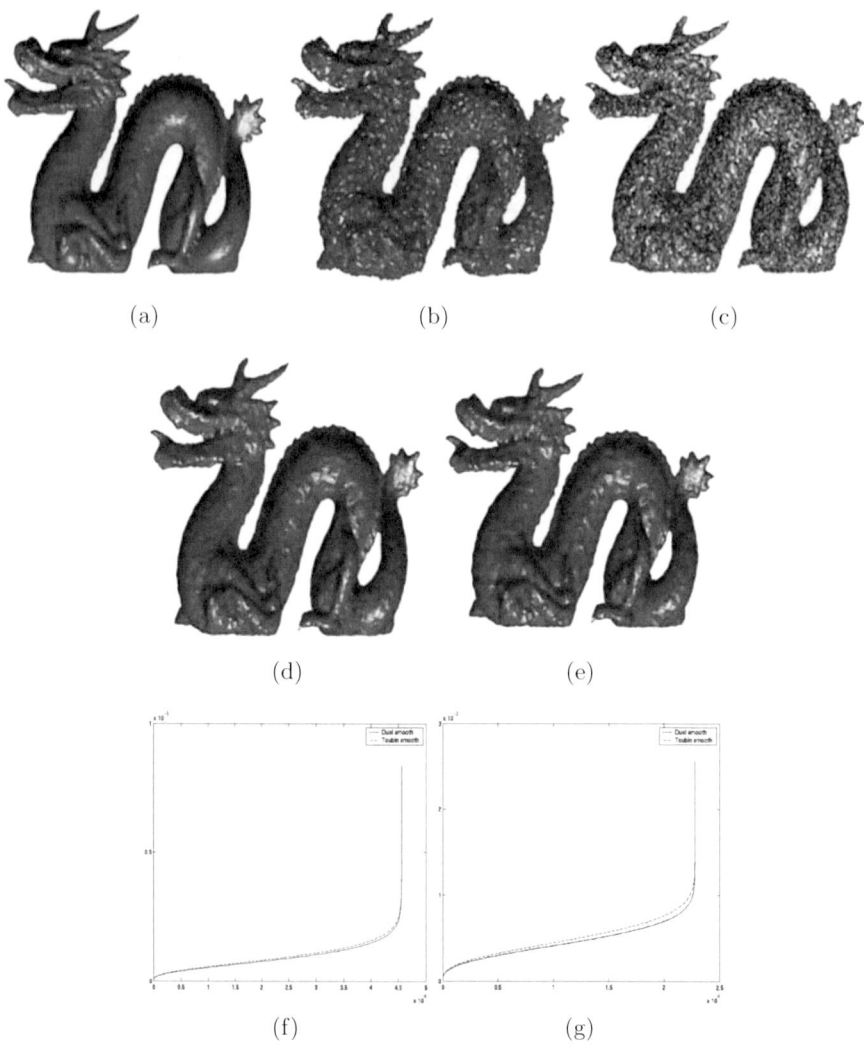

**Fig. 10.** (a) Input data set: num. vertices 22.813, num. triangles 45.626, (b) noised data set with normal error, (c) Laplacian smoothing applied to the dual mesh $\lambda = 0.6$, $\mu = -0.5640$, $k = 10$, (d) average reconstruction, (e) Taubin's smoothing with previous $\lambda$, $\mu$, $k$, (f) error evaluation on vertices, (g) error evaluation on normals.

## 5    Conclusions and Future Work

The first goal of the paper is the definition of a homeomorphism between a 2-manifold triangle mesh and its dual representation with optimal (i.e. linear) computational cost and numerical stability avoiding the least-square formulation whose solution requires $O(n_V^3)$ flops and it only achieves an approximated reconstruction of the input mesh. We have also derived two combinatorial properties (1), (2) which highlight the redundancy of the geometry stored in a triangle mesh [12], and a deeper analysis of these properties is under development. Finally, the duality analysis has been exploited for defining the dual Laplacian smoothing in order to settle a linear regularization method based on the face-face topological mask instead of the vertex-vertex one used by the Taubin's signal processing framework. This new approach has been compared with previous work highlighting its validity and optimality.

## Acknowledgements

This work has been partially supported by the National Project "MACROGeo: Metodi Algoritmici e Computazionali per la Rappresentazione di Oggetti Geometrici", FIRB grant. Thanks are given to the Computer Graphics Group of IMATI-GE, and to B. Falcidieno.

## References

1. F. Aurenhammer. Voronoi diagrams: A survey of a fundamental geometric data structure. *ACM Computing Surveys*, 23(3):345–405, Sept. 1991.
2. A. Belyaev and Y. Ohtake. A comparison of mesh smoothing methods. *To appear in Israel-Korea Bi-National Conference on Geometric Modeling and Computer Graphics*, Tel-Aviv, Fabr. 12-14, 2003.
3. M. W. Bern and D. Eppstein. Mesh generation and optimal triangulation. In F. K. Hwang and D.-Z. Du, editors, *Computing in Euclidean Geometry*, pages 23–90. World Scientific, 1992.
4. P. Cignoni, C. Rocchini, and R. Scopigno. Metro: Measuring error on simplified surfaces. *Computer Graphics Forum*, 17(2):167–174, June 1998.
5. M. de Berg, M. van Kreveld, M. Overmars, and O. Schwarzkopf. *Computational Geometry Algorithms and Applications*. Springer-Verlag, Berlin Heidelberg, 1997.
6. L. DeFloriani and P. Magillo. Multiresolution mesh representation: Models and data structures. In *Tutorials on Multiresolution in Geometric Modeling*, pages 363–417, Munich, 2002. Springer-Verlag.
7. M. Desbrun, M. Meyer, P. Schröder, and A. H. Barr. Implicit fairing of irregular meshes using diffusion and curvature flow. In A. Rockwood, editor, *Siggraph 1999*, Annual Conference Series, pages 317–324, Los Angeles, 1999. ACM Siggraph, Addison Wesley Longman.
8. H. Edelsbrunner. *Algorithms in Combinatorial Geometry*, volume 10 of *EATCS Monographs on Theoretical Computer Science*. Springer-Verlag, Nov. 1987.
9. G. Golub and G. VanLoan. *Matrix Computations*. John Hopkins University Press, 2nd. edition, 1989.
10. L. J. Guibas and J. Stolfi. Primitives for the manipulation of general subdivisions and the computation of Voronoi diagrams. In *Proceedings of the Fifteenth Annual ACM Symposium on Theory of Computing*, pages 221–234, Boston, Massachusetts, 25–27 Apr. 1983.
11. H. Hoppe, T. DeRose, T. Duchamp, J. McDonald, and W. Stuetzle. Surface reconstruction from unorganized points. *Computer Graphics*, 26(2):71–78, July 1992.

12. M. Isenburg, S. Gumhold, and C. Gotsman. Connectivity shapes. In *Visualization'01 Conference Proceedings*, pages 135–142, 2001.
13. J. Li and C. Kuo. A dual graph approach to 3D triangular mesh compression. In *IEEE 1998 International Conference on Image Processing, Chicago, Oct. 4-7, 1998.*, pages 891–894.
14. M. Mäntylä. *An Introduction to Solid Modeling*. Computer Science Press, Rockville, MD, 1987.
15. B. Mohar. The laplacian spectrum of graphs. *Graph Theory, Combinatorics and Applications.*, pages 871–898, 1991.
16. F. P. Preparata and M. Shamos. *Computational Geometry*. Springer-Verlag, New York, 1985, 1985.
17. G. Taubin. A signal processing approach to fair surface design. In R. Cook, editor, *SIGGRAPH 95 Conference Proceedings*, Annual Conference Series, pages 351–358. ACM SIGGRAPH, Addison Wesley, Aug. 1995. held in Los Angeles, California, 06-11 August 1995.
18. G. Taubin. Dual mesh resampling. In *Proceedings of Pacific Graphics 2001*, pages 180–188, 2001. Tokyo, Japan, October 2001.
19. G. Taubin. Geometric signal processing on polygonal meshes. In *Eurographics'2000, State of the Art Report*, August 2000.
20. G. Taubin, T. Zhang, and G. Golub. Optimal surface smoothing as filter design. In B. Buxton and R. Cipolla, editors, *Computer vision, ECCV '96: 4th European Conference on Computer Vision, Cambridge, UK, April 15–18, 1996: proceedings*, volume 1064–1065 of *Lecture notes in computer science*, pages 283–292 (vol. 1). Springer-Verlag, 1996.
21. D. Zorin and P. Schröder. A unified framework for primal/dual quadrilateral subdivision schemes. *Computer Aided Geometric Design*, Special issue on Subdivision Surfaces, 18, 2001., 2001.

# Hand Tracking Using a Quadric Surface Model and Bayesian Filtering

Roberto Cipolla[1], Bjorn Stenger[1],
Arasanathan Thayananthan[1], and Philip H.S. Torr[2]

[1] University of Cambridge, Department of Engineering
Trumpington Street, Cambridge, CB2 1PZ, UK
[2] Microsoft Research Ltd., 7 J J Thomson Ave, Cambridge CB3 0FB, UK

**Abstract.** Within this paper a technique for model-based 3D hand tracking is presented. A hand model is built from a set of truncated quadrics, approximating the anatomy of a real hand with few parameters. Given that the projection of a quadric onto the image plane is a conic, the contours can be generated efficiently. These model contours are used as shape templates to evaluate possible matches in the current frame. The evaluation is done within a hierarchical Bayesian filtering framework, where the posterior distribution is computed efficiently using a tree of templates. We demonstrate the effectiveness of the technique by using it for tracking 3D articulated and non-rigid hand motion from monocular video sequences in front of a cluttered background.

## 1 Introduction

Hand tracking has great potential as a tool for better human-computer interaction. Tracking hands, in particular articulated finger motion, is a challenging problem because the motion exhibits many degrees of freedom (DOF). Representing the hand pose by joint angles, the configuration space is 27 dimensional, 21 DOF for the joint angles and 6 for orientation and location. Given a kinematic hand model, one may attempt to use inverse kinematics to calculate the joint angles [19], however this problem is ill-posed when using a single view. It also requires exact feature localization, which is particularly difficult in the case of self-occlusion.

Most successful methods have followed the approach of using a geometric hand model, introduced by Rehg and Kanade [13] in the *DigitEyes* tracking system. Their hand model is constructed from truncated cylinders. The axes of these cylinders are projected into the image, and the distances to local edges are minimised using nonlinear optimisation. Heap and Hogg [9] use a deformable surface mesh model, which is constructed via principal component analysis (PCA) from example shapes obtained with an MRI scanner. This is essentially a 3D version of active shape models, and shape variation is captured by only a few principal components. The motion is not based on a physical deformation model and thus implausible finger motions can result. Wu *et al.* [20] model the articulated hand motion from data captured using a data glove. The tracker is based on

M.J. Wilson and R.R. Martin (Eds.): Mathematics of Surfaces 2003, LNCS 2768, pp. 129–141, 2003.

importance sampling, and hypotheses are generated by projecting a 'cardboard model' into the image. This model is constructed from planar patches, and thus the system is view-dependent.

It is clear that the performance of a model-based tracker depends on the type of the used model. However, there is a trade-off between accurate modelling, and efficient rendering and comparison with the image data. In fact this is generally true when modelling articulated objects for tracking, which is commonly done in the context of human body tracking (see [11] for a survey). A number of different models have been suggested in this context, using various primitives such as boxes, cylinders, ellipsoids or super-quadrics.

The next section describes the geometric hand model used in this paper. Section 3 reviews work on tree-based detection. A short introduction to Bayesian filtering is given in 4, and in section 5 we introduce filtering using a tree-based estimator. Tracking results on video sequences of hand motion are shown in section 6.

## 2    Modelling Hand Geometry

This section describes the construction of a hand model from truncated quadrics. The advantage of this method is that the object surface can be approximated with low complexity and that contours can be generated using tools from projective geometry. This hand model has previously been described in [15] but here it is used in a different tracking framework.

### 2.1    Projective Geometry of Quadrics and Conics

A quadric is a second degree implicit surface in 3D space, and it can be represented in homogeneous coordinates by a symmetric $4 \times 4$ matrix $\mathbf{Q}$ [8]. The surface is defined by all points $\mathbf{X} = [x, y, z, 1]^{\mathrm{T}}$ satisfying the equation

$$\mathbf{X}^{\mathrm{T}}\mathbf{Q}\mathbf{X} = 0. \tag{1}$$

Different families of quadrics are obtained from matrices $\mathbf{Q}$ of different ranks. Particular cases of interest are:

**ellipsoids,** represented by matrices $\mathbf{Q}$ with full rank;
**cones and cylinders,** represented by matrices $\mathbf{Q}$ with rank$(\mathbf{Q}) = 3$;
**a pair of planes $\boldsymbol{\pi}_0$ and $\boldsymbol{\pi}_1$,** represented as $\mathbf{Q} = \boldsymbol{\pi}_0\boldsymbol{\pi}_1^{\mathrm{T}} + \boldsymbol{\pi}_1\boldsymbol{\pi}_0^{\mathrm{T}}$ with rank$(\mathbf{Q}) = 2$.

Note that there are several other projective types of quadrics, such as hyperboloids or paraboloids, which like ellipsoids have a matrix of full rank. Under a Euclidean transformation $\mathbf{T} = \left[\begin{smallmatrix} \mathbf{R} & \mathbf{t} \\ \mathbf{0}^{\mathrm{T}} & 1 \end{smallmatrix}\right]$ the shape of a quadric is preserved, but in the new coordinate system $\mathbf{Q}$ is represented by $\hat{\mathbf{Q}} = \mathbf{T}^{-\mathrm{T}}\mathbf{Q}\mathbf{T}^{-1}$.

A quadric has nine degrees of freedom, corresponding to the independent elements of $\mathbf{Q}$ up to a scale factor. Given a number of point correspondences or outlines in multiple views, quadric surfaces can be reconstructed, as shown by Cross and Zisserman [6]. It is also suggested that for many objects using a piecewise quadric representation gives an accurate and compact surface representation. In order to employ quadrics for modelling such general shapes, it is necessary to truncate them. For any quadric $\mathbf{Q}$ the truncated quadric $\mathbf{Q}_\Pi$ can be obtained by finding points $\mathbf{X}$ satisfying:

$$\mathbf{X}^T\mathbf{Q}\mathbf{X} = 0 \quad \text{and} \quad \mathbf{X}^T\mathbf{\Pi}\mathbf{X} \geq 0, \tag{2}$$

where $\mathbf{\Pi}$ is a matrix representing a pair of clipping planes (see figure 1a). The image of a quadric $\mathbf{Q} = \begin{bmatrix} \mathbf{A} & \mathbf{b} \\ \mathbf{b}^T & c \end{bmatrix}$ seen from a normalised projective camera $\tilde{\mathbf{P}} = [\mathbb{I} \mid \mathbf{0}]$ is a conic $\mathbf{C}$ given by

$$\mathbf{C} = c\mathbf{A} - \mathbf{b}\mathbf{b}^T, \tag{3}$$

as shown in figure 1b. In order to obtain the image of a quadric $\mathbf{Q}$ in an arbitrary projective camera $\mathbf{P}$ it is necessary to compute the transformation $\mathbf{H}$ such that $\mathbf{P}\mathbf{H} = [\mathbb{I} \mid \mathbf{0}]$. This normalising transformation is given by the matrix $\mathbf{H} = [\mathbf{P}^\dagger \mid \mathbf{p}^\perp]$, where $\mathbf{P}^\dagger$ is the pseudo inverse of $\mathbf{P}$ and $\mathbf{p}^\perp$ is the camera centre or the null vector of $\mathbf{P}$ (see [5]).

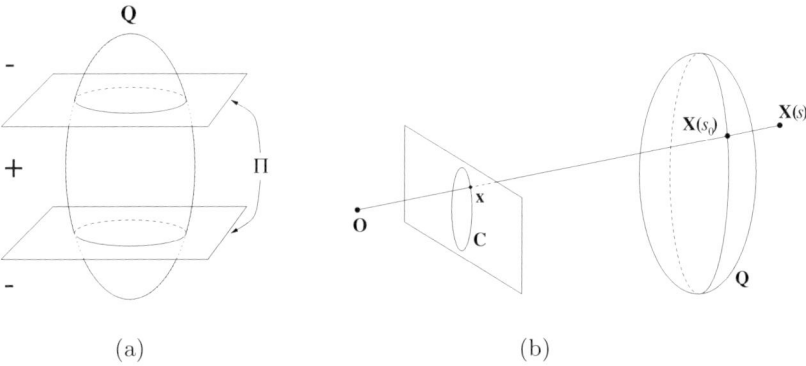

|     |     |
| --- | --- |
| (a) | (b) |

**Fig. 1. Projection of a quadric.** (a) A truncated quadric $\mathbf{Q}_\Pi$, here a truncated ellipsoid, can be obtained by finding points on quadric $\mathbf{Q}$ which satisfy $\mathbf{X}^T\mathbf{\Pi}\mathbf{X} \geq 0$. (b) The projection of a quadric $\mathbf{Q}$ into the image plane is a conic $\mathbf{C}$.

## 2.2   Description of the Hand Model

The hand model is built using a set of quadrics $\{\mathbf{Q}_i\}_{i=1}^q$, representing the anatomy of a human hand as shown in figure 2. We use a hierarchical model with 27 degrees of freedom (DOF): 6 for the global hand position, 4 for the pose of each finger and 5 for the pose of the thumb. Starting from the palm and ending at the tips, the coordinate system of each quadric is defined relative to the previous one in the hierarchy. The palm is modelled using a truncated cylinder, its top closed by a half-ellipsoid. Each finger consists of three segments of a cone, one for each phalanx. They are connected by hemispheres, representing the joints. A default shape is first obtained by taking measurements from a real hand. Given the image data, shape matching can be used to estimate a set of shape parameters, including finger lengths and a width parameter [17].

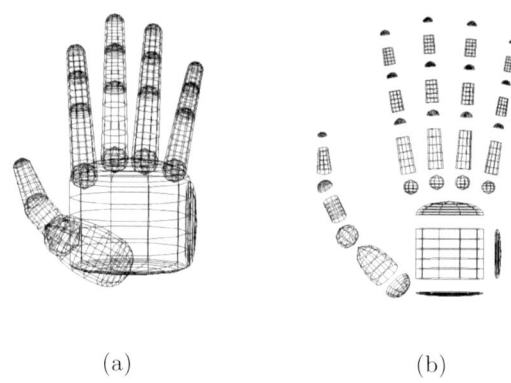

(a)                                          (b)

**Fig. 2.    Geometric hand model.** The hand model has 27 degrees of freedom is constructed using truncated quadrics as building blocks. Depicted is (a) a front view and (b) an exploded view.

## 2.3    Generation of the Contours

Each clipped quadric of the hand model is projected individually as described in section 2.1, generating a list of clipped conics. For each conic matrix $\mathbf{C}$ we use eigen-decomposition to obtain a factorisation given by

$$\mathbf{C} = \mathbf{T}^{-\mathrm{T}}\mathbf{D}\mathbf{T}^{-1}. \tag{4}$$

The diagonal matrix $\mathbf{D}$ represents a conic aligned with the $x$- and $y$-axis and centred at the origin. The matrix $\mathbf{T}$ is the Euclidean transformation that maps this conic onto $\mathbf{C}$. We can therefore draw $\mathbf{C}$ by drawing $\mathbf{D}$ and transforming the points according to $\mathbf{T}$. The drawing of $\mathbf{D}$ is carried out by different methods, depending on its rank. For rank$(\mathbf{D}) = 3$ we draw an ellipse, for rank$(\mathbf{D}) = 2$ we draw a pair of lines.

The next step is occlusion handling. Consider a point $\mathbf{x}$ on the conic $\mathbf{C}$, obtained by projecting the quadric $\mathbf{Q}$, as shown in figure 1. The camera centre and $\mathbf{x}$ define a 3D ray $L$. Each point $\mathbf{X} \in L$ is given by $\mathbf{X}(s) = [\begin{smallmatrix}\mathbf{x}\\s\end{smallmatrix}]$, where $s$ is a free parameter determining the depth of the point in space, such that the point $\mathbf{X}(0)$ is at infinity and $\mathbf{X}(\infty)$ is at the camera centre. The point of intersection of the ray with the quadric $\mathbf{Q}$ is found by solving the equation

$$\mathbf{X}(s)^{\mathrm{T}}\mathbf{Q}\mathbf{X}(s) = 0 \tag{5}$$

for $s$. Writing $\mathbf{Q} = \begin{bmatrix} \mathbf{A} & \mathbf{b} \\ \mathbf{b}^{\mathrm{T}} & c \end{bmatrix}$, the unique solution of (5) is given by $s_0 = -\mathbf{b}^{\mathrm{T}}\mathbf{x}/c$. In order to check if $\mathbf{X}(s_0)$ is visible, (5) is solved for each of the other quadrics $\mathbf{Q}_i$ of the hand-model. In the general case there are two solutions $s_1^i$ and $s_2^i$, yielding the points where the ray intersects with quadric $\mathbf{Q}_i$ . The point $\mathbf{X}(s_0)$ is visible if $s_0 \geq s_j^i \quad \forall i,j$, in which case the point $\mathbf{x}$ is drawn. Figure 3 shows examples of hand model projections.

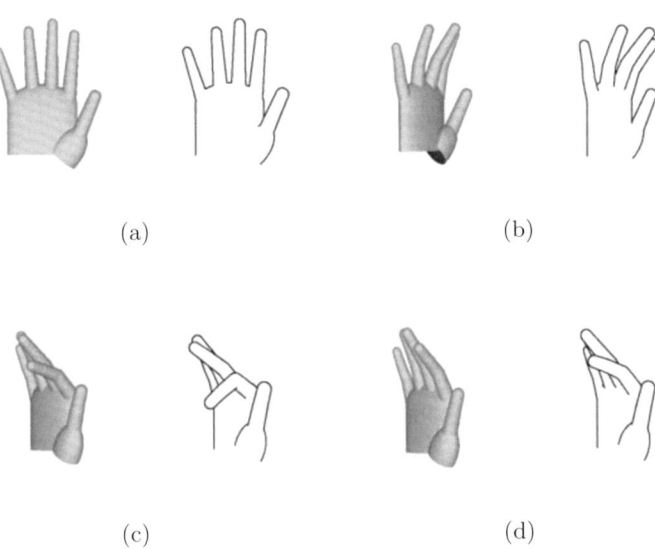

(a)                                    (b)

(c)                                    (d)

**Fig. 3. Examples of model projections.** (a)-(d) show different hand poses. For each example the 3D hand model is shown on the left and its projection into the image plane on the right. Note that self-occlusion is handled when generating the contours.

## 2.4   Learning Natural Hand Articulation

Model-based trackers commonly use a 3D geometric model with an underlying biomechanical deformation model [1,2,13]. Each finger can be modelled as a kinematic chain with 4 DOF, and the thumb with 5 DOF. Thus articulated hand motion lies in a 21 dimensional joint angle space. However, hand motion is highly constrained as each joint can only move within certain limits. Furthermore the motion of different joints is correlated, for example, most people find it difficult to bend the little finger while keeping the ring finger fully extended at the same time. Thus hand articulation is expected to lie in a compact region within the 21 dimensional angle space. We used a data glove to collect a large number of joint angles in order to capture natural hand articulation. Experiments with 15 sets of joint angles captured from three different subjects, show that in all cases 95 percent of the variance is captured by the first eight principal components, in 10 cases within the first seven, which confirms the results reported by Wu *et*

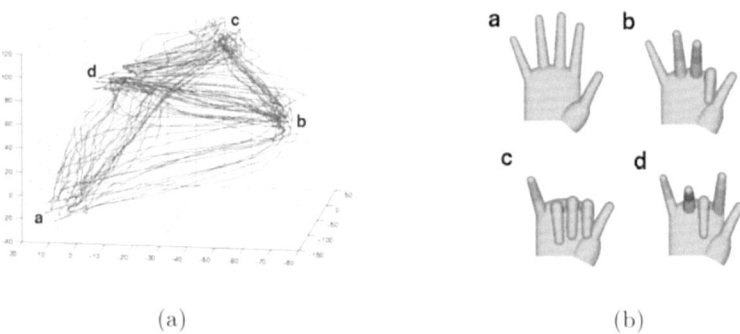

(a)                                      (b)

**Fig. 4. Paths in the configuration space found by PCA.** (a) The figure shows a trajectory of projected hand state vectors onto the first three principal components. (b) The hand configurations corresponding to the four end points in (a).

*al.* in [20]. Figure 4 shows trajectories projected onto the first three eigenvectors between a set of hand poses. As described in the next section, this lower dimensional eigen-space will be quantised into a set of discrete states. Hand motion is then modelled by a first order Markov process between these states. Given a large amount of training data, higher order models can be learned.

## 3    Tree-Based Detection

For real applications the problem of tracker initialisation, as well as the handling of self-occlusion and cluttered backgrounds remain obstacles. Current state-of-the-art systems often employ a version of particle filtering, allowing for multiple hypotheses. The use of particle filters is primarily motivated by the need to overcome ambiguous frames in a video sequence so that the tracker is able to recover. Another way to overcome the problem of losing lock is to treat tracking as object detection at each frame. Thus if the target is lost in one frame, this does not affect any subsequent frame. Template based methods have yielded good results for locating deformable objects in a scene with no prior knowledge, e.g. for hands or pedestrians [2,7,14,17]. These methods are made robust and efficient by the use of distance transforms such as the chamfer or Hausdorff distance between template and image [3,10], which were originally developed for matching a single template. A key suggestion was that multiple templates could be dealt with efficiently by building a template hierarchy and a coarse-to-fine search [7,12]. The idea is to group similar templates and represent them with a single prototype template together with an estimate of the variance of the error within the cluster, which is used to define a matching threshold. The prototype is first compared to the image; only if the error is below the threshold are the templates within the cluster compared to the image. This clustering is done at various levels, resulting in a hierarchy, with the templates at the leaf level covering the space of all possible templates.

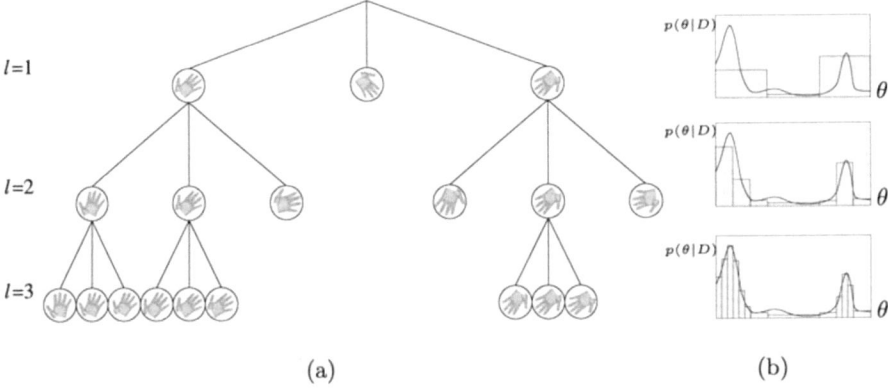

$l=1$

$l=2$

$l=3$

$p(\theta|D)$

$\theta$

$p(\theta|D)$

$\theta$

$p(\theta|D)$

$\theta$

(a)                                                                    (b)

**Fig. 5. Tree-based estimation of the posterior density.** (a) Associated with the nodes at each level is a non-overlapping set in the state space, defining a partition of the state space (here rotation angle). The posterior distribution for each node is evaluated using the centre of each set, depicted by a hand rotated by a specific angle. Sub-trees of nodes with low posterior probability are not further evaluated. (b) Corresponding posterior density (continuous) and the piecewise constant approximation using tree-based estimation. The modes of the distribution are approximated with higher precision at each level.

If a parametric object model is available, another option to build the tree is by partitioning the state space. Each level of the tree defines a partition with increasing resolution, the leaves defining the finest partition. Such a tree is depicted schematically in figure 5(a), for a single rotation parameter. This tree representation has the advantage that prior information is encoded efficiently, as templates with large distance in parameter space are likely to be in different sub-trees.

It may be argued that there is no need for a parametric model and that an exemplar-based approach could be followed, as by Toyama and Blake in [18]. However, for models with many degrees of freedom the storage space for templates becomes excessive. The use of a parametric model allows the combination of an on-line and off-line approach in the tree-based algorithm. Once the leaf level is reached more child templates can be generated for further optimisation. Hierarchical detection works well for locating a hand in images, and yet often there are ambiguous situations that could be resolved by using temporal information. The next section describes the Bayesian framework for filtering.

## 4   Bayesian Filtering

Filtering is the problem of estimating the state (hidden variables) of a system given a history of observations. Define, at time $t$, the state parameter vector as $\theta_t$, and the data (observations) as $\mathbf{D}_t$, with $\mathbf{D}_{1:t-1}$, being the set of data from time 1 to $t-1$; and the data $\mathbf{D}_t$ are conditionally independent at each time step given the $\boldsymbol{\theta}_t$. In our specific application $\boldsymbol{\theta}_t$ is the state of the hand (set of joint angles, location and orientation) and $\mathbf{D}_t$ is the image at time $t$ (or some set of features extracted from that image). Thus at time $t$ the posterior distribution of the state vector is given by the following recursive relation

$$p(\boldsymbol{\theta}_t|\mathbf{D}_{1:t}) = \frac{p(\mathbf{D}_t|\boldsymbol{\theta}_t)p(\boldsymbol{\theta}_t|\mathbf{D}_{1:t-1})}{p(\mathbf{D}_t|\mathbf{D}_{1:t-1})}, \tag{6}$$

where the normalising constant is

$$p(\mathbf{D}_t|\mathbf{D}_{1:t-1}) = \int p(\mathbf{D}_t|\boldsymbol{\theta}_t)p(\boldsymbol{\theta}_t|\mathbf{D}_{1:t-1})d\boldsymbol{\theta}_t. \tag{7}$$

The term $p(\boldsymbol{\theta}_t|\mathbf{D}_{1:t-1})$ in (6) is obtained from the Chapman-Kolmogorov equation:

$$p(\boldsymbol{\theta}_t|\mathbf{D}_{1:t-1}) = \int p(\boldsymbol{\theta}_t|\boldsymbol{\theta}_{t-1})p(\boldsymbol{\theta}_{t-1}|\mathbf{D}_{1:t-1})d\boldsymbol{\theta}_{t-1} \tag{8}$$

with the initial prior pdf $p(\boldsymbol{\theta}_0|\mathbf{D}_0)$ assumed known. It can be seen that (6) and (8) both involve integrals. Except for certain simple distributions these integrals are intractable and so approximation methods must be used. As has been mentioned, Monte Carlo methods represent one way of evaluating these integrals. Alternatively, hierarchical detection provides a very efficient way to evaluate the likelihood $p(\mathbf{D}_t|\boldsymbol{\theta}_t)$ in a deterministic manner, even when the state space is high dimensional; as the number of templates in the tree increases exponentially with the number of levels in the tree. This leads us to consider dividing up the state space into non-overlapping sets, just as the templates in the tree cover the regions of parameter space. Typically this methodology has been applied using an evenly spaced grid and is thus exponentially expensive as the dimension of the state space increases. In this paper we combine the tracking process with the empirically successful process of tree-based detection as laid out in section 3 resulting in an efficient deterministic filter.

## 5   Filtering Using a Tree-Based Estimator

Our aim is to design an algorithm that can take advantage of the efficiency of the tree-based search whilst also yielding a good approximation to Bayesian filtering. We design a grid-based filter, in which a multi-resolution partition is provided by the tree as given in Section 3. Thus we will consider a grid defined by the leaves of the tree. Because the distribution is characterised by being almost zero in large regions of the state space with some isolated peaks, many of the grid regions can be discarded as possessing negligible probability mass. The tree-based search provides an efficient way to rapidly concentrate computation on significant regions. At the lowest level of the tree the posterior distribution will be assumed to be piecewise constant. This distribution will be mostly zero for many of the leaves. At each tree level the regions with high posterior are identified and explored in finer detail in the next level (Figure 5b). It is to be expected that the higher levels will not yield accurate approximations to the posterior. However, just as for the case of detection, the upper levels of the tree can be used to discard inadequate hypotheses, for which the negative log posterior of the set exceeds a threshold value. The thresholds at the higher levels of the tree are set conservatively so as to not discard good hypotheses too soon. The equations of Bayesian filtering (6)-(8), are recast to update these states. For more details see [16].

## 5.1    Formulating the Likelihood

A key ingredient for any tracker is the likelihood function $p(\mathbf{D}_t|\boldsymbol{\theta}_t)$, which relates the observations $\mathbf{D}_t$ to the unknown state $\boldsymbol{\theta}_t$. For hand tracking finding good features and a likelihood function is challenging, as there are few features which can be detected and tracked reliably. Colour values and edges contours appear to be suitable and have been used frequently in the past [2,20]. Thus the data is taken to be composed of two sets of observations, those from edge data $\mathbf{D}_t^{edge}$ and from colour data $\mathbf{D}_t^{col}$. The likelihood function is assumed to factor as

$$p(\mathbf{D}_t|\boldsymbol{\theta}_t) = p(\mathbf{D}_t^{edge}|\boldsymbol{\theta}_t)\, p(\mathbf{D}_t^{col}|\boldsymbol{\theta}_t). \tag{9}$$

The likelihood term for edge contours $p(\mathbf{D}_t^{edge}|\boldsymbol{\theta}_t)$ is based on the chamfer distance function [3,4]. Given the set of projected model contour points $\mathcal{U} = \{\mathbf{u}_i\}_{i=1}^n$ and the set of Canny edge points $\mathcal{V} = \{\mathbf{v}_j\}_{j=1}^m$, a quadratic chamfer distance function is given by

$$d_{cham}^2(\mathcal{U}, \mathcal{V}) = \frac{1}{n}\sum_{i=1}^n d^2(i, \mathcal{V}), \tag{10}$$

where $d(i, \mathcal{V}) = \max(\min_{v_j \in \mathcal{V}} \|u_i - v_j\|, \tau)$ is the thresholded distance between the point, $u_i \in \mathcal{U}$, and its closest point in $\mathcal{V}$. Using a threshold value $\tau$ makes the matching more robust to outliers and missing edges. The chamfer distance between two shapes can be computed efficiently using a distance transform, where the template edge points are correlated with the distance transform of the image edge map. Edge orientation is included by computing the distance only for edges with similar orientation, in order to make the distance function more robust [12]. We also exploit the fact that part of an edge normal on the interior of the contour should be skin-coloured.

In constructing the colour likelihood function $p(\mathbf{D}_t^{col}|\boldsymbol{\theta}_t)$, we seek to explain all the image pixel data given the proposed state. Given a state, the pixels in the image $\mathcal{I}$ are partitioned into a set of object pixels $\mathcal{O}$, and a set of background pixels $\mathcal{B}$. Assuming pixel-wise independence, the likelihood can be factored as

$$p(\mathbf{D}_t^{col}|\boldsymbol{\theta}_t) = \prod_{k \in \mathcal{I}} p(I_t(k)|\boldsymbol{\theta}_t) = \prod_{o \in \mathcal{O}} p(I_t(o)|\boldsymbol{\theta}_t) \prod_{b \in \mathcal{B}} p(I_t(b)|\boldsymbol{\theta}_t), \tag{11}$$

where $I_t(k)$ is the intensity normalised rg-colour vector at pixel location $k$ at time $t$. The object colour distribution is modeled as a Gaussian distribution in

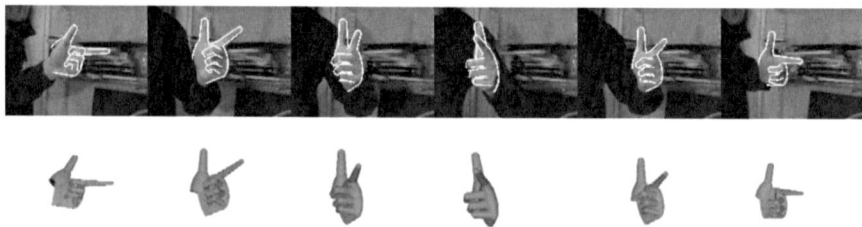

**Fig. 6. Tracking a pointing hand in front of clutter**. The images are shown with projected contours superimposed (top) and corresponding 3D avatar models (bottom), which are estimated using our tree-based algorithm. The hand is translating and rotating. A 2D deformable template would have problems coping with topological shape changes caused by self-occlusion.

the normalised colour space, and a uniform distribution is assumed for the background. For efficiency, we evaluate only the edge likelihood term while traversing the tree, and incorporate the colour likelihood only at the leaf level.

## 6   Results

We demonstrate the effectiveness of our technique by tracking both hand motion and finger articulation in cluttered scenes using a single camera. The results reveal the ability of the tree-structure to handle ambiguity arising from self-occlusion and 3D motion. In the first sequence (figure 6) we track the global 3D motion of a pointing hand. The 3D rotations are limited to a hemisphere. At the leaf level, the tree has the following resolutions: 15 degrees in two 3D rotations, 10 degrees in image rotation and 5 different scales. These 12,960 templates are then combined with a search at 2-pixel resolution in the image translation space. In the second example (figure 7) finger articulation is tracked while the hand is making transitions between different types of gestures. The tree is built by partitioning a lower dimensional eigen-space. Applying PCA to the data set shows that more than 96 percent of the variance is captured within the first four principal components, thus we partition the four dimensional eigen-space. The number of nodes at the leaf level in this case is 9,163. In the third sequence (figure 8) tracking is demonstrated for global hand motion together with finger articulation. The manifolds in section 2.4 are used to model the articulation. The articulation parameters for the thumb and fingers are approximated by the first 2 eigenvectors of the joint angle data set obtained from opening and closing of the hand. For this sequence the range of global hand motion is restricted to a smaller region, but it still has 6 DOF. In total 35,000 templates are used at the leaf level. The tree evaluation takes approximately 2 to 5 seconds per frame on a 1GHz Pentium IV machine. Note that in all cases the hand model was automatically initialised by searching the complete tree in the first frame of the sequence.

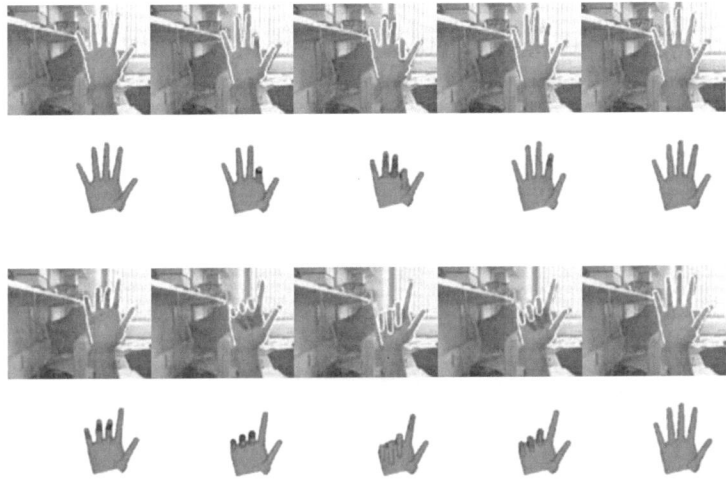

**Fig. 7. Tracking finger articulation**. In this sequence a number of different finger motions are tracked. The images are shown with projected contours superimposed (top) and corresponding 3D avatar models (bottom), which are estimated using our tree-based filter. The nodes in the tree are found by hierarchical clustering of training data in the parameter space, and dynamic information is encoded as transition probabilities between the clusters.

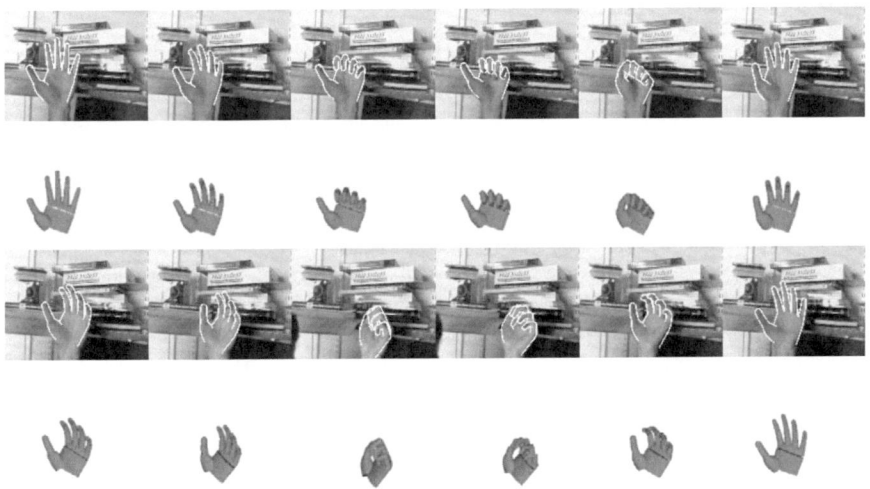

**Fig. 8. Tracking a hand opening and closing with rigid body motion.** This sequence is challenging because the hand undergoes translation and rotation while opening and closing the fingers. 6 DOF for rigid body motion plus 2 DOF using manifolds for finger flexion and extension are tracked successfully with our tree-based algorithm.

# 7   Summary and Conclusion

Within this paper we have described a model-based hand tracking system which overcomes some of the major obstacles which have limited the use of hand trackers in practical applications. These are the handling of self-occlusion, tracking in cluttered backgrounds, and tracker initialisation.

Our algorithm uses a tree of templates, generated from a 3D geometric hand model. The model is constructed from a set of truncated quadrics, and its contours can be projected into the image plane while handling self-occlusion. Articulated hand motion is learned from training data collected using a data glove. The likelihood cost function is based on the chamfer distance between projected contours and edges in the image. Additionally, edge orientation and skin colour information is used, making the matching more robust in cluttered backgrounds. The problem of tracker initialisation is solved by searching the tree in the first frame without the use of any prior information. We have tested the tracking method on a number of sequences including hand articulation and cluttered backgrounds. Furthermore within these sequences the hand undergoes rotations leading to significant topological changes in the projected contours. The tracker performs well even in these circumstances.

## Acknowledgements

The authors would like to thank the Gottlieb Daimler–and Karl Benz–Foundation, the EPSRC, the Gates Cambridge Trust, and the Overseas Research Scholarship Programme for their support.

## References

1. K. N. An, E. Y. Chao, W. P. Cooney, and R. L. Linscheid. Normative model of human hand for biomechanical analysis. *J. Biomechanics*, 12:775–788, 1979.
2. V. Athitsos and S. Sclaroff. Estimating 3D hand pose from a cluttered image. In *Proc. Conf. Computer Vision and Pattern Recognition*, Madison, USA, June 2003. to appear.
3. H. G. Barrow, J. M. Tenenbaum, R. C. Bolles, and H. C. Wolf. Parametric correspondence and chamfer matching: Two new techniques for image matching. In *Proc. 5th Int. Joint Conf. Artificial Intelligence*, pages 659–663, 1977.
4. G. Borgefors. Hierarchical chamfer matching: A parametric edge matching algorithm. *IEEE Trans. Pattern Analysis and Machine Intell.*, 10(6):849–865, November 1988.
5. R. Cipolla and P. J. Giblin. *Visual Motion of Curves and Surfaces.* Cambridge University Press, Cambridge, UK, 1999.
6. G. Cross and A. Zisserman. Quadric reconstruction from dual-space geometry. In *Proc. 6th Int. Conf. on Computer Vision*, pages 25–31, Bombay, India, January 1998.
7. D. M. Gavrila. Pedestrian detection from a moving vehicle. In *Proc. 6th European Conf. on Computer Vision*, volume II, pages 37–49, Dublin, Ireland, June/July 2000.

8. R. I. Hartley and A. Zisserman. *Multiple View Geometry in Computer Vision.* Cambridge University Press, Cambridge, UK, 2000.
9. A. J. Heap and D. C. Hogg. Towards 3-D hand tracking using a deformable model. In *2nd International Face and Gesture Recognition Conference*, pages 140–145, Killington, USA, October 1996.
10. D. P. Huttenlocher, J. J. Noh, and W. J. Rucklidge. Tracking non-rigid objects in complex scenes. In *Proc. 4th Int. Conf. on Computer Vision*, pages 93–101, Berlin, May 1993.
11. T. B. Moeslund and E. Granum. A survey of computer vision-based human motion capture. *Computer Vision and Image Understanding*, 81(3):231–268, 2001.
12. C. F. Olson and D. P. Huttenlocher. Automatic target recognition by matching oriented edge pixels. *Transactions on Image Processing*, 6(1):103–113, January 1997.
13. J. M. Rehg. *Visual Analysis of High DOF Articulated Objects with Application to Hand Tracking.* PhD thesis, Carnegie Mellon University, Dept. of Electrical and Computer Engineering, 1995. TR CMU-CS-95-138.
14. N. Shimada, K. Kimura, and Y. Shirai. Real-time 3-D hand posture estimation based on 2-D appearance retrieval using monocular camera. In *Proc. Int. WS. RATFG-RTS*, pages 23–30, Vancouver, Canada, July 2001.
15. B. Stenger, P. R. S. Mendonça, and R. Cipolla. Model based 3D tracking of an articulated hand. In *Proc. Conf. Computer Vision and Pattern Recognition*, volume II, pages 310–315, Kauai, USA, December 2001.
16. B. Stenger, A. Thayananthan, P. H. S. Torr, and R. Cipolla. Hand tracking using a tree-based estimator. Technical Report CUED/F-INFENG/TR 456, University of Cambridge, Department of Engineering, 2003.
17. A. Thayananthan, B. Stenger, P. H. S. Torr, and R. Cipolla. Shape context and chamfer matching in cluttered scenes. In *Proc. Conf. Computer Vision and Pattern Recognition*, Madison, USA, June 2003. to appear.
18. K. Toyama and A. Blake. Probabilistic tracking with exemplars in a metric space. *Int. Journal of Computer Vision*, pages 9–19, June 2002.
19. Y. Wu and T. S. Huang. Capturing articulated human hand motion: A divide-and-conquer approach. In *Proc. 7th Int. Conf. on Computer Vision*, volume I, pages 606–611, Corfu, Greece, September 1999.
20. Y. Wu, J. Y. Lin, and T. S. Huang. Capturing natural hand articulation. In *Proc. 8th Int. Conf. on Computer Vision*, volume II, pages 426–432, Vancouver, Canada, July 2001.

# Vector Transport for Shape-from-Shading

Fabio Sartori and Edwin R. Hancock

Department of Computer Science
University of York, York YO10 5DD, UK

**Abstract.** In this paper we describe a new shape-from-shading method. We show how the parallel transport of surface normals can be used to impose curvature consistency and also to iteratively update surface normal directions so as to improve the brightness error. We commence by showing how to make local estimates of the Hessian matrix from surface normal information. With the local Hessian matrix to hand, we develop an "EM-like" algorithm for updating the surface normal directions. At each image location, parallel transport is applied to the neighbouring surface normals to generate a sample of local surface orientation predictions. From this sample, a local weighted estimate of the image brightness is made. The transported surface normal which gives the brightness prediction which is closest to this value is selected as the revised estimate of surface orientation and the process is iterated until stability is reached. We experiment with the method on a variety of real world and synthetic data.

## 1 Introduction

Shape-from-shading is a problem that has been studied for over 25 years in the vision literature [2,3,11,15,21]. Stated succinctly, the problem is to recover local surface orientation information, and hence reconstruct the surface height function, from information provided by the surface brightness. Since the problem is an ill-posed one, in order to be rendered tractable, recourse must be made to strong simplifying assumptions and constraints. Hence, the process is usually specialised to matte reflectance from a surface of constant albedo, illuminated by a single point light source of known direction. To overcome the problem that the two parameters of surface slope can not be recovered from a single brightness measurement, the process is augmented by constraints on surface normal direction at occluding contours or singular points, and also by constraints on surface smoothness. An exhaustive review of the topic, which includes a detailed comparative study can be found in the recent paper of Zhang, Tsai, Cryer and Shah [22].

There have been several distinct approaches to the shape-from-shading problem. The classic approach developed by Ikeuchi and Horn [10] and, by Horn and Brooks [6], among others, is an energy minimisation one based on regularisation theory. Here the dual constraints of compliance with the image irradiance equation and local surface smoothness are captured by an error function. This has

M.J. Wilson and R.R. Martin (Eds.): Mathematics of Surfaces 2003, LNCS 2768, pp. 142–162, 2003.

distinct terms corresponding to data-closeness, i.e. compliance with the image irradiance equation, and for surface smoothness, i.e. the constraint that the local variation in the surface normal directions should be small. The shortcomings with this method are threefold. First, it is sensitive to the initial surface normal directions. Second the data-closeness and surface smoothness must be carefully balanced., Third, and finally, the solution found is invariably dominated by the smoothness model and as a result fine surface detail is lost. Ferrie and Lagarde [5] overcome some of these problems by applying the curvature consistency method of Sander and Zucker [17,18] to smooth the field of surface normals as a post-processing step in shape-from-shading. The second approach to the problem of shape-from-shading has been to adopt the apparatus of level-set theory to solve the underlying differential equation [11,12,16]. This offers two advantages. First, the recovered solution is provably correct, and second, the surface height function is recovered at the same time as the field of surface normals. The third approach was recently developed by Worthington and Hancock [21]. This method adopts the view that the image irradiance equation should be treated as a hard constraint and that curvature consistency constraints should be used in preference to local smoothing. They develop a shape-from-shading algorithm in which the surface normals are constrained to fall on the irradiance cone whose apex angle is determined by the local image brightness. The surface normals are initialised to point in the direction of the local Canny image gradient. These directions on the cone are updated by smoothing the surface normal directions in a manner which is sensitive to local surface topography. The method hence incorporates curvature consistency constraints.

The observation underpinning this paper is that although considerable effort has gone into the development of improved shape-from-shading methods, there are two areas which leave scope for further development. The first of these is the use of statistical methods in the recovery of surface normal information. The second is that relatively little effort has been expended in the use of ideas from differential geometry for surface modelling. However, the exception here is the work of Zucker and his colleagues who have explored the use of fibre bundles [4] and the relationship between the occluding boundary and the shading flow field [8,9].

Our aim in this paper is to develop a sample-based algorithm for shape-from-shading which exploits curvature consistency information. As suggested by Worthington and Hancock [21,20], we commence with the surface normals positioned on their local irradiance cone so that they are aligned in the direction of the local image gradient. From the initial surface normals, we make local estimates of the Hessian matrix. This allows us to transport neighbouring normals across the surface in a manner which is consistent with the local surface topography. The resulting sample of surface normals represent predictions of the local surface orientation which are consistent with the local surface curvature. Moreover, each transported vector can be used to make a prediction of the local image brightness.

We adopt a simple model of the distribution of brightness estimates based on the assumption that the original intensity image is subject to Gaussian measurement errors. Using this distribution, we compute the mean predicted brightness value for the sample of transported surface normals. We select a revised local surface normal direction by identifying the transported vector which gives the brightness that is closest to the mean-value.

This process may be iterated until stability is reached. From the revised surface normal directions, we make new estimates of the local Hessian matrices. These matrices in-turn are used for neighbouring surface normal transportation, and the samples of surface normals so-obtained are used to estimate mean brightness. Viewed in this way our algorithm has many features reminiscent of the EM algorithm. The surface normals may be regarded as hidden or missing data that must be recovered from the observed image brightness. In the expectation-step, we compute the mean image brightness. The maximisation step is concerned with finding the revised surface normal directions that minimise the weighted brightness error. From the perspective of differential geometry, one of the attractive features of our algorithm is that it provides a statistical framework for combining evidence for shading patterns from the Gauss map.

In this way, we facilitate a direct coupling between consistent surface normal estimation and reconstruction of the image brightness. Moreover, our method overcomes the problem of estimating surface normal directions in a natural way. This offers two advantages over existing methods for shape-from-shading. First, because it is evidence-based, unlike the Horn and Brooks [6] method, it is not model dominated and does not oversmooth the recovered field of surface normal directions. The data-closeness and surface-smoothness errors are not simply compounded in an additive way as is the case in the regularisation method. Second, and unlike the Worthington and Hancock method [21], it relaxes the image irradiance equation and hence allows for brightness errors to be corrected. Another interesting property of the method, is that we parameterise the local surface structure using the Hessian matrix, rather than quadric patch parameters. Hence we exploit the intrinsic differential geometry of the Gauss map rather than its extrinsic geometry.

The novelty of our contribution is twofold. First, we develop an evidence combining algorithm for shape-from-shading. There have been few previously documented attempts to do this in the literature. Second, is our idea of using parallel transport to ensure consistency with differential geometry. Here there are two pieces of related work. Lagarde and Ferrie [5] have shown how the Darboux smoothing idea of Sander and Zucker [17,18] can be applied to smooth extracted needle-maps as a post-processing step. Worthington and Hancock [21], on the other hand, have shown how the variance of the Koenderinck and Van Doorn [13,14] shape-index can be used to control the robust smoothing of surface normal directions.

The outline of this paper is as follows. In section 2 we review the standard regularisation approach to shape-from-shading of Horn and Brooks. Sections 3 and 4 describe a least squares procedure for extracting estimates of the Hessian matrix from the fields of surface normals delivered by shape-from-shading. In Section 5 we describe how parallel transport may be used to accumulate sets of surface normals. In Section 6 we outline a statistical method which can be used to update local surface normal direction by selecting from the sample of transported normals. In Section 7 we provide and experimental evaluation of the new shape-from-shading method, and compare it with a number of alternative algorithms. Finally, Section 8 offers some conclusions and offers directions for further research.

## 2    Shape-from-Shading

Central to shape-from-shading is the idea that local regions in an image $E(x, y)$ correspond to illuminated patches of a piecewise continuous surface, $z(x, y)$. The measured brightness $E(x, y)$ will depend on the material properties of the surface, the orientation of the surface at the co-ordinates $(x, y)$, and the direction and strength of illumination.

The *reflectance map*, $R(p, q)$ characterises these properties, and provides an explicit connection between the image and the surface orientation. Surface orientation is described by the components of the surface gradient in the $x$ and $y$ direction, i.e. $p = \frac{\partial z}{\partial x}$ and $q = \frac{\partial z}{\partial y}$. The shape from shading problem is to recover the surface $z(x, y)$ from the intensity image $E(x, y)$. As an intermediate step, we may recover the needle-map, or set of estimated local surface normals, $\mathbf{Q}(x, y)$.

Needle-map recovery from a single intensity image is an under-determined problem [15,7,2] which requires a number of constraints and assumptions to be made. The common assumptions are that the surface has ideal Lambertian reflectance, constant albedo, and is illuminated by a single point source at infinity. A further assumption is that there are no inter-reflections, i.e. the light reflected by one portion of the surface does not impinge on any other part.

The local surface normal may be written as $\mathbf{Q} = (-p, -q, 1)^T$, where $p = \frac{\partial z}{\partial x}$ and $q = \frac{\partial z}{\partial y}$. For a light source at infinity, we can similarly write the light source direction as $\mathbf{s} = (-p_l, -q_l, 1)^T$. If the surface is Lambertian the reflectance map is given by

$$R(p, q) = \mathbf{Q} \cdot \mathbf{s} \tag{1}$$

The image irradiance equation [6] states that the measured brightness of the image is proportional to the radiance at the corresponding point on the surface; that is, just the value of $R(p, q)$ for $p, q$ corresponding to the orientation of the surface. Normalising both the image intensity, $E(x, y)$, and the reflectance map, the constant of proportionality becomes unity, and the image irradiance equation is simply

$$E(x, y) = R(p, q) \tag{2}$$

Although the image irradiance equation succinctly describes the mapping between the $x, y$ co-ordinate space of the image and the $p, q$ gradient-space of the surface, it provides insufficient constraints for the unique recovery of the needle-map. To overcome this problem, a further constraint must be applied. Usually, the needle-map is assumed to vary smoothly.

The process of smooth surface recovery may be posed as a variational problem in which a global error-functional is minimised through the iterative adjustment of the needle map. Surface normals are updated with a step-size dictated by Euler's equation. Here we consider the formulation of Brooks and Horn [7] which is couched in terms of unit surface normals. Their error functional is defined to be

$$I = \int \int \underbrace{\left(E(x, y) - \mathbf{Q} \cdot \mathbf{s}\right)^2}_{Brightness Error} + \lambda \underbrace{\left(\left\|\frac{\partial \mathbf{Q}}{\partial x}\right\|^2 + \left\|\frac{\partial \mathbf{Q}}{\partial y}\right\|^2\right)}_{Regularizing Term} dx dy \qquad (3)$$

The terms $\frac{\partial \mathbf{Q}}{\partial x}$ and $\frac{\partial \mathbf{Q}}{\partial y}$ above are the directional derivatives of the needle-map in the $x$ and $y$ directions respectively. The magnitudes of these quantities are used to measure the smoothness of the surface, with a large value indicating a highly-curved region. However, it should be noted that a planar surface has $\frac{\partial \mathbf{Q}}{\partial x} = \frac{\partial \mathbf{Q}}{\partial y} = \mathbf{0}$ in this case.

The first term of Equation 3 is the brightness error, which encourages data-closeness of the measured image intensity and the reflectance function. The *regularising term* imposes the smoothness constraint on the recovered surface normals, penalising large local changes in surface orientation. The constant $\lambda$ is a Lagrange multiplier. For numerical stability, $\lambda$ must often be large, resulting in the smoothness term dominating.

Minimisation of the functional defined in Equation 3 is accomplished by applying the calculus of variations and solving the resulting Euler equation. The solution is

$$\mathbf{Q}_o^{(k+1)} = \bar{\mathbf{Q}}_o^{(k)} + \frac{\epsilon^2}{2\lambda} \left(E_o - \mathbf{Q}_o^{(k)} \cdot \mathbf{s}\right) \mathbf{s} \qquad (4)$$

where $\epsilon$ is the spacing of pixel-sites on the lattice and $\bar{\mathbf{Q}}_o$ is the local mean of the surface normals around the neighbourhood $R_o$ of the pixel at position $o$

$$\bar{\mathbf{Q}}_o = \frac{1}{|R_o|} \sum_{m \in R_o} \mathbf{Q}_o \qquad (5)$$

In practice $R_o$ is either the four or the eight pixel neighbourhood. The main criticism of this method is that in order to achieve stable algorithm behaviour, the constant $\epsilon$ must be set to be small. As a result, the smoothness term dominates the data-closeness term, with the consequence that the recovered field of surface normals is oversmoothed and in poor agreement with Lambert's law.

One way to circumvent this problem is to apply the image irradiance equation as a hard constraint. This is the idea underpinning Worthington and Hancock's method [21]. Here the surface normals are constrained to fall on cones pointing in the light source direction, and with opening angles determined by the measured image brightness. This may be over restrictive since it does not allow brightness errors to be corrected, The aim in this paper is to develop an evidence combining scheme in which we can both exploit curvature consistency constraints, and allow for brightness errors.

## 3   Differential Surface Structure

In this paper we are interested in the local differential structure of surfaces represented in terms of a field of surface normals. In differential geometry this representation is known as the Gauss map. The differential structure of the surface is captured by the second fundamental form or Hessian matrix

$$\mathcal{H} = \begin{pmatrix} \frac{\partial^2 z}{\partial x^2} & \frac{\partial^2 z}{\partial x \partial y} \\ \frac{\partial^2 z}{\partial x \partial y} & \frac{\partial^2 z}{\partial y^2} \end{pmatrix} \qquad (6)$$

where $z$ is the surface height.

The eigen-structure of the Hessian matrix can be used to gauge the curvature of the surface. The two eigen-values of $\mathcal{H}$ are the maximum $(K^{max})$ and minimum $(K^{min})$ curvatures. The orthogonal eigen-vectors of $\mathcal{H}$ are known as the principal curvature directions. The mean-curvature $K = \frac{1}{2}(K^{max} + K^{min})$ of the surface is found by averaging the maximum and minimum curvatures. Finally, the Gaussian curvature $H = K^{max}K^{min}$ is equal to the product of the two eigenvalues.

In the case when surface normal information is being used to characterise the surface, then the Hessian matrix takes on the following form

$$\mathcal{H} = \begin{pmatrix} \alpha & \beta \\ \beta & \gamma \end{pmatrix} \qquad (7)$$

The diagonal elements of the Hessian are related to the rate-of change of the surface normal components with position via the equations

$$\alpha = \left(\frac{\partial \mathbf{Q}}{\partial x}\right)_x , \gamma = \left(\frac{\partial \mathbf{Q}}{\partial x}\right)_y \qquad (8)$$

where the subscripts $x$ and $y$ on the large brackets indicate that the $x$ or $y$ components of the vector-derivative are being taken.

Treatment of the off-diagonal elements is more subtle. However, if we assume that the surface can be represented by a twice differentiable function $z = f(x,y)$, then we can write

$$\beta = \left(\frac{\partial \mathbf{Q}}{\partial y}\right)_x = \left(\frac{\partial \mathbf{Q}}{\partial x}\right)_y \qquad (9)$$

With the simplified Hessian to hand, then we can compute the maximum and minimum curvatures at locations on the surface. As mentioned earlier, the two curvatures are simply the eigenvalues of the Hessian matrix. Hence, the maximum and minimum curvatures are respectively:

$$K_o^{max} = -\frac{1}{2}(\alpha + \gamma - S) , \quad K_o^{min} = -\frac{1}{2}(\alpha + \gamma + S) \qquad (10)$$

where $S = \sqrt{(\alpha - \gamma)^2 + 4\beta^2}$. The eigenvector associated with the maximum curvature $K_o^{max}$ is the principal curvature direction. On the tangent-plane to the surface, the principal curvature direction is given by the 2-component vector

$$e_o^{max} = \begin{bmatrix} (\beta, -\frac{1}{2}(\alpha - \gamma + S))^T & \alpha \geq \gamma \\ (\frac{1}{2}(\alpha - \gamma - S), \beta)^T & \alpha < \gamma \end{bmatrix} \tag{11}$$

In the next Section we will describe how the elements of the Hessian, i.e. $\alpha$, $\beta$ and $\gamma$, can be estimated from raw surface normal data using the method of least-squares.

## 4   Estimating the Hessian

In this section we describe how to make a statistical estimate of the Hessian matrix from a sample of surface normals delivered by shape-from-shading. Specifically, we use the method of least squares to estimate the elements of $\mathcal{H}$. Here we draw on a method previously reported by Wilson and Hancock [19].

Let $\mathbf{Q}_o$ represent the surface normal at the position $(x_o, y_o)$ and let $\mathbf{Q}_m$ be a neighbouring surface normal with position $(x_m, y_m)$. If the normals are close to each other, then we can approximate the change in the components of the surface normal using a first-order Taylor expansion. Accordingly,

$$(\Delta Q_m)_x = \left(\frac{\partial \mathbf{Q}}{\partial x}\right)_x \Delta x_m + \left(\frac{\partial \mathbf{Q}}{\partial y}\right)_x \Delta y_m \tag{12}$$

$$(\Delta Q_m)_y = \left(\frac{\partial \mathbf{Q}}{\partial x}\right)_y \Delta x_m + \left(\frac{\partial \mathbf{Q}}{\partial y}\right)_y \Delta y_m \tag{13}$$

where the measured change in the components of the surface normal is given by $\mathbf{Q}_m - \mathbf{Q}_o = ((\Delta Q_m)_x, (\Delta Q_m)_y)^T$. The displacements in point co-ordinates are $\Delta x_m = x_m - x_o$ and $\Delta y_m = y_m - y_o$. We can rewrite the first-order Taylor expansion in terms of elements of the Hessian matrix, i.e.

$$(\Delta Q_m)_x = \alpha_o \Delta x_m + \beta_o \Delta y_m \tag{14}$$

$$(\Delta Q_m)_y = \beta_o \Delta x_m + \gamma_o \Delta y_m \tag{15}$$

where $\alpha_o$, $\beta_o$ and $\gamma_o$ are the elements of the Hessian matrix at the pixel indexed $o$. These equations govern the parallel transport of the vector across the curved geometry of the surface. So, to first-order, the change in the normal surface is linear in the elements of the Hessian matrix. Unfortunately, for the single neighbouring normal these equations are under-constrained and we can not recover the Hessian. However, if we have a sample of $N$ neighbouring surface normals, then there are $2N$ homogenous linear equations in the elements of $\mathcal{H}$ and the problem of recovering differential structure is no-longer under-constrained. Under these circumstances, we can estimate the elements of the Hessian matrix using the method of least-squares.

To proceed, we make the homogeneous nature of the equations more explicit by writing

$$\begin{aligned} (\Delta Q_m)_x &= \Delta x_m \cdot \alpha_o + \Delta y_m \cdot \beta_o + \quad 0 \cdot \gamma_o \\ (\Delta Q_m)_y &= \quad 0 \cdot \alpha_o + \Delta x_m \cdot \beta_o + \Delta y_m \cdot \gamma_o \end{aligned} \tag{16}$$

In order to simplify notation, we can write the full system of 2N equations in matrix form as

$$\mathbf{N} = \mathbf{X}\boldsymbol{\Phi_o} \tag{17}$$

where $\mathbf{N}$ is an aggregated column-vector of normal components

$$\mathbf{N} = \left( (\Delta Q_1)_x, (\Delta Q_1)_y, (\Delta Q_2)_x, \ldots \right)^T$$

The design matrix $\mathbf{X}$ is a matrix of co-ordinate displacements

$$\mathbf{X} = \begin{pmatrix} \Delta x_1 & \Delta y_1 & 0 \\ 0 & \Delta x_1 & \Delta y_1 \\ \Delta x_2 & \Delta y_2 & 0 \\ & \vdots & \end{pmatrix}$$

and $\Phi_o$ is the parameter vector

$$\Phi_o = (\alpha_o, \beta_o, \gamma_o)^T$$

When the system of equations is over-specified in this way, then we can extract the set of parameters that minimises the vector of error-residuals $\mathbf{N} - \mathbf{X}\Phi_o$. We pose this parameter recovery process as a least-squares estimation problem. In other words we seek the vector of estimated parameters $\hat{\boldsymbol{\Phi}}_\mathbf{o} = (\hat{\alpha}_o, \hat{\beta}_o, \hat{\gamma}_o)^T$ which satisfy the condition

$$\hat{\boldsymbol{\Phi}}_\mathbf{o} = \arg\min_{\boldsymbol{\Phi}} (\mathbf{N} - \mathbf{X}\boldsymbol{\Phi})^T (\mathbf{N} - \mathbf{X}\boldsymbol{\Phi}) \tag{18}$$

The solution-vector is found by computing the pseudo-inverse of the design matrix $\mathbf{X}$ thus

$$\hat{\Phi}_o = (\mathbf{X}^T\mathbf{X})^{-1}\mathbf{X}^T\mathbf{N} \tag{19}$$

The surface geometry described in this Section is illustrated in Figure 1(a).

## 5   Parallel Transport

In this paper we are interested in using the local estimate of the Hessian matrix to provide curvature consistency constraints for shape from-shading. Our aim is to improve the estimation of surface normal direction by combining evidence from both shading information and local surface curvature. As demonstrated by both Ferrie and Lagarde [5] and Worthington and Hancock [21], the use of curvature information allows the recovery of more consistent surface normal directions. It also provides a way to control the over-smoothing of the resulting needle maps. Ferrie and Lagarde [5] have addressed the problem by using local Darboux frame smoothing. Worthington and Hancock [21], on the other hand, have employed a curvature sensitive robust smoothing method. Here we adopt a different approach which uses the equations of parallel transport to guide the prediction of the local surface normal directions.

Our idea is as follows. At each location on the surface we make an estimate of the vector of curvature parameters. Suppose that we are positioned at the point $\boldsymbol{X}_o = (x_o, y_o)^T$ where the vector of estimated curvature parameters is $\Phi_o$ and that the resulting estimate of the Hessian matrix is $\mathcal{H}_o$. Further suppose that $\mathbf{Q}_m$ is the surface normal at the point $\boldsymbol{X}_m = (x_m, y_m)^T$ in the neighbourhood of $\boldsymbol{X}_o$. We use the local curvature parameters $\Phi_o$ to transport the vector $\mathbf{Q}_m$ to the location $\boldsymbol{X}_o$. From Equations 13 and 14 it follows that the first-order approximation to the transported vector is

$$\mathbf{Q}_m^o = \mathbf{Q}_m + \mathcal{H}_o(\boldsymbol{X}_m - \boldsymbol{X}_o) \qquad (20)$$

This procedure is repeated for each of the surface normals belonging to the neighbourhood $R_o$ of the point $o$. In this way we generate a sample of alternative surface normal directions at the location $o$. The geometry of the parallel transport procedure is illustrated in Figure 1(b).

## 6   Statistical Framework

We would like to exploit the transported surface-normal vectors to develop an evidence combining approach to shape-from-shading. To do this we require a probabilistic characterisation of the sample of available surface normals. We assume that the observed brightness $E_o$ at the point $\boldsymbol{X}_o$ follows a Gaussian distribution. As a result the probability density function for the transported surface normals is

$$p(E_o | \mathbf{Q}_m, \Phi_o) = \frac{1}{\sqrt{2\pi}\sigma} \exp\left[ -\frac{(E_o - \mathbf{Q}_m^o.\mathbf{s})^2}{2\sigma^2} \right] \qquad (21)$$

where $\sigma^2$ is the noise-variance of the brightness errors.

We exploit this simple model to develop two alternative ways of updating the surface normal direction.

**Sample Mode.** The first method involves computing the sample mode for the set of transported surface normals. To do this we use the probability density to compute the expected value of the image brightness at the location $\boldsymbol{X}_o$ for the sample of transported surface normals. The expected brightness is given by

$$\hat{E}_o = \sum_{m \in R_o} p(E_o | \mathbf{Q}_m, \Phi_o) \mathbf{Q}_m^o.\mathbf{s} \qquad (22)$$

To update the surface normal direction, we select from the sample the one which results in a brightness value which is closest to $\hat{E}_o$. This surface normal is the one for which

$$\hat{\mathbf{Q}}_o = \arg \min_{m \in R_o} \left[ \hat{E}_o - \mathbf{Q}_m^o.\mathbf{s} \right]^2 \qquad (23)$$

**Curvature Weighting.** Our second approach involves weighting the surface normals according to the curvature of the path $\Gamma_{o,m}$ from the point $m$ to the point $o$. The normal curvature at the point $o$ in the direction of the transport path is approximately

$$\kappa_{o,m} = (\boldsymbol{T}_{o,m} \cdot \boldsymbol{e}_o^{max})^2 (K_o^{max} - K_o^{min}) + K_o^{min} \tag{24}$$

where $\boldsymbol{T}_{o.m} = \frac{\boldsymbol{X}_m - \boldsymbol{X}_o}{|\boldsymbol{X}_m - \boldsymbol{X}_o|}$ is the unit vector from $o$ to $m$.

Here we adopt a model in which we assume that the sample of transported surface normals is drawn from a Gaussian prior, which is controlled by the normal curvature of the transport path. Accordingly we write

$$p(\boldsymbol{Q}_m^o) = \frac{1}{\sqrt{2\pi}\sigma_k} \exp\left[-\frac{1}{2\sigma_k^2} k_{o.m}^2\right] \tag{25}$$

With these ingredients, the weighted mean for the sample of transported surface normals

$$\hat{\boldsymbol{Q}}_o = \sum_{m \in R_o} \boldsymbol{Q}_m^o p(E_o | \boldsymbol{Q}_m^o, \Phi_o) p(\boldsymbol{Q}_m^o) \tag{26}$$

where $R_o$ is the index set of the surface normals used for the purposes of transport. Substituting for the distributions,

$$\hat{\boldsymbol{Q}}_o = \frac{\sum_{m \in R_o} \boldsymbol{Q}_{o,m} \exp\left[-\frac{1}{2}\left(\frac{(E_o - \boldsymbol{Q}_m^o \cdot \mathbf{s})^2}{\sigma_E^2} + \frac{\kappa_{o,m}^2}{\sigma_k^2}\right)\right]}{\sum_m \exp\left[-\frac{1}{2}\left(\frac{(E_o - \boldsymbol{Q}_m^o \cdot \mathbf{s})^2}{\sigma_E^2} + \frac{\kappa_{o,m}^2}{\sigma_k^2}\right)\right]} \tag{27}$$

and the predicted brightness is $\hat{E}_o = \hat{\boldsymbol{Q}}_o \cdot \mathbf{s}$.

**Update Algorithm.** The two methods for updating the surface normal direction may be iterated using the following steps:

- 1: At each location compute a local estimate of the Hessian matrix $\mathcal{H}_o$ from the currently available surface normals $\boldsymbol{Q}_o$.
- 2: At each image location $\boldsymbol{X}_o$ obtain a sample of surface normals $S_o = \{\boldsymbol{Q}_m^o | m \in R_o\}$ by applying parallel transport to the set of neighbouring surface normals whose locations are indexed by the set $R_o$.
- 3: From the set of surface normals $S_o$ compute the expected brightness value $\hat{E}_o$ and the updated surface normal direction $\hat{\boldsymbol{Q}}_o$. Note that the measured intensity $E_o$ is kept fixed throughout the iteration process and is not updated.
- 4: With the updated surface normal direction to hand, return to step 1, and recompute the local curvature parameters.

To initialise the surface normal directions, we adopt the method suggested by Worthington and Hancock [21]. This involves placing the surface normals on the irradiance cone whose axis is the light-source direction $\mathbf{S}$ and whose apex angle is $\cos^{-1} E_o$. The position of the surface normal on the cone is such that its projection onto the image plane points in the direction of the local image gradient, computed using the Canny edge detector. When the surface normals are initialised in this way, then they satisfy the image irradiance equation.

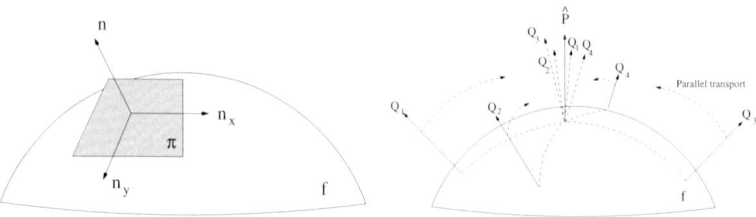

(a) Surface normal vector and its components.

(b) Parallel transport used for predicting the surface normal vector using local curvature estimation.

**Fig. 1.** Surface structure and parallel transport.

## 7    Experiments

In this section, we present some experimental evaluation of the new method. The study is divided into two parts. We commence by showing results on synthetic data with known ground truth. The second aspect of the study is concerned with real world imagery.

**Synthetic Data.** We commence by showing some results for synthetic surfaces. To do this we we have generated height data. From the height data, surface normals directions have been computed. The surface normal directions have been used to generate shading information using a Lambertian reflectance model. We have applied our shape-from-shading algorithm to the images generated in this way.

The objects studied are shown in Figure 2. The objects are a) a sphere surrounded by a torus, b) two domes intersected by a parabolic dome, c) four domes and d) four horizontal parabolic ridges intersected by a vertical parabolic ridge. The figure shows the synthetic images of the surfaces together with their ground-truth needle-maps.

In Figures 4 to 7 we show the results obtained for the synthetic images using our new shape-from-shading method and the method of Worthington and Hancock. Each figure shows the Lambertian re-illuminations computed from the recovered surface normals. The re-illuminations are obtained with varying light source direction. The three rows of each figure, from top to bottom, show the ground truth re-illuminations obtained from the original surface, the re-illuminations obtained using the surface normals delivered by the new method, and the re-illuminations obtained using the surface normals delivered by the Worthington and Hancock method.

We now present a series of plots which compare the results obtained using shape-from-shading with the corresponding ground truth for the synthetic images. We commence by considering the synthetic disk. In Figures 7, from left to right, we show the needle-map obtained using the Worthington and Hancock robust regulariser method, the result obtained using the method described in this

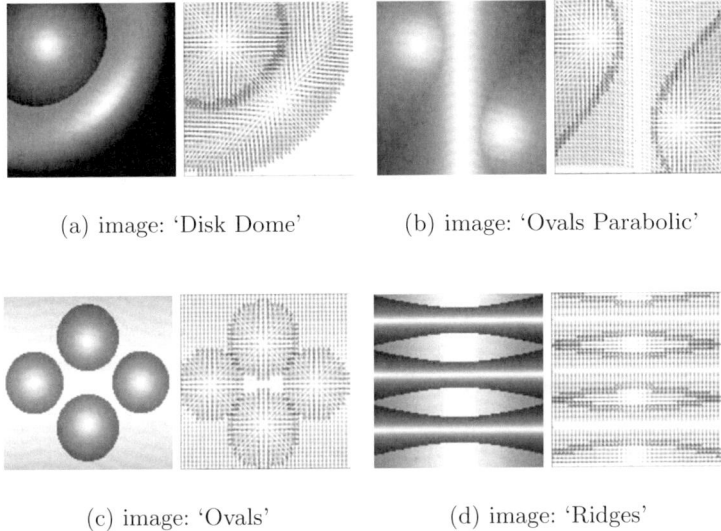

(a) image: 'Disk Dome'          (b) image: 'Ovals Parabolic'

(c) image: 'Ovals'              (d) image: 'Ridges'

**Fig. 2.** Original synthetic images and respective needle-maps.

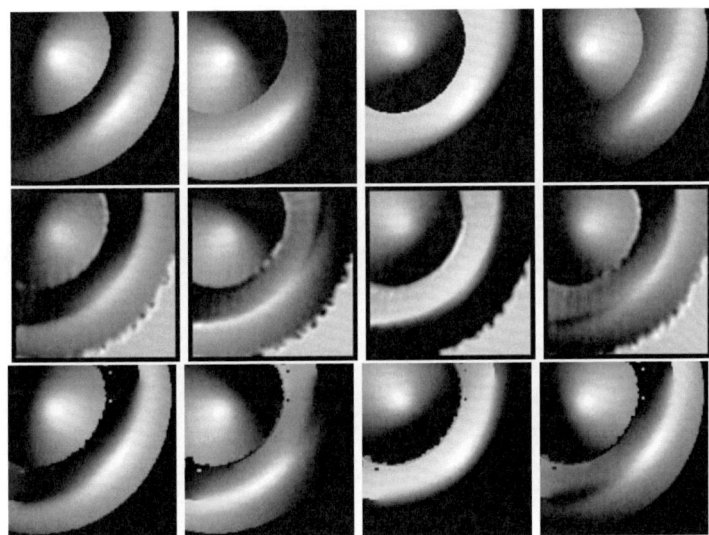

**Fig. 3.** Ground truth needle-maps re-illuminations (first row), from the new method (second row) and Worthington and Hancock method (third row).

paper, and the field of vector differences between these two needle-maps. In Figure 9, the panels are as follows. In the first and second rows we show scatter plots of estimated versus ground truth azimuth angles. The first row is for the vector transport method, while the second row is for the Worthington and Hancock method. Similarly, the third and fourth rows respectively show the scatter plots

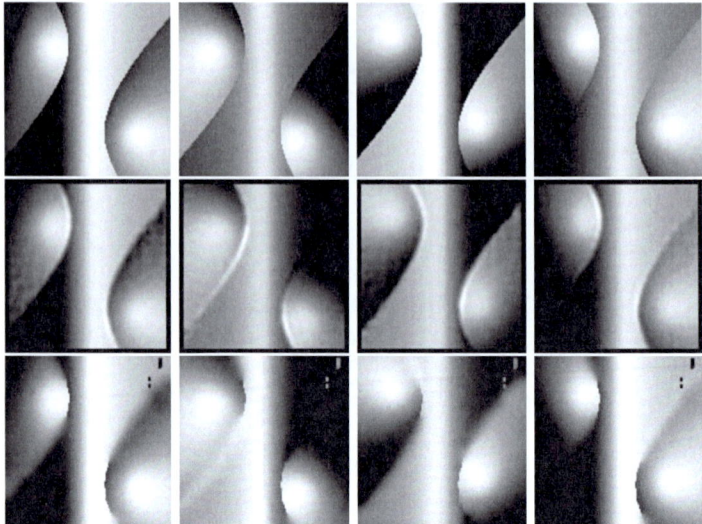

**Fig. 4.** Ground truth needle-maps re-illuminations (first row) from the new method (second row) and Worthington and Hancock method (third row).

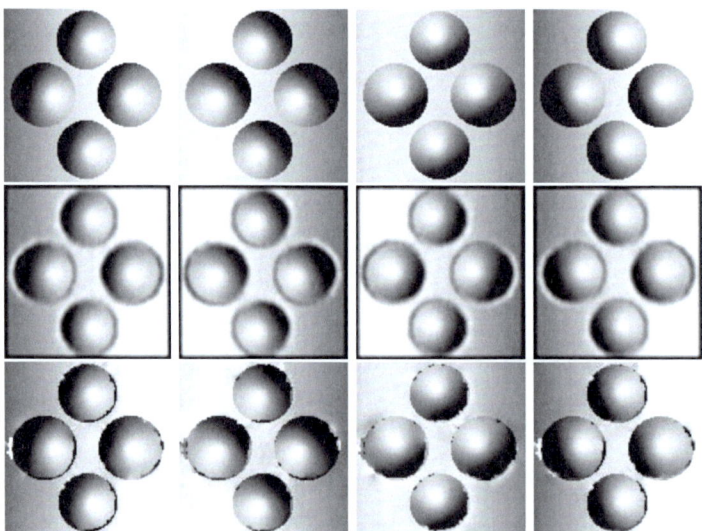

**Fig. 5.** Ground truth needle-maps re-illuminations (first row), from the new method (second row) and Worthington and Hancock method (third row).

of zenith angles for vector transport, and, the Worthington and Hancock method. The fifth and sixth rows show the shape-index $\phi_o = \frac{2}{\pi} \arctan \frac{K_o^{max}+K_o^{min}}{K_o^{max}-K_o^{min}}$ for the two methods. This is an angular measure of local surface topography, which has proved effective in the visualisation of surfaces. The first column of scatter plots

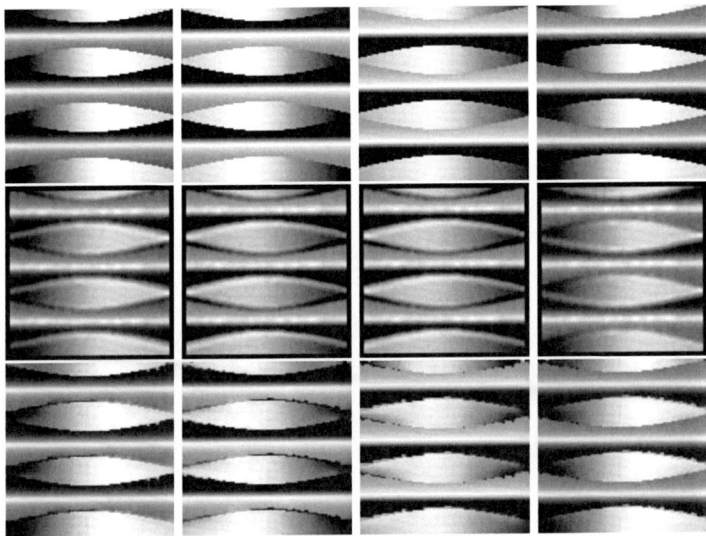

**Fig. 6.** Ground truth needle-maps re-illuminations (first row), from the new method (second row) and Worthington and Hancock method (third row).

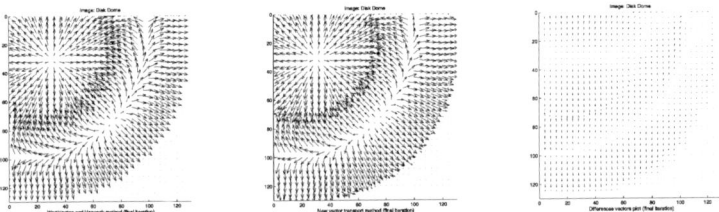

**Fig. 7.** Needle maps delivered with the two different methods. Image: Disk Dome.

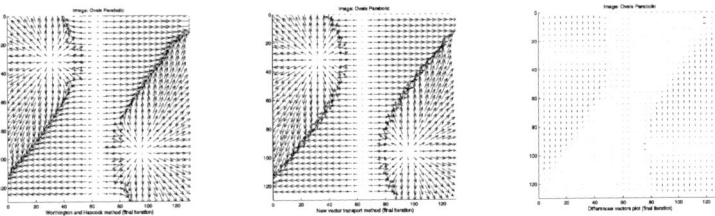

**Fig. 8.** Needle maps delivered by the two different methods for the parabolic oval.

is for the initial needle map, and this is common to both the vector transport method and the Worthington and Hancock method. The second column is for the middle iteration, and the final column is for the last iteration. Figures 8 and 10 repeat the results shown in Figures 8 and 9 for the parabolic oval.

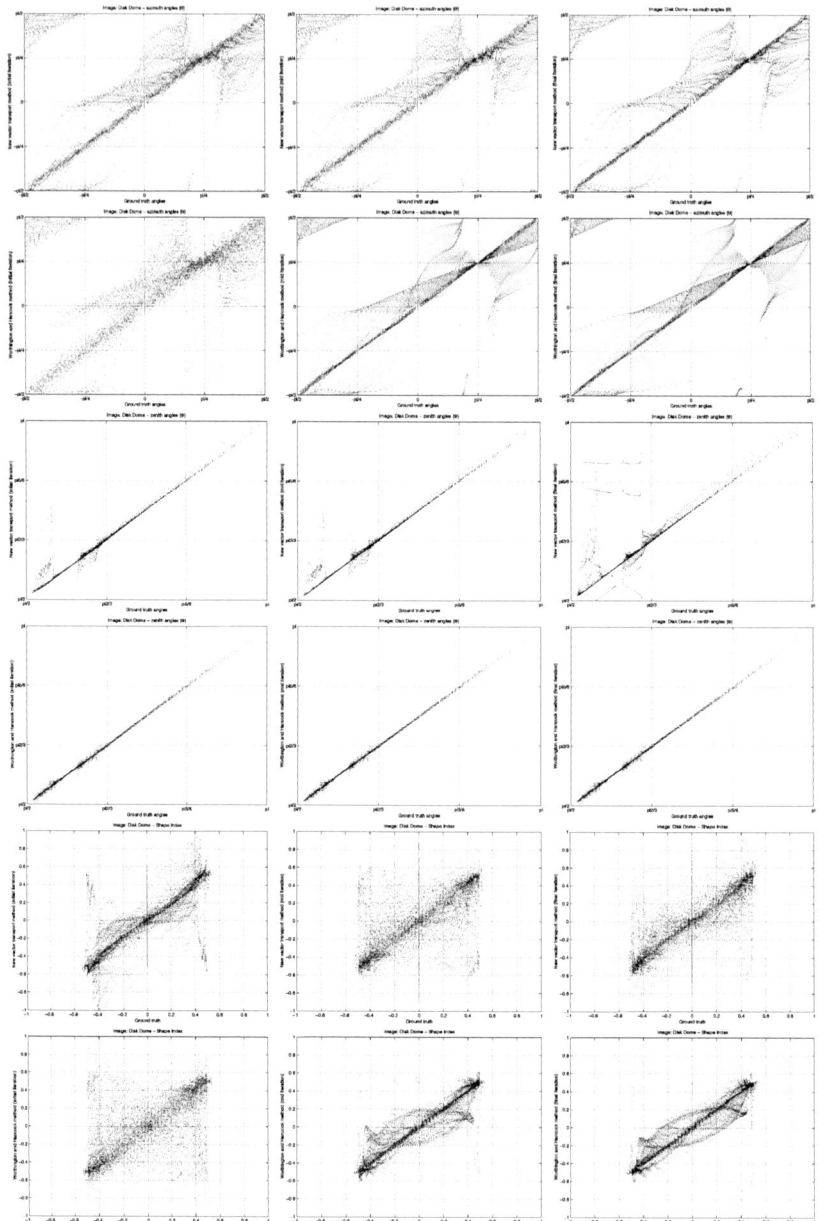

**Fig. 9.** Scatterplots of zenith and azimuth angles, shape index and curvedness for the 'disk-dome'.

The main features to note from the plots are as follows. First, there is little difference between the results obtained using the two methods. Both methods give comparable needle maps. They also give in needle maps with azimuth, zenith

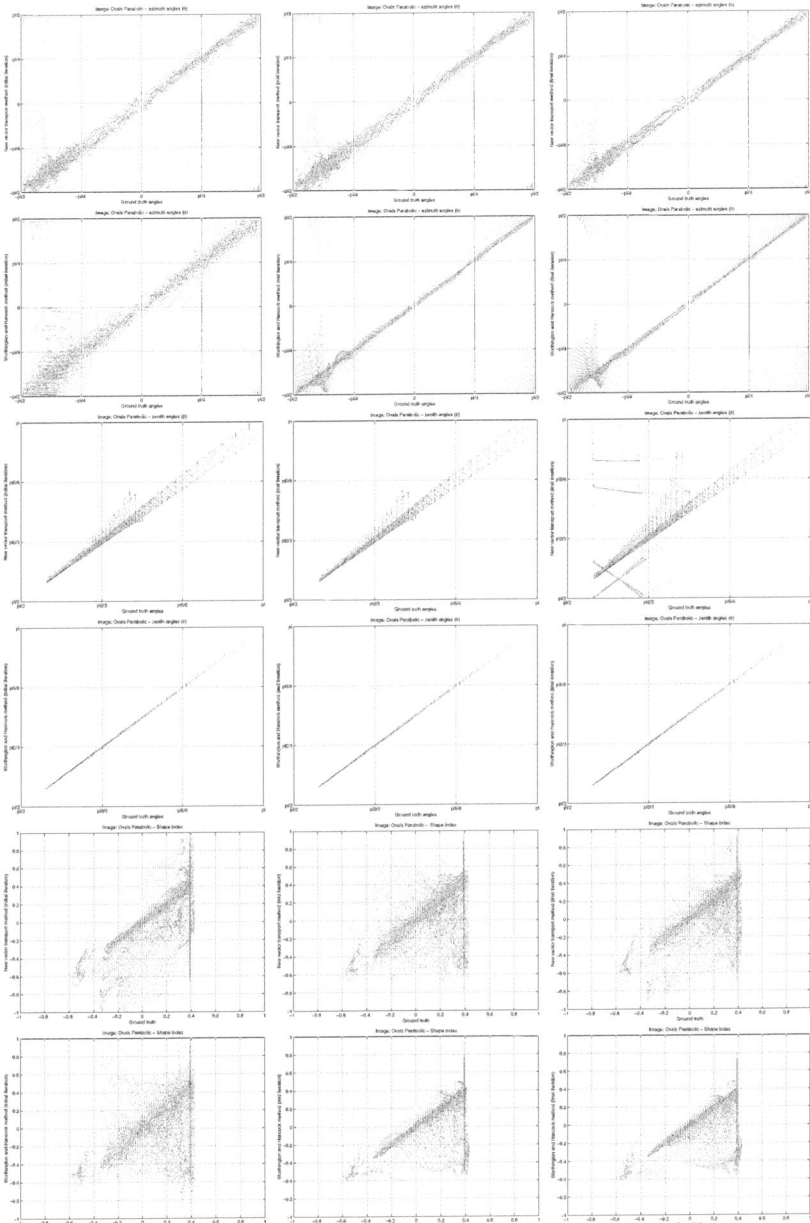

**Fig. 10.** Scatterplots of zenith and azimuth angles, shape index and curvedness for the 'ovals parabolic'.

and shape-index distributions which agree well with ground truth. The distributions in the initial scatter plots are dispersed about the diagonal. However, in the final iteration the dispersion is considerably reduced. There is some ev-

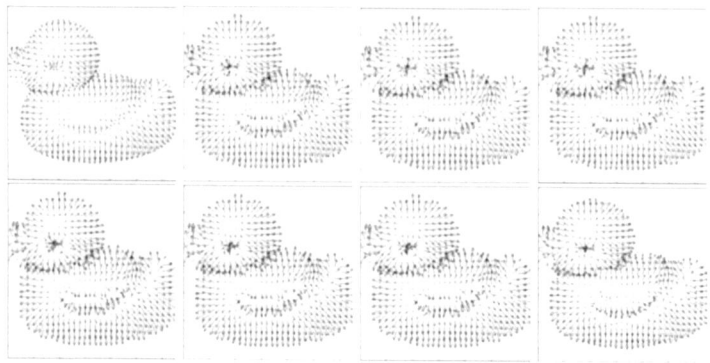

**Fig. 11.** Needle maps evolution of the toy duck image.

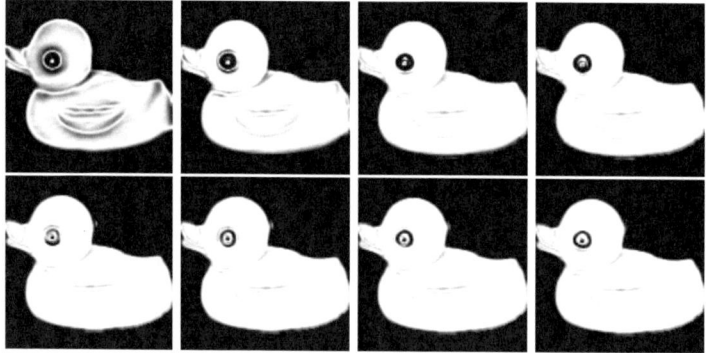

**Fig. 12.** Probability of image intensity agreement with iteration number.

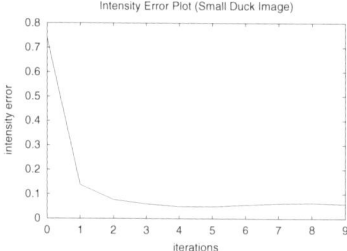

**Fig. 13.** Error plot for the reconstructed intensity.

idence of biassing and systematic error in the azimuth plots. These appear to be associated with the initialisation, and are probably attributable to quantisation effects of the initial gradient estimates on the pixel lattice. In the case of the shape-index plots it is clear that the Worthington and Hancock method introduces some systematic biases. This is not surprising since the method uses

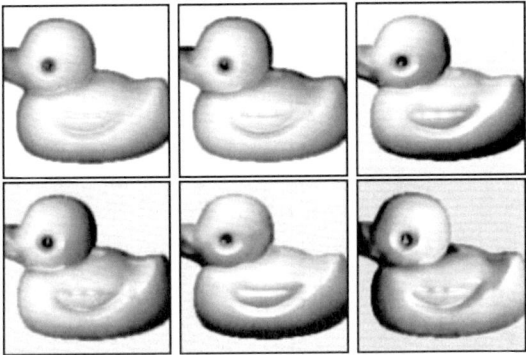

**Fig. 14.** Reconstructed image with different illumination directions.

(a) The Three Graces.    (b) Head of Canova.

**Fig. 15.** Classic statues, original images.

the shape-index to control smoothing. Our method does not appear to suffer from this problem and has dispersed but unstructured scatter plot. Finally, it is important to note that our new method takes fewer iterations than the Worthington and Hancock method. Typically, our method converges in 10 iterations, while the Worthington and Hancock method takes some 50 iterations.

**Real World Data.** We now turn our attention to real world imagery. We commence by exploring some of the iterative properties of the method. Here we experiment with an image of a toy duck from the Columbia COIL data-base. In Figure 11 we show the field of surface normal directions with iteration number. The main feature to note here is that the surface details become more marked with iteration number. In Figure 12, we plot the difference between the measured and predicted image brightness as a function of iteration number. Initially, the error is greatest on the curved regions of the surface (near the head, beak, neck,

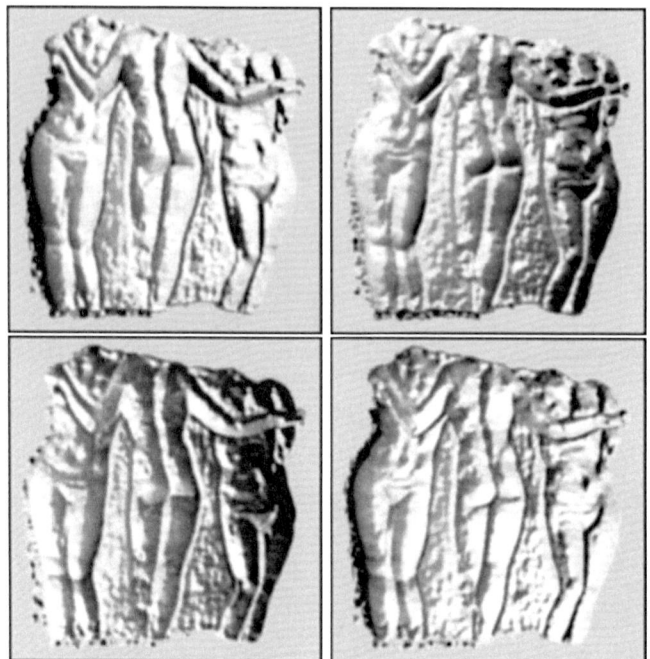

**Fig. 16.** Re-illuminations from the vector transport method.

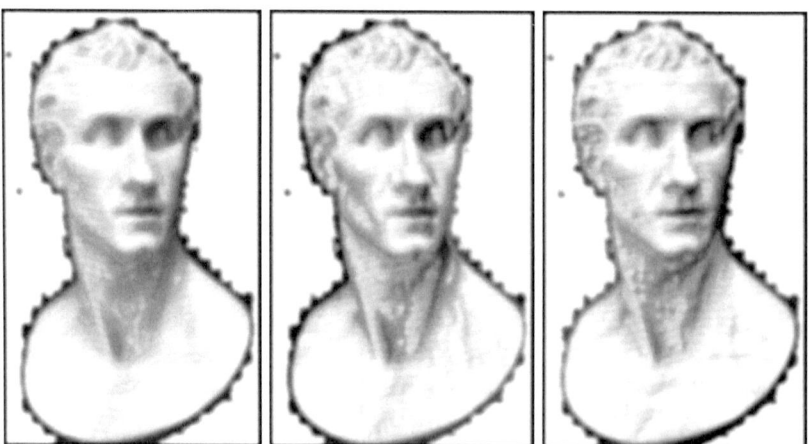

**Fig. 17.** Re-illuminations from the vector transport method.

wings and tail). After 10 iterations, the only region where there is a significant error is around the eye, where there is an albedo difference. There are some high curvature points around the neck and the wing where there is also some residual brightness error. The average brightness error is plotted as a function of iteration number in Figure 13. Figure 14 shows the effect of re-illuminating the final needle-map with different light source directions. This highlights the curvature detail on the surface, which appears to be well reconstructed.

To conclude, we show some results for the "Three Graces" relief. Figure 15(a) shows the original image while Figure 16 shows a sequence of re-illuminations generated from the surface normals delivered using our new method. The surface detail is well reconstructed and the re-illumination captures well the changes in light source direction. Figures 15(b) and 17 show a second example of re-illuminating of classical statue, the head of Canova. The re-illuminations in Figure 17 reveal important surface details including the fine structure of the hair, the indentations of the left cheek and the detail of the nose.

## 8   Conclusions

In this paper we have described a new method for shape-from-shading which relies on vector transport to accumulate evidence for surface normal directions which are consistent with the observed image brightness. The method uses a two-step iterative algorithm. First, estimates of the Hessian matrix are made using the available surface normals. These Hessian matrices are used to perform vector transport on the surrounding surface normals to accumulate a sample of orientation hypotheses. These putative directions are used to compute an expected value for the image brightness. In the second step of the algorithm, the surface normal direction is updated. The direction is taken to be that of the transported vector which yields the brightness which is closest to the expected value. The method is evaluated on a variety of real-world images where it provides promising results.

## References

1. Angelopoulou, E. (1999) Gaussian Curvature from Photometric Scatter Plots, *IEEE Workshop on Photometric Modelling for Computer Vision and Graphics.*
2. Belhumeur, P.N. and Kriegman, D.J. (1996) What is the Set of Images of an Object Under All Possible Lighting Conditions? *Proc. IEEE Conference on Computer Vision and Pattern Recognition,* pp. 270-277.
3. Bichsel, M. and Pentland, A.P. (1992) A Simple Algorithm for Shape from Shading, *Proc. IEEE Conf. on Computer Vision and Pattern Recognition,* pp. 459-465.
4. Breton, P., Iverson, L.A., Langer, M.S. and Zucker, S.W. "Shading Flows and Scenel Bundles: A New Approach to Shape from Shading", ECCV92, (135-150).
5. Ferrie, F.P. and Lagarde, J. (1990)Curvature Consistency Improves Local Shading Analysis, *Proc. IEEE International Conference on Pattern Recognition,* Vol. I, pp. 70-76.
6. Horn, B.K.P. and Brooks, M.J. (1986) The Variational Approach to Shape from Shading, *CVGIP,* Vol. 33, No. 2, pp. 174-208.
7. Horn, B.K.P. (1990) Height and Gradient from Shading, *IJCV,* Vol. 5, No. 1, pp. 37-75.
8. Huggins, P.S., Chen, H.F., Belhumeur, P.N.,and Zucker, S.W., "Finding Folds: On the Appearance and Identification of Occlusion", CVPR01, (II:718-725).

9.  Huggins, P.S., and Zucker, S.W., "Folds and Cuts: How Shading Flows Into Edges", ICCV01 (II: 153-158).
10. Ikeuchi, K. and Horn, B.K.P. (1981) Numerical Shape from Shading and Occluding Boundaries, *Artificial Intelligence*, Vol. 17, No. 3, pp. 141-184.
11. Kimmel, R. and Bruckstein, A.M. (1995) Tracking Level-sets by Level-sets: A Method for Solving the Shape from Shading Problem, *CVIU*, Vol. 62, No. 1, pp. 47-58.
12. Kimmel R., Siddiqi K., Kimia B.B., and Bruckstein A.M., "Shape from shading - Level set propagation and viscosity solutions", Int J. Computer Vision Vol 16, No 2, pp. 107-133, 1995.
13. Koenderink, J.J. (1990) *Solid Shape*, MIT Press, Cambridge MA.
14. Koenderink, J.J., van Doorn, A.J. and Kappers, A.M.L. (1992) Surface Perception in Pictures, *Perception and Psychophysics*, Vol. 52, No. 5, pp. 487-496.
15. Oliensis, J. and Dupuis, P. (1994) An Optimal Control Formulation and Related Numerical Methods for a Problem in Shape Reconstruction,*Ann. of App. Prob.* Vol. 4, No. 2, pp. 287-346.
16. Rouy E. and Tourin A., "A viscosity solutions approach to shape-from-shading", SIAM J. Numerical Analysis, vol 29, No 3, pp. 867-884, 1992.
17. Sander P.T. and Zucker S.W., "Inferring surface structure and differential structure from 3D images", IEEE PAMI, Vol. 12, pp. 833-854, 1990.
18. Sander P.T. and Zucker S.W., "Singularities in the principal direction field from 3D images", IEEE PAMI, Vol. 14, pp. 309–317, 1992.
19. Wilson, R. and Hancock., "Consistent topographic surface labeling", *PR*, 32(7), pp. 1211–1223, 1999
20. Worthington, P.L. and Hancock, E.R. (1998) Needle Map Recovery using Robust Regularizers, *Image and Vision Computing*, Vol. 17, No. 8, pp. 545-559.
21. Worthington, P.L. and Hancock, E.R. (1999) New Constraints on Data-closeness and Needle-map consistency for SFS, *IEEE Transactions on Pattern Analysis*, Vol. 21, pp. 1250-1267.
22. Zhang R., Tsai P.S., Cryer J.E. and Shah M., "Shape-from-shading: A survey", *IEEE Transactions on Pattern Analysis and Machine Intelligence*, Vol 21, pp. 690-706, 1999.

# A Graph-Spectral Method
# for Surface Height Recovery

Antonio Robles-Kelly and Edwin R. Hancock

Dept. of Comp. Science
University of York,York YO1 5DD, UK
{arobkell,erh}@cs.york.ac.uk

**Abstract.** This paper describes a graph-spectral method for 3D surface integration. The algorithm takes as its input a 2D field of surface normal estimates, delivered, for instance, by a shape-from-shading or shape-from-texture procedure. We commence by using the Mumford-Shah energy function to obtain transition weights for pairs of sites in the field of surface normals. The weights depend on the sectional curvature between locations in the field of surface normals. This curvature may be estimated using the change in surface normal direction between locations. We pose the recovery of the integration path as that of finding a path that maximises the total transition weight. To do this we use a graph-spectral seriation technique. By threading the surface normals together along the seriation path, we perform surface integration. The height increments along the path are simply related to the traversed path length and the slope of the local tangent plane. The method is evaluated on needle-maps delivered by a shape-from-shading algorithm applied to real world data and also on synthetic data. The method is compared with the height reconstruction method of Bichsel and Pentland.

## 1   Introduction

Surface integration is a process that provides the means of converting a field of surface normals (i.e. the projection of the Gauss map of a surface from a unit sphere onto an image plane) into an explicit 3D surface representation. This problem of reconstructing the surface from its Gauss map arises when attempting to infer explicit surface structure from the output of low-level vision modules such as shape-from-shading and shape-from-texture [1]. The process involves selecting a path through the surface normal locations. This may be done using either a curvature minimizing path or by advancing a wavefront from the occluding boundary or singular points. By traversing the path, the surface may be reconstructed by incrementing the height function using the known distance travelled and the local slope of the surface tangent plane.

The analysis of the literature on the topic is not a straightforward task. The reason for this is that surface recovery is frequently viewed as an integral part of the shape-from-shading or shape-from-texture process. However, with this caveat in mind we briefly discus some of the available algorithms. In the original

M.J. Wilson and R.R. Martin (Eds.): Mathematics of Surfaces 2003, LNCS 2768, pp. 163–181, 2003.

work on shape-from-shading by Horn and Brooks [2,3] the surface height recovery process is applied as a post-processing step. The process proceeds from the occluding boundary and involves incrementing the surface height by an amount determined by the distance traversed and the slope angle of the local tangent plane. In some of the earliest work, Wu and Li [4] average the surface normal directions to obtain a height estimate. A more elegant solution is proposed by Frankot and Chellappa [5] who project the surface normals into the Fourier domain to impose integrability constraints and hence recover surface height. In the level-set method of Kimmel, Bruckstein, Kimmia and Siddiqi [6,7] surface reconstruction is incorporated as an integral component into the shape-from-shading process. They use the apparatus of level-set theory to simultaneously solve the image irradiance equation and recover the associated surface height-function under constraints provided by surface integrability. Leclerc and Bobick [8] have developed a direct numerical method for height recovery from shading information which uses curvature consistency constraints. Tsai and Shah [9] describe a fast surface height recovery method, which works well except at the locations of self-shadows and numerically singular points. Dupuis and Oliensis [10] have developed a method which involves propagation in the direction of the steepest gradient from singular points. A fast variant of this algorithm is described by Bichsel and Pentland [11] who compute the relative height of the surface with respect to the highest intensity point. Jones and Taylor [12] use a Gaussian scale space to perform coarse-to-fine height recovery. Several of these methods are described in more detail and are compared in the recent comprehensive review paper of Zhang, Tsai, Cryer and Shah [13].

On the other hand, one of the problems associated with surface integration methods which propagate the height from reference contours or singular points, is that these locations need to be identified in a stable manner. In this paper we describe an eigenvector method which uses graph spectral analysis to establish a natural order for the pixel sites. Hence, the method may be less sensitive to the chosen initialisation. Despite of the effectiveness of the eigenvector approach, there may be other ways in which this problem could be addressed. For instance, the velcro mesh described in [14] is a NURBS-like surface that attaches itself to surface normals, and could be adapted for the purposes of surface integration.

As mentioned earlier, our overall aim is to pose the recovery of the surface integration path in a graph-spectral setting [15,16]. Our starting point is to use the Mumford-Shah functional [17] to compute the curvature dependant elements of a transition weight matrix between sites in the field of surface normals. The transition weights depend on the change in surface normal direction and the distance between sites. The greater the difference in surface normal direction, i.e. the sectional curvature, the smaller the weight.

The aim is to recover the integration path that maximises the sum of curvature weights across the field of surface normals. This can be viewed as a problem of graph-seriation [18], which involves ordering the set of nodes in a graph in a sequence such that strongly correlated elements are placed next to one another. The seriation problem can be approached in a number of ways. Clearly the problem of searching for a serial ordering of the nodes, which maximally preserves the edge ordering is one of exponential complexity. As a result approximate solution methods have been employed. These involve casting the problem in an optimisation setting. Hence techniques such as simulated annealing and mean field annealing have been applied to the problem. It may also be formulated using semidefinite programming, which is a technique closely akin to spectral graph theory since it relies on eigenvector methods. However, recently a graph-spectral solution has been found to the problem. Atkins, Boman and Hendrikson [18] have shown how to use the leading eigenvector of the Laplacian matrix to sequence relational data. The method has been successfully applied to the consecutive ones problem and a number of DNA sequencing tasks. There is an obvious parallel between this method and the use of eigenvector methods to locate steady state random walks on graphs. However, in the case of a random walk the path is not guaranteed to encourage edge connectivity. The spectral seriation method on the other hand does impose edge connectivity constraints on the recovered path.

Unfortunately, the analysis of the seriation problem presented in the paper by Atkins, Bowman and Hendriksen [18] is not directly applicable to our surface reconstruction problem. We hence provide an analysis which shows how the problem of recovering the optimal seriation path corresponds to maximising a Rayleigh quotient. This in turn establishes the relationship between the seriation path and the leading eigenvector of the transition weight matrix. The sites visited by this path constitute a patch on the surface.

We use the leading eigenvector property to develop a recursive algorithm for surface height reconstruction. We commence by locating the leading eigenvector of the transition weight matrix. The magnitude order of the co-efficients can be used to define the integration path, while the outer product of the vector can be used to define an adjacency matrix for the patch. We perform surface height recovery by visiting the surface normals in the order defined by the integration path. As the path is traversed, then the height function is incremented by an amount determined by the slope of the local tangent plane. Once the most significant patch has been identified and reconstructed in this way, then the elements of the associated pixel sites in the transition weight matrix are set to zero. We then repeat the procedure described above to extract each patch in turn. This process is halted when the size of the remaining patches becomes insignificant.

The outline of this paper is follows. In Section 2, we present the differential geometry which underpins the construction of our curvature-based transition weight matrix. Section 3 explains how the leading eigenvector of the transition weight matrix can be used to construct an integration path. In Section 4 we describe how patches can be extracted from the field of surface normals. Section 5 outlines the geometry necessary to reconstruct the height function of the surface from this path and the raw needle-map information. Experiments on real world image data and simulation data are described in Section 6. Finally, Section 7 offers some conclusions and suggests directions for future investigation.

## 2   Transition Weights

Stated formally, our goal is the recovery of height information from a field of surface normals. The Gauss map of an oriented surface is found by translating the surface normals onto a unit sphere. The mapping of this sphere onto an image plane is the field of surface normals taken as input to our surface reconstruction method. From a computational standpoint the aim is to find a path on the image plane along which simple trigonometry may be applied to increment the estimated height function. To be more formal suppose that the surface under study is $S$ and that the field of surface normals is mapped onto the plane $\Pi$. Our aim here is to find a curve $\Gamma$ across the plane $\Pi$ that can be used as an integration path to reconstruct the height-function of the surface $S$. The projection of the curve $\Gamma$ onto the surface $S$ is denoted by $\Gamma_S$. Further, suppose that $\kappa(s)$ is the sectional curvature of the curve $\Gamma_S$ at the point $Q$ with parametric co-ordinate $s$. Our recovery of the path $\Gamma_S$ is based on an analysis of the Mumford-Shah [17] functional

$$\mathcal{E}(\Gamma_S) = \int_{\Gamma_S} \left\{ \alpha + \beta\kappa(s)^2 \right\} ds \tag{1}$$

where $\alpha$ and $\beta$ are constants.

For the surface $S$ and the image plane plane $\Pi$ the field of unit surface normals is denoted by $\mathbf{N}$. Accordingly, and following do Carmo [19], we let $T_Q(S)$ represent the tangent plane to the surface $S$ at the point $Q$ which belongs to the curve $\Gamma_S$. To compute the sectional curvature $\kappa(s)$ we require the differential of the surface or Hessian matrix $d\mathbf{N}_Q : T_Q(S) \rightarrow T_Q(S)$. The maximum and minimum eigenvalues $\lambda_1$ and $\lambda_2$ of $d\mathbf{N}_Q$ are the principal curvatures at the point $Q$. The corresponding eigenvectors $\mathbf{e}_1 \in T_Q(S)$ and $\mathbf{e}_2 \in T_Q(S)$ form an orthogonal basis on the tangent plane $T_Q(S)$. At the point $Q$ the unit normal vector to the curve $\Gamma$ is $\mathbf{n}_\Gamma$ and the unit tangent vector is $t_Q \in T_Q(S)$. The sectional curvature of $\Gamma$ at $Q$ is given by

$$\kappa(s) = \frac{(\mathbf{t}_Q.\mathbf{e}_1)^2(\lambda_1 - \lambda_2) + \lambda_2}{\mathbf{n}_\Gamma.\mathbf{N}_Q} \tag{2}$$

where $(\mathbf{t}_Q.\mathbf{e}_1)^2(\lambda_1 - \lambda_2) + \lambda_2$ is the normal curvature and $\psi = \arccos \mathbf{n}_\Gamma.\mathbf{N}_Q$ is the angle between the curve normal and the surface normal.

In practice, we will be dealing with points which are positioned at discrete positions on the pixel lattice. Suppose that $i$ and $j$ are the pixel indices of neighbouring points sampled on the pixel lattice along the path $\Gamma_S$. With this discrete notation, the cost associated with the path is given by

$$\mathcal{E}(\Gamma_S) = \sum_{(i,j)\in\Gamma_S} \left\{ \alpha + \beta\kappa_{i,j}^2 \right\} s_{i,j} \tag{3}$$

where $\kappa_{i,j}$ is an estimate of the curvature based on the surface normal directions at the pixel locations $i$ and $j$, and $s_{i,j}$ is the path distance between these points. The energy associated with the transition between sites $i$ and $j$ is $\mathcal{E}_{i,j} = \{\alpha + \beta\kappa_{i,j}^2\}s_{i,j}$.

In order to compute the path curvature appearing in the expression for the transition energy, we make use of the surface normal directions. To commence, we note that $|\kappa_{i,j}| = \frac{1}{R_{i,j}}$, where $R_{i,j}$ is the radius of the local circular approximation to the integration curve on the surface. Suppose that the surface normal directions at the pixel locations $i$ and $j$ are respectively $\boldsymbol{N}_i$ and $\boldsymbol{N}_j$. The approximating circle connects the points $i$ and $j$, and has the path segment $s_{i,j}$ as the connecting chord. The change in direction of the radius vector of the circle is $\theta_{i,j} = \arccos \boldsymbol{N}_i.\boldsymbol{N}_j$, and hence $\cos \theta_{i,j} = \boldsymbol{N}_i.\boldsymbol{N}_j$. If the angle $\theta_{i,j}$ is small, then we can make the Maclaurin approximation $\cos \theta_{i,j} = \boldsymbol{N}_i.\boldsymbol{N}_j \simeq 1 - \frac{\theta_{i,j}^2}{2}$. Moreover, the small angle approximation to the radius of curvature of the circle is $R_{i,j} = \frac{s_{i,j}}{\theta_{i,j}}$. Hence, $\kappa_{i,j}^2 = \frac{2(1-\boldsymbol{N}_i.\boldsymbol{N}_j)}{s_{i,j}^2}$. The geometry outlined above is illustrated in Figure 1a.

As a result, we find that the cost associated with the step from the pixel $i$ to the pixel $j$ is

$$\mathcal{E}_{i,j} = \alpha s_{i,j} + \frac{2\beta}{s_{i,j}}(1 - \boldsymbol{N}_i.\boldsymbol{N}_j) \tag{4}$$

The total cost associated with the integration path $\Gamma_S$ is hence

$$\mathcal{E}_{\Gamma_S} = \alpha L_{\Gamma_S} + \sum_{(i,j)\in\Gamma_S} \frac{2\beta}{s_{i,j}}(1 - \boldsymbol{N}_i.\boldsymbol{N}_j) \tag{5}$$

where $L_{\Gamma_S}$ is the length of the path. Hence, the integration path is a form of elastica which attempts to find an energy minimising path through the field of surface normals. The energy function is a variant of the Mumford-Shah functional. It has two terms. The first encourages the integration path to be one of minimum length. The second term encourages a path which minimises the total change in surface normal direction.

With the energy function to hand, we could attempt to find the integration path $\Gamma_S$ that minimises the Mumford-Shah functional, i.e. the one that satisfies the condition $\Gamma_S = \arg\min_{\hat{\Gamma}} \mathcal{E}(\hat{\Gamma})$. There are clearly a number of ways in which the energy can be minimised. These might include expectation-maximisation [20], relaxation labelling [21] and stochastic methods [22]. However, here we choose to make use of a graph-spectral method to perform seriation on the transition weight matrix.

To pursue the graph-spectral analysis of the field of surface normals, we require a transition weight matrix which reflects the connectivity of the pixel lattice. For the pixels indexed $i$ and $j$ we define the transition weight matrix to have elements

$$W(i,j) = \begin{cases} \exp(-\mathcal{E}_{i,j}) & \text{if } j \in \mathcal{N}_i \\ 0 & \text{otherwise} \end{cases} \tag{6}$$

where $\mathcal{N}_i$ is the set of pixels-neighbours of the pixel $i$. Hence, the curvature weight is only non-zero if pixels abut one-another.

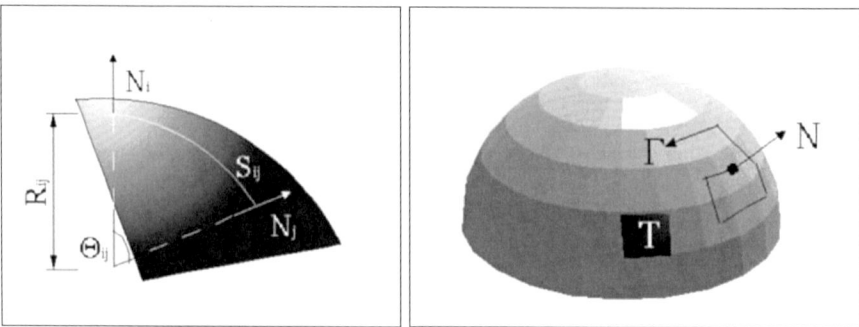

**Fig. 1.** Illustration of the curvature computation and the integration path (see text for details).

## 3   Graph Seriation

In the previous section, we used the Mumford-Shah functional to derive the transition weights between sites on the field of surface normals. The weights depend on the sectional curvature of the path on the surface. We have shown how the differential geometry of the field of surface normals can be used to compute the required curvatures. In this section, we describe how the leading eigenvector of the transition weight matrix can be used to locate an integration path that maximises the total curvature weight. We pose this as a process of graph-spectral seriation.

To commence, we pose the problem in a graph-based setting. The set of pixel sites can be viewed as a weighted graph $G = (V, E, W)$ with index-set $V$, edge-set $E = \{(i, j) | (i, j) \in V \times V, i \neq j\}$ and weight function $W : E \to [0, 1]$. Let the curvature minimising path commence at the node $j_1$ and proceed via the sequence of edge-connected nodes $\Gamma = \{j_1, j_2, j_3, ...\}$ where $(j_i, j_{i-1}) \in E$. Further, we suppose that the transition weight matrix $W(j_i, j_{i+1})$ associated with the move between the nodes $j_i$ and $j_{i+1}$ can be regarded as a pairwise similarity measure. With these ingredients, the problem of finding the path that minimises the curvature between adjacent pixel-sites can be viewed as one of seriation, subject to edge connectivity constraints.

As noted by Atkins, Boman and Hendrikson [18], many applied computational problems, such as sparse matrix envelope reduction, graph partitioning and genomic sequencing, involve ordering a set according to a permutation $\pi = \{\pi(j_1), \pi(j_2), ..., \pi(j_{|V|})\}$ so that strongly related tokens are placed next to one another. The seriation problem is that of finding the permutation $\pi$ that satisfies the condition

$$\pi(j_i) < \pi(j_k) < \pi(j_l) \Rightarrow \{W(i, k) \geq W(i, l) \wedge W(k, l) \geq W(i, l)\}$$

This task has been posed as a combinatorial optimisation problem which involves minimising the penalty function

$$g(\pi) = \sum_{i=1}^{|V|} \sum_{k=1}^{|V|} W(i,k) \left(\pi(j_i) - \pi(j_k)\right)^2$$

for a symmetric, real transition weight matrix $W$.

Unfortunately, the penalty function $g(\pi)$, as given above, does not impose edge connectivity constraints on the ordering computed during the minimisation process. Furthermore, it implies no directionality in the transition from the node indexed $j_i$ to the one indexed $j_{i+1}$. To overcome these shortcomings, we turn our attention instead to the penalty function

$$g(\pi) = \sum_{i=1}^{|V|-1} W(i,i+1) \left(\pi(j_i) - \pi(j_{i+1})\right)^2 \tag{7}$$

where the nodes indexed $j_i$ and $j_{i+1}$ are edge connected. After some algebra, it is straightforward to show that

$$g(\pi) = \sum_{i=1}^{|V|-1} W(i,i+1)(\pi(j_i)^2 + \pi(j_{i+1})^2) - 2 \sum_{i=1}^{|V|-1} W(i,i+1)\pi(j_i)\pi(j_{i+1}) \tag{8}$$

It is important to note that $g(\pi)$ does not have a unique minimiser. The reason for this is that its value remains unchanged if we add a constant amount to each of the co-efficients of $\pi$. We also note that it is desirable that the minimiser of $g(\pi)$ is defined up to a constant $\lambda$ whose solutions are polynomials in $W$. Therefore, we subject the minimisation problem to the constraints

$$\lambda\pi(j_i)^2 = \sum_{k=1}^{|V|} W(k,i)\pi(j_k)^2 \text{ and } \sum_{k=1}^{|V|} \pi(j_k)^2 \neq 0 \tag{9}$$

Since the co-efficients $\pi(j_{i+1})$ are inversely proportional to $\lambda - W(i+1,i)$, the co-efficient $\pi(j_{i+1})^2$ increase with decreasing sectional curvature (i.e. the similarity tends to one). The effect of this is to enforce edge connectivity while favouring paths of small local curvature, and also to minimise the overall cost of the path.

Combining the constraint conditions given in Equation 9 with the definition of the penalty function given in Equation 8, it is straightforward to show that the permutation $\pi$ satisfies the condition

$$\sum_{k=1}^{|V|} \sum_{i=1}^{|V|-1} \left(W(k,i) + W(k,i+1)\right)\pi(j_k)^2 = \lambda \sum_{i=1}^{|V|-1} (\pi(j_i)^2 + \pi(j_{i+1})^2) \tag{10}$$

Using matrix notation, we can write the above equation in the more compact form $\Omega W \phi = \lambda \Omega \phi$, where $\phi = \{\pi(j_1)^2, \pi(j_2)^2, \ldots, \pi(j_{|V|})^2\}^T$ and $\Omega$ is the $(N-1) \times N$ matrix

$$\Omega = \begin{bmatrix} 1 & 1 & 0 & \ldots & 0 \\ 0 & 1 & 1 & \ddots & \vdots \\ \vdots & \ddots & \ddots & \ddots & 0 \\ 0 & \ldots & 0 & 1 & 1 \end{bmatrix} \tag{11}$$

Hence it is clear that locating the permutation $\pi$ that minimises $g(\pi)$ can be posed as an eigenvalue problem, and that $\phi$ is an eigenvector of $W$. This follows from the fact Equation 3 can be obtained by multiplying both sides of the eigenvector equation $W\phi = \lambda\phi$ by $\Omega$. Furthermore, due to the norm condition of the eigenvector, the constraint $\sum_{k=1}^{|V|} \pi(j_k)^2 \neq 0$ is always satisfied. Taking this analysis one step further, we can premultiply both sides of Equation 3 by $\phi^T$ to get the matrix equation $\phi^T \Omega W \phi = \lambda \phi^T \Omega \phi$. As a result, it follows that

$$\lambda = \frac{\phi^T \Omega W \phi}{\phi^T \Omega \phi} \tag{12}$$

This expression is reminiscent of the Rayleigh quotient.

We note that the elements of the permutation $\pi$ are required to be real. Consequently, the co-efficients of the eigenvector $\phi$ are always non-negative. Since the elements of the matrices $\Omega$ and $W$ are positive, it follows that the quantities $\phi^T \Omega W \phi$ and $\phi^T \Omega \phi$ are positive. Hence, the set of solutions reduces itself to those that are determined up to a constant $\lambda > 0$. As a result, the co-efficients of the eigenvector $\phi$ are linearly independent of the all-ones vector $\mathbf{e} = (1, 1...., 1)^T$.

With these observations in mind, we focus on proving the existence of a permutation that minimises $g(\pi)$ subject to the constraints in Equation 9 and demonstrating that this permutation is unique. To this end we use the Perron-Frobenius theorem [15]. This concerns the proof of existence regarding the eigenvalue $\lambda_* = \max_{i=1,2,...,|V|}\{\lambda_i\}$ of a primitive, real, non-negative, symmetric matrix $W$, and the uniqueness of the corresponding eigenvector $\phi_*$. The Perron-Frobenius theorem states that the eigenvalue $\lambda_* > 0$ has multiplicity one. Moreover, the co-efficients of the corresponding eigenvector $\phi_*$ are all positive and the eigenvector is unique. As a result the remaining eigenvectors of $W$ have at least one negative co-efficient and one positive co-efficient. If $W$ is substochastic (i.e. $W(k, i) \geq 0$ for all $k, i \in V$ and $\sum_{i=1}^{|V|} W(k, i) \geq 1$ for all $k \in V$, with strict inequality for at least one $k$), $\phi_*$ is also known to be linearly independent of the all-ones vector $\mathbf{e}$ [23,15]. As a result, the leading eigenvector of $W$ is the minimiser of $g(\pi)$.

The elements of the leading eigenvector $\phi_*$ can be used to construct an integration path. As noted earlier, the components of $\phi_*$ decrease with increasing curvature of the seriation path. We commence from the node associated with the largest component of $\phi_*$. We then sort the elements of the leading eigenvector such that they are both in the decreasing magnitude order of the co-efficients of the eigenvector, and satisfy neighbourhood connectivity constraints on the pixel lattice. The procedure is a recursive one that proceeds as follows. At each iteration, we maintain a list of sites visited. At iteration $k$, let the list of sites be denoted by $\mathcal{L}_k$. Initially $\mathcal{L}_0 = j_o$, where $j_0 = \arg\max_j \phi_*(j)$ (i.e. $j_0$ is the component of $\phi_*$ with the largest magnitude). Next, we search through the set of 8-neighbours of $j_o$ to find the pixel associated with the largest remaining component of $\phi_*$. If $\mathcal{N}_{j_o}$ is the set of 8-neighbours of $j_0$, the second element in the list is $j_1 = \arg\max_{l \in \mathcal{N}_{j_0}} \phi_*(l)$. The pixel index $j_1$ is appended to the list of sites visited and the result is $\mathcal{L}_1$. In the $k$th (general) step of the algorithm, we are at the pixel-site indexed $j_k$ and the list of sites visited by the path so far is $\mathcal{L}_k$. We search through those 8-neighbours of $j_k$ that have not already been traversed by the path. The set of candidate pixel sites is $C_k = \{l | l \in \mathcal{N}_{j_k} \wedge l \notin \mathcal{L}_k\}$. The next site to be appended to the path list is therefore $j_{k+1} = \arg \max_{l \in C_k} \phi_*(l)$. This process is repeated until no further moves can be made. This occurs when $C_k = \emptyset$ and we denote the index of the termination of the path by $T$. The integration path $\Gamma_S$ is given by the list of pixel sites $\mathcal{L}_T$.

## 4    Extracting Patches

In practice the surface under study may have a patch structure. The patches may be identified by finding the blocks of the transition weight matrix induced under a permutation of the nodes. We find the blocks by computing the leading eigenvector of the transition weight matrix. The algorithm proceeds in an iterative fashion. The leading eigenvector of the current transition weight matrix represents a patch. The nodes with non-zero components in the leading eigenvector belong to the patch. The nodes are identified, and are then removed from further consideration by nulling their associated elements in the transition weight matrix. This process is repeated until all the principal patches are identified. This is the case when only an insignificant number of unassigned and unconnected nodes reamin.

We commence by constructing the thresholded transition weight matrix $A$ whose elements are defined as follows

$$A(i,j) = \begin{cases} 0 & \text{if } W(i,j) << 1 \\ P(i,j) & \text{otherwise} \end{cases} \tag{13}$$

The matrix $A$ is simply a thresholded version of the transition weight matrix $W$ in which the vanishingly small elements are set to zero.

Our aim is identify groups of surface normals from a potentially noisy or ambiguous adjacency matrix $A$ which correspond to surface patches. Stated formally, suppose that in an image with an adjacency matrix $A$ there are $m$ disjoint patches. Each such group should appear as a sub-block of the matrix $A$. However, as a consequence of noise or errors in the shape-from-shading method which delivers the field of surface normals, these distinct groups or patches may be merged together. In other words, their corresponding sub-blocks are no longer disjoint.

Suppose that there are $m$ distinct surface patches, each associated with an adjacency matrix $B^{(i)}$ where $i$ is the patch index. If $C$ represents a noise matrix, then the relationship between the observed transition weight matrix $A$ and the underlying block-structured transition weight matrix is $A = B + C$.

To recover the matrix $B$, we turn to the eigenvector expansion of the matrix $A$ and write

$$A = \phi_* \phi_*^T + \sum_{i=2}^{|V|} \lambda_i \phi_i \phi_i^T \tag{14}$$

where the leading eigenvalue is unity i.e. $\lambda_1 = 1$, $\phi_*$ is the leading eigenvector and the eigenvectors are normalised to be of unit length, i.e. $|\phi_i| = 1$. To identify the

patches, we use the following iterative procedure. We initialise the algorithm by
letting $A^{(1)} = A$. Further suppose that $\phi_*^{(1)}$ is the leading eigenvector of $A^{(1)}$. The
matrix $B^{(1)} = \phi_*^{(1)}\phi_*^{(1)T}$ represents the first block of $A$, i.e. the most significant
surface patch. The nodes with non-zero entries belong to the patch. These nodes
may be identified and removed from further consideration. To do this we compute
the residual transition weight matrix $A_U^{(2)} = A^{(1)} - B^{(1)}$ in which the elements of
the first patch are nulled. The leading eigenvector $\phi_*^{(2)}$ of the residual transition
weight matrix $A^{(2)}$ is used to compute the second block $B^{(2)} = \phi_*^{(2)}\phi_*^{(2)T}$. The
process is repeated iteratively to identify all of the principal blocks of $A$. At
iteration $k$, $\phi_*^{(k)}$ is the leading eigenvector of the normalised residual transition
weight matrix $A^{(k)}$, and the $k^{th}$ block is $B^{(k)} = \phi_*^{(k)}\phi_*^{(k)T}$. The patch indexed
$n$ is the set of nodes for which the components of the leading eigenvector $\phi_*^{(k)}$
are non-zero. Hence, the index-set for the $k^{th}$ patch is $S_k = \{i|\phi_*^{(k)}(i) \neq 0\}$. It is
important to stress that the patches are non-overlapping, i.e. the inner product
of the block eigenvectors for different patches is zero $\boldsymbol{b}_*^{(k)}.\boldsymbol{b}_*^{(l)} = 0$, where $k \neq l$.

## 5    Geometry

Our surface height recovery algorithm proceeds along the sequence of pixel sites
defined by the order of the co-efficients of the leading eigenvector associated
with the separate patches. For the $k^{th}$ patch, the path is $\Gamma_S^k = (j_1^k, j_2^k, j_3^k, \dots)$
where the order is established using the method outlined at the end of Section
3. As we move from pixel-site to pixel-site defined by this path we increment
the surface height-function. In this section, we describe the trigonometry of the
height incrementation process.

At step $n$ of the algorithm for the path indexed $k$, we make a transition from
the pixel with path-index $j_{n-1}^k$ to the pixel with path-index $j_n^k$. The distance
between the pixel-centres associated with this transition is

$$d_n = \sqrt{(x_{j_n^k}^2 - x_{j_{n-1}^k})^2 + (y_{j_n^k} - y_{j_{n-1}^k})^2} \tag{15}$$

This distance together with the surface normals $\boldsymbol{N}_{j_n^k}$ and $\boldsymbol{N}_{j_{n-1}^k}$ at the two
pixel-sites may be used to compute the change in surface height associated with
the transition. The height increment is given by

$$h_n = \frac{d_n}{2} \left\{ \frac{\boldsymbol{N}_{j_n^k}(x)}{\boldsymbol{N}_{j_n^k}(y)} + \frac{\boldsymbol{N}_{j_{n-1}^k}(x)}{\boldsymbol{N}_{j_{n-1}^k}(y)} \right\} \tag{16}$$

If the height-function is initialised by setting $z_{j_0^k} = 0$, then the centre-height for
the pixel with path-index $j_n^k$ is $z_{j_{n+1}^k} = z_{j_n^k} + h_n$. The geometry of this procedure
is illustrated in Figure 1b.

Once the surface normals that belong to the individual patches have been
integrated together, then we merge them together to form a global surface. Suppose that $S_k$ is the integrated surface for the $k^{th}$ patch. We compute the mean

height for the pixels belonging to this boundary. We merge the patches together by ensuring that abutting patches have the same mean boundary height along their shared perimeter. However, we acknowledge that this is crude and could result in poor patch alignment.

# 6   Experiments

The experimental evaluation of the new surface reconstruction method is divided into two parts. We commence with a sensitivity study aimed at evaluating the method on synthetic data. In the second part of the study, we focus on synthetic and real-world imagery.

At this point, it is worth pausing to note the following. First, the computation times reported throughout this section correspond to those of a 2Ghz, dual-Xeon workstation. Second, for all our experiments the constants $\alpha, \beta$ are set to unity and a threshold is used to separate the background from the foreground. As a result, only pixel-sites that have a normalised brightness greater or equal to 0.3 are considered for the height recovery process. Finally, we have normalised the surface normals in order to satisfy the condition $\mid \boldsymbol{N}_{j_n^k} \mid = 1$.

## 6.1   Performance Analysis

The aim in this section is to determine the accuracy of the surface reconstruction method. To this end we have generated synthetic surfaces. From the surfaces, we have computed the field of surface normal directions. We have then applied the graph-spectral method to the 2D field of surface normals to recover an estimate of the surface height.

In Figure 2 we show the results obtained for a series of different surfaces. In the left-hand column we show the original synthetic surface. The middle column shows the surface reconstructed from the field of surface normals. The right-hand column shows the absolute error between the ground-truth and reconstructed surface height. From top-to-bottom the surfaces studied are a dome, a sharp ridge, a torus and a volcano. In all four cases the surface reconstructions are qualitatively good. For the dome the height errors are greater at the edges of the surface where the slope is largest. In the case of the ridge, there are errors at the crest. For the volcano, there are some problems with the recovery of the correct depth of the "caldera", i.e. the depression in the centre. For the reconstructed surfaces, the mean-squared errors are 5.6% for the dome, 10.8% for the ridge, 7.8% for the torus and 4.7% for the volcano. Hence, the method seems to have greater difficulty for surfaces containing sharp creases.

We have repeated these experiments under conditions of controlled noise. To do this we have added random measurement errors to the surface normals. To do this, we sample randomly error vectors from a circularly symmetric 2D Gaussian distribution with zero mean and known variance. In the worst case, i.e. when the variance is equal to unity, the mean absolute difference between the ground truth surface normal and the noisy surface normal direction is 48.9 degrees. In the left-hand column of Figure 3, we show the field of noise-free surface

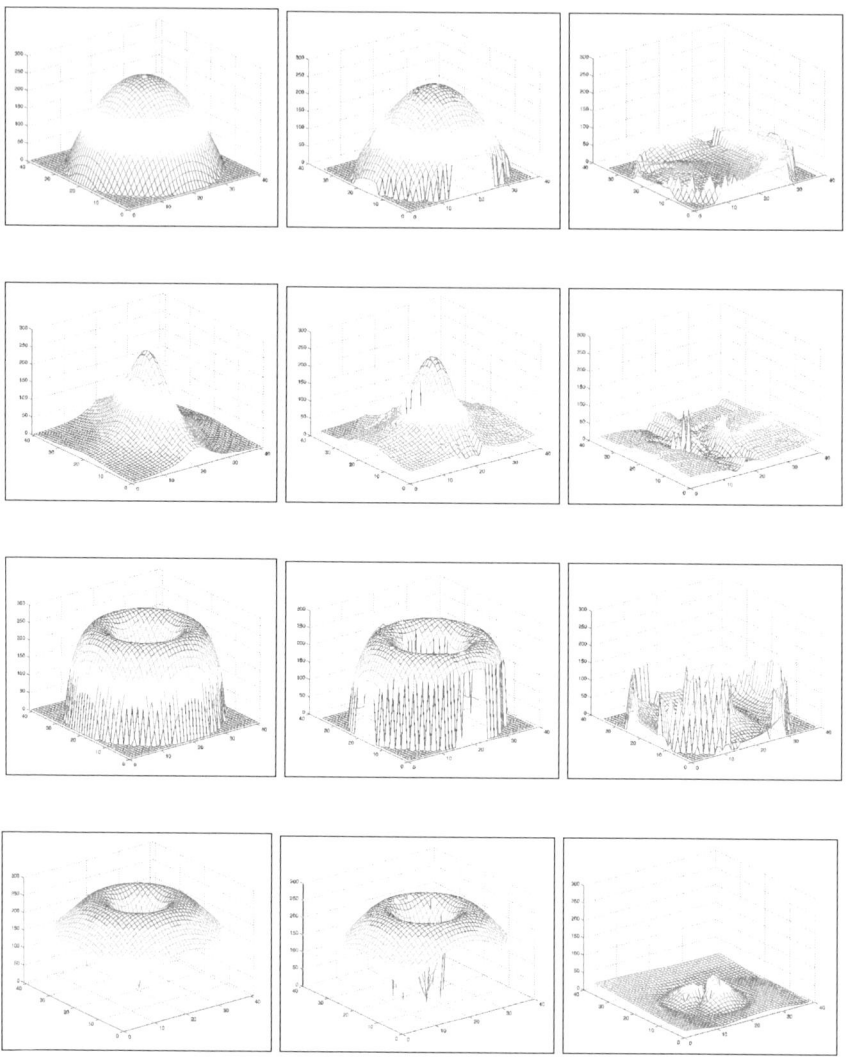

**Fig. 2.** Left-hand column: Artificially generated data; Middle Column: Reconstructed surface; Right-hand column: Error plot.

normals. In the second column, we show the noise-corrupted field of surface normals. In the third column, we show the reconstructed height-function obtained from the noisy surface normals. The fourth, i.e. rightmost, column shows the difference between the height of the surface reconstructed from the noisy surface normals and the ground-truth height function. In the case of all four surfaces, the gross structure is maintained. However, the recovered height is clearly noisy. The height difference plots are relatively unstructured. These are important observations. They mean that our graph-spectral method simply transfers errors

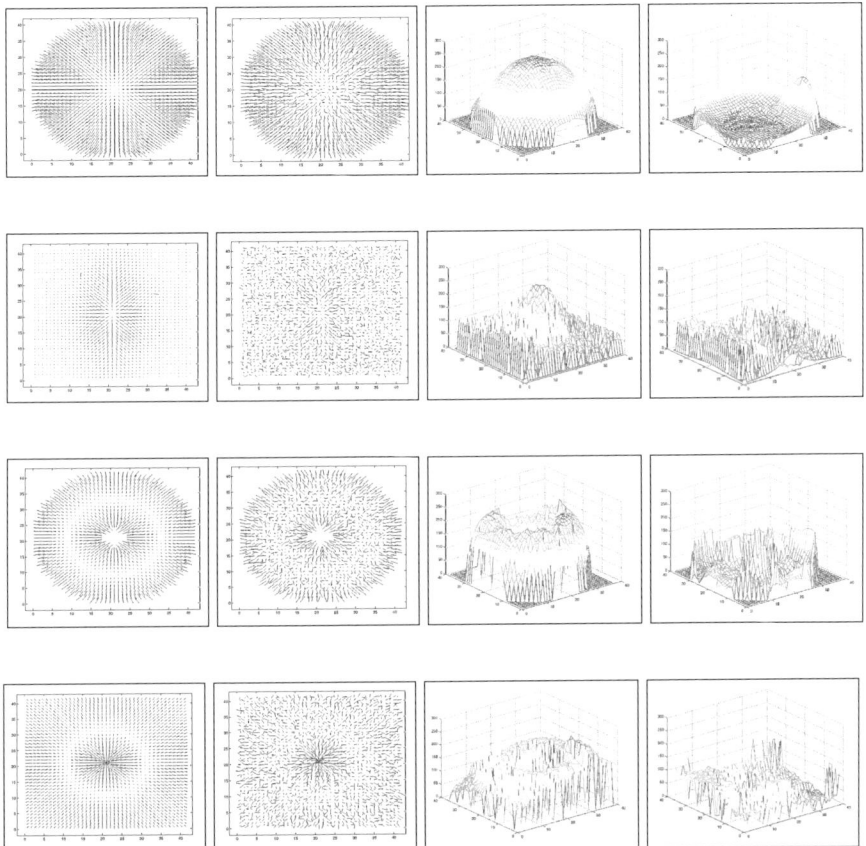

**Fig. 3.** Left-hand column: Needle-map without added noise; Second Column: Needle-map with Gaussian noise added (worst case with variance=1); Third column: Reconstructed surface; Right-hand column: Error plot.

in surface normal direction into errors in height, without producing structural noise artefacts. However, there are some large errors on the surface which may be attributed to poor patch alignment.

To investigate the effect of noise further, we plot the mean-squared error for the reconstructed surface height as a function of the standard deviation of the added Gaussian noise in Figure 4. From the plots for the different surfaces shown in Figure 3, it is clear that the mean-squared error grows slowly with increasing noise standard deviation.

## 6.2   Synthetic and Real-World Imagery

The second part of our experimental evaluation of the new surface height recovery focusses on real-world imagery. Here we have applied our surface recovery

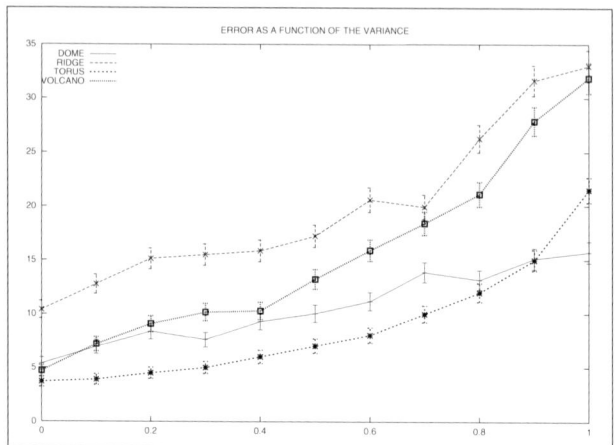

**Fig. 4.** Surface reconstruction error versus noise variance.

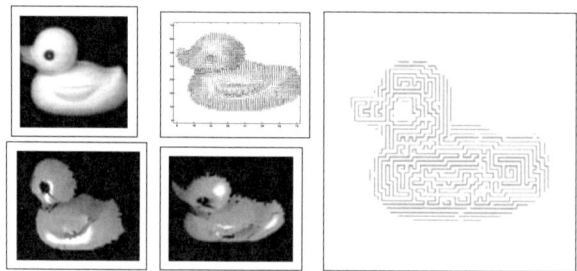

**Fig. 5.** Results on real-world imagery.

method to needle-maps extracted from real-world data using the shape-from-shading algorithm of Worthington and Hancock [24]. It should be stressed, however, that the method can be used in any situation where surface-data is presented in the form of a field of surface normals sampled on a plane. Hence, it can also be used in conjunction with shape-from-texture and motion analysis.

In Figure 5, we show our first example. The image used in this study is of a toy duck and has been taken from the Columbia University Object Image Library. Here the two panels in the top row, shows the raw image (left) and the extracted field of surface normals (right). Two views of the reconstructed surface are shown in the bottom row. The main features to note here are the well-formed beak, the spherical shape of the head, the well-defined cusp between the head and body, and, the saddle structure near the tail. However, the method fails around the wing detail. This is due to problems with the original needle-map delivered by the shape-from-shading algorithm, which has mistaken the convex wing as a concave object.

In far right-hand panel of the figure, we show the integration path, i.e. the order of the components of the leading eigenvector, for each detected patch (for

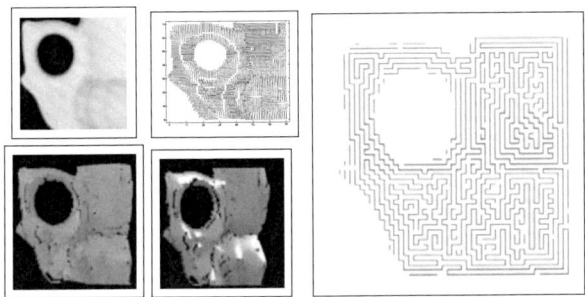

**Fig. 6.** Results on real-world imagery.

the last two images only the upper right corner detail is shown). These paths appear to follow the main height contours on the surface patches. The main patches correspond to the head, the beak, the body, the wing and the tail.

Our second example is shown in Figure 6. This is a detail from a porcelain urn. Here both the convexities and concavities on the surface are well reconstructed. The principal surface patches correspond to the handle of the urn, the lower "bulge" and the cylindrical body of the urn. In addition, some of the ribbed detail on the surface of the urn is recovered.

Finally, Figures 7 and 8 show two images from the University of Central Florida data-base which is used in the recent comparative study of Zhang et al [13]. The images are of a vase and a bust of Beethoven. The overall shape of the vase is well reconstructed and the fine detail of the face of Beethoven is well reproduced.

In Table 1 we list, for the data described above, the number of foreground pixel-sites, the number of patches extracted, the fraction of reconstruction errors and the computation time. The reconstruction errors are identified by visual inspection of the recovered surface. The errors may be due to artificial height discontinuities or locations where the surface becomes inverted (i.e. concave regions become convex, or vice versa). From the table, it is clear that only about 7-9% of the surface area of the objects studied is affected by such errors. In the case of the toy duck, the errors are dominated by the inversion of the wing.

Finally, we have compared our algorithm with two alternatives. The first of these is the shape-from-shading height recovery method of Bichsel and Pentland [11]. This is a local scheme which, from Zhang et al [13], appears to perform reasonably well on a wide variety of images. This local technique implicitly recovers a surface. Hence, comparison upon the basis of surface reconstruction is possible without introducing additional errors. The second method used in our comparison is a purely geometric surface integration method, which like the technique reported in this paper takes the needle-maps delivered by the Worthington and Hancock algorithm as input. This method uses the trapezoid rule to increment the height from equal height reference contours. The method was developed in conjunction with a study of terrain reconstruction using radar shape-from-shading and a full description can be found in [25].

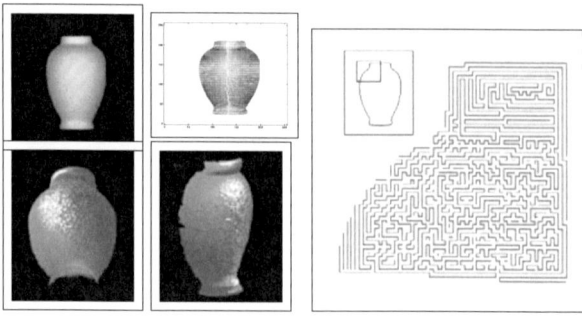

**Fig. 7.** Results on real-world imagery.

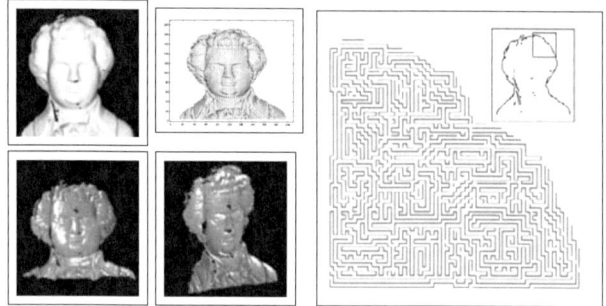

**Fig. 8.** Results on real-world imagery.

**Table 1.** Summary statistics for the graph spectral method.

| Image | No. of Foreground Pixel-sites | No. of Patches | % of Reconstruction Errors | Computation Time |
|---|---|---|---|---|
| Duck | 7512 | 60 | 7.3% | 34.5 sec. |
| Handle | 7648 | 145 | 7.8% | 36.7 sec. |
| Vase | 18256 | 76 | 2.8% | 184.4 sec. |
| Mozart | 21267 | 29 | 8.1% | 219.2 sec. |

Figures 9 and 10 respectively show views of the reconstructed surfaces obtained with the geometric integration method and the Bichsel and Pentland method. The images used are those already shown in Figures 5-8. Qualitatively, the most interesting point of comparison between the methods is the ability to recover discontinuities, and, the degree of distortion and over-smoothing. In Figure 10 the surface of the duck recovered by the Bichsel and Pentland algorithm is less detailed than that recovered using our graph-spectral method. For instance, the boundary of the neck and body is less well defined. Also most of the detail of the Beethoven bust is lost. The poorest result is obtained for the vase image. Here the handle has been distorted and over-smoothed. The results obtained by the geometric surface integration method are better than those obtained by

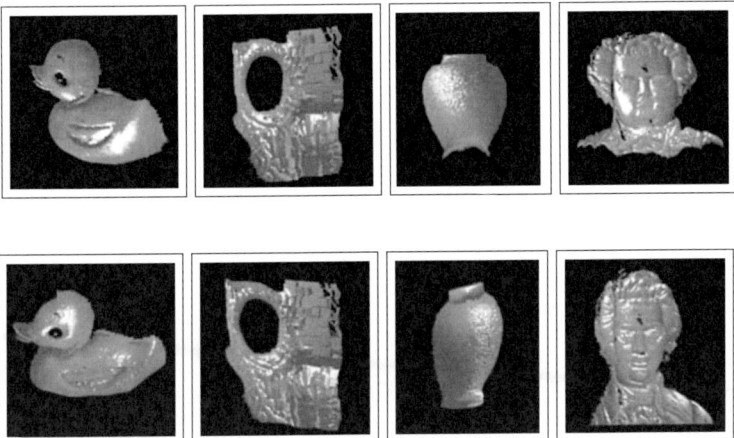

**Fig. 9.** Results of integrating the surface using a geometric approach.

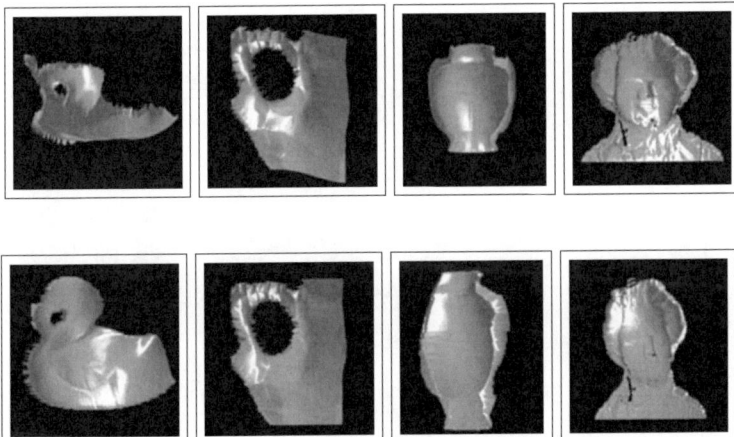

**Fig. 10.** Results of applying the Bichsel and Pentland algorithm.

Bichsel and Pentland, but do not contain the fine detail delivered by our graph spectral method. For instance, the hair and fabric details of Beethoven are not well reproduced and the rib structures on the vase are not well recovered. The fact that both integration methods outperform the Bichsel and Pentland method may be attributable to the quality of the input needle maps.

A more quantitative analysis is presented in Table 2. We list the image dimensions and the percentage of reconstruction errors for the respective surfaces. In contrast to our algorithm, the Bichsel and Pentland method delivers error rates which are as high as 48% and never less than 6%. In the case of the geometric integration method, the error rates are always in the range 7-27%.

**Table 2.** Summary statistics for the Bichsel and Pentland and the geometric integration algorithms.

| Image | Dimensions (pixels) | % of reconstruction errors (Bichsel and Pentland Algorithm) | % of reconstruction errors (Geometric Integration Algorithm) |
|-------|------------|----------------------|----------------------|
| Duck | 128x128 | 6.8% | 7.1% |
| Handle | 100x100 | 48% | 26.3% |
| Vase | 256x256 | 28.7% | 4.1% |
| Mozart | 256x256 | 37.2% | 11.1% |

# 7   Conclusions

In this paper, we have described a new surface height recovery algorithm. The method commences from a surface characterisation which is based on a transition weight matrix. The elements of this matrix are related to path curvature and are computed from the difference in surface normal direction. Based on a recently reported method for graph seriation, we use the leading eigenvector of the transition weight matrix to define an integration path. By traversing this path and applying some simple trigonometry we reconstruct the surface height function. Experiments on real world data reveal that the method is able to reconstruct quite complex surfaces. Noise sensitivity experiments on simulated data, show that the method is capable of accurate surface reconstruction.

# References

1. A. P. Rockwood and J. Winget. Three-dimensional object reconstruction from two dimensional images. *Computer-Aided Design*, 29(4):279–285, 1997.
2. B. K. P. Horn and M. J. Brooks. The variational approach to shape from shading. *CVGIP*, 33(2):174–208, 1986.
3. B. K. P. Horn and M. J. Brooks. Height and gradient from shading. *International Journal of Computer Vision*, 5(1):37–75, 1986.
4. Z. Wu and L. Li. A line-integration based method for depth recovery from surface normals. *CVGIP*, 43(1):53–66, July 1988.
5. R. T. Frankot and R. Chellappa. A method of enforcing integrability in shape from shading algorithms. *IEEE Transactions on Pattern Analysis and Machine Intelligence*, 4(10):439–451, 1988.
6. R. Kimmel, K. Siddiqqi, B. B. Kimia, and A. M. Bruckstein. Shape from shading: Level set propagation and viscosity solutions. *International Journal of Computer Vision*, (16):107–133, 1995.
7. R. Kimmel and A. M. Bruckstein. Tracking level sets by level sets: a method for solving the shape from shading problem. *Computer vision and Image Understanding*, 62(2):47–48, July 1995.
8. Y. G. Leclerc and A. F. Bobick. The direct computation of height from shading. In *Proceedings of Computer Vision and Pattern Recognition*, pages 552–558, 1991.
9. P. S. Tsai and M. Shah. Shape from shading using linear approximation. *Image and Vision Computing*, 12(8):487–498, 1994.
10. P. Dupuis and J. Oliensis. Direct method for reconstructing shape from shading. In *CVPR92*, pages 453–458, 1992.

11. M. Bichsel and A. P. Pentland. A simple algorithm for shape from shading. In *CVPR92*, pages 459–465, 1992.
12. A. G. Jones and C. J. Taylor. Robust shape from shading. *Image and Vision Computing*, 12(7):411–421, September 1994.
13. R. Zhang, Ping-Sing Tsai, J. E. Cryer, and M. Shah. Shape from shading: A survery. *IEEE Trans. on Pattern Analysis and Machine Intelligence*, 21(8):690–706, 1999.
14. W. Neuenschwander, P. Fua, G. Szekely, and O. Kubler. Deformable velcro surfaces. In *Proc. of the IEEE Int. Conf. on Comp. Vision*, pages 828–833, 1995.
15. R. S. Varga. *Matrix Iterative Analysis*. Springer, second edition, 2000.
16. L. Lovász. Random walks on graphs: a survey. *Bolyai Society Mathematical Studies*, 2(2):1–46, 1993.
17. D. Mumford and J. Shah. Optimal approximations by piecewise smooth functions and associated variational problems. *Comm. in Pure and Appl. Math.*, 42(5):577–685, 1989.
18. J. E. Atkins, E. G. Roman, and B. Hendrickson. A spectral algorithm for seriation and the consecutive ones problem. *SIAM Journal on Computing*, 28(1):297–310, 1998.
19. M. P. Do Carmo. *Differential Geometry of Curves and Surfaces*. Prentice Hall, 1976.
20. J. A. F. Leite and E. R. Hancock. Iterative curve organisation with the em algorithm. *Pattern Recognition Letters*, 18:143–155, 1997.
21. S. W. Zucker, C. David, A. Dobbins, and L. Iverson. The organization of curve detection: Coarse tangent fields and fine spline coverings. In *Proc. of the IEEE Int. Conf. on Comp. Vision*, pages 568–577, 1988.
22. L. R. Williams and D. W. Jacobs. Stochastic completion fields: A neural model of illusory contour shape and salience. *Neural Computation*, 9(4):837–858, 1997.
23. P. Bremaud. *Markov Chains, Gibbs Fields, Monte Carlo Simulation and Queues*. Springer, 2001.
24. P. L. Worthington and E. R. Hancock. New constraints on data-closeness and needle map consistency for shape-from-shading. *IEEE Transactions on Pattern Analysis and Machine Intelligence*, 21(12):1250–1267, 1999.
25. A. Bors, R. C. Wilson, and E. R. Hancock. Terrain analysis using radar imagery. To appear in the IEEE Trans. on Pattern Analysis and Machine Intelligence, 2003.

# A Robust Reconstruction Algorithm of Displaced Butterfly Subdivision Surfaces from Unorganized Points

Byeong-Seon Jeong, Sun-Jeong Kim, and Chang-Hun Kim

Dept. of Computer Science and Engineering, Korea University
1, 5-ka, Anam-dong, Sungbuk-ku, Seoul 136-701, Korea
{bsjeong,sunjeongkim,chkim}@korea.ac.kr

**Abstract.** This paper presents a more robust reconstruction algorithm to solve the genus restriction of displaced subdivision surface (DSS) from unorganized points. DSS is a useful mesh representation to guarantee the memory efficiency by storing a vertex position as one scalar displacement value, which is measured from the original mesh to its parametric domain. However, reconstructing DSS from unorganized points has some defects such as the incorrect approximation of concave region and the limited application of genus-0. Based on volumetric approach, our new cell carving method can easily and quickly obtain the shape of point clouds and preserve its genus. In addition, using interpolatory subdivision scheme, our displaced butterfly subdivision surface is also effective multiresolution representation, because it samples exclusively new odd vertices at each level, compared with previous works to resample all vertices of every level. We demonstrate that displaced butterfly subdivision surface is an effective multiresolution representation that overcome the topological restriction and preserve the detailed features nicely.

## 1 Introduction

Recently fast and accurate range scanners enable to model complex objects automatically. These scanners provide point data on the surface of an object for applications such as medicine, reverse engineering, and digital filmmaking. The computation of surfaces out of these point data is referred to as reconstruction. Surface reconstruction from the point data has been an actively researched area in computer graphics for several decades. And since subdivision mesh is very useful for the level of detail representation, editing and animation, various approaches have been proposed to remesh the irregular surfaces after the reconstruction. Especially [9] introduced the Displaced Subdivision Surfaces (DSS) as a new mesh representation to store one scalar displacement value, which is measured between the original mesh and its parametric domain, instead of three coordinates of a vertex position. It means the DSS dramatically reduces the amount of the required memory to the $1/3$ compared to the existing mesh representations. However this technique has the limitation of starting from a detailed

M.J. Wilson and R.R. Martin (Eds.): Mathematics of Surfaces 2003, LNCS 2768, pp. 182–195, 2003.

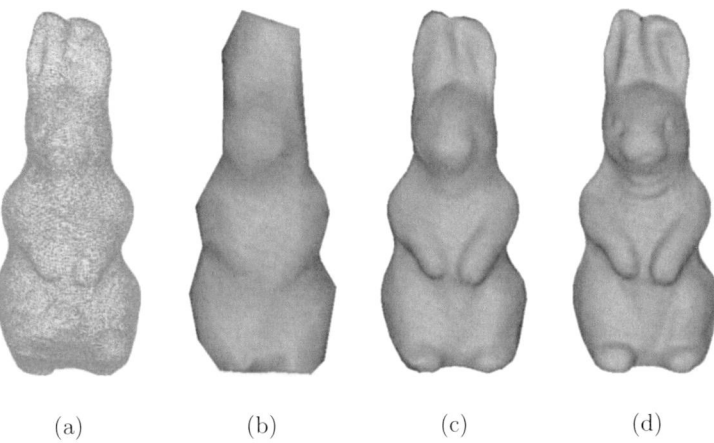

(a)    (b)    (c)    (d)

**Fig. 1.** Displaced Butterfly Subdivision Surfaces (DBSS): (a) Point cloud, (b) DBSS level 1, (c) DBSS level 2 and (d) DBSS level 3

initial mesh. In [7] a direct reconstruction algorithm of the DSS from three dimensional point clouds is proposed. The shrink-wrapping algorithm of bounding box is used for the construction of an initial mesh. However it only works on the genus-0 models and fails to reconstruct the concave shape (see Fig. 2). Also, the DSS is not a truly multiresolution representation. It is just a single level representation of the subdivision mesh. To generate the multiresolution representation with the DSS, the displaced values of all vertices are resampled at every level. Since the representation of each level has the displaced values of all vertices, the whole size of the multiresolution representations is not effective. In this paper we propose a more robust algorithm which solves the problems mentioned above and also dramatically reduces the amount of required memory for the multiresolution representation. We use the cell carving approach based on the volume grid instead of the bounding box proposed in [7]. Firstly we created a volume grid of the uniform size specified by a user containing the input point cloud. From the outmost cells, we check whether a cell contains some points or not. If points are not found inside the cell, we remove it and check the next cell in the same manner. Through this procedure, we can easily and quickly obtain the set of cells which approximates the shape of the point clouds and preserves their genus. We convert this set of cells to the initial mesh using the triangulation followed by the shrink-wrapping. The resulting initial mesh guarantees the topology and the shape of the input model to be preserved. To achieve the memory effective multiresolution representation, we use the Modified Butterfly scheme in [11], one of the interpolatory subdivision schemes, when generating the parametric domain and sampling the displaced values. The interpolatory scheme is more efficient than the approximating scheme in generating the multiresolution DSS, because the displace values can be sampled only at the odd vertices per subdivision level.

(a)                                                        (b)

**Fig. 2.** The initial mesh (Teapot model) reconstructed using the method in [7] and our method: (a) The initial mesh reconstructed using the method in [7] and (b) The one reconstructed using our method

## Contributions

- Our reconstruction algorithm is so robust that it is genus-free and feature preserving. Using the volumetric approach, i.e. Cell Carving, it can reconstruct the initial mesh approximating the point clouds of any topologies as well as concave regions. The high quality initial mesh allows the accuracy of the reconstruction to be increased.
- Our new displaced butterfly scheme makes DSS have smaller storage than previous works in the multiresolution representation. When generating a parametric domain, [7, 9] used approximating subdivision scheme, but it resulted in shrinkage effect and resampling all vertices of every subdivision level. However, Displaced Butterfly Subdivision Surface doesn't shrink from the control mesh and samples only new odd vertices at each subdivision level.

## 2   Previous Work

The general process for the multiresolution mesh representation starts with reconstructing the initial mesh from the point cloud, and then gets through the steps of remeshing for the mesh to gain the subdivision connectivity, which makes it possible to achieve the multiresolution representation. Now we will take a look at the related previous work.

### 2.1   Reconstruction

Several criteria are used to evaluate reconstruction algorithms from the point cloud scanned from three dimensional objects. Those criteria include the quality of the reconstructed model, the speed of the reconstruction process, and the robustness of the algorithm. In [5, 6], the displacement function is defined by calculating the tangent plane of the points and used for finding the voxel information which is applied to the Marching cubes algorithm that reconstructs the mesh. The quality of the resulting mesh is improved later after the mesh optimization. The high complexity and low speed are the main drawbacks of this algorithm. In [2], to reconstruct the mesh from the point cloud, a three dimensional octree containing the point cloud is created and Delaunay triangulation is applied to each cell. The speed of Delaunay triangulation algorithm is sharply dropped when the number of points begins to increase. They applied the Delaunay triangulation algorithm to each cell separately to prevent the loss of speed and to deal with the large mesh data. But the Delaunay triangulation is basically too slow. In [1], firstly the approximating function of the point cloud is constructed using the Radial Basis Function (RBF). Then, Marching cubes algorithm is used to show the visual appearance of the model. This means that the reconstruction and the rendering processes are considered as separated parts. Though the quality of the resulting mesh is considerably high, this algorithm is also too slow to get the triangle mesh. In this paper, we want to reconstruct the coarse control mesh easily and quickly which is used to generate the parametric domain. So, we need a method to reconstruct fast and correctly the initial mesh without loss of any geometrical and topological features. For this, we used the cell carving and shrink-wrapping algorithms based on the volume grid. Intuitively the cell carving algorithm is the same that an artist carves the initial wood to make a sculpture.

## 2.2   Subdivision

A subdivision mesh consists of a control mesh and the subdivision mask or rule. A subdivision rule is applied to a control mesh at every subdivision level and makes the connectivity between meshes before and after subdivision. Since the subdivision mesh has such hierarchical connectivity, there are many applications like the level of detail representation, multiresolution editing, progressive transmission, etc. However we need remeshing techniques that convert irregular meshes into regular meshes which are subdivision meshes, because ordinary irregular meshes do not have any subdivision connectivity. In [8], the Shrink-wrapping approach is proposed to make an arbitrary mesh gain the subdivision connectivity. By using the bounding sphere surrounding the input arbitrary mesh, the simple control mesh is constructed. And it is subdivided to get the hierarchical connectivity. At last a subdivision mesh is shrink-wrapped over the arbitrary mesh to approximate its original shape. This algorithm restricts the input model to have only genus-0 topology. In [3], they partitioned the input model into surface patches which are homeomorphic to the disk, and each surface is parameterized onto a corresponding triangle of the simplified model. The simplified model is subdivided and lifted back onto the original surface according to the information of parameterization. They do not restrict the topology of an input model, but the partitioning preprocess is too costly high to get the subdivision mesh. In this paper, the shrink-wrapped mesh is genus-free and is not partitioned into the disk-like patches. We just simplify the initial mesh reconstructed from the point cloud and subdivide it to get the subdivision connectivity.

## 2.3   Displaced Subdivision Surfaces

The DSS was proposed in [9] on which the concept of this paper is based. The DSS uses only one scalar distance value instead of the three scalar coordinates to represent vertex position of the mesh. That enables us to reduce the memory requirement to 1/3 and to store, edit and transmit efficiently the large mesh data of high quality. The DSS reconstruction algorithm requires the domain surface constructed by subdividing the simplified control mesh, measures the distances between the vertices on the domain and the arbitrary input mesh, and assigns it to each vertex. However, if we want to generate the multiresolution representation of the DSS, we have to subdivide the control mesh from the beginning coarsest level to the target subdivision level to get one resolution of the mesh and to sample the displacement values of all the vertices. To get several resolution meshes, we must repeat the steps mentioned. This makes us lose the memory efficiency of the DSS and its reconstruction algorithm slow. To get more efficiency in generating the multiresolution DSS, we use the Modified Butterfly subdivision [11], which is one of the interpolatory schemes. First, the control mesh is subdivided just once using the Modified Butterfly scheme followed by sampling the distances between the odd vertices on the parametric domain and the point cloud, and the odd vertices moved by the sampled distances along the vertex normal direction. To generate the next resolution, in the same manner we subdivide the resulting mesh just one time followed by the distance sampling and the vertex moving processes. It is important that the sampling and moving processes are performed only on the newly introduced vertices at every subdivision level. That enables us to make the true multiresolution representation with far less memory requirement compared to the existing methods.

## 3   Robust Reconstruction Algorithm

Our algorithm consists of two steps. The first step is generating the control mesh by reconstructing the initial mesh from the point cloud followed by simplifying and projecting to the point cloud. The next is sampling the displacement values between the subdivided control mesh, i.e. the parametric domain and the original input point cloud.

### 3.1   Control Mesh Generation

First of all, some definitions have to be clarified. Let the volume grid surrounding the point cloud be the set G which consists of the whole cells of the grid. The size of the volume grid is specified by the user like 4 x 4 x 4 (see Fig. 3(a)). According to the complexity of the shape of the point cloud, the size of the volume grid can be various. Each cell in the set G has its six neighbors which are adjacent to its six faces (see Fig. 3(b)). By carving the volume grid, we can extract the shape of the point cloud which the remaining cells have. We call the carving procedure Cell Carving algorithm. It tests whether the cells have the points or not. The

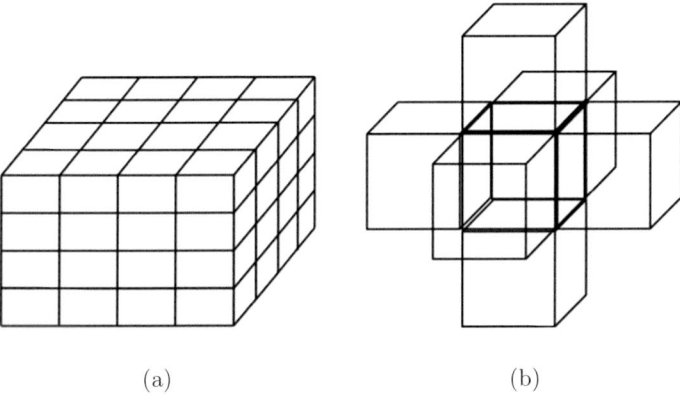

(a)                                          (b)

**Fig. 3.** The volume gird and the six neighbors of each cell in the grid: (a) Volume grid and (b) Six neighbor cells

set C consists of these candidate cells. At the beginning of the algorithm, the set C contains the outmost cells of the volume grid. After the Cell Carving, we can get the cells approximating the surface of the point cloud. The surface set S consists of these cells. And the other sets used in the Cell Carving algorithm are just temporary. Now the detail procedure of the Cell Carving algorithm is as follows (also see Fig. 4).

**Step 1** Insert the outmost cells into the set $C$.
**Step 2** Repeat the following substeps on each cell in the set $C$ until the set $C$ becomes empty.
   **Step 2.1** If a cell contains some points, then insert it into the set $T$ and remove it from the set $C$.
   **Step 2.2** If a cell does not contain any points, then perform the following substeps on its neighbor cells and remove it from the set $C$ and the set $G$.
      **Step 2.2.1** If a neighbor cell contains some points, then insert init to set $T$, and remove it from set $C$ if it is in the set $C$.
      **Step 2.2.2** If a neighbor cell does not contain any points and is not in the set $C$ either, then insert it into the set $N$.
**Step 3** If there are some elements in the set $N$, then move all the elements in the set $N$ to the set $C$ and return to the step 2.
**Step 4** For each cell in the set $T$, perform the following substeps.
   **Step 4.1** For each neighbor cell of this cell, perform the following substeps.
      **Step 4.1.1** If the neighbor cell contains some points and is not in the set $T$ or $S$, then insert it into the set $W$.
**Step 5** If there are some cells in the set $W$, then move all the elements in the set $T$ to the set $S$, move all the elements in the set $W$ to the set $T$, and return to the step 4. Otherwise terminate the algorithm.

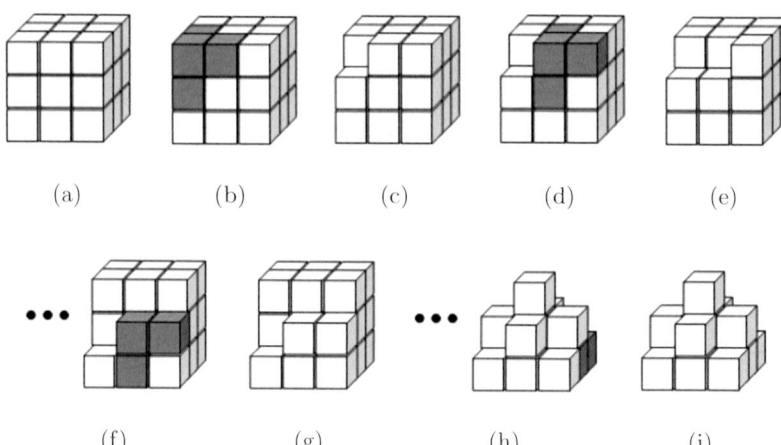

**Fig. 4.** The Cell Carving algorithm: (a) the initial volume grid, (b) the test with the first candidate cell and its neighbors, (c) the first cell is removed because it is the empty cell, (d) the test with the second candidate cell and its neighbors, (e) the second cell is removed because it is empty too, (f) the test with the fifth candidate cell and its neighbors, (g) the fifth cell is not removed because it is the non empty cell. (h) the test with the last candidate cell and its neighbors, and (i) the final result of the Cell Carving algorithm

After carving the volume grid, the cells in the set S represent the surface of the point cloud. Then we triangulate the quad faces on the cells in the set S which aren't adjacent to any other cells to make the triangle mesh. We call those quad faces the air faces and call the air vertices the vertices of the cell such that are not adjacent to any other cells. There are four types of triangulation:

Type 1 If there are five air faces in a cell, then we collapse the four air vertices into one vertex. The collapsed vertex will be located at the averaged position of the four vertices. As a result, there are five vertices, four from the non-air faces and one from the air faces (see Fig. 5(a)).

Type 2 If there are four air faces which are not arranged in the shape of the ribbon in a cell, then we collapse the two air vertices into one vertex. The collapsed vertex will be located at the averaged position of the two vertices. As a result, there are seven vertices, six from the non-air faces and one from the air faces (see Fig. 5(b)).

Type 3 If there are three air faces in a cell, which means that there is only one air vertex, then we triangulate three air faces (see Fig. 5(c)).

Type 4 Otherwise, we perform the general triangulation (see Fig. 5(d)).

This process guarantees the good aspect ratio of the triangles in the initial mesh. Consequently, we can minimize the distortion which could be raised at the displacement sampling stage. Now we apply the shrink-wrapping algorithm to the triangulated mesh over the point cloud. The shrink-wrapping process is

- non-air vertex
○ air vertex

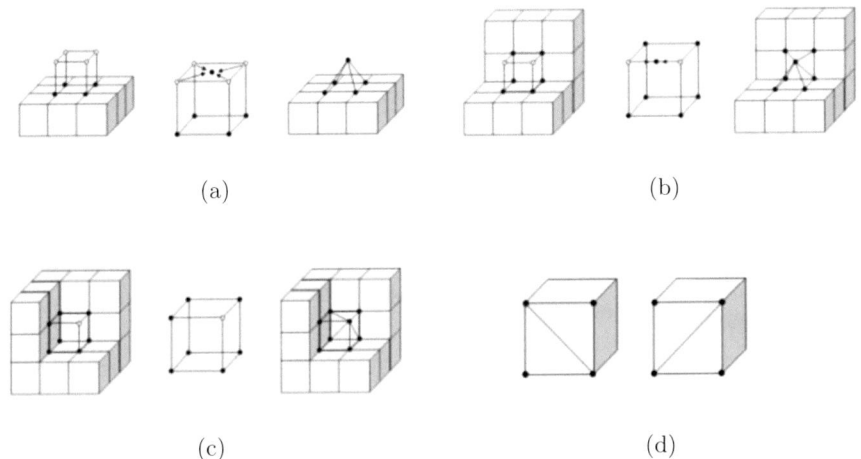

(a)

(b)

(c)

(d)

**Fig. 5.** The types of the triangulation: (a) Four air vertices, (b) Two air vertices, (c) One air vertices, and (d) General type

the same as [7]. The projection and smoothing operators are repeatedly applied in shrinking the mesh over the point cloud. The projection operator moves each vertex in the direction of the closest point to it. The smoothing operator relaxes the vertices to be distributed equally in order to minimize the distortion produced by the projection. Now the shrink-wrapped mesh is approximating the shape of the point cloud and eventually it becomes the initial mesh. The next step is the simplification of the initial mesh up to the user specified level to get the simpler one. If the initial mesh is sufficiently coarse, then the simplification is not necessary. In [7], to simplify the initial mesh they used a point based QEM method that is more time consuming than the original QEM in [4]. This is because the quality of the initial mesh was not good enough to guarantee them the adequate quality of the simplified one. However the accuracy of the initial mesh reconstructed by our algorithm is higher than the previous one and even enough to assure the quality of the simplified mesh so that we can use the original QEM in this paper. Finally, we need to fit the simplified mesh to the point cloud to generate the well-shaped parametric domain. To generate the parametric domain from the control mesh, Loop subdivision scheme was used in both [9] and [7]. Loop scheme in [10] which is one of approximating subdivision scheme causes the shrinkage effect, so the global energy minimization and local subdivision surface fitting have to be performed on the simplified mesh before subdividing it for the better approximation of the resulting parametric domain to the input model. For this, in [7], they use the local subdivision surface fitting algorithm which is faster than the global fitting in [9]. In [7], they computed the limit positions of the vertices of the control mesh which the subdivision scheme allows, and found the closest points to those positions. Then, they adjust the

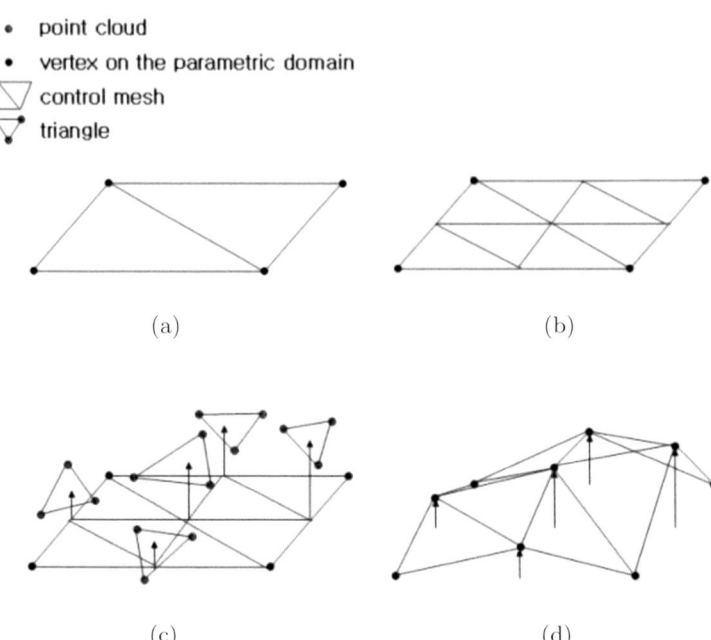

- point cloud
- vertex on the parametric domain
- control mesh
- triangle

(a)     (b)

(c)     (d)

**Fig. 6.** The parametric domain generation and the displacement sampling: (a) Control mesh, (b) Parametric domain, (c) Displacement sampling, and (d) Vertex moving

position of the vertices of the control mesh so that the limit position may be close to those points. But we need not to fit the surfaces like that. We only need to project the vertices of the control mesh to the point cloud because we use the interpolatory subdivision scheme not to have the shrinkage effect. After projecting the simplified mesh to the point cloud we have the control mesh (see Fig. 9).

### 3.2   Parametric Domain Generation and Displacement Sampling

The parametric domain is a subdivision mesh constructed from the control mesh by subdivision. In [7], they use Loop subdivision to generate the parametric domain. So they can only get the single resolution of the DSS because Loop scheme repositions all the vertices whenever it subdivides the control mesh. But, we use the interpolatory subdivision scheme which does not reposition the even vertices. Therefore we can construct the multiresolution mesh through a single process using the interpolatory subdivision scheme, i.e. Modified Butterfly scheme in [11]. Here are the details (also see Fig. 6):

Step 1 Subdivide the control mesh using the Modified Butterfly scheme.
Step 2 Measure the distances between the newly generated vertices on the subdivided mesh and the point cloud in the vertex normal direction. To measure the distances we use the triangle intersection test as [7] but the different method to gather the sampling points (see Fig. 7).

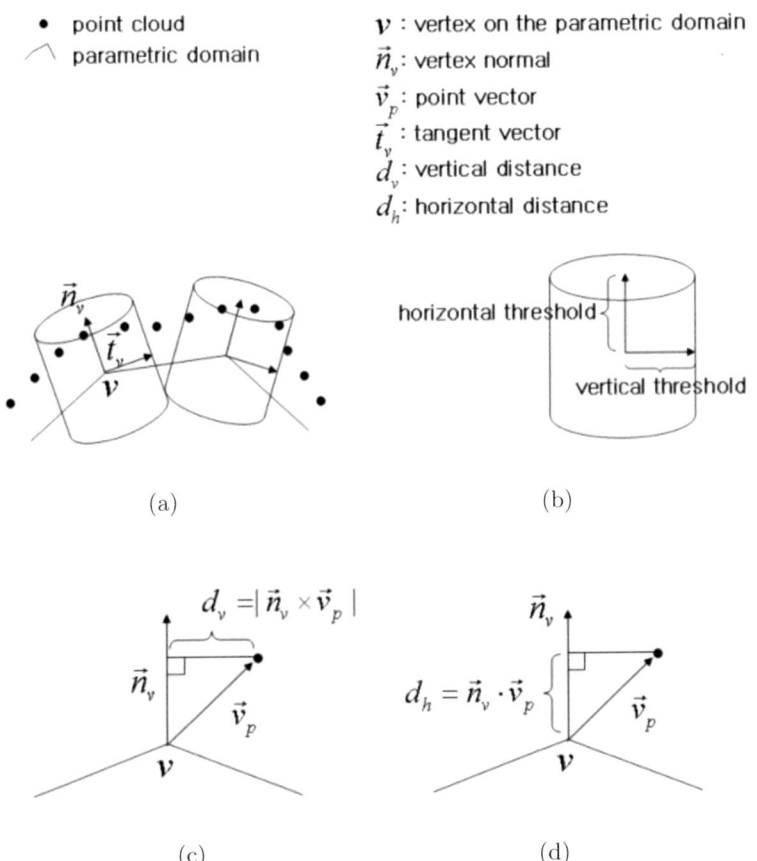

- point cloud
- parametric domain

$v$ : vertex on the parametric domain
$\vec{n}_v$ : vertex normal
$\vec{v}_p$ : point vector
$\vec{t}_v$ : tangent vector
$d_v$ : vertical distance
$d_h$ : horizontal distance

horizontal threshold

vertical threshold

(a)

(b)

$$d_v = |\vec{n}_v \times \vec{v}_p|$$

$$d_h = \vec{n}_v \cdot \vec{v}_p$$

(c)

(d)

**Fig. 7.** The cylindrical distances for sampling points: (a) Cylindrical point sampling, (b) Cylindrical thresholds, (c) Vertical distance, and (d) Horizontal distance

**Step 3** Move each odd vertex in the vertex normal direction according to the measured distance respectively.

**Step 4** If the subdivision level is fine enough, then terminate the procedure. Otherwise, subdivide the resulting mesh once again and return to the step 2.

To gather the sampling points, we must compute two distances. At first, we calculate the vertical distance of the point as Fig. 7(c) and the horizontal distance of the point as Fig. 7(d). Therefore the point sampling area is the cylinder whose height axis is along the vertex normal. If these distances of the point are smaller than the thresholds, then they becomes sampling points. Then we can sample the displacement between the candidate triangles which consist of three sampling points and the odd vertices using the triangle intersection method in [7]. After sampling the displacements, the odd vertices are moved up to the amount of the displacements along the normal of each vertex. Compared with our approach, the

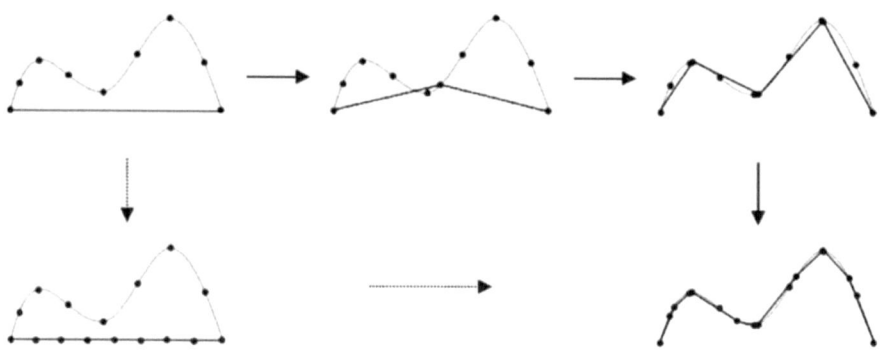

**Fig. 8.** Our method and the previous one in [7, 9] in sampling the displacements

method in [7] sampled the displaced values of the whole vertices after subdividing the control mesh up to the user specified level (see Fig. 8). According to their method, to construct multi-level representation, resampling of all the vertices is necessary at every subdivision level. Therefore its multiresolution representation has the redundancy in the number of the displaced values. However our Displaced Butterfly Subdivision Surfaces (DBSS) can get more subdivision levels in a given memory space than their DSS, because it consists of the control mesh and the displaced values of the newly introduced vertices only at every subdivision level.

## 4   Results

We implemented our algorithm in Pentium IV 1.4 GHz CPU, 512MB memory and GeForce2 MX400 Graphic card. We can generate the control mesh fast and easily using the proposed Cell Carving algorithm. The bounding box algorithm used in [7] fails to reconstruct the non genus-0 models (see Fig. 2). It also fails to preserve the details of the complicated and highly curved models. With the proposed algorithm, we can reconstruct both the genus and the detail features of the model. For the multiresolution representation, the existing method requires to sample the displacement of all the vertices at every subdivision level. The proposed algorithm needs the displacement sampling only on the newly introduced vertices at each subdivision level. So, our DBSS can represent any levels of the multiresolution mesh (see Fig. 1, 9, and 10, and Table 1). We can easily infer that our approach to build the multiresolution representation is about more efficient than the previous one in the memory by 25%, because we store the scalar values of the odd vertices at each subdivision level (see Table 1).

## 5   Conclusion and Future Work

In this paper we proposed a robust algorithm that directly reconstructs a multiresolution Displaced Butterfly Subdivision Surfaces from three dimensional

**Table 1.** The number of times of the displacement sampling in ours and the previous methods

| Schemes | Models | Level 1 | Level 2 | Level 3 |
|---------|--------|---------|---------|---------|
| Our method | Torus (control mesh #v = 50) | 150 | 600 | 2,400 |
| | Rabbit (control mesh #v = 107) | 315 | 1,260 | 5,040 |
| Previous method | Torus | 200 | 800 | 3,200 |
| | Rabbit | 422 | 1,682 | 6,722 |

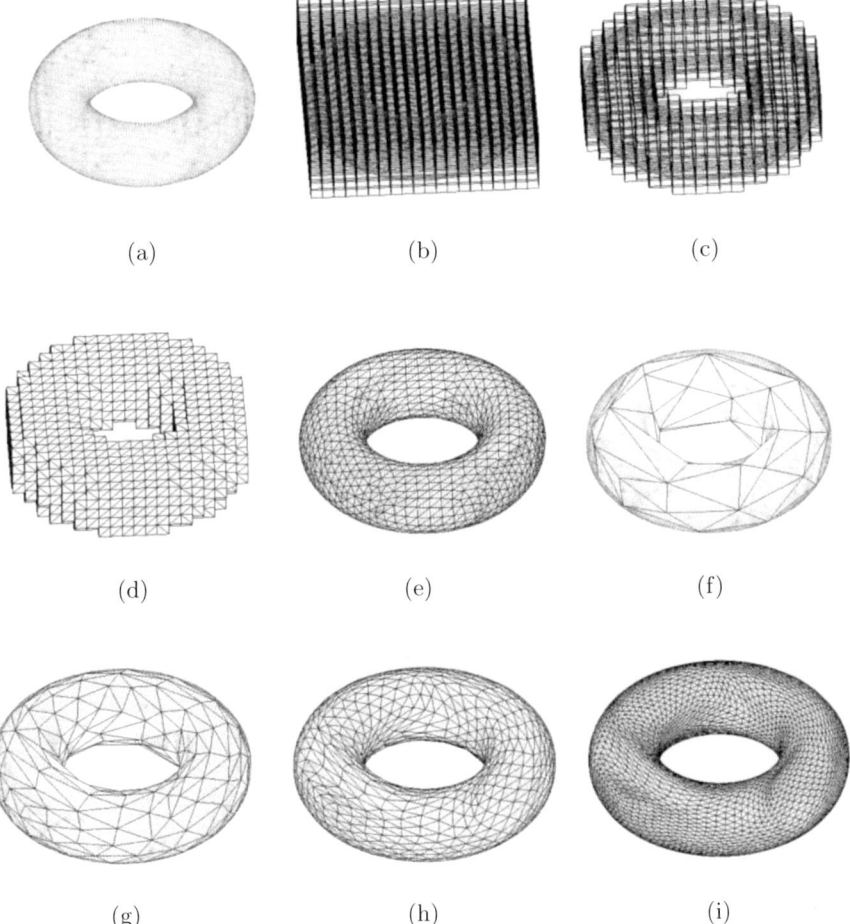

(a)          (b)          (c)

(d)          (e)          (f)

(g)          (h)          (i)

**Fig. 9.** The whole process of our algorithm (Torus model): (a) Point cloud, (b) Volume grid, (c) After cell carving, (d) Triangulated cells, (e) Initial mesh, (f) Control mesh, (g) DBSS level 1, (h) DBSS level 2, and (i) DBSS level 3

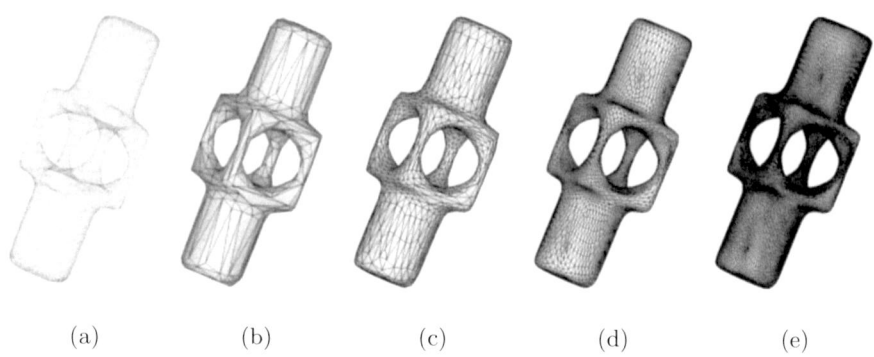

(a)            (b)            (c)            (d)            (e)

**Fig. 10.** The resulting Displaced Butterfly Subdivision Surfaces of the Mechanical part model: (a) Point cloud, (b) Control mesh, (c) DBSS level 1, (d) DBSS level 2, (e) DBSS level 3

point clouds. Our Cell Carving algorithm is very simple but reconstructs robustly an initial mesh, because it preserves the genus and features of the input points. Since the displaced values are sampled at the odd vertices per subdivision level, our algorithm is effective in representing the multiresolution mesh, compared to the previous DSS, which must resample the displaced values of all the vertices of every subdivision level. Therefore our algorithm has the efficiency in term of time and space on the multiresolution representation. Our Displaced Butterfly Subdivision Surfaces can be widely used in computer graphics applications. However, since the CAD models with the sharp edges can hardly be represented using the DSS with the relatively small number of triangles, a new sampling scheme must be developed.

# References

1. J. C. Carr, R. K. Beatson, J.B. Cherrie T. J. Mitchell, W. R. Fright, B. C. McCallum and T. R. Evans.: Reconstruction and Representation of 3D Objects with Radial Basis Functions. ACM SIGGRAPH 2001 Proceedings, pp.67-76, August 2001

2. T. K. Dey, J. Giesen, and J. Hudson.: Delaunay Based Shape Reconstruction from Large Data. IEEE Symposium on Parallel and Large Data Visualization and Graphics (PVG 2001), pp.19-27, October 2001

3. M. Eck, T. DeRose, T. Duchamp, H. Hoppe, M. Lounsbery, and W. Stuetzle.: Multiresolution Analysis of Arbitrary Meshes. ACM SIGGRAPH 95 Proceedings, pp. 173-182. August 1995

4. M. Garland and P. Heckbert.: Surface Simplification Using Quadric Error Metrics. Proceedings of SIGGRAPH 97, pp.209-216, August 1997

5. H. Hoppe, T. DeRose, T. Duchamp, M. Halstead, H. Jin, J. McDonald, J. Schweitzer, and W. Stuetzle.: Piecewise smooth surface reconstruction. ACM SIGGRAPH 94 Proceedings, pp.295-302, July 1994

6. H. Hoppe, T. DeRose, T. Duchamp, J. McDonald, and W. Stuetzle.: Surface Reconstruction from Unorganized Points. Computer Graphics (SIGGRAPH 92 Proceedings), 26(2):71-78, July 1992
7. W.-K. Jeong and C.-H. Kim.: Direct Reconstruction of Displaced Subdivision Surface from Unorganized Points. Graphical Models, 64(2):78-93, March 2002
8. L. Kobbelt, J. Vorsatz, U. Labsik, and H.-P. Seidel.: A Shrink Wrapping Approach to Remeshing Polygonal Surfaces. Computer Graphics Forum (EUROGRAPHICS 99 Proceedings), 18(3):119-130, September 1999
9. A. Lee, H. Moreton, and H. Hoppe.: Displaced Subdivision Surfaces. ACM SIGGRAPH 2000 Proceedings, pp.85-94, July 2000
10. C. Loop.: Smooth Subdivision Surfaces Based on Triangles. Masters thesis, Department of Mathematics, University of Utah, August 1987
11. D. Zorin, P. Schröder, and W. Sweldens.: Interpolating Subdivision for Meshes with Arbitrary Topology. ACM SIGGRAPH 96 Proceedings, pp.189-192, August 1996

# Filling Holes in Point Clouds

Pavel Chalmoviansky[1] and Bert Jüttler[2]

[1] Johannes Kepler University, Spezialforschungsbereich SFB 013
Freistädter Str. 313, 4040 Linz
Pavel.Chalmoviansky@jku.at
http://fractal.dam.fmph.uniba.sk/~chalmo
[2] Johannes Kepler University, Dept. of Applied Geometry
Altenberger Str. 69, 4040 Linz
bert.juettler@jku.at
http://www.ag.jku.at

**Abstract.** Laser scans of real objects produce data sets (point clouds) which may have holes, due to problems with visibility or with the optical properties of the surface. We describe a method for detecting and filling these holes. After detecting the boundary of the hole, we fit an algebraic surface patch to the neighbourhood and sample auxiliary points. The method is able to reproduce technically important surfaces, such as planes, cylinders, and spheres. Moreover, since it avoids the parameterization problem for scattered data fitting by parametric surfaces, it can be applied to holes with complicated topology.

**Keywords:** Reverse engineering, scattered data, algebraic surface fitting, meshless methods

## 1 Introduction

Since the advent of advanced laser scanners, even complicated objects can be digitized with impressive accuracy. This led to the technology of reverse engineering [18], which has been developed into a valuable alternative to the traditional top–down construction process in CAD. Instead of designing a CAD model from scratch, the model is (semi–) automatically created from the cloud of measurement data.

The data acquired by the scanning device, however, may have various problems. For instance, some parts of the objects can be missing, due to problems with accessibility/visibility, or due to the special physical properties of the scanned surface (e.g. transparentness, reflectivity, etc.). This produces holes in the data set, which do not correspond to any holes in the object (cf. [5,17]).

There are several possible ways to address this problem. For instance, one may try to combine several views, i.e., several scanned point sets of the same object. As another approach, the holes can be filled with auxiliary, artificially generated points. The latter approach will work also in regions where the object is difficult to scan, due to visibility and/or optical properties. Also, it can be used even if the problems have not been realized immediately during the scanning process.

M.J. Wilson and R.R. Martin (Eds.): Mathematics of Surfaces 2003, LNCS 2768, pp. 196–212, 2003.
© Springer-Verlag Berlin Heidelberg 2003

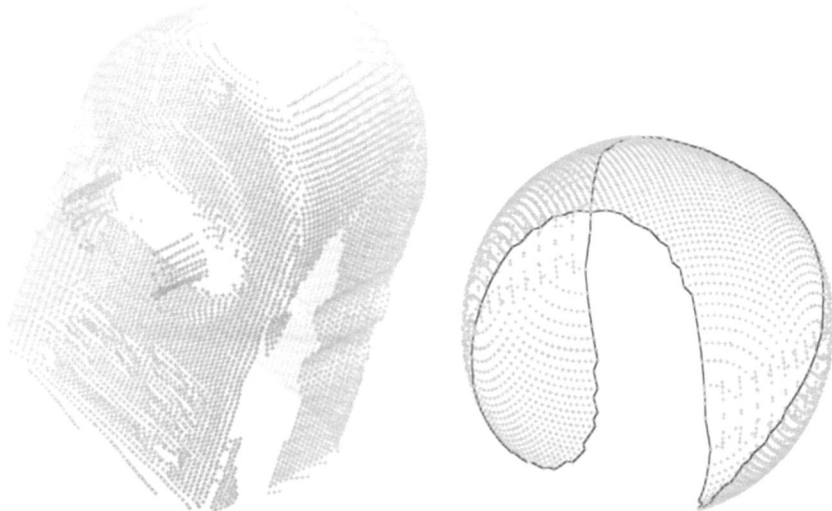

**Fig. 1.** Left: babyphone data, generated by a 3D Laser scanner. Right: tennis ball data (artificially generated).

In this paper, we use algebraic surfaces for filling the hole by generating auxiliary points. After detecting the hole, a surface is fitted to the neighbourhood of its boundary. The method is able to reproduce spheres, planes and circular cylinders. Finally, by sampling points from the part of the algebraic surface which corresponds to the hole, the missing points are constructed.

This paper is organized as follows. The next section describes basic notions and facts used throughout the paper. Section 3 summarizes the algorithms for estimating normals associated with the points. Techniques for detecting boundary points and for constructing polygonal boundaries for the holes are outlined in section 4. Section 5 describes the algebraic surface fitting. The sampling process used for generating the new points is introduced in section 6. Finally, we conclude this paper. Technical details can be found in the appendices.

## 2   Preliminaries

The data generated by the scanner form a set $P = \{\mathbf{p}_0, \ldots, \mathbf{p}_N\} \subset \mathbb{R}^3$, where $N \in \mathbb{Z}_+$ is the number of points. Typically, $N$ is in the order of tens of thousands for objects of a size of coffee mug, see Figure 1.

For the convenience of the reader, we summarize some facts about the Bernstein–Bézier representation of trivariate polynomials (see [6] for more details).

Let $\mathbf{v}_i \in \mathbb{R}^3$ for $i \in \{0, 1, 2, 3\}$ be four non-coplanar points with coordinates $\mathbf{v}_i = (v_{i1}, v_{i2}, v_{i3})^\top$. They span a simplex $\mathbf{V} = \triangle(\mathbf{v}_0, \mathbf{v}_1, \mathbf{v}_2, \mathbf{v}_3)$. Hence, each point $\mathbf{p} \in \mathbb{R}^3$ can be expressed as a unique linear combination

$$\mathbf{p} = \sum_{i=0}^{3} p_i \mathbf{v}_i \quad \text{with} \quad \sum_{i=0}^{3} p_i = 1. \tag{1}$$

The quadruple $\tilde{\mathbf{p}} = (p_0, p_1, p_2, p_3)$ are the *barycentric coordinates of the point* $\mathbf{p}$ *with respect to the simplex* $\mathbf{V}$. They can be computed from

$$p_i = \frac{[\mathbf{v}_0, \ldots, \{\mathbf{v}_i, \mathbf{p}\}, \ldots, \mathbf{v}_3]}{[\mathbf{v}_0, \mathbf{v}_1, \mathbf{v}_2, \mathbf{v}_3]}, \quad \text{for} \quad i \in \{0, 1, 2, 3\} \tag{2}$$

with

$$[\mathbf{a}, \mathbf{b}, \mathbf{c}, \mathbf{d}] = \begin{pmatrix} 1 & 1 & 1 & 1 \\ a_1 & b_1 & c_1 & d_1 \\ a_2 & b_2 & c_2 & d_2 \\ a_3 & b_3 & c_3 & d_3 \end{pmatrix}, \tag{3}$$

where $[\mathbf{v}_0, \ldots, \{\mathbf{v}_i, \mathbf{p}\}, \ldots, \mathbf{v}_3]$ indicates that we replace the column containing the Cartesian coordinates of the point $\mathbf{v}_i$ with those of the point $\mathbf{p}$.

Let $\mathbf{i} = (i_0, i_1, i_2, i_3) \in \mathbb{Z}_+^4$, $\tilde{\mathbf{x}} = (x_0, x_1, x_2, x_3) \in \mathbb{R}^4$ and $|\mathbf{i}| = i_0 + i_1 + i_2 + i_3$. The *Bernstein-Bézier polynomials of degree* $n$ are

$$B_{\mathbf{i}}^n(\mathbf{x}) = \binom{n}{\mathbf{i}} \tilde{\mathbf{x}}^{\mathbf{i}} = \frac{n}{i_0! i_1! i_2! i_3!} x_0^{i_0} x_1^{i_1} x_2^{i_2} x_3^{i_3} \tag{4}$$

for all $\mathbf{i}$ such that, $|\mathbf{i}| = n$. Clearly, such a polynomial is homogeneous in the barycentric coordinates. It is well known that the $\{B_{\mathbf{i}}^n(\mathbf{x}): |\mathbf{i}| = n\}$ form a basis of the linear space $\Pi_n(\mathbb{R}^3)$ of all trivariate polynomials of degree $n$.

Let $f \in \Pi_n(\mathbb{R}^3)$ be a trivariate polynomial of degree $n$. After introducing barycentric coordinates for its argument and homogenization, $f$ can be uniquely written as

$$f(\mathbf{x}) = \sum_{|\mathbf{i}|=n} B_{\mathbf{i}}^n(\mathbf{x}) b_{\mathbf{i}}, \tag{5}$$

where the $b_{\mathbf{i}} \in \mathbb{R}$ are called the *coefficients* of the polynomial $f(\mathbf{x})$.

The derivatives of the polynomial $f$ and their expression in the Bernstein-Bézier basis can be easily calculated using the polar form of the polynomial. This concept from multilinear algebra was introduced by L. Ramshaw to geometric modeling as the *blossoming principle* (see [11] and, more recently, [12]). See appendices A and B.

## 3   Normal Estimation

This part is devoted to the estimation of normals for each point of a given point cloud. The normals are used later during the patch fitting (see section 5). We estimate the normal using the plane of regression for certain neighborhood of each points in the data set.

We assume that the set $P$ of points is uniformly distributed on the surface of the solid[1]. The estimation of normals from scattered data is a standard technique for data processing (see [7,9]). The algorithm is as follows.

---

[1] More precisely, we assume, that the eigenvalues calculated during the PCA are well separated.

## Algorithm 1 (Normal Estimation)

1. Find all points within a certain neighbourhood of each point $\mathbf{p}_i \in P$.
2. Estimate the direction of the normal $\mathbf{n}_i$ for $\mathbf{p}_i$ via PCA (see below).
3. Orient the generated normals $\mathbf{n}_i$ consistently, via region growing.

The three steps will now be discussed in more detail.

*Step 1.* The neighbourhood of each point $\mathbf{p}_i \in P$ consists of the $k$ nearest neighbours in the data set,

$$P_i = \{\mathbf{p}_{i,0}, \ldots, \mathbf{p}_{i,k-1}\} \tag{6}$$

It can be computed using suitable algorithms from computational geometry, such as hashing and $k$D-trees, see [14]. We suppose that $\mathbf{p}_{i,0} = \mathbf{p}_i$ and the points in $P_i$ are sorted by ascending distance to the point $\mathbf{p}_i$.

*Step 2.* The estimation of the unit direction $\mathbf{n}_i$ of the normal, relies on Principal Component Analysis (PCA) applied on the above computed neighbourhood $P_i$ of the point. Let

$$\mathbf{a}_i = \frac{1}{k} \sum_{j=0}^{k-1} \mathbf{p}_{i,j} \tag{7}$$

be the centroid of the neighbourhood $P_i$. Consider the quadratic form

$$q(\mathbf{x}) = (\mathbf{x} - \mathbf{a}_i)^\top Q(\mathbf{x} - \mathbf{a}_i) \quad \text{with} \quad Q = \sum_{j=0}^{k-1} (\mathbf{p}_{i,j} - \mathbf{a}_i)(\mathbf{p}_{i,j} - \mathbf{a}_i)^\top. \tag{8}$$

The eigenvector of $Q$ which is associated with the smallest eigenvalue is used as an estimate of the normal. It can be shown to be the normal of the plane of regression of the points $P_i$. The normal is normalized in order to obtain a unit vector. In the sequel, we will often refer to the estimated normal at $\mathbf{p}_i$ shortly as the normal at $\mathbf{p}_i$.

*Step 3.* The estimated normals may not be oriented consistently, since normals at neighbouring points may have different orientations (cf. Figure 2). We use a region–growing–type algorithm for generating a consistent orientation via systematic reorientation ("swapping") of the normals. It is based on the quantity

$$\cos \gamma_{ij} = \langle \mathbf{n}_i, \mathbf{n}_j \rangle \tag{9}$$

which is compared with a given threshold. As an example, we applied this technique to the tennis ball data, see Figure 2.

Clearly, the estimated tangent plane of the original surface at a point $\mathbf{p}_i$ has the equation

$$\langle \mathbf{x} - \mathbf{p}_i, \mathbf{n}_i \rangle = 0 \tag{10}$$

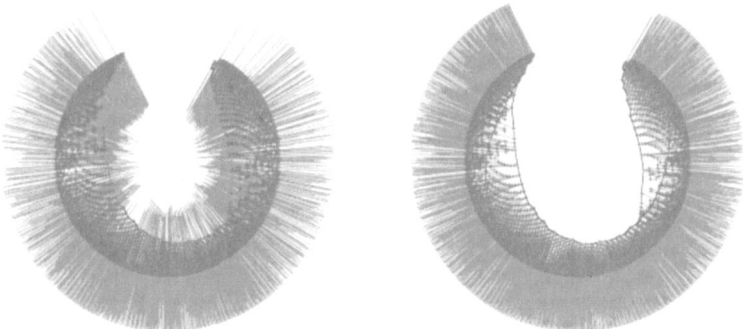

**Fig. 2.** Inconsistent(left) and consistent(right) orientation of normals.

*Implementation Issues.* In order to define the neighbourhood of a point $\mathbf{p}_i$, we used all points within a ball of a certain constant radius, and among those we picked the $k = 35$ closest ones. Note that the number of points in the neighbourhood might be smaller than $k$, especially in regions with a low sampling rate. It is recommended either to use advanced techniques of computational geometry for larger sets $P$ to find the $k$ nearest neighbours, or to split the input data into smaller subsets whenever possible.

Currently, our region–growing algorithm for orientation is a semi-automatic one, requiring some user interaction (e.g., adjusting the threshold). For all our examples it worked without problems.

## 4   Constructing Boundaries

Boundary detection for triangulated point clouds is well understood. In this paper, however, we describe a meshless method, which does not assume the existence of a triangulation[2].

The detection of boundaries in the discrete set of points is a subtle issue, which strongly depends on measurement errors in the point cloud and the distribution of the points. It consists of two major steps:

### Algorithm 2 (Boundary Building)

1. Detect the candidate boundary points.
2. Construct boundary polygons from boundary points.

*Step 1.* The points on the boundary of the set $P$ are characterized by a having a bigger distance from the centroid of their neighbourhood than the points within the inner part of the surface, see Figure 3. Consequently, the first criterion for the boundary points is

---

[2] Although methods for triangulations of point clouds have recently made some progress, they are still difficult and computationally expensive. Also, they may not give the desired results, especially for complicated and/or noisy data. Therefore, meshless methods are a valuable alternative [15].

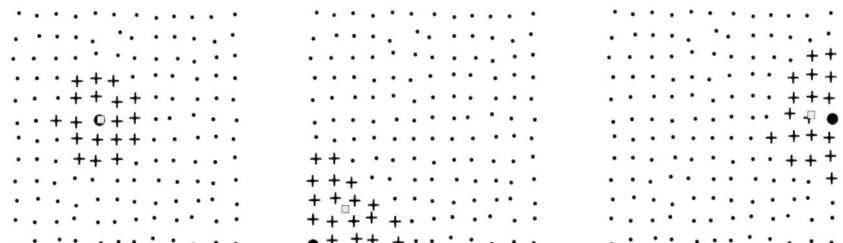

**Fig. 3.** Points and centroids in a point cloud for inner (left) and boundary points (middle and right). The crosses indicate the closest neighbors of the big black dot. The centroid is shown as a square.

$$\|\mathbf{p}_i - \mathbf{a}_i\| \geq \varepsilon_{\text{bdist}} \tag{11}$$

for an appropriate threshold $\varepsilon_{\text{bdist}}$. Let $P_B \subseteq P$ be the set of all such points in $P$.

More sensitive criteria can be constructed based on the use of higher moments. This has recently been explored for feature detection [3]. Other information about the local behaviour of the point cloud can be obtained by analyzing the distribution of the eigenvalues of the matrix (8).

According to our experience, the criterion (11) works reasonably well to generate candidate points both for points on boundary and at sharp edges of the object. A finer classification can then be obtained by a local analysis in the estimated tangent plane (10), as follows.

Consider the neighbourhood $P_i$ of a boundary point $\mathbf{p}_i$. In addition to $\mathbf{p}_i$, it contains the points $\{\mathbf{p}_{i,1}, \ldots, \mathbf{p}_{i,k-1}\}$. By projecting it into the estimated tangent plane $T_{p_i}(S)$ we get the points

$$\mathbf{R} = \{\mathbf{r}_1, \ldots, \mathbf{r}_{k-1}\}. \tag{12}$$

Let

$$w_{\mathbf{p}_i}(\mathbf{r}_l, \mathbf{r}_j) = \text{wedge}(\mathbf{r}_l \mathbf{p}_i \mathbf{r}_j) \tag{13}$$

be the wedge spanned by points $\mathbf{r}_l$ and $\mathbf{r}_j$ with the apex $\mathbf{p}_i$, and

$$\max_{l,j}\{\angle w_{\mathbf{p}_i}(\mathbf{r}_l, \mathbf{r}_j): \text{ int } w_{\mathbf{p}_i}(\mathbf{r}_l, \mathbf{r}_j) \cap \mathbf{Q} = \emptyset\} \tag{14}$$

be the biggest angle of all those wedges which do not contain any other point from $\mathbf{R}$ (see Figure 4, left).

In the limit, if the neighbourhood became infinitesimal small, and the sampling density were arbitrarily high, the angle would be equal to $\pi$ for all regular points on the boundary. For regular points in the inner part, the corresponding limit is zero.

In order to detect the boundary points $\mathbf{p}_i$ among the candidate points $P_B$, we use the following test:

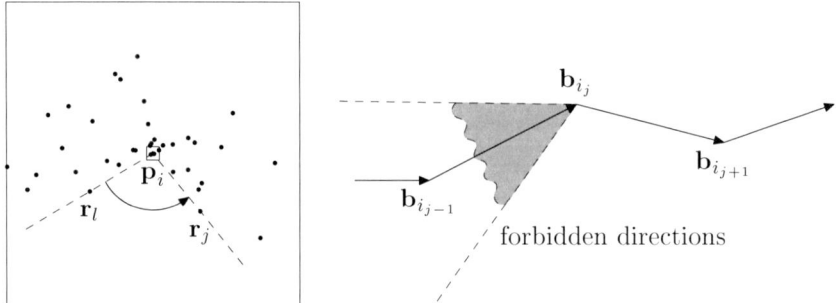

**Fig. 4.** Left: the maximal angle between the consecutive points. Right: boundary polygon construction. The colored wedge zone is forbidden for the next point $\mathbf{c}_{i+1}$.

- Project the neighbouring points orthogonally into the local plane of regression (the estimated tangent plane) $T_{\mathbf{p}_i}$.
- Sort them according to the polar coordinates around the point $\mathbf{p}_i$.
- Calculate the (oriented) angles of the wedges spanned by any two consecutive points with apex at $\mathbf{p}_i$ (see Figure 4), left. Let $\alpha_i$ be the maximum of these angles.
- Delete the point $\mathbf{p}_i$ from $P_B$ if $\alpha_i < \alpha_0$, where $\alpha_0$ is a user–defined threshold.

Clearly, this test will also delete vertices of the boundary curve with sharp inner angles. This, however, is not a problem for the hole filling application. The complexity of this algorithm is $O(Nk \log k)$.

*Step 2.* Now we are ready to find the boundary polygons. For the sake of simplicity, we first suppose that there are no sharp edges (no singular points and the curvature has an upper bound) along the boundary of the surface. (Such points are handled by glueing together several branches of the boundary polygon.) We use the following greedy algorithm:

Assume the set $B = \{\mathbf{b}_0, \ldots, \mathbf{b}_q\}$ contains all boundary points of set $P$. Points of $B$ which are not in any constructed polygon are called *free*.
- Take a free point $\mathbf{b}_{i_0}$ and its closest free neighbour $\mathbf{b}_{i_1}$ from $B$.
- Extend the polygon by adding the nearest free point from $B$ which belongs to the allowed wedges as shown in Figure 4, right. The polygon is allowed to grow in both directions.
- If there is no point to extend the polygon, or if the polygon is already closed, start another one until all the points in $B$ are used up.

This algorithm produces the set of polygonal boundaries. If the boundary of a hole makes sharp turns, then the method will produce different polygonal segments, one for each edge. In this case, one has – as a postprocessing step – to glue the different segments together.

Clearly, the complexity of this step is $O(q^2)$. An example of the whole process is shown in the Figure 5.

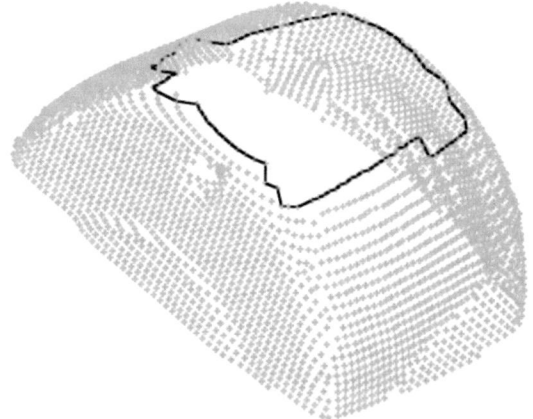

**Fig. 5.** Building the boundary of a hole in scanned data set.

Unfortunately, there is no guarantee of the topological correctness for the boundary obtained by this algorithm. The result strongly depends on the distribution of the points and errors in $P$. Topologically correct solutions could be guaranteed by using so–called "snakes" (active curves) from computer vision [10]. These techniques, however, always need an initialization, which has to be provided by the user. Also, since we do not assume that the given data are triangulated, the rules for the evolution of a "snake" on a point cloud are not fully obvious. Still, this method may have some potential, and we intend to look into it in the future.

*Implementation Issues.* Several user–defined parameters appear in the above algorithms. They strongly depends on statistical characteristics of the set $P$ such as the average distance of closest neighbours, curvature bounds of the scanned surface, etc. They can be locally adapted according to these parameters.

The parameters are estimated according to statistical properties. The parameter $\varepsilon_{\mathrm{bdist}}$ is set to a quarter of the average distance of the closest neighbour in the set. The parameter $\varepsilon_{\mathrm{bangle}} \in [-1, 1]$ was chosen equal to $-0.5$ in our examples. It controls the size of the feasible wedge during the boundary point detection.

Note that "outliers" on the boundary of a hole may sometimes not be detected as boundary points, since its neighbourhood may not contain sufficiently many points to get reliable estimates. However, outliers have never been a problem for the further processing, and it may sometimes even be better to ignore them.

## 5   Algebraic Surface Fitting

The fitting of algebraic surfaces to given data have been discussed in several publications, see [9] and references cited therein. The method in [9] can be used to solve the following problem:

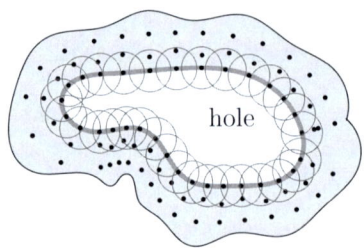

**Fig. 6.** Tubular neighbourhood of a boundary.

Consider a set of points $\{\mathbf{q}_0, \ldots, \mathbf{q}_r\}$ inside a tetrahedron $\triangle(\mathbf{v}_0\mathbf{v}_1\mathbf{v}_2\mathbf{v}_3)$. Let $\mathbf{n}_i$ be a unit vector (representing the normal) associated with every point $\mathbf{q}_i$. Find an algebraic surface of degree $n$ matching simultaneously the points and the associated normal vectors.

Considering simultaneously points and associated normals has two major advantages. First, the solution can be found by solving a linear system of equations. (Other methods require solving an eigenvalue problem instead [16].) Second, the use of the normals helps to keep unwanted branches away from the region of interest, since the resulting surface can be expected to be regular in the neighbourhood of the data. (Alternatively, regularity can always be achieved by adding suitable "tension terms" to the objective function, or by imposing monotonicity conditions or sign conditions, see [2]. While the first approach tends to "flatten" the shape, the second one requires more sophisticated solvers.)

We represent the algebraic surface as the zero set of a polynomial $f$ of degree $n$ in Bernstein-Bézier form over a simplex $\mathbf{V}$, see (5). The simplex $\mathbf{V}$ is chosen such that it contains the given points.

In order to fill the hole(s) in the data, we want to extend the shape of the point cloud according to the shape of the vicinity of the boundary. Thus, in order to match the boundary data by a surface

$$\mathbf{S} = \{\mathbf{x} \in \mathbf{V} : f(\mathbf{x}) = 0\}, \tag{15}$$

we take into account not only the points of $P$ from the detected boundary polygon $Q$, but also the points in a certain tubular neighbourhood $N(Q)$. The tubular neighbourhood is approximated by a union of balls,

$$N(Q) = \bigcup_{i=0}^{r} B_{\varepsilon_{\text{tube}}}(\mathbf{q}_i), \tag{16}$$

where $B_\varepsilon(\mathbf{p})$ is the ball centered at $\mathbf{p}$ with radius $\varepsilon$ (see Figure 6 for 2d analogon).

To each point we assign the weight $\mu_i$ for $i \in \{0, \ldots, r\}$. It is proportional to number of balls $B_{\varepsilon_{\text{tube}}}(\mathbf{q}_j)$ in (16), which contain the point $\mathbf{q}_i$. For points further away from the hole, this weight decreases. As another possibility, one may choose these weights $\mu_i$ according to the distance $d(\mathbf{p}_i, Q)$ from the boundary polygon $Q$. Clearly, this is more expensive to compute.

The objective function has the form

$$F(\boldsymbol{b}) = w_0 D(\boldsymbol{b}) + w_1 N(\boldsymbol{b}) + w_2 T_1(\boldsymbol{b}) \ \{+w_3 T_2(\boldsymbol{b})\} \tag{17}$$

where

$$D(\boldsymbol{b}) = \sum_{k=0}^{r} \mu_k f(\mathbf{q}_k)^2, \tag{18}$$

$$N(\boldsymbol{b}) = \sum_{k=0}^{r} \mu_k \|\nabla f(\mathbf{q}_k) - \mathbf{n}_k\|^2, \tag{19}$$

$$T_1(\boldsymbol{b}) = \int_{\mathbf{V}} f_{xx}^2 + f_{yy}^2 + f_{zz}^2 + 2f_{xy}^2 + 2f_{xz}^2 + 2f_{yz}^2 \, dV, \tag{20}$$

$$T_2(\boldsymbol{b}) = \int_{\mathbf{V}} f_{xxx}^2 + \cdots + f_{zzz}^2 \, dV \tag{21}$$

and $\boldsymbol{b} = (b_{\mathbf{i}})_{|\mathbf{i}|=n}$ are the coefficients in (5) and $w_0, w_1, w_2, w_3 \in \mathbb{R}_+$ are constant weights. The term $T_1(\boldsymbol{b})$ ($T_2(\boldsymbol{b})$) is a "tension term", which can be used to pull the solution towards a plane (quadric surface). It should be used only if it is needed for avoiding singularities (unwanted branches of the algebraic surface). Otherwise, one may set $w_2 = 0$ ($w_3 = 0$). Clearly, the value of the tension term also depends on the shape of the simplex $\mathbf{V}$.

Since the objective function is a quadratic positive definite function of the coefficients $\boldsymbol{b}$, we can find its minimum by solving the linear system

$$\nabla F(\boldsymbol{b}) = \mathbf{0}. \tag{22}$$

The resulting formulas have been gathered in Appendix C.

As an example, we applied the fitting procedure to the upper part of the babyphone point cloud (see Figure 1, left), in order to fill it with an algebraic surface of degree 4. The choice of the degree is a useful compromise between flexibility and number of degrees of freedom (shape parameters). According to our experience, degree 4 was sufficient in most cases. The part of the surface corresponding to the inner part of the hole is shown in Figure 7. It was produced with the weights $w_0 = 1000$, $w_1 = 550.0$ and $w_2 = 1.0$.

*Remark 1.* If the weight $w_2 \neq 0$, then the algebraic surface fitting *reproduces planes.* If $w_2 = 0$, but $w_3 \neq 0$, then it *reproduces spheres and circular cylinders.* Indeed, if both the points and the associated normals are sampled from an algebraic surface, whose gradients have the same length everywhere, then this algebraic surface is the unique minimizer of the objective function. Clearly, planes, circular cylinders and spheres enjoy this property. Other quadrics (such as cones, ellipsoids, etc.) are therefore generally not preserved.

*Remark 2.* If $n \geq 4$ and $w_2 = w_3 = 0$, the data the data taken from a sphere (or circular cylinder) produces a singular system (22), since any product of the equation of the sphere (or cylinder) with a cocentric sphere (or coaxial cylinder) would be a solution.

For holes with more complicated geometry, it is necessary to use algebraic spline surfaces (instead of single patches), and/or to split the data.

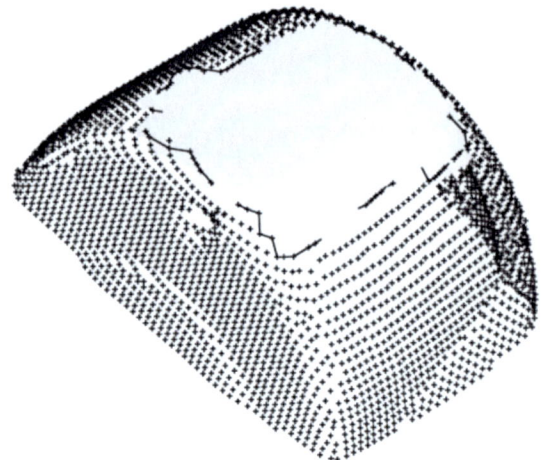

**Fig. 7.** Fitting a surface to the boundary data of the upper hole in the babyphone data set.

## 6   Generating the Points

In order to finish the filling of the hole, we need to generate sample points from the surface, which has been fitted to the boundary data. Clearly, we need only points from the "inner part" of the hole. More precisely, let $S \subset \mathbb{R}^3$ be an algebraic surface (or a part of it) defined by the equation $f(\mathbf{x}) = 0$, with a chosen orientation. Let $\mathbf{c} \colon [0, 1] \mapsto S$ be a positively oriented simple closed curve on the surface $S$. We have to find a finite subset of points $P' \subset S$ such that all points are inside the region of $S$ enclosed by the curve $\mathbf{c}$.

Our algorithm is based on a triangulation of the surface $S$, and on approximate geodesic offsets of the boundary curve $\mathbf{c}$.

**Algorithm 3 (Generation of Points)**

1. Approximate the given surface $S$ in region of interest by a triangulation $\mathcal{T}$.
2. Project the curve $\mathbf{c}$ onto the triangulation and find the region $\mathcal{R} \subset \mathcal{T}$ enclosed by the projected curve.
3. Generate a set $P_{\mathcal{R}}$ of points in the $\mathcal{R}$ and compute the set $P_S$ of their corresponding footpoints on $S$.
4. Choose a uniform subset of $P_S$ by geodesic offsetting, in order to get $P'$.

*Step 1.* We have used the algorithm of "marching triangles". The global strategy can be found in [1]. In general, this algorithm provides relatively nice triangulations; no post-processing is needed.

The original algorithm had to be modified in order to avoid thin triangles during the approximation as far as possible. Similarly, the final phase of the algorithm ("connecting the cracks") has been improved.

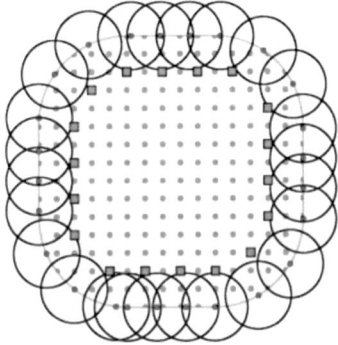

**Fig. 8.** Generating of points approximating geodesic offsets.

*Step 2.* The polygon $\mathbf{c}_0\mathbf{c}_1 \ldots \mathbf{c}_r$ which approximates the boundary curve $\mathbf{c}$ is projected onto the triangulation $\mathcal{T}$. We detect all triangles of $\mathcal{T}$ which are enclosed by the projected boundary curve; they form the domain $\mathcal{R}$.

*Step 3.* The generation of the points on the $\mathcal{R}$ and the their projection back to the $\mathbf{S}$ is straightforward. Since the triangles have approximately the same size, we generate – for each one – the same number of points.

*Step 4.* We approximate the geodesic offsets of the boundary curve. (For more detail on numerical methods of computing geometric offsets, see e.g. [13].) It is assumed that we have a "sufficiently dense" set of points on $\mathbf{S}$. In addition, the user has to specify a radius $\varepsilon_r$ for offsetting; it should be equal to the average point distance in the point cloud.

- Delete from $P_{\mathbf{S}}$ all the points in a tubular neighbourhood of the boundary curve $\mathbf{c}$.
- Repeat until $P_{\mathbf{S}}$ is empty:
  For every point on the boundary find the closest point in $P_{\mathbf{S}}$, and add it to the set of boundary points, see Figure 8. Delete all points within a tubular neighbourhood of the new set from $P_{\mathbf{S}}$.

The method has been applied to the two data sets from Figure 1, see Figures 9, 10. As one can see from the tennis ball example, the method for algebraic surface fitting is able to reproduce spheres. We used the degree $n = 4$ and weights $w_0 = 10,000$, $w_1 = 1.0$ and $w_2 = 0.0001$ [3].

*Implementation Issues.* All the tubular neighbourhoods have the radius $\varepsilon_r$. They are always approximated by the union of balls around the data.

---

[3] Due to $w_2 \ll w_1 \ll w_0$, the influence of $T_2(\boldsymbol{b})$ is relatively small. Consequently, the method almost exactly reproduces sphere, even if the condition of Remark 1 is violated.

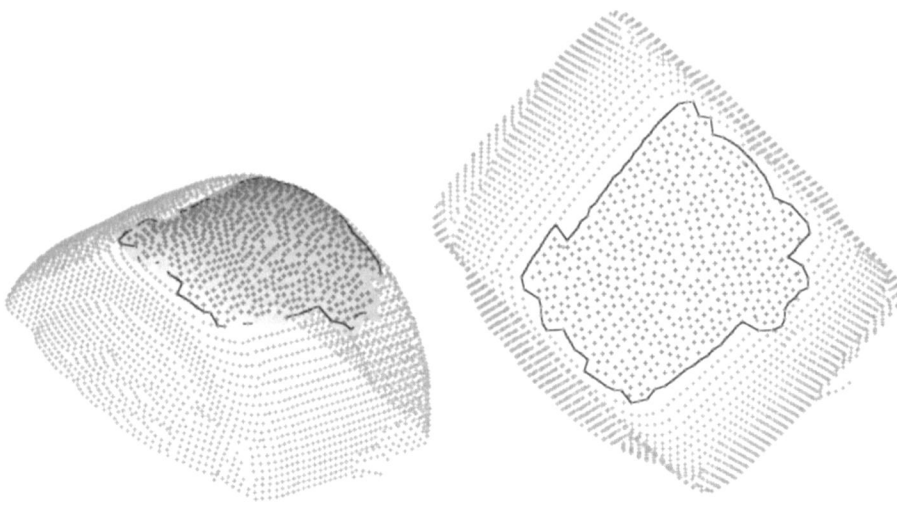

**Fig. 9.** Filling the hole in the babyphone data.

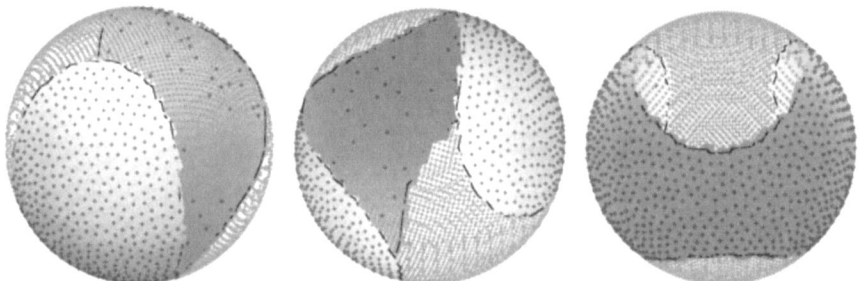

**Fig. 10.** Reconstructing the tennis ball.

The size of the triangle should be smaller than the offsetting radius $\varepsilon_r$. If this is satisfied, then the first part of Step 4 excludes the points which are on the wrong side of the boundary.

*Remark 3.* The number of generated points is proportional to the number of triangles. Hence, it is appropriate to find a balance between the number of points generated per triangle and the size of generic triangle chosen for the approximation of the surface. For medium-sized holes, the number of generated points is reasonable to deal with. In the case of bigger holes, it is recommended to use efficient methods of computational geometry (such as hashing) to delete unwanted points from the generated set.

# 7   Conclusions

We described a method for filling holes in point clouds. After detecting the holes and their boundaries, we fitted an algebraic surface patch to the neighbour-hood of the boundary. Finally, auxiliary points filling the hole were generated by approximating the geodesic offsets of the boundary.

The use of an algebraic fit has several advantages. First, the surface can be obtained without assuming the existence of a suitable parameterization of the data. Second, the method can be applied to holes of more general shapes (such as the tennis ball data, see Figure 10), and more general topologies. For instance, holes with a cylindrical shape can be handled by our algorithm. Finally, the method is able to reproduce planes, circular cylinders and spheres, which are clearly important for engineering applications.

Currently, we implemented only a single patch of an algebraic surface. In order to fit more complicated shapes, that cannot be covered by a single al-gebraic patch, the use of algebraic splines (such as defined by Clough-Tocher, Powell-Sabin or Bajaj macroelements [2,8]) will be more appropriate. In this case, however, the use of suitable tension terms will become more important, since no geometric information is present in the inner part of the hole.

# A   Blossoming

We recall some results associated with the blossoming principle (polar forms). There is a unique bijective mapping between polynomials of degree $n$ and mul-tilinear, symmetric functions. Let $f$ be a polynomial of degree $n$. Then, there is a multilinear symmetric function

$$f_* : \underbrace{\mathbb{R}^4 \times \cdots \times \mathbb{R}^4}_{n\text{-times}} \mapsto \mathbb{R} \tag{23}$$

such that

$$D_{\boldsymbol{\eta}_1,\ldots,\boldsymbol{\eta}_q} f(\mathbf{u}) = \frac{n!}{(n-q)!} f_*(\tilde{\boldsymbol{\eta}}_1 \ldots \tilde{\boldsymbol{\eta}}_q \hat{\mathbf{u}}^{n-q}) \tag{24}$$

where $\hat{\mathbf{u}} = (1, u_1, \ldots, u_d)^\top$ and $\tilde{\boldsymbol{\eta}}_i = (0, \eta_{i1}, \ldots, \eta_{id})^\top$ are the projective exten-sions of points and vectors. Note that the derivative of order zero is the value of the polynomial itself; the formula is valid in this case, too. Hence, the polar form can be used to evaluate all directional derivatives of polynomial $f$. Clearly, the existence of a polynomial from a given multilinear symmetric function follows directly.

A multilinear function (23) is determined by $\binom{n+3}{3}$ coefficients $b_{\mathbf{i}}$ which can be seen as the values of $f_*$ on the $e_0^{i_0} e_1^{i_1} e_2^{i_2} e_3^{i_3}$ for an arbitrary but fixed basis $\{e_0, \ldots, e_3\} \in \mathbb{R}^4$.

# B   Multiplying Multivariate Polynomials in Bernstein-Bézier Form

In the sequel we will need multiplication formulas for polynomials in Bernstein-Bézier form. We briefly recall the algorithm, see [4] for a more general approach (also covering the composition of polynomials in Bernstein–Bézier form). Let

$$f(\mathbf{x}) = \sum_{|\mathbf{i}|=n} B_{\mathbf{i}}^n(\mathbf{x}) b_{\mathbf{i}} \quad \text{and} \quad g(\mathbf{x}) = \sum_{|\mathbf{j}|=m} B_{\mathbf{j}}^m(\mathbf{x}) c_{\mathbf{j}} \tag{25}$$

be two polynomials of degree $n$ and $m$. Their product in the corresponding Bernstein-Bézier basis,

$$h(\mathbf{x}) = \sum_{|\mathbf{k}|=m+n} B_{\mathbf{k}}^{n+m}(\mathbf{x}) d_{\mathbf{k}}, \tag{26}$$

has the coefficients

$$d_{\mathbf{k}} = \sum_{\mathbf{i}+\mathbf{j}=\mathbf{k}, |\mathbf{i}|=n, |\mathbf{j}|=m} \frac{\binom{|\mathbf{i}|}{\mathbf{i}}\binom{|\mathbf{j}|}{\mathbf{j}}}{\binom{|\mathbf{k}|}{\mathbf{k}}} b_{\mathbf{i}} c_{\mathbf{j}}. \tag{27}$$

Finally, we recall the identity

$$\int_{\mathbf{V}} f(\mathbf{x}) \mathrm{d}V = \frac{\mathrm{vol}(\mathbf{V})}{\binom{n+3}{3}} \sum_{|\mathbf{i}|=n} b_{\mathbf{i}}, \quad \text{where} \quad \mathrm{vol}(\mathbf{V}) = \frac{1}{3!}[\mathbf{v}_0, \mathbf{v}_1, \mathbf{v}_2, \mathbf{v}_3], \tag{28}$$

which is needed for evaluating the tension terms.

# C   Minimizing the Objective Function

The objective function (17) is minimized by solving (22). More precisely, this leads to

$$\frac{\partial}{\partial b_{\mathbf{i}}} F(\boldsymbol{b}) = w_0 A_0 + w_1 A_1 + w_2 A_2, \tag{29}$$

where

$$A_0 = \sum_{k=0}^r 2\mu_k f(\mathbf{q}_k) B_{\mathbf{i}}^n(\mathbf{q}_k), \tag{30}$$

$$A_1 = \sum_{k=0}^r 2\mu_i \langle \nabla f(\mathbf{q}_k) - \mathbf{n}_k, \frac{\partial}{\partial b_i} \nabla f(\mathbf{q}_k) \rangle \tag{31}$$

and

$$A_2 = \int_{\mathbf{V}} \left( 2f_{xx}(\mathbf{x}) \frac{\partial}{\partial b_{\mathbf{i}}} f_{xx}(\mathbf{x}) + 2f_{yy}(\mathbf{x}) \frac{\partial}{\partial b_{\mathbf{i}}} f_{yy}(\mathbf{x}) + 2f_{zz}(\mathbf{x}) \frac{\partial}{\partial b_{\mathbf{i}}} f_{zz}(\mathbf{x}) + \right.$$
$$\left. 4f_{xy}(\mathbf{x}) \frac{\partial}{\partial b_{\mathbf{i}}} f_{xy}(\mathbf{x}) + 4f_{xz}(\mathbf{x}) \frac{\partial}{\partial b_{\mathbf{i}}} f_{xz}(\mathbf{x}) + 4f_{yz}(\mathbf{x}) \frac{\partial}{\partial b_{\mathbf{i}}} f_{yz}(\mathbf{x}) \right) \mathrm{d}V, \tag{32}$$

Equation (30) can be rewritten as

$$A_0 = \sum_{|\mathbf{j}|=n} \left( \sum_{k=0}^{r} 2\mu_k B_{\mathbf{j}}^n(\mathbf{q}_k) B_{\mathbf{i}}^n(\mathbf{q}_k) \right) b_{\mathbf{j}}. \tag{33}$$

In order to simplify (31), we write – according to (24) –

$$D_{\boldsymbol{\eta}} f(\mathbf{x}) = n f_*(\tilde{\boldsymbol{\eta}} \hat{\mathbf{x}}^{n-1}) \tag{34}$$

$$= \sum_{|\mathbf{j}|=n} n \underbrace{(\eta_0 B_{\mathbf{j}-\mathbf{e}_0}^{n-1}(\mathbf{x}) + \eta_1 B_{\mathbf{j}-\mathbf{e}_1}^{n-1}(\mathbf{x}) + \eta_2 B_{\mathbf{j}-\mathbf{e}_2}^{n-1}(\mathbf{x}) + \eta_3 B_{\mathbf{j}-\mathbf{e}_3}^{n-1}(\mathbf{x}))}_{D_{\boldsymbol{\eta}} B_{\mathbf{j}}^n(\mathbf{x})} b_{\mathbf{j}},$$

where $\boldsymbol{\eta}$ is the representation of any vector in barycentric coordinates with respect to the tetrahedron $\mathbf{V}$ and polynomials with negative indices vanish identically. Consequently,

$$A_1 = 2n \sum_{k=0}^{r} \mu_k \left( n \sum_{|\mathbf{j}|=n} \langle \nabla B_{\mathbf{j}}^n(\mathbf{q}_k), \nabla B_{\mathbf{i}}^n(\mathbf{q}_k) \rangle b_{\mathbf{j}} - \langle \mathbf{n}_{\mathbf{j}}, \nabla B_{\mathbf{i}}^n(\mathbf{q}_k) \rangle \right) \tag{35}$$

with

$$\nabla B_{\mathbf{i}}^n(\mathbf{q}_k) = (D_{\mathbf{x}} B_{\mathbf{i}}^n(\mathbf{q}_k), D_{\mathbf{y}} B_{\mathbf{i}}^n(\mathbf{q}_k), D_{\mathbf{z}} B_{\mathbf{i}}^n(\mathbf{q}_k)). \tag{36}$$

Similarly as in (34) one gets

$$D_{\boldsymbol{\eta}_1 \boldsymbol{\eta}_2} f(\mathbf{x}) = \sum_{|\mathbf{j}|=n} D_{\boldsymbol{\eta}_1 \boldsymbol{\eta}_2} B_{\mathbf{j}}^n(\mathbf{x}) b_{\mathbf{j}}$$

where by (24)

$$D_{\boldsymbol{\eta}_1 \boldsymbol{\eta}_2} B_{\mathbf{j}}^n(\mathbf{x}) = n(n-1) f_{\mathbf{j}*}(\tilde{\boldsymbol{\eta}}_1 \tilde{\boldsymbol{\eta}}_2 \hat{\mathbf{x}}^{n-2}).$$

Further, if we denote

$$\Delta_{\mathrm{diag}} B_{\mathbf{i}}^n(\mathbf{x}) = (D_{\mathbf{xx}} B_{\mathbf{i}}^n(\mathbf{x}), D_{\mathbf{yy}} B_{\mathbf{i}}^n(\mathbf{x}), D_{\mathbf{zz}} B_{\mathbf{i}}^n(\mathbf{x}))^\top \tag{37}$$

and

$$\Delta_{\mathrm{offdiag}} B_{\mathbf{i}}^n(\mathbf{x}) = (D_{\mathbf{xy}} B_{\mathbf{i}}^n(\mathbf{x}), D_{\mathbf{xz}} B_{\mathbf{i}}^n(\mathbf{x}), D_{\mathbf{yz}} B_{\mathbf{i}}^n(\mathbf{x}))^\top, \tag{38}$$

the term (32) can be rewritten as

$$A_2 = 2 \int_{\mathbf{V}} \sum_{|\mathbf{j}|=n} \left( \langle \Delta_{\mathrm{diag}} B_{\mathbf{j}}^n(\mathbf{x}), \Delta_{\mathrm{diag}} B_{\mathbf{i}}^n(\mathbf{x}) \rangle \right.$$
$$\left. + 2\langle \Delta_{\mathrm{offdiag}} B_{\mathbf{j}}^n(\mathbf{x}), \Delta_{\mathrm{offdiag}} B_{\mathbf{i}}^n(\mathbf{x}) \rangle \right) b_{\mathbf{j}} \ dV \tag{39}$$

and evaluated using (25)–(28).

## Acknowledgements

This research was supported by the Austrian Science Fund (FWF) through the SFB F013 "Numerical and Symbolic Scientific Computing" at Linz, project 15. The authors wish to thank the referees for their comments which have helped to improve the paper.

# References

1. S. Akkouche and E. Galin. Adaptive implicit surface polygonization using marching triangles. *Computer Graphics Forum*, 20(2):67–80, 2001.
2. C. L. Bajaj. Implicit surface patches. In J. Bloomenthal, editor, *Introduction to implicit surfaces*. Morgan Kaufmann, San Francisco, 1997.
3. U. Clarenz, M. Rumpf, and A. Telea. Robust feature detection and local classification for surfaces based on moment analysis. *IEEE Transactions on Visualization and Computer Graphics*, 2003. submitted, available online at `http://numerik.math.uni-duisburg.de/research/publications.htm`.
4. T. DeRose, R.N. Goldman, H. Hagen, and S. Mann. Functional composition algorithms via blossoming. *ACM Trans. Graph.*, 12(2):113–135, 1993.
5. B. Curless et al. The Digital Michelangelo Project.
`http://graphics.stanford. edu/projects/mich/`, 2000.
6. G. Farin, J. Hoschek, and M.-S. Kim, editors. *Handbook of computer aided geometric design*. North-Holland, Amsterdam, 2002.
7. H. Hoppe, T. DeRose, T. Duchamp, J. McDonald, and W. Stuetzle. Surface reconstruction from unorganized points. *Computer Graphics*, 26(2):71–78, 1992.
8. J. Hoschek and D. Lasser. *Fundamentals of Computer Aided Geometric Design*. AK Peters, 1993.
9. B. Jüttler and A. Felis. Least-squares fitting of algebraic spline surfaces. *Adv. Comput. Math.*, 17(1-2):135–152, 2002.
10. M. Kass, A. Witkin, and D. Terzopoulos. Snakes: Active contour models. *International Journal of Computer Vision*, 1(4):321–331, 1987.
11. L. Ramshaw. Blossoms are polar forms. *Comput. Aided Geom. Des.*, 6(4):323–358, 1989.
12. L. Ramshaw. On multiplying points: The paired algebras of forms and sites. SRC Research Report #169, COMPAQ Corp., 2001.
13. T. Rausch, F.-E. Wolter, and O. Sniehotta. Computation of medial curves on surfaces. In T. Goodman and R. Martin, editors, *The mathematics of surfaces VII*, pages 43–68. Information Geometers, Ltd., 1997.
14. J.-R. Sack and J. Urrutia, editors. *Handbook of computational geometry*. North-Holland, Amsterdam, 2000.
15. R. Schaback. Remarks on meshless local construction of surfaces. In R. Cipolla and R. Martin, editors, *The mathematics of surfaces IX. Proceedings of the 9th IMA conference Cambridge.*, pages 34–58. Springer, London, 2000.
16. R. Taubin. Estimation of planar curves, surfaces, and non-planar space curves defined by implicit equations with applications to edge and range image segmentation. *IEEE Trans. Pattern Anal. Mach. Intelligence*, 13:1115–1139, 1991.
17. K. Tucholsky. Zur soziologischen Psychologie der Löcher. In *Zwischen Gestern und Morgen*. Rohwolt, Hamburg, 1952.
18. T. Varady, R.R. Martin, and J. Cox. Reverse engineering of geometric models - an introduction. *Comput.–Aided Des.*, 29(4):255–268, 1997.

# Trimming Local and Global Self-intersections in Offset Curves Using Distance Maps[*]

Gershon Elber

Computer Science Department, Technion
Haifa 32000, Israel
gershon@cs.technion.ac.il

**Abstract.** The problem of detecting and eliminating self-intersections in offset curves is a fundamental question that has attracted numerous researchers over the years. The interest has resulted in copious publications on the subject.

Unfortunately, the detection of self-intersections in offset curves, and more so, the elimination of these self-intersections are difficult problems with less than satisfactory answers.

This paper offers a simple, and equally important robust, scheme to detect and eliminate local as well as global self-intersections in offsets of freeform curves. The presented approach is based on the derivation of an analytic distance map between the original curve and its offset.

## 1 Introduction and Background

Offsets of curves and surfaces are crucial in many applications from robotic navigation, through CNC machining, to geometric design. This basic operation could be found in virtually any contemporary geometric modeling system or environment. The offset curve $C_d^o(t)$ by amount $d$ to a given rational curve $C(t)$ equals,

$$C_d^o(t) = C(t) + N(t)d, \qquad (1)$$

where $N(t)$ is the unit normal of $C(t)$. The offset of rational curve $C(t)$ is not rational, in general, due to the necessary normalization of $N(t)$.

The fact that $C_d^o(t)$ is not rational introduces an enormous difficulty into the computation of $C_d^o(t)$, as virtually all geometric design systems support only rationals. Hence, $C_d^o(t)$ must be approximated. Numerous publications can be found on this topic of approximating offsets of rational curves as rationals [5,7,10,11,12].

The problem of proper offset computation is reflected in a more gloomy light if one considers why offsets are so useful. The offset to a border line can offer the path to follow at a fixed distance from that border, for robot navigation. Successive offsets from a boundary of a pocket could similarly serve as the toolpath

[*] The research was supported in part by the Fund for Promotion of Research at the Technion, IIT, Haifa, Israel.

M.J. Wilson and R.R. Martin (Eds.): Mathematics of Surfaces 2003, LNCS 2768, pp. 213–222, 2003.
© Springer-Verlag Berlin Heidelberg 2003

along which to drive a CNC cutter. In geometric modeling, the offset of some shape is a simple way of building a constant thickness wall out of the shape.

Given a general curve as is presented in Figure 1, in all these applications, be it for robotic navigation, CNC machining, or geometric modeling, the offset result shown in Figure 1 is not the one typically desired. Nonetheless and while it is the accurate offset, the self-intersections in the resulting computed offset are expected to be trimmed away.

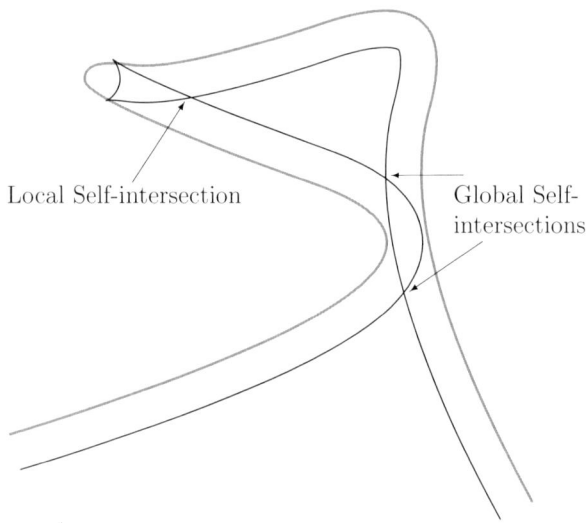

Local Self-intersection

Global Self-intersections

**Fig. 1.** A general curve (in gray) and its accurate offset. Both local and global self-intersecting could occur in the offset curve, and should be trimmed away.

The self-intersections in offsets are typically classified into two types. Let $\kappa$ be the curvature of curve $C(t)$. *Local self-intersections* in offset curves are intersections that are due to regions in $C(t)$ that present a radius of curvature, $1/\kappa$, that is smaller than $d$, the offset distance.

Additionally, two unrelated points in $C(t)$, $C(t_0)$ and $C(t_1)$, that are close, could intersect in the offset curve. If

$$C_d^o(t_0) = C(t_0) + N(t_0)d = C(t_1) + N(t_1)d = C_d^o(t_1),$$

$C_d^o(t_0)$ and $C_d^o(t_1)$ are at the same location and a *global self-intersection* will result.

The detection and trimming of local, and more so, global self-intersections in offset curves are considered to be difficult problems. A monotone curve cannot self-intersect, which leads to a conceptually simple, yet computationally complex approach. We can subdivide $C_d^o(t)$ into monotone regions, and then intersect each of the regions against all other regions, in order to detect all the self-intersection locations. A somewhat simplified approach, taken in [10], converts the curves to a piecewise linear approximation and processes all the (monotone) linear segments using a plane sweep [1] scheme.

Solutions to eliminate the self intersecting regions in closed curves that are piecewise-lines and -arcs were also proposed in the past, by computing the Voronoi map of the curves, [13,2,8]. The bisectors forming the Voronoi diagram directly hint on the distance from the boundary (the closed curve). As one moves along the bisectors starting from the boundary, the distance to the boundary increases monotonically. Together, all the bisectors' arrangment that form the Voronoi diagram, could be viewed as a distance map (proximity map in [8]). Then, the proper offset could be extracted as the iso-surface of the Voronoi diagram seen as height field with the distance mapped to height. The extension of this approach for freeform shapes is difficult. The bisectors between points, lines and arc are all conics. Unfortunately, the bisectors of rational planar curves are not rational, in general [6], and forming the complete Voronoi diagram of freeform planar shape is still an open question.

Let $T(t)$ and $T_d^o(t)$ be the tangent fields of $C(t)$ and $C_d^o(t)$, respectively. For a self-intersection-free offset curve, $T(t)$ and $T_d^o(t)$ are parallel, for all $t$. In [3], the mutual characteristic of the tangent fields of $C(t)$ and $C_d^o(t)$ is used to detect local self-intersections. Consider,

$$\delta(t) = \langle T(t), T_d^o(t) \rangle. \tag{2}$$

$\delta(t)$ is negative only at the neighborhood of a local self-intersection as the tangent field of $C_d^o(t)$ flips its direction, for $1/\kappa$ that is smaller than $d$.

Global self-intersections are more difficult to detect. Nevertheless, the need to robustly detect and eliminate self-intersections stems from the simple fact that a failure in the detection and/or the elimination process would take the robot into a collision path or will gouge into the part on the CNC machine.

In this paper, we present a simple yet robust scheme to detect and eliminate self-intersections in offsets of freeform planar curves. While we do employ the concept of a distance maps, we do attempt to build the complete Voronoi diagram for the given curve. Instead, the distance function is computed only for a small neighborhood of the computed offsets. The rest of this paper is organized as follows. In Section 2, the proposed offset trimming approach is discussed, an approach that is based on an analytic distance function computation. In Section 3, some examples are presented and finally, we conclude in Section 4.

## 2   The Offset Trimming Approach

Consider the rational offset approximation of rational curve $C(t)$ by amount $d$, $C_{d_\epsilon}^o(t)$, where $\epsilon \in \mathbb{R}^+$ denotes the accuracy of the approximation. That is, the offset distance is bound to be between $d \pm \epsilon$. The presented trimming process of self-intersections is independent of the offset approximation scheme. Henceafter and unless otherwise stated, we assume $C_{d_\epsilon}^o(t)$ is parameterized independently of $C(t)$ as $C_{d_\epsilon}^o(r)$. Consider the new distance square function of,

$$\Delta_d^2(r,t) = \langle C(t) - C_{d_e}^o(r), C(t) - C_{d_e}^o(r) \rangle. \tag{3}$$

If no self-intersection occurs in $C_{d_e}^o(r)$, then $\Delta_d^2(r,t) \geq (d-\epsilon)^2$. In contrast, if $C_{d_e}^o(r)$ is self-intersecting, there exist points in $C_{d_e}^o(r)$ that are closer than $d-\epsilon$ to $C(t)$. Therefore, any pair of points $C_{d_e}^o(r)$ and $C(t)$ such that $\Delta_d^2(r,t) < (d-\epsilon)^2$ hints at a self-intersection. Moreover, any point $C_{d_e}^o(r)$ for which there exists a point $C(t)$ such that $C(t) - C_{d_e}^o(r) < d-\epsilon$, for some $t$, must be trimmed away.

Let $\rho \in \mathbb{R}^+$ be another small positive real value and let $\mathcal{D}(r,t) = \Delta_{d_e}^2(r,t) - (d-\epsilon-\rho)^2$. Any point in the zero set of $\mathcal{D}(r_0,t_0)$ represents two points, $C(t_0)$ and $C_{d_e}^o(r_0)$, that are $(d-\epsilon-\rho)$ apart. Every such point $C_{d_e}^o(r_0)$ must be purged away as a self-intersecting point. We denote this trimming process of $\rho$ below the offset approximation, a $\rho$-accurate trimming or $\rho$-trimming, for short. Hence, we now offer the following algorithm to detect and eliminate the self-intersection regions:

**Algorithm 1**
**Input:**
  $C(t)$, A rational curve;
  $C_{d_e}^o(r)$, A rational approximation offset of $C(t)$ by distance $d$
$\qquad\qquad\qquad\qquad\qquad\qquad\qquad$ and tolerance $\epsilon$;
  $\rho$, a trimming tolerance for the self-intersections.

**Output:**
  $\overline{C}_{d_e}^o(r)$, A rational approximation offset of $C(t)$ by distance $d$,
$\qquad\qquad\qquad\qquad\qquad$ tolerance $\epsilon$, and $\rho$-trimming;

**Begin**
  $\Delta_d^2(r,t) \Leftarrow \langle C(t) - C_{d_e}^o(r), C(t) - C_{d_e}^o(r) \rangle$;
  $\mathcal{D}(r,t) \Leftarrow \Delta_d^2(r,t) - (d-\epsilon-\rho)^2$;
  $\mathcal{Z} \Leftarrow$ the zero set of $\mathcal{D}(r,t)$;
  $\mathcal{Z}_r \Leftarrow$ the projection of $\mathcal{Z}$ onto the $r$ axis;
  $\overline{C}_{d_e}^o(r) \Leftarrow$ the $r$ domain(s) of $C_{d_e}^o(t)$ not included in $\mathcal{Z}_r$;
**End.**

$\Delta_d^2(r,t)$ and $\mathcal{D}(r,t)$ are clearly piecewise rational, provided $C(t)$ and $C_{d_e}^o(t)$ are, and that they are the result of products and differences of piecewise rational functions. See, for example, [4].

With $\mathcal{D}(r,t)$ as a piecewise rational, the zero set, $\mathcal{Z}$, could be derived by exploiting the convex hull and subdivision properties, yielding a highly robust divide and concur zero set computation that is reasonably efficient. Nevertheless, we are not really interested in the zero set of $\mathcal{D}(r,t)$, but merely in all $r$ such that $\overline{C}_{d_e}^o(r)$ is closer to $C(t)$ more than $(d-\epsilon-\rho)$, for some $t$. Hence, $\mathcal{Z}$ is projected onto the $r$ axis, as $\mathcal{Z}_r$. The domain of $r$ covered by this projection prescribes the regions of $C_{d_e}^o(r)$ that must be purged away.

At this point, it is crucial to emphasize the total separation between the two stages: the offset approximation step and the self-intersection trimming stage.

The result of the offset approximation step, $C^o_{d_\epsilon}(r)$, can be an arbitrary offset curve approximation of $C(t)$ that is accurate to within $\epsilon$. Furthermore, $C^o_{d_\epsilon}(r)$ can assume any regular parameterization. Algorithm 1 assumes nothing of the parameterization of the curve and its offset, in contrast, for example, to the local self-intersection detection and elimination method proposed in [3].

The outcome of Algorithm 1 is a subset of $C^o_{d_\epsilon}(r)$. The latter comprises of curve segments that have no point closer than $(d - \epsilon - \rho)$ to $C(t)$. Clearly, the offset path should be computed conservatively, to be a bit more than $d$. $\epsilon$ and $\rho$ could be added to $d$, computing an offset approximation to a distance of $(d + \epsilon + \rho)$, only to ensure a minimal distance constraint of $d$.

A point on $C^o_{d_\epsilon}(r_0)$ is said to be *a match to point* $C(t_0)$ if it matches location $C(t_0) + N(t_0)d$. Denote by $\Delta^2_d(t)$ the distance square between the matched point on $C^o_{d_\epsilon}(r)$ to $C(t)$ and $C(t)$. The need to exploit a positive $\rho$ value stems from the fact that $d + \epsilon \geq \Delta^2_d(t) \geq d - \epsilon$. For numerical stability, $\rho$ should be as large as possible, reducing the chance of detecting matched points as self-intersections. In contrast, the larger $\rho$ is, the bigger the likelihood that we will miss small self-intersections. In practice, $\rho$ was selected to be between 95% and 99% of $d$.

By selecting $\rho > 0$, the curve segments that result from Algorithm 1 are not exactly connected. Instead, a sequence of curve segments is output with end points that are very close to each other. Numeric Newton Raphson marching steps at each such close pair of end points could very quickly converge to the exact self-intersection location. Very few steps are required to converge to the highly precise self-intersection location. In Section 3, we present several examples that demonstrate this entire procedure, including the aforementioned numerical marching stage.

## 3   Examples and Extensions

We now present several examples of trimming of both local and global self-intersections in offset curves, via the (square of the) distance map, $\Delta^2_d(r,t)$. In Figure 2, the example from Figure 1 is presented again. In Figure 2 (a), the original curve and its offset are presented. In Figure 2 (b), the result of trimming the curve using the $\Delta^2_d(r,t)$ functions is shown while Figure 2 (c) presents the same trimmed offset curve after the numerical marching stage.

In Figure 3, the log of the distance (square) function, $\Delta^2_d(r,t)$, is presented for the curve in Figure 2. The minimal distance is in the order of the offset distance itself whereas the maximal distance is in the order of the diameter of the curve and hence, the figure is shown in a logarithmic scale. Also presented in Figure 3 are the zero set, $\mathcal{Z}$, of $\Delta^2_d(r,t) - (d - \epsilon - \rho)^2$ and its projection, $\mathcal{Z}_r$, on the $t = 0$ axis. In this case, the $r$ axis is divided into four valid domains by three sub-regions that are self-intersecting. The first and third black sub-regions along the $r$ axis are due to the global self-intersection of the curve in Figure 2, whereas the middle large black sub-region is due to the local self-intersection in the curve. Extracting the four valid sub-regions, we see the resulting curve segments in Figure 2 (b). A numerical marching step completes the computation in Figure 2 (c).

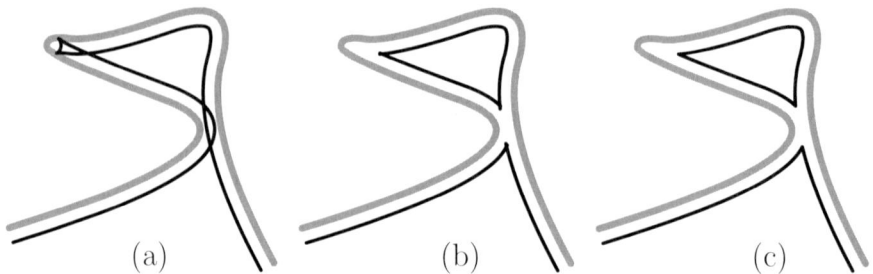

**Fig. 2.** A simple curve and its offset, from Figure 1. The curve and its offset are presented in (a). (b) is the result of $\rho$-trimming the curve using $\Delta_d^2(r,t)$ whereas, in (c), the result is improved using numerical marching.

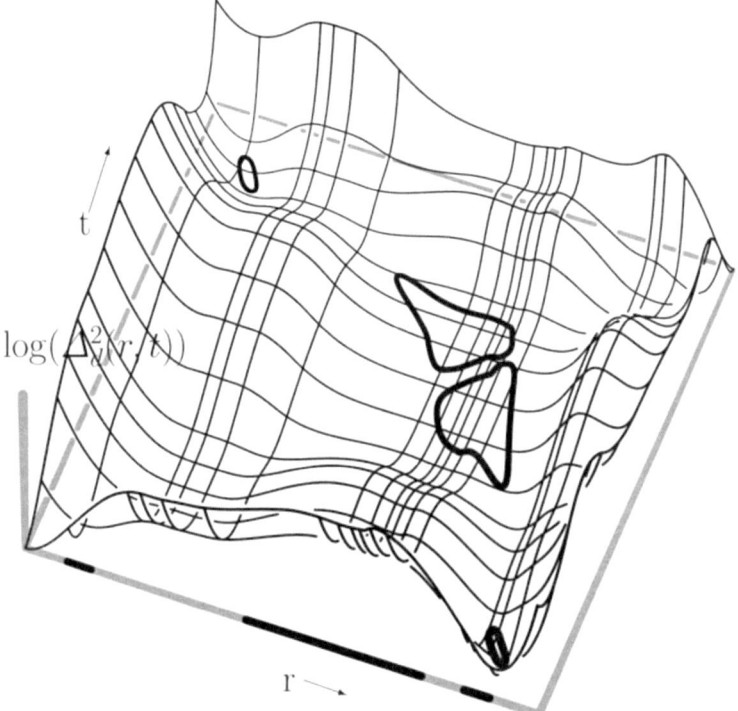

**Fig. 3.** The distance function, $\Delta_d^2(r,t)$, of the curve in Figure 2, on a logarithmic scale. Also shown, in thick lines, are the zeros, $\mathcal{Z}$, of $\Delta_d^2(r,t) - (d - \epsilon - \rho)^2$ as well as the projection of $\mathcal{Z}$ on the $t = 0$ axis.

Figure 4 presents another example of a curve with several offsets to both directions. In Figure 4 (a), the original curve is shown (in gray) with the offsets. With the aid of the distance function square, $\Delta_d^2(r,t)$, the self-intersections are $\rho$-trimmed in (b), whereas the result of applying the numerical marching stage is presented in Figure 4 (c).

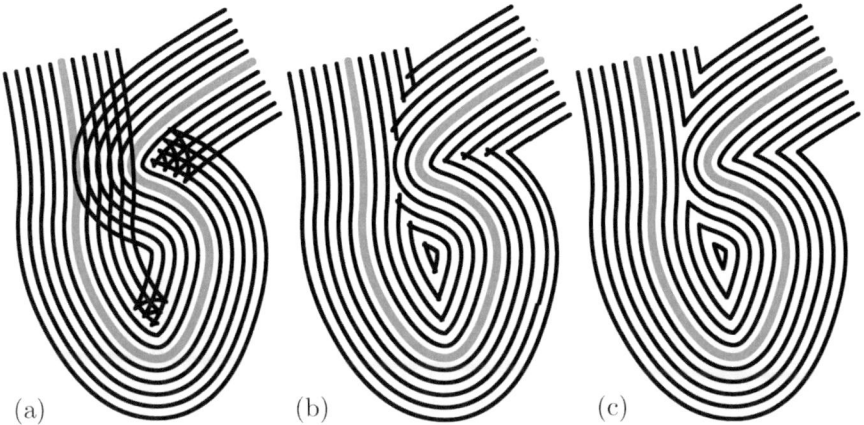

(a)                          (b)                          (c)

**Fig. 4.** Another example of a curve (in gray) and its offsets is shown in (a). (b) is the result of $\rho$-trimming the curve using $\Delta_d^2(r, t)$ and (c) is the result of applying the numerical marching stage.

Figures 5 and 6 present two more complex examples. Here (a) is the original curve (in gray) and its offsets, and (b) is the result of $\rho$-trimming, $\rho = 95\%$, of the self-intersections with the aid of the $\Delta_d^2(r, t)$ function. In all the examples presented in this work, the trimming distance $\rho$ was from 95% to 99% of the offset distance, with an offset tolerance about ten times better (i.e. offset accuracy of 99% to 99.9% of the offset distance). (c) and (d) in Figures 5 and 6 present the result of trimming at $\rho = 95\%$ and $\rho = 99\%$ of the offset distance, respectively. Small local self-intersections escape the $\rho$-trimming step at $\rho = 95\%$ but are properly trimmed at $\rho = 99\%$.

These small local self-intersections could clearly appear at any percentage of the $\rho$-trimming distance, below 100%. In many applications, such as robotics and CNC machining, local and arbitrary small self-intersections will enforce large accelerations along the derived path, and hence are highly undesired. One could employ the local self-intersection test presented in Equation (2) as another filtering step that should completely resolve such small local events.

The computation of the offset curves as well as the trimming of the curves in Figures 5 and 6 took about one minute on a modern PC workstation.

## 4    Conclusions and Future Work

We have presented a robust and reasonably efficient scheme to trim both local and global self-intersections in offsets of freeform planar curves. No complete distance map was defined for the plane, which is highly complex computationally. Instead, the distance function was examined only for the neighborhood of the computed offset curve.

The presented scheme is robust in the sense that the trimmed offset curve that result is at least trimming distance apart, $(d - \epsilon - \rho)$, from the original

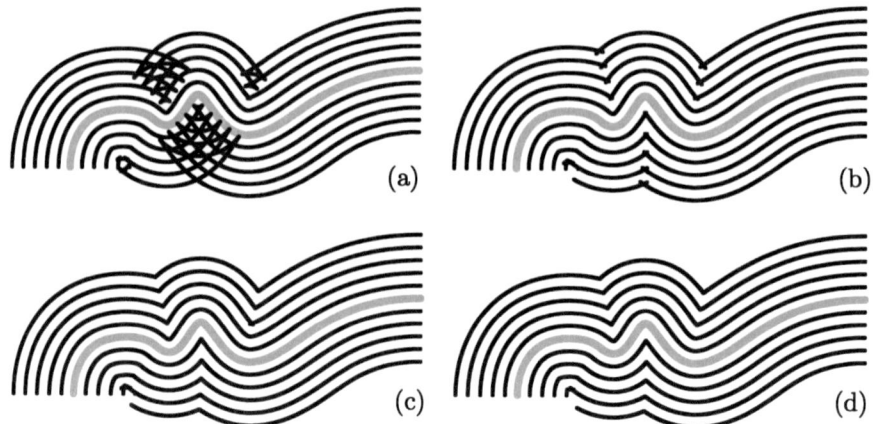

**Fig. 5.** (a) presents the original curve (in gray) and its offsets, while (b) is the result of $\rho$-trimming, $\rho = 95\%$, of the offset with the aid of $\Delta_d^2(r,t)$. (c) and (d) present two different $\rho$-trimming percentages of 95% and 99%, respectively, after a numerical marching stage.

curve. Nevertheless, since the trimming distance must be some finite distance smaller than the offset approximation, small loops such as those from local self-intersections might still exist and a combined scheme that employs an approach similar to that of [3] to detect and eliminate local self-intersection, using Equation (2), could provide the complete solution.

The potential of extending this trimming offset approach to surfaces is an immediate issue that needs to be considered, in this context. While trimming of self-intersection of curves is considered a difficult problem, the question of trimming self-intersections in offset surfaces is far more complex. An even more challenging question is the issue of trimming self-intersections in related applications such as the computation of bisector sheets. The topology of the self-intersections in offsets of surfaces is exceptionally complex, which makes this problem extremely difficult to handle. Yet, we are hopeful that the distance map could aid simplifying this problem by globally examining the distance function. Let $S(u,v)$ and $S_{d_\epsilon}^o(u,v)$ be a freeform rational surface and its rational offset approximation with tolerance $\epsilon$, and let

$$\Delta_d^2(u,v,r,t) = \langle S(u,v) - S_{d_\epsilon}^o(r,t), S(u,v) - S_{d_\epsilon}^o(r,t) \rangle. \tag{4}$$

The zero set of $\Delta_d^2(u,v,r,t) - (d - \epsilon - \rho)^2$, projected onto the $(u,v)$ domain, will prescribe the domains of $S(u,v)$, which need to be purged. Trimming curves could then be constructed in the parametric domain of $S(u,v)$, representing the resulting $\rho$-trimmed offset as a trimmed tensor product surface. The question of numerically improving the $\rho$-trimming for offset surfaces is more difficult, and while probably feasible, will have to be dealt with more cautiously.

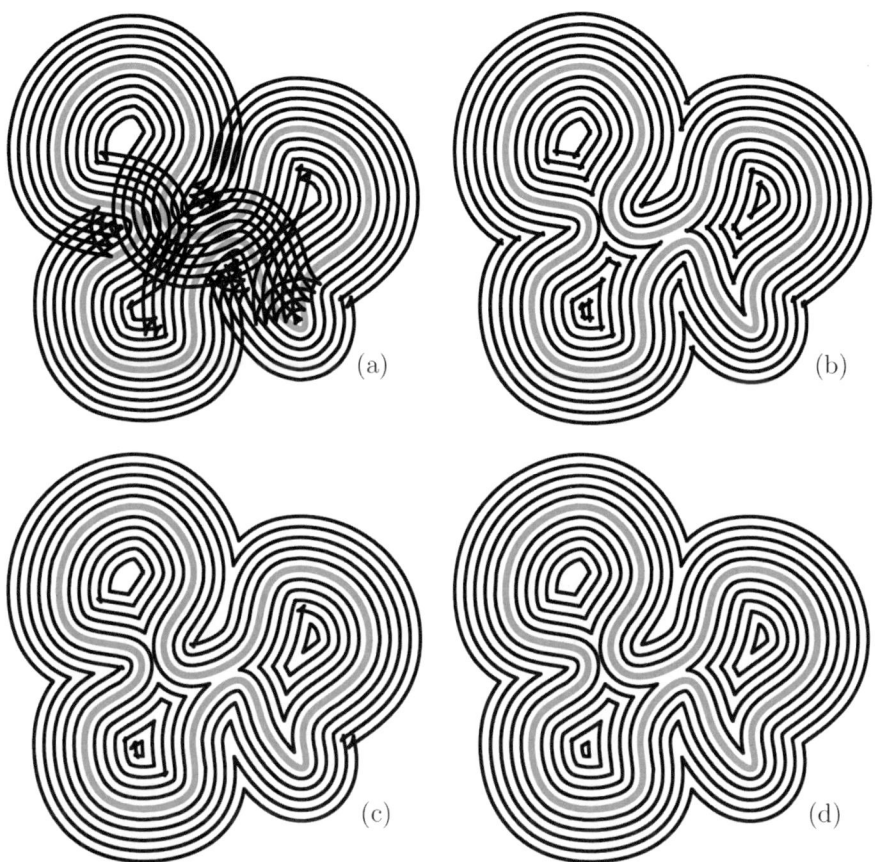

**Fig. 6.** (a) presents the original curve (in gray) and its offsets, while (b) is the result of $\rho$-trimming, $\rho = 95\%$, of the offset with the aid of $\Delta_d^2(r, t)$. (c) and (d) present two different trimming percentages of $95\%$ and $99\%$, respectively, with a numerical marching stage.

This same approach could also be employed to handle the trimming of self-intersections in *variable distance* offsets of curves and surfaces where we now seek the zero set of $\Delta_d^2(r, t) - (d(t) - \epsilon - \rho)^2$ and $\Delta_d^2(u, v, r, t) - (d(u, v) - \epsilon - \rho)^2$, with the distance being a function of the parametric location.

## Acknowledgment

The implementation of the presented algorithm was conducted with the aid of the IRIT [9] modeling environment.

# References

1. M. de Berg, M. van Kreveld, M. Overmars, O. Schwarzkopf. "Computational Geometry, Algorithms, and Applications (2nd ed.)", Springer-Verlag, Berlin, 2000.

2. J. J. Chou. "NC Milling Machine Toolpath Generation for Regions Bounded by Free Form Curves and Surfaces." PhD thesis, Department of Computer Science, University of Utah, April 1989.

3. G. Elber and E. Cohen. "Error Bounded Variable Distance Offset Operator for Free Form Curves and Surfaces." International Journal of Computational Geometry & Applications, Vol 1, No 1, pp 67-78, March 1991.

4. G. Elber and E. Cohen. "Second Order Surface Analysis Using Hybrid Symbolic and Numeric Operators." Transactions on Graphics, Vol 12, No 2, pp 160-178, April 1993.

5. G. Elber, I. K. Lee, and M. S. Kim. "Comparing Offset Curve Approximation Methods." CG&A, Vol 17, No 3, pp 62-71, May-June 1997.

6. G. Elber and M. S. Kim. "Bisector Curves of Planar Rational Curves." Computer Aided Design, Vol 30, No 14, pp 1089-1096, December 1998.

7. R. T. Farouki, Y. F. Tsai and G. F. Yuan. "Contour machining of free-form surfaces with real-time PH curve CNC interpolators", Computer Aided Geometric Design, No 1, Vol 16, pp 61-76, 1999.

8. "Pocket Machining Based on Countour-Parallel Tool Paths Generated by Means of Proximity Maps." Computer Aided Design. Vol 26, No 3, pp 189-203. 1994.

9. IRIT 8.0 User's Manual. The Technion—IIT, Haifa, Israel, 2000. Available at `http://www.cs.technion.ac.il/~irit`.

10. I. K. Lee, M. S. Kim, and G. Elber. "Planar Curve Offset Based on Circle Approximation." Computer Aided Design, Vol 28, No 8, pp 617-630, August 1996.

11. Y. M. Li and V. Y. Hsu. "Curve offsetting based on Legendre series." Computer Aided Geometric Design, Vol 15, No 7, pp 711-720, 1998.

12. M. Peternell and H. Pottmann. "A Laguerre geometric approach to rational offsets." Computer Aided Geometric Design, Vol 15, No 3, pp 223-249, 1998.

13. H. Persson. "NC machining of arbitrary shaped pockets". Computer Aided Design. Vol 10, No 3, pp 169-174. 1978.

# Using Line Congruences
# for Parameterizing Special Algebraic Surfaces

## Dedicated to the Memory of Professor Dr. Josef Hoschek

Bert Jüttler and Katharina Rittenschober

Johannes Kepler University, Dept. of Applied Geometry
Altenberger Str. 69, 4040 Linz
bert.juettler@jku.at
http://www.ag.jku.at

**Abstract.** Surfaces in line space are called line congruences. We consider several special line congruences forming a fibration of the three–dimensional space. These line congruences correspond to certain special algebraic surfaces. Using rational mappings associated with the line congruences, it is possible to generate rational curves and surfaces on them. This approach is demonstrated for quadric surfaces, cubic ruled surfaces, and for Veronese surfaces and their images in three–dimensional space (quadratic triangular Bézier surfaces).

## 1   Introduction

Line Geometry – the geometry of lines in three–dimensional space – is a classical part of geometry, whose origins can be traced back to works of Plücker in the 19th century. Differential line geometry studies line manifolds using the techniques provided by differential geometry [11,12]. Recently, computational line geometry [17] has been demonstrated to be useful for various branches of applied geometry, ranging from robot kinematics to computer aided geometric design.

Two–dimensional manifolds of lines are called line *congruences*. A simple example is the system of normals of a surface. Via the Klein mapping, which identifies each line with a point on a hyperquadric in a five–dimensional real projective space, line congruences correspond to surfaces. We are mainly interested in special line congruences, which are equipped with an associated rational mapping.

Linear congruences are the simplest class of line congruences. They have been used for parameterizing the various types of quadric surfaces [8,6,16]. The parameterization is based on the quadratic mapping which is associated with them.

After summarizing this approach from the viewpoint of line geometry, we generalize it to other classes of algebraic surfaces. By using other, more sophisticated line congruences, we derive similar results for the various types of cubic ruled surfaces, and for Veronese surfaces.

M.J. Wilson and R.R. Martin (Eds.): Mathematics of Surfaces 2003, LNCS 2768, pp. 223–243, 2003.
© Springer-Verlag Berlin Heidelberg 2003

The paper is organized as follows. First we summarize some fundamental concepts from line geometry. Section 3 discusses line models of quadric surfaces, which are related to the generalized stereographic projection. Section 4 is devoted to cubic ruled surfaces, which are shown to be closely connected with a certain class of line congruences. Similarly, section 5 deals with Veronese surfaces. Finally, we conclude this paper.

## 2   Line Geometry

In this section we summarize the fundamentals of the geometry of lines in three–dimensional space. Due to space limitations, this section can give only an outline of this fascinating branch of geometry. For further information, the reader should consult suitable textbooks, such as [11,12,17].

### 2.1   Homogeneous Coordinates

Throughout this paper, points in three–dimensional space will be described by *homogeneous coordinate vectors*

$$\mathbf{p} = (p_0, p_1, p_2, p_3)^\top \in \mathbb{R}^4 \setminus \{(0,0,0,0)^\top\}. \tag{1}$$

Any two linearly dependent vectors correspond to the same point. The associated Cartesian vectors of points satisfying $p_0 \neq 0$ are

$$\underline{\mathbf{p}} = (\frac{p_1}{p_0}, \frac{p_2}{p_0}, \frac{p_3}{p_0})^\top. \tag{2}$$

Points with $p_0 = 0$ are called ideal points or *points at infinity;* they can be identified with the $\infty^2$ equivalence classes of parallel lines in three–dimensional space.

### 2.2   Plücker's Line Coordinates

The line $\mathcal{L}$ spanned by two different points $\mathbf{p}$, $\mathbf{q}$ (i.e., with linearly independent homogeneous coordinate vectors) consists of all points

$$\mathbf{x} = \lambda \mathbf{p} + \mu \mathbf{q}, \quad (\lambda, \mu) \in \mathbb{R}^2 \setminus \{(0,0)\}. \tag{3}$$

Consider the $2 \times 2$ determinants $l_{ij} = p_i q_j - p_j q_i$. They produce essentially six different numbers

$$\mathbf{L} = \mathbf{p} \wedge \mathbf{q} = (l_{01}, l_{02}, l_{03}, l_{23}, l_{31}, l_{12})^\top. \tag{4}$$

The components of the vector $\mathbf{L} \in \mathbb{R}^6$, which are called the *Plücker coordinates,* are homogeneous coordinates of the line $\mathcal{L}$. They do not depend on the choice of the points $\mathbf{p}$ and $\mathbf{q}$. Indeed, the two points $\mathbf{p}' = \lambda_0\mathbf{p} + \mu_0\mathbf{q}$ and $\mathbf{q}' = \lambda_1\mathbf{p} + \mu_1\mathbf{q}$ lead to the modified Plücker coordinates $\mathbf{L}'$ which are linearly dependent on $\mathbf{L}$,

$$\mathbf{L}' = \mathbf{p}' \wedge \mathbf{q}' = \det \begin{pmatrix} \lambda_0 & \lambda_1 \\ \mu_0 & \mu_1 \end{pmatrix} \mathbf{L}. \tag{5}$$

The Plücker coordinates $l_{ij}$ of a line satisfy the *Plücker's identity*

$$l_{01}l_{23} + l_{02}l_{31} + l_{03}l_{12} = 0. \tag{6}$$

If $\mathbf{p}$, $\mathbf{q}$ and $\mathbf{r}$, $\mathbf{s}$ are two pairs of points spanning two lines $\mathbf{L}$ and $\mathbf{M}$, respectively, then the determinant of the $4 \times 4$ matrix $[\mathbf{p}, \mathbf{q}, \mathbf{r}, \mathbf{s}]$ can be expanded to

$$\langle \mathbf{L}, \mathbf{M} \rangle = l_{01}m_{23} + l_{02}m_{31} + l_{03}m_{12} + m_{01}l_{23} + m_{02}l_{31} + m_{03}l_{12} = 0. \tag{7}$$

Consequently, the two lines intersect if and only if $\langle \mathbf{L}, \mathbf{M} \rangle = 0$. Plücker's identity is obtained the special case $\mathbf{L} = \mathbf{M}$, i.e., $\frac{1}{2}\langle \mathbf{L}, \mathbf{L} \rangle = 0$.

*Remark 1.* If $\mathbf{q} = (0, v_1, v_2, v_3)$ is chosen as an ideal point, and $\mathbf{p} = (1, \underline{q}_1, \underline{q}_2, \underline{q}_3)$ is the vector of Cartesian coordinates, homogenized by adding a leading 1, then the Plücker coordinates

$$\mathbf{L} = (\ \underbrace{v_1, v_2, v_3}_{\text{direction vector}}\ , \underbrace{(v_1, v_2, v_3) \times (\underline{q}_1, \underline{q}_2, \underline{q}_3)}_{\text{momentum vector}})^{\top}. \tag{8}$$

are the so–called momentum vector (which is perpendicular to the plane spanned by the line, and whose length is equal to the distance from the origin) and the direction vector of the line, see Figure 1. Plücker's identity (6) is satisfied, since direction and momentum vector are mutually perpendicular.

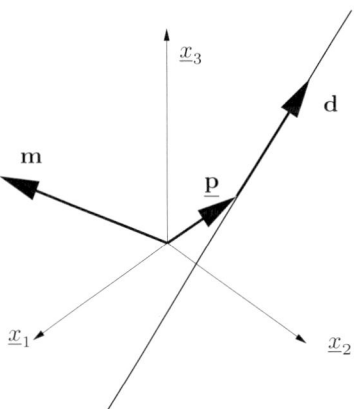

**Fig. 1.** Momentum vector $\mathbf{m}$ and direction vector $\mathbf{d}$ of a line $\mathcal{L}$.

*Remark 2.* An alternative notation is based on three–dimensional vectors from $\mathbb{R}^3 + \varepsilon\mathbb{R}^3$, where the so–called *"dual unit"* $\varepsilon$ satisfies $\varepsilon^2 = 0$. Dual unit vectors $\mathbf{u} = \mathbf{d} + \varepsilon\mathbf{m}$ satisfying $\langle \mathbf{u}, \mathbf{u} \rangle = \mathbf{d}^2 + \varepsilon\mathbf{d}, \mathbf{m}$ correspond to the line with direction vector $\mathbf{d}$ and the momentum vector $\mathbf{m}$. Two lines intersect if and only if the dual part of the inner product vanishes. The inner product can be used to determine both the distance and the angle between two lines. This notation is frequently used in space kinematics [2], where it leads to the dual quaternion representation of rigid body motions.

## 2.3   Klein's Mapping

Using Plücker coordinates, any line $\mathcal{L}$ in three–dimensional space is identified with the point

$$\mathbf{L} = (l_{01}, l_{02}, l_{03}, l_{23}, l_{31}, l_{12})^\top \qquad (9)$$

in the five–dimensional real projective space $P^5(\mathbb{R})$, which is contained in the hyperquadric $M$ given by Plücker's identity (6). On the other hand, any such point $\mathbf{L}$ corresponds to a unique line. Points with $l_{01} = l_{02} = l_{03} = 0$ correspond to points at infinity. This bijective mapping

$$\text{line } \mathcal{L} \subset P^3(\mathbb{R}) \quad \mapsto \quad \text{point } \mathbf{L} \in M \subset P^5(\mathbb{R}) \qquad (10)$$

is called the *Klein mapping.* The point $\mathbf{L}$ is called the *Klein image* of the line $\mathcal{L}$.

The polar hyperplane of a point $\mathbf{L}$ with respect to the hyperquadric $M$,

$$\Pi_{\mathbf{L}} = \{ \mathbf{X} \,|\, \langle \mathbf{X}, \mathbf{L} \rangle = 0 \}, \qquad (11)$$

intersects the hyperquadric $M$ in the Klein images of all lines which intersect the given line $\mathcal{L}$.

*Remark 3.* The homogeneous coordinates of points in five–dimensional real projective space $P^5(\mathbb{R})$ will also be indexed as usual,

$$\mathbf{P} = (p_0, p_1, p_2, p_3, p_4, p_5)^\top. \qquad (12)$$

If such a point corresponds to the Plücker coordinates $\mathbf{L}$ of a line $\mathcal{L}$, then the coordinates are identified according to $p_0 = p_{01}$, $p_1 = p_{02}$, $p_2 = p_{03}$, $p_3 = p_{23}$, $p_4 = p_{31}$, $p_5 = p_{12}$.

## 2.4   Curves in Line Space

Any *curve* $\mathbf{L}(u)$, $u \in (a, b) \subseteq \mathbb{R}$, which is fully contained in the hyperquadric $M$, is the Klein image of a one–parameter family of lines, i.e., of a *ruled surface.* The Klein images of the generators are the points of the curve.

As an example, we consider all lines which intersect three given lines $\mathcal{L}_1$, $\mathcal{L}_2$, $\mathcal{L}_3$. They form a so–called *regulus*, which is one of the two systems of straight lines on a ruled quadric surface.

The Klein image of a regulus is the intersection of the three polar hyperplanes $\Pi_{\mathbf{L}_1}$, $\Pi_{\mathbf{L}_2}$, $\Pi_{\mathbf{L}_3}$ of the three points $\mathbf{L}_1$, $\mathbf{L}_2$, $\mathbf{L}_3$ with the hyperquadric $M$. The three polar hyperplanes intersect in a two–dimensional plane. Consequently, the Klein image of the regulus is simply a conic on the hyperquadric $M$.

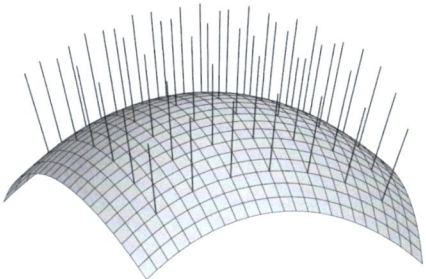

**Fig. 2.** Normal congruence of a surface.

## 2.5    Surfaces in Line Space

Now consider a (two-dimensional) surface $\mathbf{L}(u, v)$ which is fully contained in the hyperquadric $M$. It is the Klein image of a two–parameter family of lines. Such a system of lines is called a *line congruence*.

As an example we consider the *normal congruence* of a surface $\underline{\mathbf{x}}(u, v)$, $(u, v) \in \Omega \subset \mathbb{R}^1$, which consists of all normals

$$\mathbf{x}(u, v) + \lambda \mathbf{n}(u, v), \tag{13}$$

where $\mathbf{n}(u, v)$ is the field of the normal vectors of the given surface, see Figure 2. The Klein image is the surface

$$\mathbf{L}(u, v) = (\ \mathbf{n}(u, v)^\top,\ [\mathbf{n}(u, v) \times \underline{\mathbf{x}}(u, v)]^\top\ )^\top. \tag{14}$$

Normal congruences of surfaces have been used in order to detect the shortest distance between free–form surfaces [24]. In line space, this task can be formulated as a problem of surface–surface intersection.

A line congruence is said to have the *space–filling property*, if any point is contained in exactly one line, except for the points on finitely many curves. With other words, the line congruence forms a *fibration* of the three–dimensional space. In this situation, the exceptional curve(s) will be called the *focal curve(s)* of the line congruence.

Such congruences can be used for defining rational mappings on algebraic surfaces. Several examples will be discussed in the remainder of this paper.

*Remark 4.*   1. Space filling line congruences without exceptions (i.e., without focal curves) are called *spreads*; they have been analyzed in the field of Foundations of Geometry (see e.g. [18]). Line congruences and spreads are also of recent interest in Computer Vision.

2. Using a notion from the classical theory of algebraic line geometry, space filling line congruences are characterized by having the *bundle degree* 1 – the number of lines passing through a generic point equals one.

## 3   Line Models of Quadric Surfaces

Consider two skew lines $\mathcal{F}_1$ and $\mathcal{F}_2$ in three–dimensional space. For any point **p**, which does not belong to one of these lines, the two planes spanned by **p** and either line intersect in a unique line $\mathcal{L}(\mathbf{p})$. Clearly $\mathcal{L}(\mathbf{p})$ passes through **p** and intersects both $\mathcal{F}_1$ and $\mathcal{F}_2$.

The two–parameter family of lines obtained in this way is called a *linear congruence of lines*. It consists of all lines connecting arbitrary points on the lines $\mathcal{F}_1$ and $\mathcal{F}_2$. Clearly, this line congruence has the space–filling property with the two focal lines $\mathcal{F}_1$ and $\mathcal{F}_2$.

The Klein image of the line congruence is the intersection of the Klein quadric $M$ with the two polar hyperplanes $\Pi_{\mathbf{F}_1}$ and $\Pi_{\mathbf{F}_2}$. Since the intersection of two hyperplanes in a five–dimensional space is three–dimensional, we obtain a quadric surface in a three–dimensional space. Depending on the choice of the focal lines, we get the two different types of non–singular quadric surfaces: ruled quadrics, which are projectively equivalent to the hyperboloid of revolution, and oval quadrics, which are projectively equivalent to a sphere. As observed in [16], this leads to an alternative approach to the so–called *generalized stereographic projection* [8,9], which has been shown to be a useful tool for generating rational curves and surfaces on oval and ruled quadric surfaces.

### 3.1   Ruled Quadrics

We consider the two real focal lines

$$\mathcal{F}_1 = \{\mathbf{p} \mid \mathbf{p} = (0,0,\lambda,\mu)^\top, \lambda,\mu \in \mathbb{R}\} \quad \text{and}$$
$$\mathcal{F}_2 = \{\mathbf{p} \mid \mathbf{p} = (\lambda,\mu,0,0)^\top, \lambda,\mu \in \mathbb{R}\}. \tag{15}$$

These lines are the infinite line which is shared by all planes parallel to the $(\underline{x}_2, \underline{x}_3)$–plane, and the $\underline{x}_1$–axis, respectively. The resulting line congruence $\widehat{\mathcal{R}}$ is shown in Figure 3 (top right). The Plücker coordinates of the focal lines are

$$\mathbf{F}_1 = (0,0,0,1,0,0)^\top \quad \text{and} \quad \mathbf{F}_2 = (1,0,0,0,0,0)^\top \tag{16}$$

Hence, due to the intersection condition (7), the Klein image of the congruence satisfies the two linear equations $\langle \mathbf{L}, \mathbf{F}_1 \rangle = \langle \mathbf{L}, \mathbf{F}_2 \rangle = 0$, or, equivalently, $l_{01} = l_{23} = 0$, and Plücker's identity simplifies to

$$l_{02}l_{31} + l_{03}l_{12} = 0. \tag{17}$$

This is the equation of a *ruled quadric surface* $\mathcal{R}$ in the three–dimensional subspace of $P^5(\mathbb{R})$, which is given by $l_{01} = l_{23} = 0$.

For any point $\mathbf{p} = (p_0, p_1, p_2, p_3)^\top$, the plane spanned by **p** and $\mathcal{F}_1$ (resp. $\mathcal{F}_2$) intersects $\mathcal{F}_2$ (resp. $\mathcal{F}_1$) in the point $\mathbf{p}^* = (p_0, p_1, 0, 0)^\top$ (resp. $\mathbf{p}_* = (0, 0, p_2, p_3)^\top$). Consequently, the Plücker coordinates of the line of the congruence through **p** are

$$\mathbf{L}_{\widehat{\mathcal{R}}}(\mathbf{p}) = \mathbf{p}^* \wedge \mathbf{p}_* = (0, p_0p_2, p_0p_3, 0, -p_3p_1, p_1p_2)^\top \tag{18}$$

The mapping $\mathbf{p} \mapsto \mathbf{L}_{\widehat{\mathcal{R}}}(\mathbf{p})$ is equivalent to the *generalized stereographic projection* onto the hyperbolic paraboloid, as introduced in [8,9].

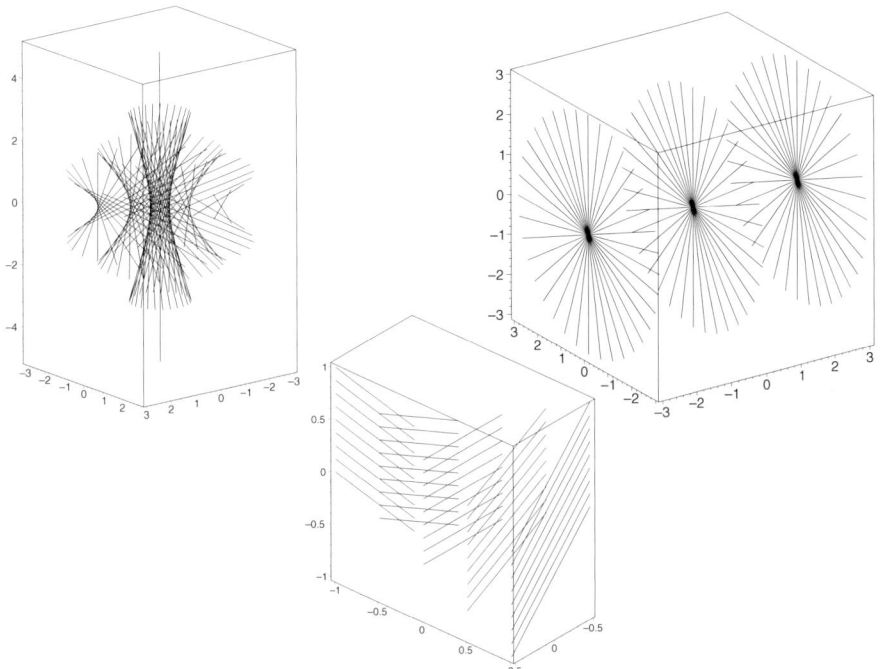

**Fig. 3.** Elliptic, hyperbolic and parabolic linear congruence.

## 3.2    Oval Quadrics

In the complex extension of the real projective 3–space, we consider the two conjugate–complex focal lines

$$\mathcal{F}_1 = \{\mathbf{p} \mid \mathbf{p} = (\lambda, \mu, -i\mu, i\lambda)^\top, \lambda, \mu \in \mathcal{C}\} \quad \text{and}$$
$$\mathcal{F}_2 = \{\mathbf{p} \mid \mathbf{p} = (\lambda, \mu, i\mu, -i\lambda)^\top, \lambda, \mu \in \mathcal{C}\}. \tag{19}$$

The resulting line congruence $\widehat{\mathcal{O}}$ is shown in Figure 3 (top left). The Plücker coordinates of the focal lines are

$$\mathbf{F}_1 = (1, -i, 0, -1, i, 0)^\top \quad \text{and} \quad \mathbf{F}_2 = (1, i, 0, -1, -i, 0)^\top. \tag{20}$$

Hence, again due to the intersection condition (7), the Klein image of the congruence satisfies the two linear equations $\langle \mathbf{L}, \mathbf{F}_1 + \mathbf{F}_2 \rangle = \langle \mathbf{L}, \mathbf{F}_1 - \mathbf{F}_2 \rangle = 0$, which lead to the two conditions $l_{01} = l_{23}$ and $l_{02} = l_{31}$. Consequently, Plücker's identity simplifies to

$$l_{01}^2 + l_{02}^2 + l_{03}l_{12} = 0. \tag{21}$$

This is the equation of a *oval quadric surface* $\mathcal{O}$ in the three–dimensional subspace of $P^5(\mathbb{R})$, which is given by $l_{01} = l_{23}$ and $l_{02} = l_{31}$. In fact, by introducing Cartesian coordinates according to

$$1 : X : Y : Z = (-l_{12}) : l_{01} : l_{02} : l_{03}, \tag{22}$$

equation (21) is transformed into the elliptic paraboloid $Z = X^2 + Y^2$.

For any real point $\mathbf{p} = (p_0, p_1, p_2, p_3)^\top$, the plane spanned by $\mathbf{p}$ and $\mathcal{F}_1$ (resp. $\mathcal{F}_2$) intersects $\mathcal{F}_2$ (resp. $\mathcal{F}_1$) the point

$$
\begin{aligned}
\mathbf{p}^+ &= (p_0 + ip_3, p_1 - ip_2, p_2 + ip_1, p_3 - ip_0)^\top \\
(\text{resp. } \mathbf{p}_+ &= (p_0 - ip_3, p_1 + ip_2, p_2 - ip_1, p_3 + ip_0)^\top).
\end{aligned}
\tag{23}
$$

These two intersections are conjugate complex. A real point on the congruence line through the point $\mathbf{p}$ can be generated by taking the linear combination

$$
\mathbf{p}^\perp = \frac{i}{2}(\mathbf{p}_+ - \mathbf{p}^+) = (-p_3, p_2, -p_1, p_0)^\top.
\tag{24}
$$

Consequently, the Plücker coordinates of the line of the congruence through $\mathbf{p}$ are

$$
\begin{aligned}
\mathbf{L}_{\widehat{\mathcal{O}}}(\mathbf{p}) &= \mathbf{p} \wedge \mathbf{p}^\perp \tag{25}\\
&= (p_0 p_2 + p_1 p_3, p_2 p_3 - p_0 p_1, p_0^2 + p_3^2, p_0 p_2 + p_1 p_3, p_2 p_3 - p_0 p_1, -p_1^2 - p_2^2)^\top
\end{aligned}
$$

The mapping $\mathbf{p} \mapsto \mathbf{L}_{\widehat{\mathcal{O}}}(\mathbf{p})$ is equivalent to the *generalized stereographic projection* onto the unit sphere, as introduced in [8,9].

### 3.3   Images of Lines

Consider a line $\mathcal{C}$ in 3–space, which does not intersect both focal lines. All lines of the congruence which pass through $\mathcal{C}$ form a regulus of lines. The Klein image of this regulus is a conic section on the quadric. Consequently, the images of the points of $\mathcal{C}$ under the generalized stereographic projections $L_{\widehat{\mathcal{R}}}, L_{\widehat{\mathcal{O}}}$ form conics.

If the line $\mathcal{C}$ intersects one of the focal lines, say $\mathcal{F}_1$, then the regulus degenerates into a pencil of lines, since all its lines pass through the intersection of the plane spanned by $\mathcal{C}$ and $\mathcal{F}_1$ with the second focal line $\mathcal{F}_2$. In this case, the images of the points of $\mathcal{C}$ under the generalized stereographic projection (18) form a line. Clearly, only real focal lines lead to real lines on the quadric, hence hyperbolic (resp. elliptic) linear congruences correspond to ruled (resp. oval) quadric surfaces.

The Klein image of any line $\mathcal{L}$ of the linear congruence is a point $\mathbf{L}$ on the quadric surface. Those lines of the congruence, which are contained in the two planes spanned by the line $\mathcal{L}$ and either one of the two focal lines are mapped to the two generating lines of the quadric $\mathcal{R}$ through that point. In the ruled quadric case, these two lines are real, otherwise they are conjugate complex.

### 3.4   Cones and Cylinders

The Klein images $\mathbf{F}_1$, $\mathbf{F}_2$ of the two focal lines span a line in $P^5(\mathbb{R})$ which intersects the hyperquadric $M$ in two real points (hyperbolic case) or in two conjugate-complex points (elliptic case). The Klein image of the line congruence is the intersection of the polar 4–plane of this line with the hyperquadric $M$. If the line in $P^5(\mathbb{R})$ touches the hyperquadric $M$, then one obtains the Klein

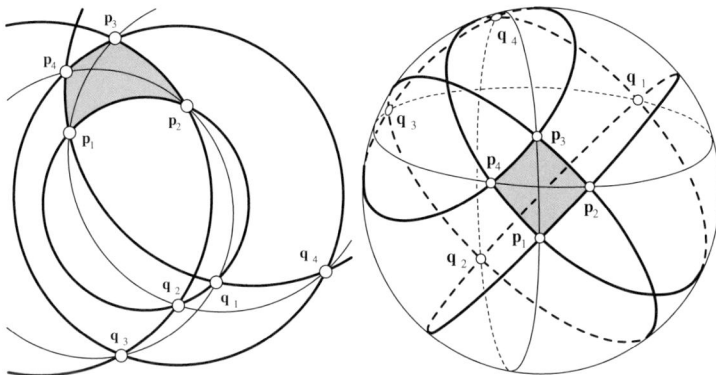

**Fig. 4.** Miquel's theorem in the plane and on the sphere: The circle through $\mathbf{p}_1, \mathbf{p}_3, \mathbf{q}_2, \mathbf{q}_4$ exists if and only if the circle through $\mathbf{p}_2, \mathbf{p}_4, \mathbf{q}_1, \mathbf{q}_3$ exists too. Four spherical arcs can act as the boundaries of a biquadratic patch iff both additional circles exist.

image of a *parabolic line congruence* (see Figure 3, bottom), which is a singular quadric (cone or cylinder). Similar to the case of oval and ruled quadrics, an associated generalized stereographic projection can be obtained, see [6] for more information.

### 3.5 Additional Algebraic Properties

As shown in [8], the mappings $L_{\widehat{\mathcal{O}}}$ and $L_{\widehat{\mathcal{R}}}$ have a very useful algebraic property: *any irreducible[1] rational parametric representation of a curve or surface can be obtained by applying these mappings to an irreducible rational curve or surface.* Without going into detail, we mention two consequences.

Any quadratic triangular Bézier patch on an oval quadric is the image of a linear patch. Consequently, the three boundary curves of the quadratic patch intersect in a single point[2]. This point is the Klein image of the unique line of the congruence, which is contained in the plane spanned by the linear patch.

Any biquadratic tensor–product Bézier patch on an oval quadric is the image of a bilinear patch. Consequently, the four boundary curves of the quadratic patch belong to the configuration of Miquel's theorem, see Figure 4. The boundary curves two additional circles are the images of the edges of the tetrahedron which is spanned by the four control points of the bilinear patch. See [7,8] for further information and related references.

Note that the generalized stereographic projection onto the unit sphere is also closely related to quaternion calculus and the kinematical mapping of spherical

---

[1] A rational parametric representation of a curve or surface is said to be irreducible, if the components of the homogeneous coordinates do not share any polynomial factors.

[2] This fact had already been observed in [20].

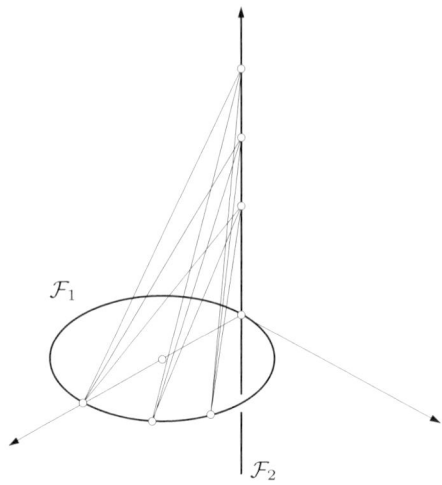

**Fig. 5.** The line congruence $\widehat{\mathcal{C}}$.

kinematics [17]. Recently, this connection has been exploited for generating spatial Pythagorean hodograph curves [4,10].

## 4    Line Models of Cubic Ruled Surfaces

We generalize this line–geometric approach to another, more complicated class of line congruences. After analyzing its Klein image, we obtain the various types of cubic ruled surfaces by projecting it back into three–dimensional space.

### 4.1    The Line Congruence

The line congruence $\widehat{\mathcal{C}}$ has the two focal curves

$$
\begin{aligned}
\mathcal{F}_1 &= \{\, \mathbf{p} \mid \mathbf{p} = (\lambda, 0, 0, \mu)^\top,\ \lambda, \mu \in \mathbb{R} \,\} \\
\mathcal{F}_2 &= \{\, \mathbf{p} \mid \mathbf{p} = (p_0, p_1, p_2, 0)^\top,\ p_1^2 - 2p_0 p_1 + p_2^2 \,\} \\
&= \{\, \mathbf{p} \mid \mathbf{p} = (s^2 + t^2, (s+t)^2, s^2 - t^2, 0)^\top,\ s, t \in \mathbb{R} \,\}.
\end{aligned}
\tag{26}
$$

The second focal curve is the circle in the plane $\underline{x}_3 = 0$ with radius 1 and center $\underline{\mathbf{c}} = (1, 0, 0)$. The first one is the $\underline{x}_3$–axis.

Clearly, the metric properties of the focal curves are not important. As a projectively equivalent choice one may take any non–degenerate conic section and any line which intersects it, but which is not contained in the same plane.

**Lemma 1.**  *The line congruence $\widehat{\mathcal{C}}$ with the focal curves (26) has the space–filling property: any point $\mathbf{p} = (p_0, p_1, p_2, p_3)^\top$ ($\mathbf{p} \notin \mathcal{F}_1 \cup \mathcal{F}_2$) lies on exactly one line through the two focal curves. This line has the Plücker coordinates*

$$
\begin{aligned}
\mathbf{L}_{\widehat{\mathcal{C}}} = (&4p_0 p_1^2 - 2p_2^3 - 2p_1 p_2^2,\ 4p_0 p_1 p_2 - 2p_1^2 p_2 - 2p_2^3, \\
&-p_3(2p_1^2 + 2p_2^2),\ -4p_1 p_2 p_3,\ 4p_1^2 p_3,\ 0)^\top.
\end{aligned}
\tag{27}
$$

**Proof.** The plane spanned by $\mathbf{p}$ and the $x_3$-axis intersects the circle $\mathcal{F}_2$ in the origin of the Cartesian coordinate system $(1,0,0,0)^{\top}$ and in the point

$$\mathbf{q}(\mathbf{p}) = (2p_1^2 + 2p_2^2, 4p_1^2, 4p_1p_2, 0)^{\top}. \tag{28}$$

The Plücker coordinates of the line are $\mathbf{L}_{\widehat{\mathcal{C}}} = \mathbf{p} \wedge \mathbf{q}(\mathbf{p})$. $\square$

**Proposition 1.** *The Klein image of the line congruence $\widehat{\mathcal{C}}$ is a cubic ruled surface, which is contained in a four–dimensional subspace of $P^5(\mathbb{R})$.*

**Proof.** The line congruence $\widehat{\mathcal{C}}$ is projectively equivalent to the congruence generated by the two focal curves $(1,0,0,s)^{\top}$, $s \in \mathbb{R}$ ($x_3$–axis), and $(1,t,t^2,0)^{\top}$, $t \in \mathbb{R}$ (a parabola in the plane $x_3 = 0$, which will be called the focal parabola). The Plücker coordinates of the lines are

$$\mathbf{L}(s,t) = (t, t^2, -s, -st^2, st, 0)^{\top}, \quad s, t \in \mathbb{R} \tag{29}$$

The area of the Newton polygon of this surface in $P^5(\mathbb{R})$ equals $3/2$, hence it is a cubic surface (see [14]). On the other hand, it is a ruled surface, since the parameter lines $t = $ constant are lines. $\square$

The Klein images of all lines which pass through a fixed point of the circle $\mathcal{F}_2$ are the generators of the cubic ruled surface. In the framework of projective differential geometry, the cubic ruled surfaces in four–dimensional space have been studied by Weitzenböck and Bos [25].

## 4.2    Projection Back into Three–Space

The Klein image of the congruence $\widehat{\mathcal{C}}$ is a surface in a four–dimensional subspace of $P^5(\mathbb{R})$. By mapping it back into three–space, we obtain a cubic ruled surface. This mapping is described by a projective transformation

$$\pi : \mathbf{L} \mapsto \pi(\mathbf{L}) = A \, (l_{01}, l_{02}, l_{03}, l_{23}, l_{31})^{\top} \tag{30}$$

where $A$ is a $4 \times 5$ matrix. We assume that $A$ has maximal rank. Otherwise, the image of the surface is contained in a plane. The kernel of $A$ is called the *center* $\mathbf{C}$ of this mapping.

*The Types of Cubic Ruled Surfaces in Three–Space.* Recall that there are three types of cubic ruled surfaces in three–dimensional space [17], see Figure 6. All are equipped with a unique double line. Each generator intersects the double line. The double line consists of singular points (in the sense of algebraic geometry), and the osculating cone (the zero set of the associated Hessian matrix) degenerates into two (possibly conjugate–complex) planes. For one or two points along the double line, these two planes degenerate into a double plane. These two points are called the cuspidal points of the surface. Depending on their nature, one either obtains a Plücker conoid (two real points), a Zindler conoid (two conjugate–complex points) or Cayley's cubic ruled surface (Cayley surface for short, one cuspidal point).

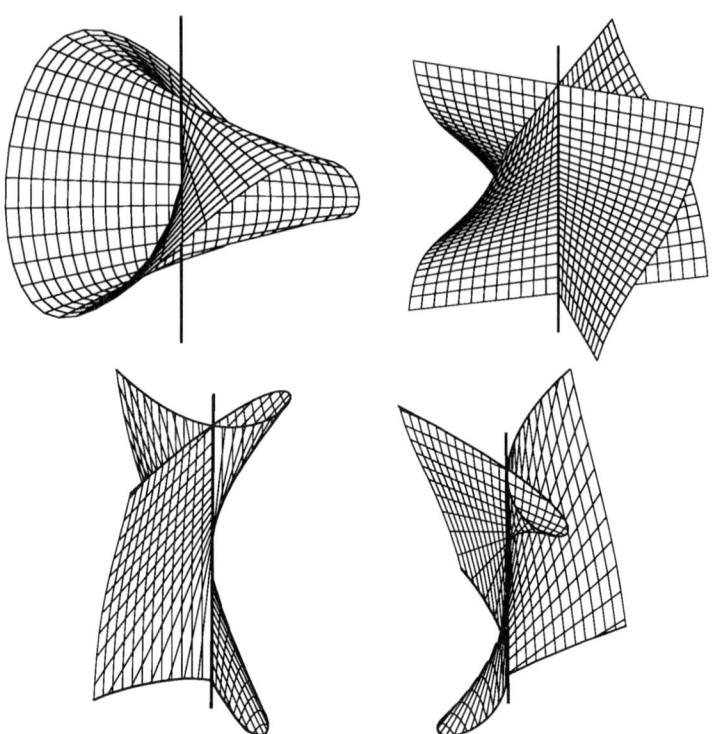

**Fig. 6.** The three types of cubic ruled surfaces. Top left: Plücker conoid, top right: Zindler conoid, bottom: Cayley surface (two views).

Clearly, the class of cubic ruled surfaces may degenerate in various ways (cubic cones, quadric surfaces, etc.). For the sake of simplicity, we restrict ourself to the generic case. Up to projective mappings, any cubic ruled surface is equivalent to one of the surfaces shown in Figure 6.

*Projecting Space Cubics into Planar Ones.* Before proceeding to four–dimensional space, we discuss a similar situation in three–dimensional space.

Any planar rational cubic curve can be obtained by applying a projective transformation $\pi : P^3(\mathbb{R}) \to P^2(\mathbb{R})$ to the space cubic $\mathbf{c}(t) = (1, t, t^2, t^3)^\top$. The mapping $\pi$ has a unique center, which is the kernel of the corresponding $3 \times 4$–matrix. The location of the center governs the shape of the result. If the center is located on one of the tangents of the space cubic, the image is a planar cubic with a cusp. If the center is even on the curve itself, the image is a conic section (conic for short). Otherwise, the planar cubic has either a double point or an isolated singular point (in the algebraic sense).

Any line connecting two points on the curve is called a chord. A double point is generated by a chord of the curve which passes through the center. A cusp is generated by a tangent of the curve through the center. A isolated singular

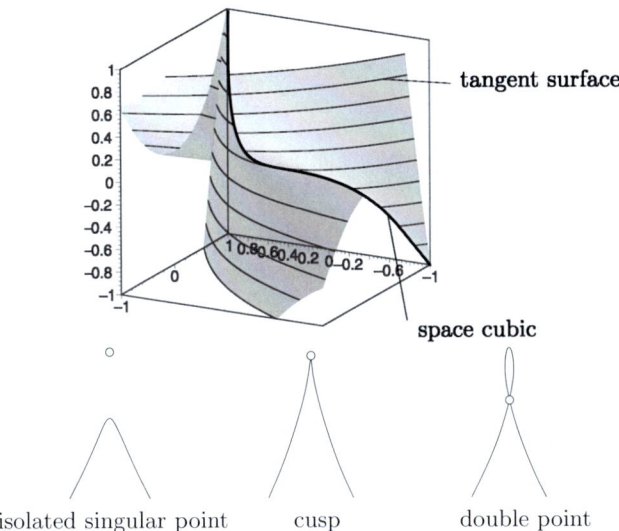

**Fig. 7.** The three types of planar rational cubics (bottom) can be obtained by projecting a space cubic into the plane. The shape of the result depends on the location of the center of projection with respect to the tangent surface (top) of the cubic. The tangent surface is visualized by level curves.

point is generated by a chord connecting two conjugate–complex points of the curve which passes through the center.

For all centers which are on the same side of the tangent surface (see Figure 7), the type of the singularity is the same[3].

*Projecting 4D Cubic Ruled Surfaces into 3D Ones.* In order to simplify the calculations, we use again the representation (29) of the Klein image of the line congruence which was used in the proof of Proposition 1. This line congruence has a the focal line $(1, 0, 0, s)^\top$ and the focal parabola $(1, t, t^2, 0)^\top$. Moreover, since the Klein image of the line congruence is contained in the hyperplane $l_{12} = 0$, we omit the last coordinate throughout the remainder of this paper, i.e.,

$$\mathbf{L}(s, t) = (t, t^2, -s, -st^2, st)^\top, \quad s, t \in \mathbb{R} \tag{31}$$

Consequently, we deal with a 2–surface in real projective 4–space.

Any point of a ruled surface has a 2–dimensional tangent plane. Along each generating line, the union of the tangent planes forms a hyperplane, which will be called the *tangent hyperplane*.

**Lemma 2.** *The one–parameter family of tangent hyperplanes covers an open subset of $P^4(\mathbb{R})$ twice, while the interior of the complementary subset is not covered. The two subsets of $P^4(\mathbb{R})$ are separated by the hyperquadric $T$,*

---

[3] This fact has been exploited for deriving an alternative approach to earlier results [21] on a so–called characterization diagram for planar cubics in Bernstein–Bézier form [16].

$$T = \{ \mathbf{q} = (q_0, q_1, q_2, q_3, q_4)^\top \mid q_2 q_3 = q_4^2 \}. \tag{32}$$

**Proof.** The tangent hyperplanes are spanned by the four points

$$\mathbf{L}\bigg|_{s=0,t=t_0}, \quad \frac{\partial}{\partial s}\mathbf{L}\bigg|_{s=0,t=t_0}, \quad \frac{\partial}{\partial t}\mathbf{L}\bigg|_{s=0,t=t_0}, \quad \frac{\partial}{\partial t}\mathbf{L}\bigg|_{s=1,t=t_0}, \tag{33}$$

for any generating line $\mathbf{x}(s, t_0)$, $t_0 \in \mathbb{R}$, constant, $s \in \mathbb{R}$. A short calculation leads to their homogeneous coordinates $(0, 0, t^2, 1, 2t)^\top$. Their envelope can be shown to be the quadric $T$. □

*Remark 5.*   1. The situation is similar for the tangents of a conic in the plane: Points in the exterior part of the conic are covered twice, while points in the interior part are not covered, see Figure 8.
   2. The hyperquadric $T$ is a degenerate quadric with two–dimensional generating planes. It can be thought of as a cylinder, but with two–dimensional rulings.

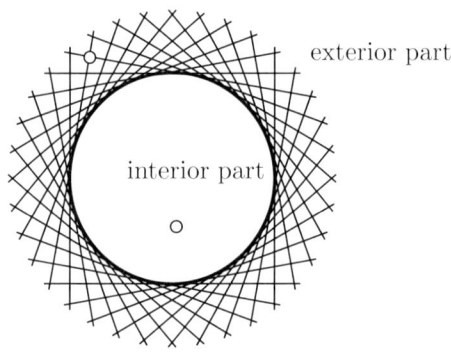

exterior part

interior part

**Fig. 8.** Tangents of a conic in the plane.

As the next step, we analyze the conic sections on the Klein image of the congruence. These conic section are the Klein images of the reguli which are contained in the congruence.

**Lemma 3.** *There exists a two–parameter family of conic sections on the Klein image of the congruence. Any conic can be obtained by substituting $s = t/(ct+d)$ in (31), with constants $c, d \in \mathbb{R}$.*

*Consider the associated two–parameter family of 2–planes, which are spanned by the conics. Except for the points in the hyperquadrics $M$ and $T$, each point in four–space belongs to exactly one of those 2–planes.*

**Proof.** Clearly, any curve $\mathbf{L}(t/(ct + d), t)$ is a quadratic rational curve, i.e., a conic section. It remains to be shown that any conic can be obtained in this way.

Any hyperplane intersects the Klein image in a cubic curve. In order to obtain a conic, this cubic has to factor into a conic section and a line. The line has to be one of the generating lines of the Klein image.

The hyperplanes through the generator $\mathbf{L}(s, t_0)$ ($t_0 \in \mathbb{R}$ constant, $s \in \mathbb{R}$) of the Klein image have the homogeneous coordinates

$$\mathbf{H} = (-t_0 k_1, \ k_1, \ t_0(-k_3 t_0 + k_4), \ k_3, \ k_4)^\top, \tag{34}$$

They form a 2–parameter family, since they depend on the three homogeneous parameters $k_1, k_3, k_4 \in \mathbb{R}$.

Indeed, the intersection between hyperplane and surface leads to

$$\mathbf{L}(s, t)^\top \mathbf{H} = - \underbrace{(t_0 - t)}_{\text{generator}} \underbrace{(-s k_3 t_0 + s k_4 + t k_1 - s k_3 t)}_{\text{conic}}. \tag{35}$$

The second factor can be solved for $s$, which leads to the parametric representations of the conics,

$$s(t) = \frac{t k_1}{(t + t_0) k_3 - k_4} = \frac{t}{ct + d} \quad (k_1 \neq 0). \tag{36}$$

This gives the following representation of the conics on the Klein image of the congruence (which are Klein images of reguli)

$$\mathbf{R}(t) = \mathbf{L}(s(t), t) = ((ct + d), \ (ct + d)t, \ -1, \ -t^2, \ t)^\top, \quad t \in \mathbb{R}. \tag{37}$$

The system of conics depends on 2 parameters $c, d \in \mathbb{R}$.

A point $\mathbf{Q} = (q_0, q_1, q_2, q_3, q_4)$ lies in the same 2–plane as one of these conics if and only if the rank of the $5 \times 4$ matrix $(\mathbf{R}(t), \dot{\mathbf{R}}(t), \ddot{\mathbf{R}}(t), \mathbf{Q})$ is less than 4. This condition leads to unique solutions for the parameters $c$ and $d$,

$$c = \frac{q_0 q_4 + q_1 q_2}{q_4^2 - q_2 q_3}, \quad d = \frac{q_0 q_3 + q_1 q_4}{q_4^2 - q_2 q_3}. \tag{38}$$

Thus, except for the points $Q$ on the quadric surface $T$ which is characterized by the equation $q_4^2 = q_2 q_3$, there is always a unique conic section lying in a 2–plane through it. $\square$

*Remark 6.*   1. If $q_0 q_3 + q_1 q_4 = 0$, i.e., $d = 0$, then this conic is the Klein image of the cone spanned by the focal parabola and by the point $(1, 0, 0, 1/c)$ on the focal line. This equation characterizes the intersection of the hyperquadric $M$ with the hyperplane $q_5 = q_{12} = 0$.

2. The Lemma can also be concluded from the fact, that a regulus is generated by a projective mapping between a conic and a line intersecting the conic, where the intersection point is a fixed point of the mapping. The mapping is given by the bilinear transformation $t \mapsto s(t)$ (see (36)), which keeps the intersection, due to $s(0) = 0$.

3. The two quadric surfaces $M$ and $T$ intersect in the Klein image $\mathbf{L}(s, t)$ of the line congruence.

**Corollary 1.** *If the center of the projective mapping $\pi : P^4(\mathbb{R}) \to P^3(\mathbb{R})$ does not belong to the hyperquadric $T$, then the unique conic on the Klein image $\mathbf{L}(s, t)$, which shares a 2–plane with the center, is mapped to the double line* of the cubic ruled surface. The two tangents of the conic through the center touch the conic in two points. These points are mapped to the cuspidal points of the cubic ruled surface. The two points are either real (Plücker conoid) or conjugate complex (Zindler conoid). See Figure 9 for a sketch of the situation in 4–dimensional space.

**Proof.** The first part of the corollary is an immediate consequence of the previous Lemma. The second part results from the fact that the two tangents to the conic are also tangents to the Klein image of the line congruence. Consequently, the image surface under $\pi$ has a singularity. □

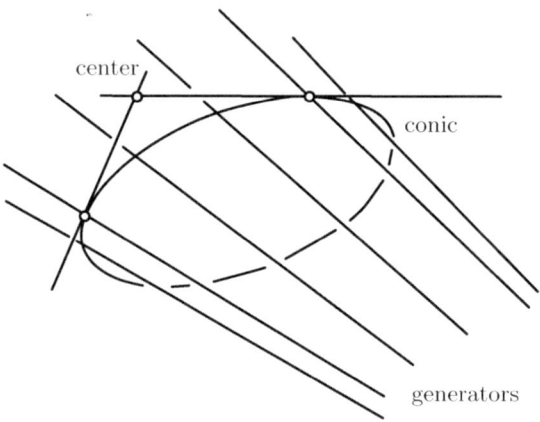

**Fig. 9.** Sketch of the situation in 4D.

Summing up, we have the following results.

**Theorem 1.** *Assume that the center* $\mathbf{C} = (c_0, c_1, c_2, c_3, c_4)^\top$ *of the projective mapping is not contained in the Klein image of the congruence, i.e., in* $\mathbf{C} \notin M \cap T$. *If* $\mathbf{C}$ *is contained in* $T$, *then the image of* $\mathbf{L}(s, t)$ *is a Cayley surface. If the coordinates of* $\mathbf{C}$ *satisfy* $c_4^2 - c_2 c_3 > 0$, *then it is a Plücker conoid, otherwise it is a Zindler conoid.*

**Proof.** If one of the hyperplanes passes through the center $\mathbf{C}$, then it contains one tangent of the surface through $\mathbf{C}$, leading to a cuspidal point. Depending on the number of cuspidal points, we get the three different types of cubic ruled surfaces. □

*Remark 7.* If the center of the projective mapping is on the Klein image itself, then the image surface degenerates in various ways. For instance, one may obtain various types of ruled quadric surfaces. A more detailed discussion of these cases is beyond the scope of this article. Further results will be presented in [19].

| point **p** | $\longrightarrow$ | line $\mathcal{L}_{\widehat{\mathcal{C}}}(\mathbf{p})$ through **p** which is contained in the congruence | $\longrightarrow$ | Klein image (Plücker coordinates $\mathbf{L}_{\widehat{\mathcal{C}}}(\mathbf{p})$) | $\longrightarrow$ | point $\pi(\mathbf{L}_{\widehat{\mathcal{C}}}(\mathbf{p}))$ |
|:---:|:---:|:---:|:---:|:---:|:---:|:---:|
| ⋒ | | ⋒ | | ⋒ | | ⋒ |
| 3D space | $\longrightarrow$ | line space (4D) | $\longrightarrow$ | hyperquadric $M$ in 5D space | $\longrightarrow$ | cubic ruled surface $\pi(\widehat{\mathcal{C}})$ in 3D space |

**Fig. 10.** Constructing a rational mapping onto a cubic ruled surface.

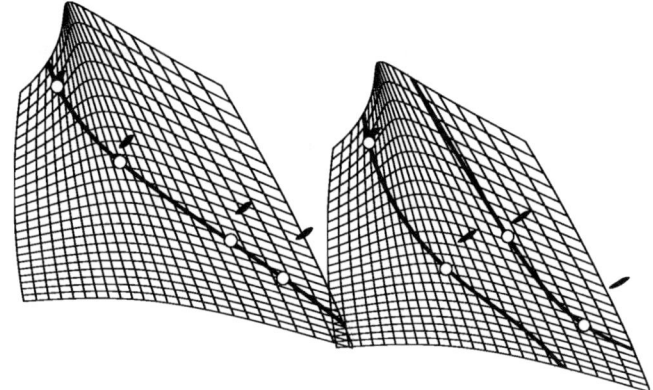

**Fig. 11.** Bicubic tensor-product patch on a cubic ruled surface. Interpolation of four points with a cubic (2 examples).

## 4.3   Constructing Curves and Surfaces

In order to generate rational curves and surfaces on a given cubic ruled surface, one may now construct a suitable rational mapping, as follows. The mapping can be found by composing the mapping $\mathbf{p} \mapsto \mathbf{L}_{\widehat{\mathcal{C}}}(\mathbf{p})$ (see (27)), which maps each point in three–space to the Klein image of the unique congruence line through it, with a suitable projective mapping $\pi : P^5(\mathbb{R}) \rightarrow P^3(\mathbb{R})$, which maps the Klein image of the congruence into the desired cubic ruled surface. This process is summarized in Figure 10.

Since $\mathbf{L}_{\widehat{\mathcal{C}}}$ is a cubic rational mapping, the image of a curve of degree $n$ is a curve of degree $3n$ on the cubic ruled surface. Similarly, the image of a tensor–product patch of degree $(n, n)$ is a tensor–product patch of degree $(3n, 3n)$.

As an example, Figure 11 shows a bicubic patch on cubic ruled surface. In addition, four points are interpolated by a cubic curve. Two different examples (one of them with a point at infinity) are shown.

The cubic curve has been constructed by applying the rational mapping to a line $\mathcal{I}$ in three–dimensional space. The preimages of the four given points are four lines which belong to the congruence $\widehat{\mathcal{C}}$. The line $\mathcal{I}$ has to intersect these four preimage lines.

Generally, there may exist two lines which intersect all four lines. They can be found by intersecting the ruled quadric surface through the first three lines with the last one. This leads to two solutions, which may be conjugate–complex, or identical. In our situation, however, one of them is the focal line of the congruence, and the other line is the desired preimage of the cubic curve.

Following the ideas presented in [8,9], the rational mapping can be used to construct rational curves and surfaces on the three types of cubic ruled surfaces.

## 5   Line Models of Veronese Surfaces

Any line intersecting a space curve in two points is said to be a *chord* of it. We consider the line congruence $\widehat{\mathcal{V}}$ that consists of all chords of the twisted cubic curve

$$\mathcal{F}_1 = \mathcal{F}_2 = \{\, \mathbf{p} \mid \mathbf{p} = (1, t, t^2, t^3)^\top, \ t \in \mathcal{C} \,\}. \tag{39}$$

Real lines are obtained either by connecting to real points on the curve, or by connecting two conjugate–complex ones. As already discussed in the second paragraph of section 4.2, the system of chords has the space–filling property: any point in three–dimensional space (except for the points on the cubic itself) belongs to exactly one chord. If the point is used as a center of a projective mapping into a plane, then corresponding two points on the cubic are mapped to the singular point of the resulting planar cubic.

As a well–known fact from advanced geometry, the system of chords is closely related to the Veronese surface [1].

**Proposition 2.** *The Klein image of the line congruence is a Veronese surface* $\widehat{\mathcal{V}}$, *which is contained in the hyperquadric M of five–dimensional real projective space.*

**Proof.** Recall that the Veronese surface in five–dimensional projective space is given by

$$(1, u, v, u^2, uv, v^2)^\top, \quad , u, v \in \mathbb{R}. \tag{40}$$

The Plücker coordinates of the chords connecting two points $(1, t, t^2, t^3)$ and $(1, s, s^2, s^3)$ of the twisted cubic can be shown to be equal to

$$(1, u, u^2 - v, v^2, -uv, v)^\top, \quad \text{with} \quad u = s + t, v = st, \tag{41}$$

where the common factor $(s - t)$ has been factored out. The Klein image of the chords is projectively equivalent to the Veronese surface (40). □

The Veronese surface can be seen as the generic triangular quadratic Bézier surface in five–dimensional space. By projecting it into three–dimensional space it is possible to obtain any triangular quadratic Bézier surface. This fact has been exploited in [5] in order to classify these surfaces. Similar to the discussion of the cubic ruled surfaces, the type of the result depends on the location of the (here one–dimensional) center of the mapping.

**Proposition 3.** *The Plücker coordinates of the line in $\widehat{\mathcal{V}}$ through a given point* **p** *are*

$$
\begin{aligned}
\mathbf{L}_{\widehat{\mathcal{V}}}(\mathbf{p}) = (\ &(p_1^2 - p_0\,p_2)^2, \quad (p_1^2 - p_0\,p_2)(p_1\,p_2 - p_0\,p_3), \\
&(p_0 p_2^3 + p_0^2 p_3^2 - 3\,p_0\,p_1\,p_2\,p_3 + p_1^3 p_3), \quad (p_2^2 - p_1\,p_3)^2, \\
&(p_1\,p_3 - p_2^2)(p_1\,p_2 - p_0\,p_3), \quad (p_1\,p_3 - p_2^2)(p_0\,p_2 - p_1^2)\ )^\top.
\end{aligned}
\tag{42}
$$

**Proof.** This can be shown by computing the singular point of the planar cubic obtained by projecting the space cubic into the plane, where **p** serves as the center. The details are omitted. □

Similar to the ideas discussed in Section 4.3, the mapping

$$
\mathbf{p} \mapsto \mathbf{L}_{\widehat{\mathcal{V}}}(\mathbf{p})
\tag{43}
$$

can be used for parameterizing the images of the Veronese surface under projective mappings into three–dimensional space (which are all types of quadratic triangular Bézier surfaces).

## 6 Concluding Remarks

Line congruences have been shown to be useful for the construction of rational curves and surfaces on special algebraic surfaces. In the case of quadric surfaces, this leads to an additional geometrical approach (which had already been outlined in [16]) to the generalized stereographic projection. Originally, this technique had earlier been derived mainly relying on algebraic results [8]. As shown in this paper, similar techniques are available for cubic ruled surfaces, and for Veronese surfaces (triangular Bézier surfaces).

As an obvious question, one may ask which line congruences provide the space–filling property, and – related to it – an associated rational mapping. This is related to the *bundle degree* (the number of lines passing through a generic points) of these congruences. The classification of line congruences with respect to their bundle degree has been studied mainly in the 19th century, in classical texts on algebraic line geometry. We mention the following results:

Generally, the lines connecting two different algebraic space curves of order $m$ and $n$ with $s$ intersections (counted with multiplicities) form a congruence of bundle degree $mn - s$ [22]. Consequently, the construction of the line congruence $\widehat{\mathcal{C}}$ can immediately be generalized to congruences with two focal curves, where one of them is a straight line. For instance, the line congruence spanned by a space cubic and one of its chords (or tangents) has the space–filling property, since any plane through the chord (or tangent) intersects the cubic in exactly one additional point. Clearly, the Klein images of the congruences generated in this way are ruled surfaces. Other choices of $m$ and $n$ with $1 \notin \{m, n\}$ do not produce congruences with bundle degree 1, since the algebraic space curve cannot intersect in sufficiently many points without becoming identical.

The bundle degree $b$ of the chords of an irreducible algebraic space curve of order $n$ satisfies

$$
\lfloor (n-1)^2/4 \rfloor \le b \le (n-1)(n-2)/2,
\tag{44}
$$

see [3,23]. It depends on the types of singularities of the space curve. Only cubic curves ($n = 3$) give congruences with $b = 1$.

Note that there are other possibilities to define space–filling line congruences than the two possibilities described in this paper. For instance, one may take all lines which touch a given surface and pass through a curve. These lines form a congruence of bundle degree $rn$, where $r$ is the rank of the surface (i.e., the algebraic order of its tangent cones) and $n$ is the order of the space curve [22].

Further research will be devoted to possible generalizations of this approach, which may include systems of linear subspaces in spaces of dimension higher than three. In addition, we plan to develop computational techniques for generating rational curves and surfaces on these special algebraic surfaces, such as techniques for interpolation and approximation (see [6] for the quadric case). Also, we will analyze the relation to Müller's results on universal parameterizations of special cubic surfaces, which also cover the case of ruled ones [15]. Last but not least, we plan to complete the results on cubic ruled surfaces by analyzing the degenerate situations.

## Acknowledgements

This research was supported by the Austrian Science Fund (FWF) through the SFB F013 "Numerical and Symbolic Scientific Computing" at Linz, project 15. The authors wish to thank Dr. Josef Schicho (RISC Linz) and Dr. Martin Peternell (Vienna) for helpful discussions.

## References

1. M. Bertolini and C. Turrini, On the automorphisms of some line congruences in $P^3$, Geom. Dedicata 27.2 (1988), 191-197.
2. O. Bottema and B. Roth, Theoretical Kinematics, New York, Dover 1990.
3. H. Brauner, Über die Projektion mittels der Sehnen einer Raumkurve 3. Ordnung, Monatsh. Math. 59 (1955), 258–273.
4. H.I. Choi, D.S. Lee and H.P. Moon, Clifford algebra, spin representation, and rational parameterization of curves and surfaces, Adv. Comput. Math. 17 (2002), 5–48.
5. W.L.F. Degen, The types of triangular Bézier surfaces, The mathematics of surfaces VI (G. Mullineux, ed.), Oxford Univ. Press., 153-170 (1996).
6. R. Dietz, Rationale Bézier-Kurven und Bézier-Flächenstücke auf Quadriken, Ph.D. thesis, TU Darmstadt; Shaker, Aachen 1995.
7. R. Dietz, G. Geise and B. Jüttler, Zur verallgemeinerten stereographischen Projektion, Mathematica Pannonica 6/2 (1995), 181-197.
8. R. Dietz, J. Hoschek and B. Jüttler, An algebraic approach to curves and surfaces on the sphere and on other quadrics, Comput. Aided Geom. Design 10 (1993), 211-229.
9. R. Dietz, J. Hoschek and B. Jüttler, Rational patches on quadric surfaces, Computer-Aided Design 27 (1995), 27-40.

10. R.T. Farouki, M. al–Kandari and T. Sakkalis, Hermite interpolation by rotation-invariant spatial Pythagorean-hodograph curves, Adv. Comp. Math. 17 (2002), 369-383.
11. V. Hlavatý, Differential Line Geometry, Noordhoff, Groningen-Holland, 1953.
12. J. Hoschek, Liniengeometrie, B.I.-Hochschulskripten 733a/b, Bibliographisches Institut, Zürich, 1971.
13. J. Hoschek and D. Lasser, Fundamentals of Computer Aided Geometric Design, AK Peters, Wellesley Mass., 1996.
14. R. Krasauskas, Toric surface patches, Adv. Comput. Math. 17 (2002), 89–133.
15. R. Müller, Universal parametrization and interpolation on cubic surfaces, Comput. Aided Geom. Design 19 (2002), 479-502.
16. H. Pottmann, Studying NURBS curves and surfaces with classical geometry, in: M. Dæhlen et al. (eds.), Mathematical methods for curves and surfaces, Vanderbilt University Press, Nashville 1995, 413-438.
17. H. Pottmann and J. Wallner, Computational Line Geometry, Springer, Berlin 2001.
18. R. Riesinger, Beispiele starrer, topologischer Faserungen des reellen projektiven 3–Raums, Geom. Dedicata 40 (1991), 145-163.
19. K. Rittenschober, Constructing curves and surfaces on rational algebraic surfaces, Ph.D. thesis, Johannes Kepler University of Linz, Austria, in preparation.
20. T. Sederberg and D. Anderson, Steiner Surface Patches, IEEE Comp. Graphics Appl. (1985), 23–26.
21. M.D. Stone and T.D. DeRose, A geometric characterization of parametric cubic curves, ACM Trans. Graph. 8.3 (1989), 147-163.
22. R. Sturm, Liniengeometrie (1. Theil), Teubner, Leipzig 1892.
23. R. Sturm, On some new theorems on curves of double curvature, British Association for the advancement of science, London 1881, 440.
24. K.-A. Sun, B. Jüttler, M.-S. Kim and W. Wang, Computing the distance between two surfaces via line geometry, in: Proceedings of the 10th Pacific Conference on Computer Graphics and Applications (S. Coquillart et al., eds.), IEEE Press, Los Alamitos 2002, 236–245.
25. R. Weitzenböck and W.J. Bos, Zur projektiven Differentialgeometrie der Regelflächen im $R_4$, 6. Mitt., Akad. Wetensch. Amsterdam Proc. 44 (1941), 1052–1057.

# Boundary Conditions
# for the 3-Direction Box-Spline

Malcolm A. Sabin[1] and Aurelian Bejancu[2]

[1] Numerical Geometry Ltd
[2] University of Leeds

**Abstract.** In their seminal paper Boehm et al[3] show how box splines over regular bivariate grids are defined by coefficients (control points) associated with centres outside the region being defined ("phantom points"), as well as with those inside.

If we apply the pure subdivision rules derived from the box splines, this means that the configuration shrinks at every step from the original coarse lattice, which includes the phantoms, to the final limit surface.

This is inconvenient: we much prefer to have the boundary curve of the limit surface defined as a univariate subdivision curve which is interpolated by the limit surface, and so the practical schemes of importance (Loop[9] and Catmull-Clark[6]) have a 'fix' at the boundary, whereby new vertices associated with the boundary are added at every stage to those defined by the box-spline rules inside the configuration.

Unfortunately this leads in the Catmull-Clark case to the equivalent of the 'natural' end-condition, which 'in spite of its positive sounding name has little to recommend it from an approximation-theoretic point of view' (quoted from de Boor [4]), because the second derivative at the end of each isoline crossing the boundary is zero. In the case of Loop subdivision the second derivative across the edge is always one quarter of that along the edge, and so the limit surface tends to have positive Gaussian curvature at the edge whether or not this is desired.

This paper explores the idea that the coefficients of Boehm et al's "phantom points" can be related to those in the domain by use of better boundary conditions. More precisely, we introduce and study extensions of the univariate 'not-a-knot' end-conditions (called 'uniform' by Kershaw in [8]) to 3-direction box-splines, generating boundary conditions whose templates turn out to have simple, elegant and interesting structure. The work is closely related to that of Bejancu and Sabin [2] on the approximation order of semi-cardinal approximation, but is presented here fore a CAGD, rather than approximation theory, audience.

**Keywords:** subdivision; boundary conditions; not-a-knot; curvature.

We first introduce the problem, that the current standard subdivision methods of Catmull-Clark and Loop have boundary conditions that are equivalent to the natural end-condition of the cubic spline. This means that points at the boundary of a Catmull-Clark surface cannot have positive Gaussian curvature,

M.J. Wilson and R.R. Martin (Eds.): Mathematics of Surfaces 2003, LNCS 2768, pp. 244–261, 2003.
© Springer-Verlag Berlin Heidelberg 2003

and those on the boundary of a Loop surface will tend to have positive Gaussian curvature unless the surface is seriously twisted.

In section 2 we repeat standard results in univariate theory in order to introduce our notations: in particular the idea of **phantom coefficients** and our use of **tableaux** as a notational device. Then in section 3 we apply the same approach to the bivariate cases of tensor product splines and the 3-direction box-spline.

We describe everything in terms of functions. However, the motivation is in the realm of representation of homogeneously three-dimensional objects, whose shape is invariant under solid body transformations, and the ideas are then applied in the parametric sense where the ordinate is a 3-vector describing position and the abscissa is an auxiliary variable (or pair of variables) usually known as parameter(s). In this context the coefficients become control points.

# 1   The Problem

Despite its elegant derivation from a minimization principle and its simplicity of implementation, the 'natural' end condition for the cubic spline has the severe disadvantage that at the ends the second derivative vanishes. This is a catastrophe for the fitting of parts of curves, as can be seen in the examples of figure 1.

Closely related bad effects occurs in two of the standard subdivision methods, Catmull-Clark and Loop. The direction perpendicular (in the parameter plane) to the boundary always has a zero second derivative vector in a Catmull Clark surface, while in a Loop surface that direction always has a second derivative exactly a quarter of the second derivative along the edge, and of the same sign.

This is not related to the extraordinary points, but to the fact that these schemes do not give enough control at the boundaries even of regular regions.

# 2   The Univariate Case

A cubic spline function is a map from a connected subset of $\mathbb{R}$ to $\mathbb{R}^n$ made of cubic polynomials over **spans**, which meet at their junctions with continuity of second derivative. We call the abscissae of the junctions 'knots'. If the knots are equally spaced in the abscissa space we call it an 'equal interval spline'. This is the focus of this paper, since unequal intervals merely complicate the presentation without bringing extra insight.

A cubic spline function is expressed in the form

$$f(x) = \sum_i y_i f_i(x)$$

where the $f_i(x)$ are **basis functions** and the $y_i$ are **coefficients**.

There are two bases of importance for the space of such functions

1) the **cardinal basis**, also called the **fundamental** or **Lagrange** basis, where each basis function takes the value 1 at one of the knots, and zero at all the others. Thus the coefficients of the spline are exactly the values taken at the knots.

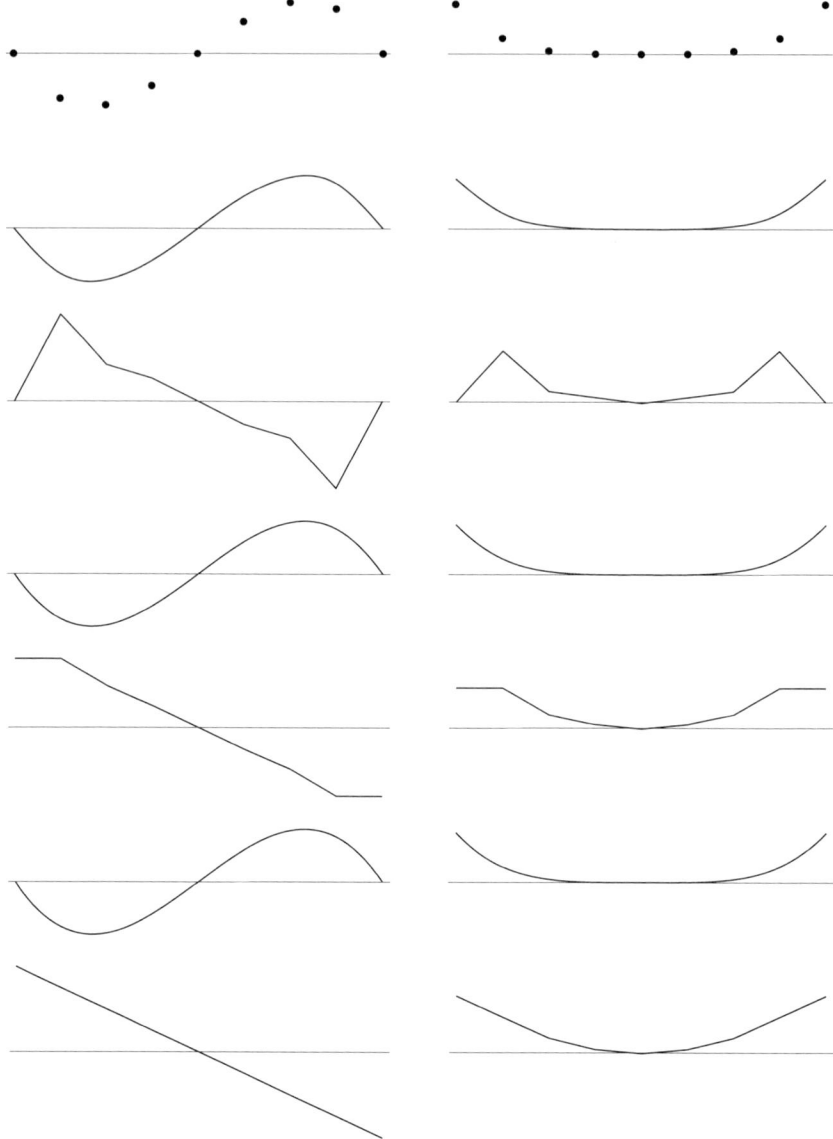

**Fig. 1.** On the left are shown a set of points lying at equal abscissa intervals on a cubic polynomial. Beneath it are shown cubic splines with the natural end condition, the constant curvature end condition and the not-a-knot end condition. Beneath each curve is shown the plot of its second derivative. On the right are shown points lying on a quartic curve and the corresponding interpolants with their second derivative plots. The curves themselves do not look very different at this scale, but the curvature plots indicate very clearly the higher quality of the fit with the not-a-knot end-condition.

2) the **B-spline basis**, where, for an equal interval spline, all of the basis functions are translates of the same basic function, and that basic function has a support of exactly four spans. It is identically zero outside that support, and so the curve in any span depends on the values of only four of the coefficients.

In both cases the basis functions have strong localisation properties, and we can associate them 1:1 with the knots. In the first case we associate each basis function with the knot at which it has non-zero value: in the second with the knot at the centre of its support. This induces an association of the coefficients with the knots.

The total number of basis functions whose supports overlap a given span in the real line is two greater than the number of knots in, and at the end of, that interval. This means that we have two more freedoms in choosing the coefficients for the curve than the number of knot points. It is therefore necessary to define end-conditions which specify how those extra freedoms are to be determined. We call the coefficients of those basis functions which overlap the domain in only one span **phantom coefficients**, and end conditions can be articulated in terms of how they are determined from the coefficients within the domain.

The four end conditions of most significance in cubic spline interpolation and approximation are:-

1) the '**natural**' end condition where the second derivative goes to zero at the end of the curve. This is the case which minimizes the 'bending energy' (the integral of the square of the second derivative) of the function.

2) the '**constant curvature**' end condition where the second derivative is constant over the end span, so that the end span is a quadratic, not a cubic.

3) the '**Not-a-knot**' end condition where there is no discontinuity of third derivative at the penultimate knot, so that the two spans at an end are part of the same cubic.

4) the **pseudo-Bézier** end condition where there is an extra control point used to define the first derivative at the end. This is equivalent to the **complete** or **clamped** conditions where the first derivatives at the end knots are specified as well as the values at all the knots. We could view this as being an unequal interval not-a-knot condition, because there is a coefficient where there is no discontinuity. Because we are concerned here with equal intervals we do not explore this case in this paper.

Now, using the B-spline basis, we can express the conditions of zero second derivative at a knot, constant second derivative over a span and zero change of third derivative at a knot in terms of linear combinations of the local coefficients. These combinations are conveniently derived by writing the Bézier coefficients of the pieces as tableaux.

The B-spline basic function and its tableau are shown in figure 2.

It is a property of the Bézier basis for the individual spans that we can determine the derivatives by taking differences of the Bézier coefficients, and this is exploited as shown in figure 3, where the first, second and third derivatives and their tableaux are displayed.

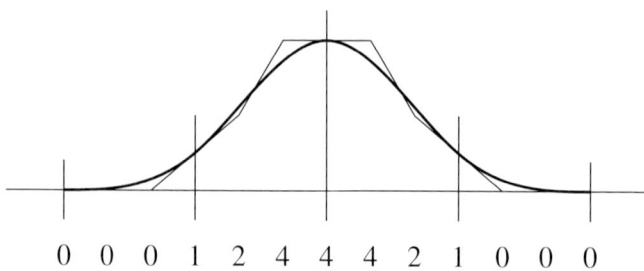

0 0 0 1 2 4 4 4 2 1 0 0 0

**Fig. 2.** The univariate cubic B-spline basis function. Also shown are the Bézier polygons of the individual pieces and the ordinates of the Bézier control points. The normalising factor of 1/6 is ignored because we are only concerned with relative magnitudes within the function.

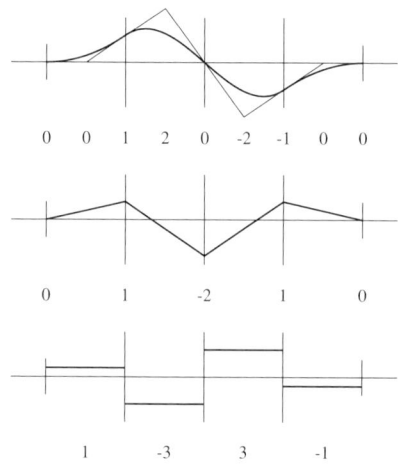

0    0    1    2    0    -2    -1    0    0

0        1        -2        1        0

1        -3        3        -1

**Fig. 3.** The first, second and third derivatives of the univariate cubic B-spline basis function. The ordinates of their Bézier polygons, shown beneath the graphs of the functions are determined by taking differences and then eliding the values either side of a knot when these are the same. In the third difference case elision is not possible because the original function is not C3 continuous. Again we ignore the normalising factors.

The values actually at the knots can be abstracted from these tableaux, using the property that the Bézier representation interpolates its end control values.

$$
\begin{array}{lrrr}
\text{value} & 1 & 4 & 1 \\
\text{first derivative} & 1 & 0 & -1 \\
\text{second derivative} & 1 & -2 & 1
\end{array}
$$

These give the values of a derivative of one basis function at the three interior knots. They are **masks**. The same numbers, differently arranged, give the values of a derivative at a particular knot in terms of the three local coefficients. These are **stencils**.

Let the symbols **A**, **B** and **C**... denote the knots and $A$, $B$, $C$... their corresponding B-spline coefficients.

The value and derivatives at **C** are given by

$$
\begin{array}{lcc}
\text{value} & B + 4C + D & /6 \\
\text{first derivative} & -B + D & /2 \\
\text{second derivative} & B - 2C + D &
\end{array}
$$

The cubic B-spline does not have a third derivative at the knots, because the spans on the two sides have different values of third derivative, but we can create expressions for the third derivative in, for example, the spans **BC** and **CD**, as

$$
\begin{array}{l}
\text{within } \mathbf{BC} \; -A + 3B - 3C + D \\
\text{within } \mathbf{CD} \; -B + 3C - 3D + E
\end{array}
$$

so that the discontinuity of third derivative at **C** is

$$
(-B + 3C - 3D + E) - (-A + 3B - 3C + D) = A - 4B + 6C - 4D + E
$$

Thus if our domain is the part of the curve from **B** to the right, we can express the conditions for the three end-conditions as equations between the B-spline coefficients

$$
\begin{array}{rl}
\text{natural:} & A - 2B + C = 0 \\
\text{constant curvature:} & -A + 3B - 3C + D = 0 \\
\text{not a knot:} & A - 4B + 6C - 4D + E = 0
\end{array}
$$

## 2.1  Interpolation

The selected end-condition from these three can be combined with the equations for the values at the knots in terms of the coefficients, to give a linear system which can be solved to determine the appropriate coefficients to make the function interpolate specific values at the knots.

This is always well-conditioned except in the case that the domain is too short. For the natural end-condition it is necessary to have at least one span in the domain, for the constant curvature end-condition, two, and for the not-a-knot end-condition at least three.

The order of approximation is 2 for the natural end-condition, 3 for constant curvature, and 4 for not-a-knot.

## 2.2  End-Control

Alternatively we may seek a scheme, reminiscent of the way Bernstein polynomials are applied in the definition of Bézier curves, in which the end values are specified, while B-spline coefficients are specified internally to the domain. In this context we may solve a $2 \times 2$ linear system to determine $A$ and $B$ from the desired end condition and the interpolation condition at **B**. Let $b$ denote the value that we require to interpolate at **B**.

natural:
$$A - 2B + C = 0$$
$$A + 4B + C = 6b$$
$$\text{giving } B = b$$
$$\text{and } A = 2b - C$$

constant curvature:
$$-A + 3B - 3C + D = 0$$
$$A + 4B + C = 6b$$
$$\text{giving } B = (6b + 2C - D)/7$$
$$\text{and } A = (18b - 15C + 4D)/7$$

not a knot:
$$A - 4B + 6C - 4D + E = 0$$
$$A + 4B + C = 6b$$
$$\text{giving } B = (6b + 5C - 4D + E)/8$$
$$\text{and } A = (6b - 7C + 4D - E)/2$$

Once $A$ and $B$ are determined, if regular subdivision is applied, the polygon will contract back at each refinement until the limit curve has **B** as end abscissa and $b$ as end value.

## 3    Bivariate Cases

The tensor product case is a trivial application of the univariate results, and we shall not spend further time on it. The interested reader may easily develop the theory and it is set as an exercise at the end of the paper.

Of much more interest is the box-spline on a three-direction triangular grid. This was first described by Frederickson[7] but is probably best known in the CAGD community as the foundation in regular regions of the Loop[9] subdivision scheme. The theory of box-splines is well described in the book by de Boor, Höllig and Riemenschneider[5]. The natural boundary condition for this box-spline was also described by Sabin in [10] (pp101-105) which covers both the edge and the corner condition.

The function consists of triangular pieces in a regular triangular grid, each piece being a quartic polynomial in the two abscissa variables. These pieces meet with continuity of second derivative across all boundaries.

The 'B-spline' basic function is 6-fold rotationally symmetric about its centre, and is non-zero over a hexagon containing 24 triangles. The coefficients of translates of this function are associated with the vertices of the triangles in a regular grid.

The triangles immediately within a finite domain are influenced by the coefficients within the domain and those one row outside.

The tableau showing the Bézier coefficients of the basic function is shown in figure 4 and those for the derivatives in the directions across and along the triangle edges in figures 5 and 6 respectively.

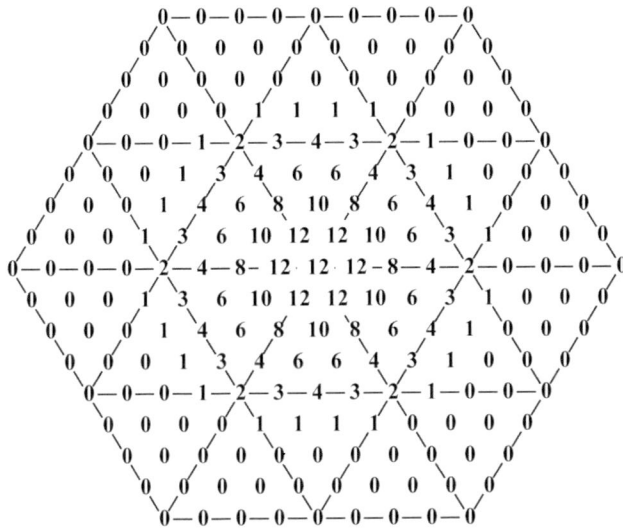

**Fig. 4.** Tableau of the box-spline Bézier coefficients.

The directional differences used to give the values in figure 5 are given by convolution with the stencils $\begin{smallmatrix} 2 & \\ -1 & -1 \end{smallmatrix}$ and $\begin{smallmatrix} 1 & 1 \\ & -2 \end{smallmatrix}$ in up- and down-pointing triangles respectively, and those in figure 6 by the stencils $\begin{smallmatrix} 0 & \\ -1 & 1 \end{smallmatrix}$ and $\begin{smallmatrix} & -1 & 1 \\ 0 & \end{smallmatrix}$.

From these tableaux we can determine the behaviour of the edge conditions.

## 3.1  Natural Edge-Conditions

The usual Loop scheme creates new vertices on the boundary from univariate cubic B-spline subdivision applied to the boundary vertices.

To express this in terms of phantom vertices means that we have to position the phantom vertices in such a way that the total contributions to the new edge vertices add up to the right values. The cross-section of the Loop subdivision mask is not the same as the univariate B-spline mask, and so specific contributions from the phantoms' functions, as well as those from the vertices one row in, need to be added in to give this condition.

This gives the condition that the phantom vertices beyond the edge are chosen to make the 'diamonds' which cross the boundary planar.

It is also possible to deduce the cubic-B-spline boundary from the flat diamond condition, a route which is taken by Bejancu in [1], where this property was first noted.

This is an exact analogue for the $A = 2b - C$ condition given by the univariate natural edge condition.

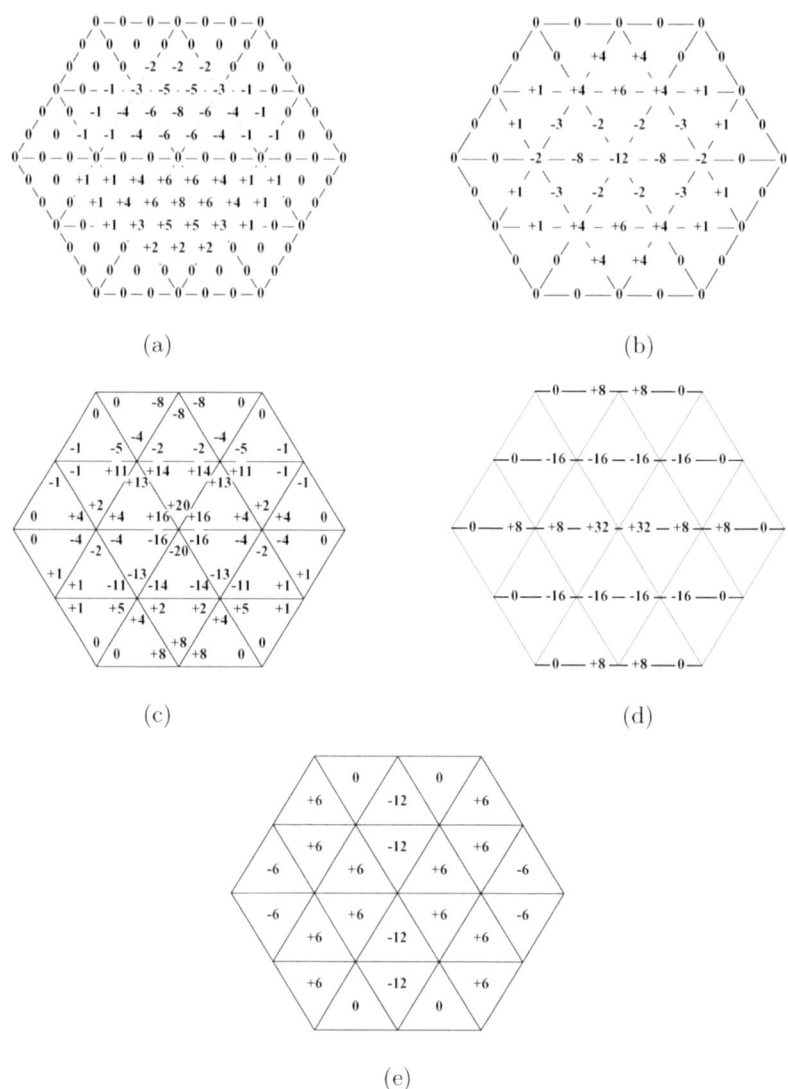

**Fig. 5.** Tableaux of derivatives in the vertical direction of the box-spline function. (a) shows $d/dy$, (b) $d^2/dy^2$, (c) $d^3/dy^3$ and (d) the mismatch of $d^3/dy^3$ across horizontal edges. At the bottom, (e) shows the tableau for $d^4/dy^4$. It is evident that there is no mismatch across any of the horizontal edges, and so if the mismatch of $d^3/dy^3$ is cancelled across a particular horizontal edge, the two adjacent triangles become part of the same polynomial, and the edge is not a knot.

In fact the condition that the boundary should be a cubic spline is not quite 'flat diamonds': the condition that the edge should be a cubic spline is essentially that the fourth derivative along the edge should be zero. Figure 7 shows that the

**Fig. 6.** Tableaux of derivatives in the horizontal direction of the box-spline function. Top left is $d/dx$, top right $d^2/dx^2$, bottom left $d^3/dx^3$ and bottom right $d^4/dx^4$.

fourth derivative zero condition is that the convolution of the diamond with the second difference along the edge should be zero, i.e., that the out-of-planarity of the diamonds must vary linearly. This does not give any local freedom, although in principle it gives two extra freedoms for an entire edge of the domain. However, what is actually imposed is that it should be the cubic B-spline of which the given boundary values are the coefficients. This gives the flat-diamonds condition.

The second derivative across an edge is in principle a quadratic function given by convolving figure 5b with the control values centred on the edge vertices, and on the vertices on the two sides of the edge. If, however, we impose the 'flat-diamond' edge condition, it becomes piecewise linear, and the shape of the variation from movement of one boundary value is exactly the same, though only one fourth as large, as the longitudinal second derivative.

## 3.2   Not-a-Knot Edge-Conditions

There is no obvious analogue in the box-spline context of the constant-curvature condition. We would certainly not expect quartic pieces to become quadratic, and forcing the triangles meeting edge vertices to become cubics perpendicular to the boundary would give two conditions per boundary vertex rather than one. We therefore move directly to the not-a-knot condition which does respond to analysis.

The not-a-knot edge condition imposes a zero discontinuity of $d^3/dy^3$ across the edges in the next row parallel to the boundary. As can be seen from figure 8,

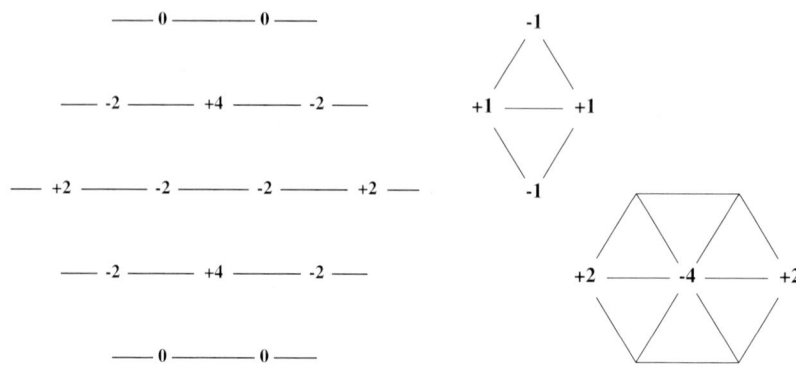

**Fig. 7.** (left) Mask of $d^4/dx^4$ values along horizontal edges, taken from figure 6d. (right) A factorisation of this mask. If we require a boundary to reduce to a cubic spline, all fourth derivatives along that edge must reduce to zero, and this figure shows that for a complete edge of the domain to reduce to a cubic spline the second derivative along that edge must vary linearly over the whole edge.

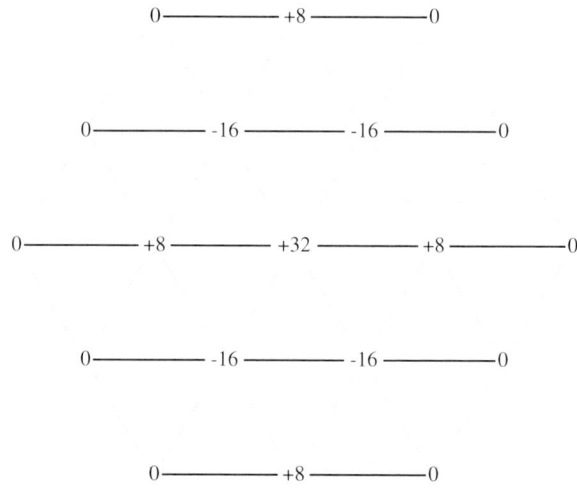

**Fig. 8.** Mask of discontinuities of $d^3/dy^3$ across horizontal edges, expressed as a partial function on the horizontal edges only. Each row is continuous and so the repeated values from figure 5d have been elided. Because these discontinuity functions are only of first degree in $x$, we can assure zero discontinuity of $d^3/dy^3$ across a line everywhere along it by applying a set of vertex conditions.

the variation of this discontinuity is first degree with respect to movement along the edge, and so it can be set to zero by enforcing zero value at its vertices. This involves applying, at each vertex on the first row in, the stencil corresponding to the values in figure 8, which gives an equation for a single phantom coefficient in terms of the coefficients within the domain. It is interesting to observe that this stencil is the square under convolution of the flat diamond stencil.

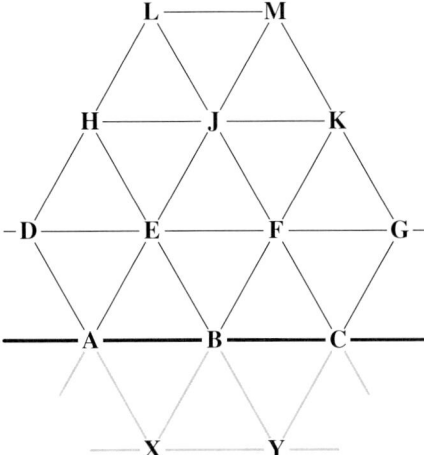

**Fig. 9.** Point labels for a not-a-knot edge. The bold line is the edge of the domain. Points **A** to **M** are actual points whose values are either to be interpolated eventually or to be used as control points. Points **X** and **Y** are 'phantom points' to which we associate values in order to span the region.

Thus the conditions which have to be satisfied so that the discontinuities of horizontal third derivative at **E** and **F** in figure 9 should be zero are

$$X + D + F + L + 4E = 2A + 2B + 2H + 2J$$
$$Y + E + G + M + 4F = 2B + 2C + 2J + 2K$$

respectively, which we can solve for $X$ and $Y$ to give

$$X = 2A + 2B - D - 4E - F + 2H + 2J - L$$
$$Y = 2B + 2C - E - 4F - G + 2J + 2K - M$$

### 3.3   Interpolation

We denote by lower case letters the corresponding actual surface values. The actual value, $b$, at **B** will be given by

$$\begin{aligned}
12b &= 6B + X + Y + A + E + F + C \\
&= 6B + (2A + 2B - D - 4E - F + 2H + 2J - L) \\
&\quad + (2B + 2C - E - 4F - G + 2J + 2K - M) + A + C + E + F \\
&= 3A + 10B + 3C - D - 4E - 4F - G + 2H + 4J + 2K - L - M
\end{aligned}$$

There is one of these equations for each edge vertex, and these may be included in a linear system along with all the equations giving the interior values in terms of the coefficients. This is a large system, which does not reduce by tensor products to a set of small systems, but is still soluble for reasonable size systems.

## 3.4   Edge Control

The large problem here in setting up a scheme analogous to end-control in the univariate case, is the question of 'what is the edge supposed to be?' and 'how is it to be specified'. In the natural case this is straightforward, but we have seen above that if the flat diamond condition is not met, we cannot have a cubic B-spline curve for the boundary, and so there is no easy equivalent for not-a-knot.

One option is to settle merely for interpolating specific boundary points, while retaining the use of the interior coefficients as normal B-spline control points.

In this case we can use the equations for $b$ (and the other boundary points) from above, to make a univariate band system, which can be solved to give the edge coefficients in terms of the interior ones. This is diagonally dominant (3,10,3), if not quite as strongly as the (1,4,1) row of the interpolating cubic spline.

The values $b$ can be given as the positions at the knots of a B-spline. In this case the influence of one of the edge control points actually specified will be very strongly focussed on the corresponding coefficient of the box-spline. However, the boundary will not be a cubic B-spline, but a piecewise quartic. What is more, the detailed shape of the actual boundary will depend to a small extent on the control values in the interior of the domain, which implies that it will not be easy to define a pair of surfaces whose intersection is a given curve. This is an important aspect for future research.

## 4   Corners

A **corner** is a place on the boundary of the domain at which the direction of the boundary turns. It becomes a place where end-conditions apply to the boundary curves.

The quad lattice deals with corners trivially as the tensor product of two edges, so we do not elaborate it here.

A triangular lattice supports two kinds of convex corner, a blunt one, with an angle of turn of 60 degrees, and a sharp one with an angle of turn of 120 degrees.

### 4.1   Blunt Corners

The abscissae points near a blunt corner are labelled as shown in figure 10.

**The Natural Case.** At a corner which is dealt with in the obvious way during subdivision by just using the old position of a control point as the new position, we have the following equations:

$$W = B + C - H$$
$$Y = C + D - H$$
$$10C + B + H + D + W + X + Y = 16C$$

which reduce to

$$X = 4C + H - 2B - 2D$$

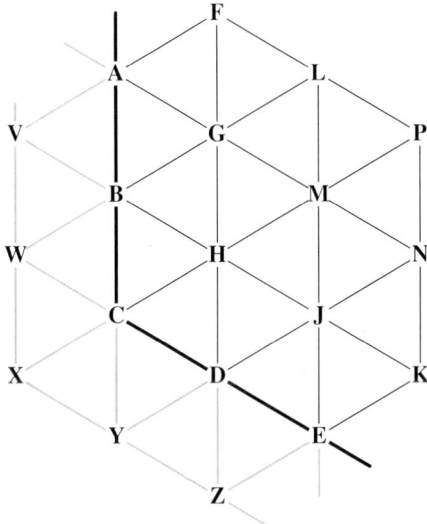

**Fig. 10.** Point labels for the blunt corner.

If $B + D = H + C$ this reduces to $X = 2C - H$, a familiar form. Notice that if this condition does not hold, the dihedral angle across **CH** does not change during refinement, and so the subdivision result has unbounded curvature at **C**. In this respect the subdivision differs from the box-spline.

**The Not-a-Knot Case.** The values of $V$, $W$, $Y$ and $Z$ are directly derived from the required conditions across the not-a-knot lines at **G**, **H** and **J**.

across **GH**:
$$V = 2(A + B + L + M) - F - 4G - H - P$$
$$W = 2(B + C + M + J) - G - 4H - D - N$$
across **HJ**:
$$Y = 2(C + D + G + M) - B - 4H - J - L$$
$$Z = 2(D + E + M + N) - H - 4J - K - P$$

We can derive the value of $X$ in either of two ways, from the condition across **BH** at **B** or that across **HD** at **D**.

$$X = 2(W + C + A + G) - V - 4B - H - F$$
$$\text{or } X = 2(C + Y + J + E) - H - 4D - Z - K$$

It is not obvious that these two equations are compatible, but in fact both of them, on substitution of the other phantom vertices, give the same symmetric form

$$X = -2B + 6C - 2D + 4G - 8H + 4J - 2L + 2M - 2N + P$$

which has the stencil

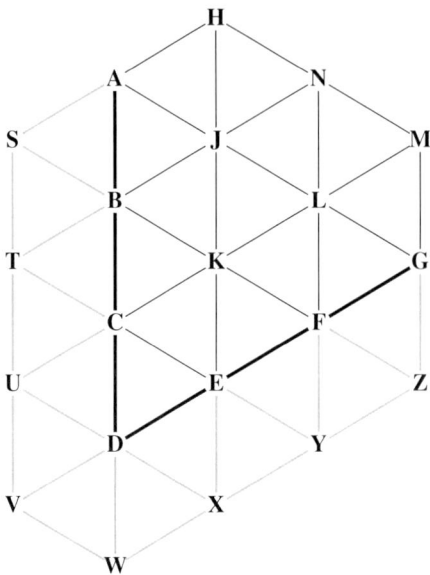

**Fig. 11.** Point labels for the sharp corner.

$$
\begin{array}{ccccc}
 & & -2 & & \\
 & 4 & & 1 & \\
-2 & & 2 & & \\
 & -8 & & -2 & \\
6 & & 4 & & \\
* & & -2 & &
\end{array}
$$

## 4.2  Sharp Corners

The abscissa points near a sharp corner are labelled as shown in figure 11.

**The Natural Case.** The relevant equations are

$$U = C + D - E$$
$$X = D + E - C$$
$$C + E + U + X + V + W = 6D$$

These tell us that

$$V + W = 4D - C - E$$

To determine $V$ and $W$ independently we need to note that the first derivative at **D** in the direction **UDX** needs to be equal to $E - C$ because it lies in the plane given by the derivatives in the directions **DC** and **DE**. It is also given by a stencil derived from figure 6a, and equating these two gives the equation

$$V - W = C - E$$

and thence $V = 2D - E$, $W = 2D - C$.

This is exactly the pattern we expect and it means that the configuration round **D** is to be coplanar. The curvature of the limit surface at **D** is zero in all directions.

**The Not-a-Knot Case.** Here the conditions for $S$, $T$, $Y$ and $Z$ are all nicely independent

$$S = 2(A + B + N + L) - H - 4J - K - M$$
$$T = 2(B + C + L + F) - J - 4K - E - G$$
$$Y = 2(E + F + B + J) - C - 4K - L - A$$
$$Z = 2(F + G + J + N) - K - 4L - M - H$$

The interesting point is that the equations for $U$ and for $X$ are both essentially equations for $U + X$.

$$U + X = 2(C + D + F + Y) - K - 4E - Z$$
$$U + X = 2(D + E + T + B) - K - 4C - S$$

It turns out that substitution of the other phantom vertices into either of these leads to the same symmetric equation

$$U + X = -2A + 4B + 2D + 4F - 2G + H + 2J - 8K + 2L + M - 2N$$

The question becomes, '*How do we determine the relative values of $U$ and $X$ ?*'

This is resolved by the use of precision set arguments. In particular, if the coefficients are **extruded**, so that all the coefficients in any given row in a chosen direction are the same, the box-spline surface turns out to be extruded also. We would like to maintain this property, and this leads to a set of equations satisfied by the stencils

$$U = -B + 4C + D + 2J - 4K - 4E - N + 5F + M - 2G$$
$$X = -F + 4E + D + 2L - 4K - 4C - N + 5B + H - 2A$$

In fact there is a univariate set of possible stencils with the required symmetry, the required sum and not merely the extrusion property, but also the linearly varying extrusion property, whereby if the coefficients along a row vary linearly in a given direction this property is inherited by the surface. This particular choice has fewest non-zero coefficients. Deeper analysis may yet give a good reason for choosing a different member of the family, but this will certainly do for now.

It then becomes straightforward to determine $V$ and $W$. We have no other conditions, and so we might as well turn the edge **CE** into a not-a-knot line as well. This means that the whole triangle **BDF** is a single polynomial, as is the quadrilateral **BJLF**.

$$V = 2(U + D + B + K) - T - 4C - E - J$$
$$W = 2(D + X + K + F) - C - 4E - Y - L$$

## 4.3   Interpolation

As before, the equation for the value of the corner point itself can be determined from the equations for the surrounding phantom coefficients. This equation can then be included in the linear system to be solved for interpolation.

## 4.4   Corner Control

In the context where only boundary values are given, together with the required values for coefficients for interior vertices, there is no additional complication from the corners. Indeed, the coupling between the linear systems being solved for the edge coefficients along adjacent edges is very small for the blunt corner case. Not zero, however, and so the entire loop of boundary has to be solved with cyclic end conditions. This is not significantly more expensive than solving the separate edges separately.

# 5   Exercises for the Reader

The Catmull-Clark subdivision scheme is, in the regular regions, a bicubic B-spline. However, the standard subdivision rules generate new vertices only in the interior of the surface, so that the surface shrinks at every step. In order to be able to constrain the boundary of the surface being described, it is usual to provide additional rules for appending new vertices around the boundary so that the surface converges to one which has a cubic B-spline boundary whose control points are the boundary vertices of the initial polyhedron. The rule is very simple: the appended vertices are exactly those created by univariate cubic subdivision of the boundary at each stage.

## 5.1   Exercise 1

Confirm that this construction is equivalent to applying the natural end-condition.

## 5.2   Exercise 2

Determine an algorithm to give the univariate cubic B-spline coefficients whose function interpolates given end values and with given B-spline coefficients in the interior of the domain, and which has not-a-knot end-conditions. Hint: only worry about one end: the other will be just the same. Then finally tidy up the case of short domains.

## 5.3    Exercise 3

Determine a pre-process for the Catmull-Clark scheme which determines a control polyhedron which will contract during refinement to a limit surface whose boundary is the cubic B-spline given by the original boundary control polygon, and whose edges and corners have the not-a-knot condition.

# Acknowledgements

The second author has received support from a Nuffield Foundation research grant (NUF-NAL-02).

# References

1. A.Bejancu, 2002, Semi-cardinal interpolation and difference equations: from cubic B-splines to a three-direction box-spline construction., Technical report, University of Leeds, Department of Applied Mathematics, submitted for publication
2. A.Bejancu and M.A.Sabin, 2003, Maximal approximation order for a box-spline semi-cardinal interpolation scheme on the three-direction mesh, Technical report, University of Leeds, Department of Applied Mathematics, submitted for publication
3. W.Boehm, G.Farin and J.Kahmann, 1984, A survey of curve and surface methods in CAGD, CAGD vol 1 no 1, pp 1–60, July 1984.
4. C. de Boor, 1978, A Practical Guide to Splines, Springer, ISBN 3-540-90356-9
5. C.de Boor, K.Höllig and S.Riemenschneider, 1993, Box Splines, Springer Verlag, ISBN 3-540-94101-0
6. E.Catmull and J.Clark, 1978, Recursively generated B-spline surfaces on arbitrary topological meshes, CAD vol 10 pp350–355
7. P.O.Frederickson, 1971, Quasi-interpolation, extrapolation and approximation on the plane, pp159–167 in Proceedings of Manitoba conference on Numerical Mathematics, Winnipeg
8. D.Kershaw, 1973, Two interpolatory cubic splines, J.inst Maths Applics vol 11 pp329–333
9. C.T.Loop, 1987, Smooth subdivision surfaces based on triangles, Master's Thesis, University of Utah
10. M.A.Sabin, 1977, The use of piecewise forms for the numerical representation of shape, Tanulmanyok 60/1977, Hungarian Academy of Sciences, ISBN 963 311 035 1

# The Plateau-Bézier Problem

Juan Monterde

Dpt. de Geometria i Topologia, Universitat de València
Avd. Vicent Andrés Estellés, 1, E-46100-Burjassot (València), Spain
monterde@uv.es

**Abstract.** We study the Plateau problem restricted to polynomial surfaces using techniques coming from the theory of Computer Aided Geometric Design. The results can be used to obtain polynomial approximations to minimal surfaces. The relationship between harmonic Bézier surfaces and minimal surfaces with free boundaries is shown.

## 1 Introduction

This work deals with polynomial surfaces of minimal area. It seems to be a very simple question but we will try to study such kind of surfaces from a non usual point of view: the Bézier description of polynomial surfaces.

People working on minimal surface theory know very well that S. Bernstein was a prolific researcher on this subject at the beginning of the XXth century. The same people probably know that he found an alternative proof of the theorem of Weiertrass about the approximation of arbitrary functions with polynomial functions. What is possibly unknown is that the basis of polynomial functions he used in their proof, nowadays called Bernstein polynomials, is a fundamental component of CAD (Computer Aided Design).

From the very beginning of CAD, polynomial functions are considered the most easy way to construct curves and surfaces from the point of view of computer science. Nevertheless, the coefficients of a polynomial function in the usual basis of powers of the variable have no geometrical meaning. It is hard to control the shape of a polynomial curve or surface just from this set of coefficients.

The alternative basis of Bernstein polynomials solves this drawback because now the coefficients, called control points, have a very intuitive and clear geometric information. It is easy to control the shape of the designed objects just by variations of the control points.

In particular, the end points of a Bézier curve are two of the control points, and the border curves of a Bézier surface can be controlled by a subset of control points.

Like discrete surfaces (see [10]), Bézier surfaces have finite dimensional spaces of admissible variations, therefore the study of linear differential operators on the variation spaces reduces to the linear algebra of matrices.

We start by stating the corresponding Plateau problem for this kind of surfaces: Given the border, or equivalently, the boundary control points, of a Bézier surface, the Bézier-Plateau problem consists in finding the inner control points

M.J. Wilson and R.R. Martin (Eds.): Mathematics of Surfaces 2003, LNCS 2768, pp. 262–273, 2003.

in such a way that the resulting Bézier surface be of minimal area among all other Bézier surfaces with the same boundary control points.

As it also happens in the theory of minimal surfaces, the area functional is highly nonlinear, so we start by studying instead the Dirichlet functional. We obtain the existence and uniqueness of the Dirichlet extremals and we show how to obtain a sequence of Dirichlet extremals whose areas converge to the area of a previously given minimal surface.

Nevertheless, the Bézier Dirichlet extremals are not harmonic charts. So, in a second part, we give the conditions that a Bézier surface must fulfill in order to be harmonic. The result is very surprising because it has a close relation with the theory of minimal surfaces with free boundaries. We show that harmonic Bézier surfaces are totally determined by the first and last rows of control points. As these two rows determine two of the boundary curves, what we get is that given two opposed boundary curves of a harmonic Bézier surface, the whole surface is fully determined.

## 2    The Dirichlet Functional for Bézier Surfaces

Let $\mathcal{P} = \{P_{ij}\}_{i,j=0}^{n,m}$ be the control net of a Bézier surface. Let us denote by $\vec{\mathbf{x}} : [0,1] \times [0,1] \to \mathbb{R}^3$, the chart of the Bézier surface.

$$\vec{\mathbf{x}}(u,v) = \sum_{i=0}^{m} \sum_{j=0}^{n} B_i^m(u) B_j^n(v) P_{ij},$$

being $B_i^n(t)$ the ith-Bernstein polynomial of degree $n$

$$B_i^n(t) = \binom{n}{i} t^i (1-t)^{n-i}, \qquad \text{for } i \in \{0, \ldots, n\},$$

otherwise $B_i^n(t) = 0$.

The area of the Bézier surface is

$$A(\mathcal{P}) = \int_R ||\vec{\mathbf{x}}_u \wedge \vec{\mathbf{x}}_v|| du\ dv = \int_R (EG - F^2)^{\frac{1}{2}} du\ dv,$$

where $R = [0,1] \times [0,1]$ and $E, F, G$ are the coefficients of the first fundamental form of $\vec{\mathbf{x}}$.

Since the border of a Bézier surface is determined by the exterior control points we can state a kind of Plateau problem, that we will call the Bézier-Plateau problem: Given the exterior control points, $\{P_{ij}\}$ with $i = 0, n$ or $j = 0, m$, of a Bézier surface, find the inner ones in such a way that the area of the resulting Bézier surface be a minimum among all the areas of all Bézier surfaces with the same exterior control points.

The first non trivial example of polynomial minimal surface is the Enneper's surface. For its description as a bicubical Bézier surface, ie., its control net, we address the reader to the references [4] or [1]. Note that the Bézier surface defined by such control net is not an approximation of the Enneper's surface, like it happens with the discrete Enneper surface (see [10]), it is the same Enneper's surface.

In general, just a few configurations of the border points will produce a polynomial minimal surface. So, for arbitrary configurations we need to develop general methods for obtaining the extremal of the area functional. Nevertheless, as usual, we do not try to minimize directly the area functional due to its high nonlinearity. We shall work instead with the Dirichlet functional

$$D(\mathcal{P}) = \frac{1}{2} \int_R (||\vec{\mathbf{x}}_u||^2 + ||\vec{\mathbf{x}}_v||^2) du\ dv. \tag{1}$$

Let us recall the following fact relating the area and Dirichlet functionals:

$$(EG - F^2)^{\frac{1}{2}} \leq (EG)^{\frac{1}{2}} \leq \frac{E+G}{2}. \tag{2}$$

Therefore, for any control net, $\mathcal{P}$, $A(\mathcal{P}) \leq D(\mathcal{P})$. Moreover, equality in (2) can occur only if $E = G$ and $F = 0$, i.e., for isothermal charts.

Anyway, both functionals have a minimum in the Bézier case due to the following facts: first, they can be considered as continuous functions defined on $\mathbb{R}^{3(n-1)(m-1)}$. Indeed, the functions depend on the inner control points and its number is $(n-1) \times (m-1)$ and each inner control point belongs to $\mathbb{R}^3$. For example, if $n = m = 2$ there is just one free control point. So, the area functional, or the Dirichlet functional, are just real functions defined on $\mathbb{R}^3$.

Second, as a consequence of $E > 0, G > 0$ and $EG - F^2 > 0$, both functionals are bounded from below.

Third, the infima are attained: when looking for a minimum, we can restrict both functions to a suitable compact subset. If a control point goes far away, then the same happens with a portion of the surface and then, the area, and then the sum of $E$ and $G$, increases. So, we can choose a compact subset such that, if one of the inner control points is outside the compact subset, then the area functional, and then the Dirichlet functional too, are greater than some bound. Finally, if we restrict both continuous functions to a compact subset we can affirm that the infima exist and they are attained.

## 2.1   Extremals of the Dirichlet Functional

The next result translates the condition "a control net $\mathcal{P}$ is an extremal of the Dirichlet problem" into a system of linear equations in terms of the control points. Let us say that we are not computing the Euler-Lagrange equations of the Dirichlet functional. We will simply compute the points where the gradient of a real function defined on $\mathbb{R}^{3(n-1)(m-1)}$ vanishes.

**Proposition 1.** *A control net, $\mathcal{P} = \{P_{ij}\}_{i,j=0}^{n,m}$, is an extremal of the Dirichlet functional with prescribed border if and only if*

$$0 = \frac{n^2}{2(2n-2)m} \binom{n-1}{i}\binom{m}{j} \sum_{k,\ell=0}^{n-1,m} A_{ni}^k \frac{\binom{m}{\ell}}{\binom{2m}{j+\ell}} \Delta^{10} P_{k\ell}$$

$$+ \frac{m^2}{2(2m-2)n} \binom{n}{i}\binom{m-1}{j} \sum_{k,\ell=0}^{n,m-1} \frac{\binom{n}{k}}{\binom{2n}{i+k}} A_{mj}^\ell \Delta^{01} P_{k\ell}, \tag{3}$$

*for any $i \in \{1, \dots, n-1\}$ and $j \in \{1, \dots, m-1\}$ where $A_{ni}^k$ is defined by*

$$\frac{ni - nk - i}{(n-i)(2n-1-i-k)} \frac{\binom{n-1}{k}}{\binom{2n-2}{i+k-1}}.$$

**Proof:** Let us compute the gradient of the Dirichlet functional with respect to the coordinates of a control point $P_{ij} = (x_{ij}^1, x_{ij}^2, x_{ij}^3)$. For any $a \in \{1,2,3\}$, $i \in \{1,\ldots,n-1\}$ and any $j \in \{1,\ldots,m-1\}$

$$\frac{\partial D(\mathcal{P})}{\partial x_{ij}^a} = \int_R (< \frac{\partial \overrightarrow{\mathbf{x}}_u}{\partial x_{ij}^a}, \overrightarrow{\mathbf{x}}_u > + < \frac{\partial \overrightarrow{\mathbf{x}}_v}{\partial x_{ij}^a}, \overrightarrow{\mathbf{x}}_v >) du\, dv$$

Let us compute now the partial derivatives

$$\frac{\partial \overrightarrow{\mathbf{x}}_u}{\partial x_{ij}^a} = \frac{\partial}{\partial x_{ij}^a} \frac{\partial}{\partial u} \overrightarrow{\mathbf{x}} = \frac{\partial}{\partial u} \frac{\partial}{\partial x_{ij}^a} \overrightarrow{\mathbf{x}}$$

$$= \frac{\partial}{\partial u} B_i^n(u) B_j^m(v) e^a = n(B_{i-1}^{n-1}(u) - B_i^{n-1}(u)) B_j^m(v) e^a,$$

where $e^a$ denotes the $a$-th vector of the canonical basis, i.e. $e^1 = (1,0,0), e^2 = (0,1,0), e^3 = (0,0,1)$. Analogously

$$\frac{\partial \overrightarrow{\mathbf{x}}_v}{\partial x_{ij}^a} = m B_i^n(u)(B_{j-1}^{m-1}(v) - B_j^{m-1}(v)) e^a.$$

Therefore

$$\frac{\partial D(\mathcal{P})}{\partial x_{ij}^a} = \int_R \big( n(B_{i-1}^{n-1}(u) - B_i^{n-1}(u)) B_j^m(v) < e^a, \overrightarrow{\mathbf{x}}_u >$$

$$+ m B_i^n(u)(B_{j-1}^{m-1}(v) - B_j^{m-1}(v)) < e^a, \overrightarrow{\mathbf{x}}_v > \big)\, dudv$$

$$= \int_R (n(B_{i-1}^{n-1}(u) - B_i^{n-1}(u)) B_j^m(v) < e^a, n \sum_{k,\ell=0}^{n-1,m} B_k^{n-1}(u) B_\ell^m(v) \Delta^{10} P_{k\ell} >$$

$$+ m B_i^n(u)(B_{j-1}^{m-1}(v) - B_j^{m-1}(v)) < e^a, m \sum_{k,\ell=0}^{n,m-1} B_k^n(u) B_\ell^{m-1}(v) \Delta^{01} P_{k\ell} >) dudv.$$

Applying now that for any $n \in \mathbb{N}$ and for any $i = 0,\ldots,n$, $\int_0^1 B_i^n(t) dt = \frac{1}{n+1}$, we get

$$\frac{\partial D(\mathcal{P})}{\partial x_{ij}^a} = \frac{n^2}{2(2n-2)m} \sum_{k,\ell=0}^{n-1,m} \left( \frac{\binom{n-1}{i-1}\binom{n-1}{k}}{\binom{2n-2}{i+k-1}} - \frac{\binom{n-1}{i}\binom{n-1}{k}}{\binom{2n-2}{i+k}} \right) \frac{\binom{m}{\ell}\binom{m}{j}}{\binom{2m}{j+\ell}} < e^a, \Delta^{10} P_{k\ell} >$$

$$+ \frac{m^2}{2(2m-2)n} \sum_{k,\ell=0}^{n,m-1} \frac{\binom{n}{k}\binom{n}{i}}{\binom{2n}{i+k}} \left( \frac{\binom{m-1}{j-1}\binom{m-1}{\ell}}{\binom{2m-2}{j+\ell-1}} - \frac{\binom{m-1}{j}\binom{m-1}{\ell}}{\binom{2m-2}{j+\ell}} \right) < e^a, \Delta^{01} P_{k\ell} >$$

$$= \frac{n^2}{2(2n-2)m} \binom{n-1}{i}\binom{m}{j} \sum_{k,\ell=0}^{n-1,m} A_{ni}^k \frac{\binom{m}{\ell}}{\binom{2m}{j+\ell}} < e^a, \Delta^{10} P_{k\ell} >$$

$$+ \frac{m^2}{2(2m-2)n} \binom{n}{i}\binom{m-1}{j} \sum_{k,\ell=0}^{n,m-1} \frac{\binom{n}{k}}{\binom{2n}{i+k}} A_{mj}^\ell < e^a, \Delta^{01} P_{k\ell} > . \square$$

*Remark 1.* A simple computation shows that the second derivatives $\frac{\partial^2 D(\mathcal{P})}{\partial x_{ij}^a x_{k\ell}^b}$ are constant. Indeed, it is a consequence of the fact that the chart depends linearly on the control points.

Let us recall that the Weiertrass approach to the theory of minimal surfaces points out that any minimal surface is related with some complex functions on a variable $z = u + \mathbf{i}v$, being $u, v$ the parameters of the surface. When dealing with polynomial complex functions, the degrees of the resulting minimal surface in the variables $u$ and $v$ are the same. So, this seems to indicate that Bézier surfaces with a squared control net are more suitable in this setting. In the squared case, equations (3) are simpler.

**Corollary 1.** *A squared control net,* $\mathcal{P} = \{P_{ij}\}_{i,j=0}^{n,n}$, *is an extremal of the Dirichlet functional with prescribed border if and only if*

$$0 = \sum_{k,\ell=0}^{n-1,n} \frac{\binom{n}{\ell}}{\binom{2n}{j+\ell}} C_{ni}^k \Delta^{10} P_{k\ell} + \sum_{k,\ell=0}^{n,n-1} \frac{\binom{n}{k}}{\binom{2n}{i+k}} C_{mj}^\ell \Delta^{01} P_{k\ell}, \tag{4}$$

*for any* $i, j = 1, \ldots, n-1$, *where* $C_{ni}^k = \frac{(n-1)i - nk}{i+k} \frac{\binom{n-1}{k}}{\binom{2n-2}{i+k}}$.

Let us recall that, as we have said before, a minimum of the Dirichlet functional with prescribed border always exists. So, fixing the exterior control points and taking as unknowns the inner control points, the linear system (3) and, in particular, the linear system (4), always is compatible and it can be solved in terms of the exterior control points.

## 2.2  Uniqueness of the Dirichlet Extremal

We have seen that the extremals of the Dirichlet functional always exists. Let us now prove the uniqueness.

**Theorem 1.** *The Dirichlet extremal is unique.*

**Proof.** We know that Dirichlet extremals are computed as solutions of the linear system (1). Let us write it as

$$A \cdot P = B. \tag{5}$$

where $B$ is a column vector computed through the boundary control points, $P$ is the column vector of the inner control points, and $A$ is a square matrix whose entries are independent of the control points, they just depend on the dimensions of the control net.

Let us check that the rank of $A$ is maximal, i.e., the linear system has a unique solution.

A well known theorem of Bernstein ([9], page 38) affirms that the only minimal surface being the graph of a function is a plane. So, let us choose the control net

$$\mathcal{P} := \{(\frac{i}{n}, \frac{j}{m}, 0)\}_{i,j=0}^{n,m}.$$

The associated Bézier chart is $\overrightarrow{\mathbf{x}}(u, v) = (u, v, 0)$. This is the so called linear precision property of the Bézier surfaces. It is minimal and isothermal, therefore, $\mathcal{P}$ is a Dirichlet extremal for the same boundary conditions.

Let us check that it is the unique Dirichlet extremal. Any other configuration $\mathcal{P}_0$ of the inner control points with at least a control point with nonzero third coordinate will produce a non planar associated Bézier surface, and then

$$\mathcal{D}(\mathcal{P}) = \mathcal{A}(\mathcal{P}) < \mathcal{A}(\mathcal{P}_0) \le \mathcal{D}(\mathcal{P}_0).$$

Any other configuration of the inner control points with zero third coordinate will produce a planar surface but a non isothermal parametrization, and then

$$\mathcal{D}(\mathcal{P}) = \mathcal{A}(\mathcal{P}) = \mathcal{A}(\mathcal{P}_0) < \mathcal{D}(\mathcal{P}_0).$$

Therefore, $\mathcal{P}$ is the only Dirichlet extremal. This implies that Eq. (5) has an unique solution for the boundary conditions given by $\mathcal{P}$. Therefore the matrix $A$ is of maximal rank. □

Note that the situation in the Bézier case, and for the Dirichlet functional, is rather different than in the discrete surface case for the area functional. In this case, as in the differentiable surface case, given a prescribed border, there could exist more than just one surface with minimal area.

## 3   Convergence Results

In this section, we will study how to reach the minimal area with prescribed boundary by a sequence of Bézier surfaces which are Dirichlet extremals.

**Theorem 2.** *Let $\overrightarrow{\mathbf{x}} : [0, 1] \times [0, 1] \to \mathbb{R}^3$ be an isothermal chart of a surface of minimal area among all surfaces with the same boundary.*

*Let $\overrightarrow{\mathbf{y}}_n$ be the Dirichlet extremal of degree $n$ with boundary defined by the exterior control points of the control net $\mathcal{P}_n = \{\overrightarrow{\mathbf{x}}(\frac{i}{n}, \frac{j}{n})\}_{i,j=0}^n$.*
*Then,*

$$\lim_{n \to \infty} \mathcal{A}(\overrightarrow{\mathbf{y}}_n) = \mathcal{A}(\overrightarrow{\mathbf{x}}).$$

**Proof:** Let $\overrightarrow{\mathbf{x}}_n$ be the associated Bézier chart to the control net $\mathcal{P}_n$. The Bernstein's proof of the Weiertrass theorem indicates that the sequence $\{\overrightarrow{\mathbf{x}}_n\}_{n=1}^\infty$ is uniformly convergent to $\overrightarrow{\mathbf{x}}$. Moreover,

$$\lim_{n \to \infty} \mathcal{A}(\overrightarrow{\mathbf{x}}_n) = \mathcal{A}(\overrightarrow{\mathbf{x}}) = \mathcal{D}(\overrightarrow{\mathbf{x}}) = \lim_{n \to \infty} \mathcal{D}(\overrightarrow{\mathbf{x}}_n), \tag{6}$$

where the equality $\mathcal{A}(\overrightarrow{\mathbf{x}}) = \mathcal{D}(\overrightarrow{\mathbf{x}})$ is a consequence of the fact that the chart is minimal and isothermal.

Let $\overrightarrow{\mathbf{z}}_n$ be a chart of the surface with minimal area and with the same boundary than $\overrightarrow{\mathbf{y}}_n$. Therefore, $\mathcal{A}(\overrightarrow{\mathbf{z}}_n) \le \mathcal{A}(\overrightarrow{\mathbf{y}}_n)$.

Moreover, due to the fact that area functional is always lesser than the Dirichlet functional, we have that for any $n \in \mathbb{N}$, $\mathcal{A}(\overrightarrow{\mathbf{y}}_n) \le \mathcal{D}(\overrightarrow{\mathbf{y}}_n)$.

And, recalling that $\vec{\mathbf{y}}_n$ is the Dirichlet extremal of degree $n$ and that it has the same boundary than the polynomial chart $\vec{\mathbf{x}}_n$, we have that $\mathcal{D}(\vec{\mathbf{y}}_n) \leq \mathcal{D}(\vec{\mathbf{x}}_n)$.

Resuming, we have for any $n \in \mathbb{N}$

$$\mathcal{A}(\vec{\mathbf{z}}_n) \leq \mathcal{A}(\vec{\mathbf{y}}_n) \leq \mathcal{D}(\vec{\mathbf{y}}_n) \leq \mathcal{D}(\vec{\mathbf{x}}_n). \tag{7}$$

Now, according to the results on the boundary behaviour of minimal surfaces (see [8], paragraph 327), as the boundary of $\vec{\mathbf{z}}_n$ converges uniformly to the boundary of $\vec{\mathbf{x}}$ then

$$\lim_{n \to \infty} \mathcal{A}(\vec{\mathbf{z}}_n) = \mathcal{A}(\vec{\mathbf{x}}). \tag{8}$$

On the other hand, due to the fact that $\vec{\mathbf{x}}_n$ $C^1$-converges to $\vec{\mathbf{x}}$, (see [6], Th. 1.8.1) then

$$\lim_{n \to \infty} \mathcal{D}(\vec{\mathbf{z}}_n) = \mathcal{D}(\vec{\mathbf{x}}) = \mathcal{A}(\vec{\mathbf{x}}). \tag{9}$$

Therefore, the result follows from Eqs. (7), (8) and (9). □

Nevertheless, the previous result is not useful to obtain good approximations of low degree. The reason of this is a consequence of a property of Bézier curves that has a correspondence in Bézier surfaces. For example, if we take the control points on a circle, the resulting Bézier curve is a bad approximation to the circle. In order to obtain better approximations with the same degree one has to solve a least square problem.

A good approximation to the catenoid can be obtained with a degree 7 Bézier surface solution of the Dirichlet problem. In Figure 1, the depicted Bézier surface has an area exceeding the area of the corresponding half catenoid in 0.05%.

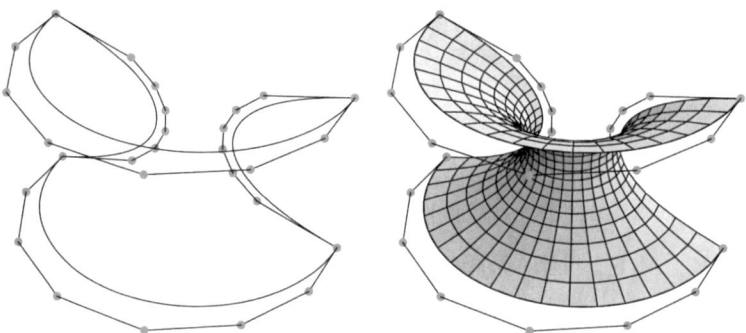

**Fig. 1.** If we take suitable control points such that the corresponding Bézier curves are a good approximation of the border of a half catenoid, then the resulting Bézier surface obtained as the minimum of the Dirichlet problem is a good approximation to catenoid. Left, the border conditions. Right, the degree 7 Bézier surface.

A degree $8 \times 8$ Bézier surface can be built resembling the minimal surface obtained by Schwarz (see [8], page 75) by placing the exterior control points on some of the edges of a cube (Fig. 2).

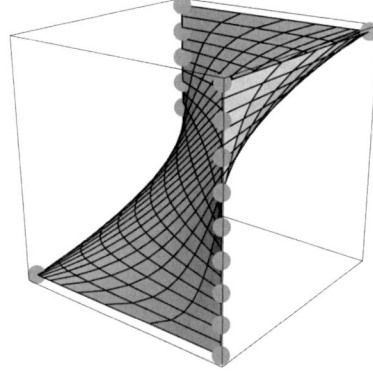

**Fig. 2.** An approximation to a Schwarz's surface. Note that the control points $(0,0,0)$ and $(1,1,1)$ are repeated 7 times.

## 4   Bézier Harmonic Charts

The Euler-Lagrange equations defined by the Dirichlet functional (1) are $\Delta \vec{x} = 0$. So, in the unrestricted case, extremals of the Dirichlet functional are harmonic charts. But harmonic charts are not polynomial in general and then, they can not be solutions of the Bézier-Plateau problem.

The *Dirichlet principle* for domains bounded by a Jordan curve says that harmonic functions are the infima of the Dirichlet functional among all functions defined on the same domain and with the same values on the border. (See [8], paragraph 229). In our case the Jordan curve is the square $[0,1] \times [0,1]$, therefore, if a polynomial chart is harmonic, then it is an extremal of the Dirichlet functional for Bézier surfaces with prescribed border.

A natural question is then to ask for the conditions that a Bézier surface must fulfill in order to be harmonic. To answer this question, we will compute the Laplacian of a Bézier chart .

**Theorem 3.** *Given a control net in* $\mathbb{R}^3$, $\{P_{ij}\}_{i,j=0}^{n,m}$, *the associated Bézier surface,* $\vec{x} : [0,1] \times [0,1] \to \mathbb{R}^3$, *is harmonic, i.e,* $\Delta \vec{x} = 0$ *if and only if for any* $i \in \{1, \ldots, n\}$ *and* $j \in \{1, \ldots, n\}$

$$
\begin{aligned}
0 = {} & n(n-1)(P_{i+2,j}a_{in} + P_{i+1,j}(b_{i-1,n} - 2a_{in}) \\
& \qquad + P_{i-1,j}(b_{i-1,n} - 2c_{i-2,n}) + P_{i-2,j}c_{i-2,n}) \\
& +m(m-1)(P_{i,j+2}a_{jm} + P_{i,j+1}(b_{j-1,m} - 2a_{jm}) \\
& \qquad + P_{i,j-1}(b_{j-1,m} - 2c_{j-2,m}) + P_{i,j-2}c_{j-2,m}) \\
& +P_{ij}((a_{in} - 2b_{i-1,n} + c_{i-2,n})n(n-1) \\
& \qquad + (a_{jm} - 2b_{j-1,m} + c_{j-2,m})m(m-1)),
\end{aligned}
$$

*where, for* $i \in \{0, \ldots, n-2\}$

$$a_{in} = (n-i)(n-i-1), \quad b_{in} = 2(i+1)(n-i-1), \quad c_{in} = (i+1)(i+2),$$

and $a_{in} = b_{in} = c_{in} = 0$ otherwise.

For the proof, see [1].

In the case of a quadratic net $(n = m)$ we can state the following corollary

**Corollary 2.** *Given a quadratic net of points in $\mathbb{R}^3$, $\{P_{ij}\}_{i,j=0}^n$, the associated Bézier surface, $\vec{\mathbf{x}} : [0,1] \times [0,1] \to \mathbb{R}^3$, is harmonic, i.e, $\Delta\vec{\mathbf{x}} = 0$ if and only if for any $i \in \{1,\dots,n\}$*

$$
\begin{aligned}
0 = {}& P_{i+2,j}a_{in} + P_{i+1,j}(b_{i-1,n} - 2a_{in}) + P_{i-1,j}(b_{i-1,n} - 2c_{i-2,n}) \\
& + P_{i-2,j}c_{i-2,n} + P_{i,j+2}a_{jn} + P_{i,j+1}(b_{j-1,n} - 2a_{jn}) \\
& + P_{i,j-1}(b_{j-1,n} - 2c_{j-2,n}) + P_{i,j-2}c_{j-2,n} \\
& + P_{ij}(a_{in} - 2b_{i-1,n} + c_{i-2,n} + a_{jn} - 2b_{j-1,n} + c_{j-2,n}).
\end{aligned}
\tag{10}
$$

An analysis of Equations (10) for degree $n = 3$ (the control net has sixteen points) shows that all the equations can be reduced to a system of just eight independent linear equations. Moreover, it is possible to show that the linear system can be solved by expressing the eight control points in the two middle rows as functions of the other eight control points in the first and last rows. This was done in [1]. So, our aim in the rest of the section is to show that this is true for any dimension, i.e., that the first and last rows of control points fully determine the harmonic Bézier surface.

In order to do that, it is better to come back to the usual basis of polynomials.

**Lemma 1.** *Let $f(u,v) = \sum_{k,\ell=0}^n a_{k\ell}u^k v^\ell$ be a harmonic polynomial function of degree $n \geq 2$, then,*

1. *If $n$ is odd, then all coefficients $\{a_{k\ell}\}_{k=2,\ell=0}^n$ are totally determined by the coefficients $\{a_{0\ell}, a_{1\ell}\}_{\ell=0}^n$.*
2. *If $n$ is even, then all coefficients $\{a_{k\ell}\}_{k=2,\ell=0}^n$ and also the coefficient $a_{1n}$ are totally determined by the coefficients $\{a_{0\ell}\}_{\ell=0}^n$ and $\{a_{1\ell}\}_{\ell=0}^{n-1}$.*

**Proof:** The harmonic condition $\Delta f = 0$ can be translated into a system of linear equations in terms of the coefficients $\{a_{k\ell}\}_{k,\ell=0}^n$

$$(k+2)(k+1)a_{k+2,\ell} + (\ell+2)(\ell+1)a_{k,\ell+2} = 0, \quad k,\ell = 0,\dots,n,$$

but with the convention $a_{n+1,\ell} = a_{n+2,\ell} = a_{n,\ell+2} = a_{n,\ell+1} = 0$.

This means that any coefficient $a_{k\ell}$ with $k > 1$ can be related with $a_{k-2,\ell+2}$ and so on until the first subindex is 0 or 1, or until the second subindex is greater than $n$. In this second case, $a_{k\ell}$ is directly 0. Indeed, if $\ell + 2k > n$ then $a_{2k,\ell} = a_{2k+1,\ell} = 0$, otherwise

$$a_{2k,\ell} = (-1)^k \binom{2k+\ell}{\ell} a_{0,2k+\ell}, \quad a_{2k+1,\ell} = (-1)^k \frac{1}{2k+1}\binom{2k+\ell}{\ell} a_{1,2k+\ell}. \tag{11}$$

So, when $n$ is odd, the result is proved. When $n$ is even, we have that, in addition, coefficient $a_{1n}$ vanishes.    □

As we have said before, the next result was conjectured and checked for low dimensions in [1]. We can now give the general result.

**Proposition 2.** Let $\vec{x}(u,v) = \sum_{k,\ell=0}^{n} B_k^n(u)B_\ell^n(v)P_{k\ell}$ be a harmonic Bézier chart of degree $n$ with control net $\{P_{k\ell}\}_{k,\ell=0}^{n}$, then

1. If $n$ is odd, control points in the inner rows $\{P_{k\ell}\}_{k=1,\ell=0}^{n-1,n}$ are determined by the control points in the first and last rows, $\{P_{0\ell}\}_{\ell=0}^{n}$ and $\{P_{n\ell}\}_{\ell=0}^{n}$.

2. If $n$ is even, control points in the inner rows $\{P_{k\ell}\}_{k=1,\ell=0}^{n-1,n}$ and also the corner control point $P_{nn}$ are determined by the control points in the first and last rows, $\{P_{0\ell}\}_{\ell=0}^{n}$ and $\{P_{n\ell}\}_{\ell=0}^{n-1}$.

**Proof:** Let us write the Bézier chart in the usual basis of polynomials

$$\vec{x}(u,v) = \sum_{k,\ell=0}^{n} u^k v^\ell (a_{k\ell}, b_{k\ell}, c_{k\ell}).$$

Let us consider the case $n$ odd. Note that the first and last rows of control points determine the two opposed border curves $\vec{x}(0,v)$, $\vec{x}(1,v)$, $v \in [0,1]$. The first border curve is

$$\vec{x}(0,v) = \sum_{\ell=0}^{n} v^\ell (a_{0\ell}, b_{0\ell}, c_{0\ell}), \tag{12}$$

and the second one is

$$\vec{x}(1,v) = \sum_{\ell=0}^{n} v^\ell \sum_{k=0}^{n} (a_{k\ell}, b_{k\ell}, c_{k\ell}). \tag{13}$$

From Eq. (12) we can obtain coefficients $(a_{0\ell}, b_{0\ell}, c_{0\ell})$ for $\ell = 0, \ldots, n$. By the previous lemma, all coefficients $(a_{k\ell}, b_{k\ell}, c_{k\ell})$ are determined by the coefficients $(a_{0\ell}, b_{0\ell}, c_{0\ell})$ and $(a_{1\ell}, b_{1\ell}, c_{1\ell})$. In particular, thanks to Eq. (11), we can reduce Eq. (13) to just a system of linear equations involving the coefficients $(a_{1\ell}, b_{1\ell}, c_{1\ell})$. Moreover, the matrix of coefficients of this system is triangular and with the unit in the diagonal entries. Therefore, the knowledge of the first and last rows of control points, implies the knowledge of the coefficients $(a_{0\ell}, b_{0\ell}, c_{0\ell})$ and $(a_{1\ell}, b_{1\ell}, c_{1\ell})$ and then, the knowledge of all the coefficients, i.e., of the whole harmonic chart, or equivalently, of the whole control net.

For the even case, the arguments are similar.    □

## 5   The Gergonne Problem Revisited

The result shown in Proposition 2 is analogous to what happens with problems about minimal surfaces with free boundaries: To find minimal surfaces the boundary of which (or part of it) is left free on supporting manifolds. With Bézier surfaces we have seen that given two disjoint border curves, i.e., given two border lines of control points, the other lines are determined thanks to Eqs. (10), and then, the whole Bézier surface is determined.

A typical problem of minimal surfaces with free boundaries is the well known Gergonne problem giving raise to a surface (see [8], page 79, or [5]) that should be no confused with another surface called the Gergonne surface. The original problem was stated as follows: "Couper un cube en deux parties, de telle manière que la section vienne se terminer aux diagonales inverses de deux faces opposées, et que l'aire de cette section, terminée à la surface du cube, soit un minimum".

The solution was finally found by Schwarz in 1872 (see Fig. 3, right). What is remarkable is that given the inverse diagonals of two opposed faces of a cube, the Gergonne surface is fully determined.

In the Bézier case, given two opposed lines of border control points, the harmonic Bézier surface is fully determined. A degree 6 harmonic Bézier approximation of the this surface can be seen in Fig. 3, left.

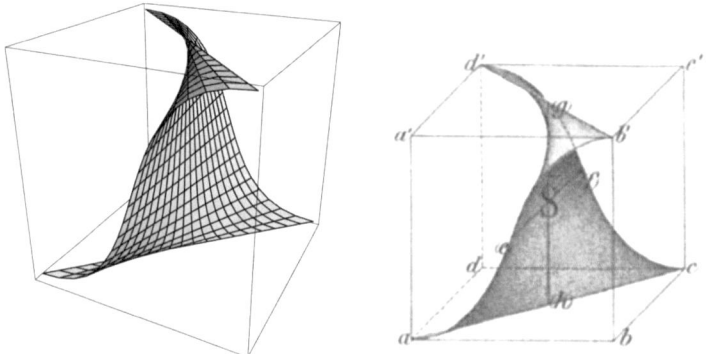

**Fig. 3.** Right, the Schwarz solution to the Gergonne problem. Left, an approximation found as a harmonic Bézier surface of degree $6 \times 6$.

The border control points to generate such a surface has been chosen as follows:

The bottom row is

$$P_{00} = (0,0,0), \qquad P_{01} = (a,a,0), \qquad P_{02} = (b,b,0),$$
$$P_{03} = (1-b,1-b,0), \; P_{04} = (1-a,1-a,0), \; P_{05} = (1,1,0),$$

and the top row,

$$P_{50} = (1,0,1), \qquad P_{51} = (1-a,a,1), \; P_{52} = (1-b,b,1),$$
$$P_{53} = (b,1-b,1), \; P_{54} = (a,1-a,1), \; P_{55} = (0,1,1).$$

The choice of the parameters $a$ and $b$ can be made according to different principles. For example, we can ask for a uniform distribution of the control points by taking $a = \frac{1}{5}, b = \frac{2}{5}$. The resulting Bézier surface is a hyperbolic paraboloid and its area is 1.28079. Or we can ask for isothermality of the Bézier surface on the corners. Then the values are ($a = \frac{\sqrt{257}-9}{32} \sim 0.22, b = \frac{32a+9}{32} \sim 0.40$.) and the resulting Bézier surface is an approximation to a portion of helycoid. The area of the restriction of the helycoid to the cube $[0,1]^3$ is 1.25364.

But there is a choice of the parameters $a$ and $b$ minimizing the area of surface inside the cube. An approximation of these values is $a = 0.41, b = 0.51$ and the resulting Bézier surface, plotted in Figure 4, left, shows a shape resembling the Gergonne surface. The area is 1.24294.

## Acknowledgments

This work has been partially supported by DGICYT grant BFM2002-00770.

## References

1. C. Cosín, J. Monterde, *Bézier surfaces of minimal area*, Proceedings of the International Conference of Computational Science, ICCS'2002, Amsterdam, eds. Sloot, Kenneth Tan, Dongarra, Hoekstra, Lecture Notes in Computer Science, 2330, vol II, pages 72–81, Springer-Verlag, (2002).
2. do Carmo, M. P., *Differential Geometry of curves and surfaces*, Prentice- Hall Inc., (1976).
3. G. Farin, *Curves and surfaces for computer aided geometric design. A practical guide*, Morgan Kaufmann, 5th. ed., (2001).
4. J. Gallier, *Curves and surfaces in geometric modeling. Theory and algorithms*, Morgan Kaufmann Publ., (2000).
5. S. Hildebrandt, *Some results on minimal surfaces with free boundaries*, in Nonlinear Analysis and Optimization, LNM 1007, eds. A. Dold and B. Eckmann, pags. 115-134, (1983).
6. G. G. Lorentz, *Bernstein polynomials*, 2nd. ed., Chelsea Publishing Co., New York, (1986).
7. D. Mars, *Applied Geometry for computer graphics and CAD*, Springer Verlag, London, (1999).
8. J. C. C. Nitsche, *Lectures on minimal surfaces*, vol. 1, Cambridge Univ. Press, Cambridge (1989).
9. R. Osserman, *A survey of minimal surfaces*, Dover Publ., (1986).
10. K. Polthier, W. Rossman, *Discrete constant mean curvature surfaces and their index*, J. Reine Angew. Math. **549**, (2002) 47–77.

# A Functional Equation Approach to the Computation of the Parameter Symmetries of Spline Paths

Helmut E. Bez

Department of Computer Science
Loughborough University
Leicestershire LE11 3TU, UK

**Abstract.** The paper considers a particular general class of parametrised path functions used in computer graphics, geometric modeling and approximation theory for the construction of curves and surfaces. General methods are developed for the identification of the conditions under which parameter transformations preserve the path geometry. The determination of these 'parameter symmetries' is shown to be equivalent to the identification of the solution space of a functional equation.

The main results of the paper are the determination of the parameter symmetries of $C^1$ and $C^2$ cubic parametric splines; in particular a complete answer to the following question for cubic splines with natural end conditions is given:

For any given set of interpolation points, under what conditions do different sets of knots determine the same geometry?

## 1   Introduction

The simplest parametric paths used in approximation theory, computer graphics and geometric design take the generic form

$$\sum_{k=0}^{n} b_{n,k}(t)v_k, \quad \text{for} \ \ 0 \le t \le 1 \ ,$$

for some predefined $v_k$-independent 'basis' functions $b_{n,k}$. The most familiar example of this type being the Bézier-Bernstein paths, for which $b_{n,k}(t) = \binom{n}{k}t^k(1-t)^{n-k}$. Such paths are completely specified by the control points $v_i \in \mathbb{R}^2$. If $V$ is the set of all $(n+1)$-tuples of vectors in $\mathbb{R}^n$ and $\mathcal{P}$ denotes the set of all paths in $\mathbb{R}^n$ then the paths may be regarded as being defined by a function of the form

$$p : V \to \mathcal{P} \ ,$$

where for $v \in V$, $p[v]$ denotes the path defined by $v$ and $p[v](t)$ is the point in $\mathbb{R}^2$ on the path at parameter value $t$.

Other parametrically defined paths, including those defined by splines, are not of this simple type. For many types it is necessary to specify both a control

M.J. Wilson and R.R. Martin (Eds.): Mathematics of Surfaces 2003, LNCS 2768, pp. 274–296, 2003.

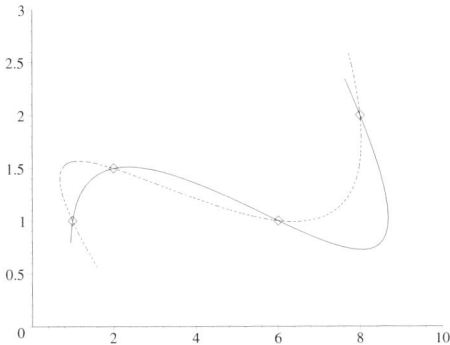

**Fig. 1.** In general, different $\tau$ produce different paths for functions of the class $p : V \times T \to \mathcal{P}$

point set and $v \in V$ and a set $\tau = (\tau_0, \ldots, \tau_n)$ of 'parameters'. The parameters $\tau$ are, in the simplest case, real numbers with the monotone property $\tau_0 < \cdots < \tau_n$ and the paths can be written in the generic form

$$\sum_{k=0}^{n} b_{n,k}(\tau, t) v_k, \quad \text{for} \quad \tau_0 \leq t \leq \tau_n \ .$$

It is clear that $V$ is not the appropriate domain for these paths - rather they should be defined on the extended domain $V \times T$ where $T$ is the set of all $(n+1)$−tuples of real numbers $\tau_i$ with $\tau_0 < \cdots < \tau_n$. The Lagrange paths, for which

$$b_{n,k}(\tau, t) = \prod_{j=0, j \neq k}^{n} \frac{(t - \tau_j)}{(\tau_k - \tau_j)} \ ,$$

are of this type.

For splines, the parameters $\tau_0, \ldots \tau_n$ are known as knots. Cubic splines can be written as the join $c_1[v, \tau] \vee c_2[v, \tau] \vee \cdots \vee c_n[v, \tau]$ of a number of cubic spans $c_i$, on $[\tau_{i-1}, \tau_i]$, that meet with suitable smoothness conditions at the knots and where each span can be expressed as $c_i[v, \tau](t) = \sum_{k=0}^{n} b_{k,i}(\tau, t) v_k$ - in this paper this is referred to as the canonical representation of the spline.

The shape of the paths determined by functions of the generic type $p : V \times T \to \mathcal{P}$ can be changed by keeping the geometric data, or 'control points', $v$ fixed and modifying the auxiliary parameters $\tau \in T$; the latter is illustrated in Figure 1 where Lagrange cubics on $v = \{(1,1), (2, \frac{3}{2}), (6,1), (8,2)\}$ are shown for both $\tau = (0,1,2,3)$ and $\tau = (0,1,3,5)$. However, distinct $\tau$ can give rise to paths having the same shape. This paper is concerned with investigating the precise conditions under which the graphs of paths defined by parametric cubic spline functions remain invariant under a change of parameters.

## 2   Scope of the Paper

The invariance problem identified in the introduction leads naturally to the question of invariance transformations of the $T$ component of the domain of path functions of the generic class $p : V \times T \to \mathcal{P}$; i.e., transformations $g : T \to T$ with the property $p[v, g\tau] = p[v, \tau]$ for all $v \in V$ and $\tau \in T$. We refer to invariant transformations of the $T$ domain as the 'parameter symmetries' of $p$.

In general if $\Theta$ is a mathematical object of a particular category and $g$ is an invertible mapping such that $g * \Theta$ is an object of the same category then $g$ is said to be a symmetry of $\Theta$ if $g*\Theta = \Theta$; i.e., if $\Theta$ is a fixed point (or invariant) of $g$. The set of all symmetries of an entity forms a group. The topic of this paper can therefore be regarded as the determination of a class of symmetries of path functions of the class $p : V \times T \to \mathcal{P}$ where $g * p$ is defined by $(g * p)[v, \tau] = p[v, g^{-1}\tau]$.

The paper establishes a generic functional equation for the parameter symmetries. The computation of the parametric symmetries is shown to be, essentially, the determination of the complete solution space of an instance of a generic functional equation. The equation is shown to be tractable for Lagrange polynomials, parametric $C^1$ cubic splines and parametric $C^2$ cubic splines with natural end conditions and the general solutions obtained provide complete information on the invariant parameter transformations of these paths. The paper deals with paths in the plane, but the fundamentals generalise readily to paths in vector spaces of higher finite dimension.

Apart from theoretical interest, a knowledge of these symmetries is of practical value. They enable, for example,

- embedded tests to be incorporated in application software to indicate when specified parameter modifications are geometry modifying

- a greater understanding of the behaviour of curves and surfaces, constructed using splines, under geometric transformation of the input data - as discussed in [1] and [2] for global polynomial constructions.

## 3   The Fundamental Functional Equation
   for Parameter Symmetries

In this section of the paper the generic functional equation for the determination of parameter symmetries of the generic class $p : V \times T \to \mathcal{P}$ is given. The equation is, fundamentally, a statement of equivalence for the class - where paths are equivalent if they have the same geometry.

The general conditions under which distinct paths have the same geometry are as follows: a path in $\mathbb{R}^2$ is a $C^1$ function $q : I \to \mathbb{R}^2$ where $I$ is an interval of $\mathbb{R}$ and $q' \neq 0$ on $I$. Distinct paths $q_1, q_2$ define the same geometry if there is a $C^1$ function $\phi : I_1 \to I_2$ such that $\phi' \neq 0$ on $I_1$ and $q_1 = q_2 \circ \phi$. The condition $\phi' \neq 0$ means that $\phi$ is strictly monotone; $\phi$ is therefore invertible and has $C^1$

inverse $\phi^{-1}$ with $(\phi^{-1})' \neq 0$. The condition $\phi' \neq 0$ implies either $\phi' > 0$ or $\phi' < 0$, i.e., $\phi$ either strictly increasing or strictly decreasing. We can, without loss of generality assume that $\phi$ is strictly increasing; $M$ denotes the set of all strictly increasing $C^1$ functions on $\mathbb{R}$.

It follows that for an arbitrary path function of the class $p : V \times T \to \mathcal{P}$ elements $\tau, \tau^*$ of the auxiliary parameter domain $T$ determine paths with the same geometry if and only if there exists a function $\phi : T \times T \to M$ such that

$$p[v, \tau] = p[v, \tau^*] \circ \phi_{\tau,\tau^*} \text{ for all } v \in V \ .$$

Hence if all the solutions of this functional equation for $\tau$, $\tau^*$ and $\phi_{\tau,\tau^*}$ are known, then all the parameter symmetry transformations of the path function $p : V \times T \to \mathcal{P}$ are also known. The symmetries may be characterised as triples $(\tau, \tau^*, \phi_{\tau,\tau^*})$ that satisfy the functional equation. Computing the complete set of symmetries requires the complete solution space of $p[v, \tau] = p[v, \tau^*] \circ \phi_{\tau,\tau^*}$ to be determined. The solution strategy is to first determine the nature of $\phi$ directly from the functional equation; with $\phi$ known the functional equation becomes, essentially, a relationship between $\tau$ and $\tau^*$ - from which the symmetries of $p$ can be determined.

## 4  Three Case Studies

The following simple Lemma, the proof of which is omitted, is required in all the cases studies.

**Lemma 1.** *If $\phi(t) = at + b$ where $a > 0$, then*

$$[1, \phi(t), \dots, \phi(t)^n] = [1, t, \dots, t^n] L_\phi \ ,$$

*where $L_\phi$ is the invertible, upper-triangular, $(n+1)$-square matrix*

$$L_\phi = \begin{bmatrix} 1, & b, & b^2, & b^3, & \cdots & , & b^n \\ 0, & a, & 2ba, & 3b^2a, & \cdots & , & nb^{n-1}a \\ 0, & 0, & a^2, & 3ba^2, & \cdots, & n(n-1)b^{n-2}a^2 \\ \vdots & & & & & \vdots \\ 0, & 0, & 0, & 0, & \cdots & , & a^n \end{bmatrix} .$$

### 4.1  Case Study 1: Parametric Lagrange Paths

The computation of the symmetries for Lagrange paths has been presented previously by the author [1,2]; it is repeated here as the simplest non-trivial demonstration of the solution of the functional equation for parameter symmetries. The explicit form of the Lagrange path function $p_L : V \times T \to \mathcal{P}$ was given earlier; it can be expressed as

$$p_L[v, \tau](t) = [1, t, t^2, \dots, t^n] D_\tau^{-1} [v_0, v_1, \dots, v_n]^T$$

where

$$D_\tau = \begin{bmatrix} 1, & \tau_0, & \ldots, & \tau_0^n \\ 1, & \tau_1, & \ldots, & \tau_1^n \\ & \vdots & & \\ 1, & \tau_n, & \ldots, & \tau_n^n \end{bmatrix}$$

is the Vandermonde matrix associated with $\tau$.

**Proposition 1.** *If $\tau, \tau^* \in \mathcal{T}$, and $p_L : V \times \mathcal{T} \to \mathcal{P}$ is the Lagrange curve function; then the paths $p_L[v, \tau]$ and $p_L[v, \tau^*]$ are equivalent for every $v$ if and only if the parameter sets $\tau$ and $\tau^*$ are affine related; i.e., if there are functions $a, b : \mathcal{T} \times \mathcal{T} \to \mathbb{R}$ with $a(\tau, \tau*) > 0$, such that $\tau_i^* = a(\tau, \tau*)\tau_i + b(\tau, \tau*)$ for all $i$.*

*Proof.* It is necessary to determine the solution space of the functional equation

$$p_L[v, \tau] = p_L[v, \tau^*] \circ \phi_{\tau, \tau^*} \text{ for all } v \in V ;$$

where $\phi_{\tau, \tau^*}$ is $C^1$ everywhere and strictly monotone increasing. The equation may be written, with $\phi$ denoting $\phi_{\tau, \tau^*}$, as

$$[1, t, \ldots, t^n]D_\tau^{-1}[v_0, v_1, \ldots, v_n]^T = [1, \phi(t), \ldots, \phi(t)^n]D_{\tau^*}^{-1}[v_0, v_1, \ldots, v_n]^T$$

for all $v_0, \ldots, v_n$ and where $[v_0, v_1, \ldots, v_n]^T$ denotes the transpose of $[v_0, v_1, \ldots, v_n]$. It follows that

$$[1, t, \ldots, t^n]D_\tau^{-1} = [1, \phi(t), \ldots, \phi(t)^n]D_{\tau^*}^{-1} ,$$

i.e.,

$$[1, \phi(t), \phi(t)^2, \ldots, \phi(t)^n] = [1, t, t^2, \ldots, t^n]D_\tau^{-1}D_{\tau^*} .$$

The matrix $D_\tau^{-1}D_{\tau^*}$ is independent of t and hence, by equating components, it follows that $\phi(t)$ is a polynomial in $t$ of degree $\leq n$, $\phi(t)^2$ is a polynomial in $t$ of degree $\leq n$ ,..., $\phi(t)^n$ is a polynomial in $t$ of degree $\leq n$. It follows that $\phi(t)$ is a polynomial in $t$ of degree $\leq 1$; i.e., $\phi$ takes the affine form $\phi(t) = a(\tau, \tau^*)t + b(\tau, \tau^*)$.

Hence, using Lemma 1, we have $[1, t, t^2, \ldots, t^n]L_\phi = [1, t, t^2, \ldots, t^n]D_\tau^{-1}D_{\tau^*}$ or equivalently $D_{\tau^*} = D_\tau L_\phi$. Substituting the explicit forms of $D_{\tau^*}$, $D_\tau$ and $L_\phi$ into this relationship gives

$$\tau_i^* = \phi(\tau_i)$$
$$= a(\tau, \tau^*)\tau_i + b(\tau, \tau^*) \text{ for all } 0 \leq i \leq n .$$

The explicit form of the functions $a$ and $b$ follow from these relations; i.e.,

$$\phi_{\tau, \tau^*}(t) = [\frac{\tau_n^* - \tau_0^*}{\tau_n - \tau_0}]t + [\frac{\tau_0^* \tau_n - \tau_n^* \tau_0}{\tau_n - \tau_0}] .$$

As $\tau_n > \tau_0$ and $\tau_n^* > \tau_0^*$ the degree of $\phi_{\tau, \tau^*}$ is 1 and Proposition 1 is proven.

## 4.2  Case Study 2: Parametric $C^1$ Cubic Spline Paths

The path function $p_S$ associated with parametric $C^1$ cubic spline paths takes the form $p_S : V \times \mathcal{T} \to \mathcal{P}$ and is determined locally on $\tau_i < t < \tau_{i+1}$ by $s_i : V \times \mathcal{T} \to \mathcal{P}$, where

$$s_i[v,\tau](t) = a_{0i} + a_{1i}t + a_{2i}t^2 + a_{3i}t^3$$
$$s_i[v,\tau](\tau_{i-1}) = v_{i-1}$$
$$s_i[v,\tau](\tau_i) = v_i$$

and $1 \le i \le n$. It follows that $p_S$ can be expressed on the interval $\tau_{i-1} < t < \tau_i$ as

$$s_i[v,\tau](t) = [1, t, t^2, t^3] D_{\tau,i}^{-1} [v_{i-1}, v_i, w_{i-1}, w_i]^T$$

where $D_{\tau,i}$ is the matrix

$$D_{\tau,i} = \begin{bmatrix} 1, & \tau_{i-1}, & \tau_{i-1}^2, & \tau_{i-1}^3 \\ 1, & \tau_i, & \tau_i^2, & \tau_i^3 \\ 0, & 1, & 2\tau_{i-1}, & 3\tau_{i-1}^2 \\ 0, & 1, & 2\tau_i, & 3\tau_i^2 \end{bmatrix}$$

and $w_i$ denotes the tangent vector at $v_i$. The spline is completely determined when the vectors $w_0, \ldots, w_n$ are specified. For example we can define $w_i$, for $1 \le i \le (n-1)$ by

$$w_i = \frac{v_{i+1} - v_{i-1}}{\tau_{i+1} - \tau_{i-1}}.$$

A simple way to define $w_0$ and $w_n$ is to introduce 'ghost' vectors $v_{-1}$ and $v_{n+1}$ and associated parameters $\tau_{-1}$ and $\tau_{n+1}$, with $\tau_{-1} < \tau_0$ and $\tau_{n+1} > \tau_n$. The tangent vectors $w_0$ and $w_n$ are then defined to be

$$w_0 = \frac{v_1 - v_{-1}}{\tau_1 - \tau_{-1}}$$
$$w_n = \frac{v_{n+1} - v_{n-1}}{\tau_{n+1} - \tau_{n-1}}.$$

In this way, the initial and final cubics, $s_1$ and $s_n$, take the same functional form as $s_2, \ldots, s_{n-1}$. With $w_i$, for $0 \le i \le n$ specified as above we have

$$[v_{i-1}, v_i, w_{i-1}, w_i]^T = E_{\tau,i} [v_{i-2}, v_{i-1}, v_i, v_{i+1}]^T$$

where

$$E_{\tau,i} = \begin{bmatrix} 0, & 1, & 0, & 0 \\ 0, & 0, & 1, & 0 \\ -\frac{1}{\tau_i - \tau_{i-2}}, & 0, & \frac{1}{\tau_i - \tau_{i-2}}, & 0 \\ 0, & -\frac{1}{\tau_{i+1} - \tau_{i-1}}, & 0, & \frac{1}{\tau_{i+1} - \tau_{i-1}} \end{bmatrix}$$

and the complete spline $p_S[v, \tau] = s_1[v, \tau] \vee \cdots \vee s_n[v, \tau]$ is expressed locally in canonical form as

$$s_i[v, \tau](t) = [1, t, t^2, t^3] C_{\tau,i}{}^{-1} [v_{i-2}, v_{i-1}, v_i, v_{i+1}]^T$$

where $C_{\tau,i} = E_{\tau,i}^{-1} D_{\tau,i}$ and $1 \le i \le n$. The symmetry result for $C^1$ parametric cubic spline paths may now be shown.

**Proposition 2.** *If $\tau, \tau^* \in \mathcal{T}$, and $p_S : V \times \mathcal{T} \to \mathcal{P}$ is the $C^1$ parametric cubic spline path function constructed above; then the paths $p_S[v, \tau]$ and $p_S[v, \tau^*]$ are equivalent for every $v$ if and only if the parameter sets $\tau$ and $\tau^*$ are affine related; i.e., if there are functions $a, b : \mathcal{T} \times \mathcal{T} \to \mathbb{R}$ with $a(\tau, \tau*) > 0$, such that $\tau_i^* = a(\tau, \tau*)\tau_i + b(\tau, \tau*)$ for all $i$.*

*Proof.* It is necessary to determine the solution space of the functional equation

$$p_S[v, \tau] = p_S[v, \tau^*] \circ \phi_{\tau,\tau^*} \text{ for all } v \in V ;$$

where $\phi_{\tau,\tau^*}$ is $C^1$ everywhere and strictly monotone increasing. Any solution to the above for $\phi_{\tau,\tau^*}$ must satisfy the system of functional equations defined by

$$s_i[v, \tau] = s_i[v, \tau^*] \circ \phi_{\tau,\tau^*} \text{ for all } v \in V .$$

If $\phi_{i,\tau,\tau^*}$ be $C^1$ is strictly monotone on the interval $\tau_{i-1} \le t \le \tau_i$ we define a system of 'local' functional equations

$$s_i[v, \tau] = s_i[v, \tau^*] \circ \phi_{i,\tau,\tau^*} \text{ for all } v \in V ;$$

where $1 \le i \le n$. It follows that the solution space to the 'global' functional equation

$$p_S[v, \tau] = p_S[v, \tau^*] \circ \phi_{\tau,\tau^*} \text{ for all } v \in V ;$$

is a subspace of the solution space of the local system. The proof determines the solutions of the local system and identifies the subset applicable to the global equation.

The local system may be written, writing $\phi_i$ for $\phi_{i,\tau,\tau^*}$, as

$$[1, t, t^2, t^3] C_{\tau,i}^{-1} [v_{i-2}, v_{i-1}, v_i, v_{i+1}]^T$$
$$= [1, \phi_i(t), \phi_i(t)^2, \phi_i(t)^3] C_{\tau^*,i}^{-1} [v_{i-2}, v_{i-1}, v_i, v_{i+1}]^T$$

for all $v \in V$ and for $0 \le i \le n$; Following the proof for Lagrange paths, the cancellation of the column of $v$'s is valid. This produces the reduced functional equations

$$[1, t, t^2, t^3] C_{\tau,i}^{-1} = [1, \phi_i(t), \phi_i(t)^2, \phi_i(t)^3] C_{\tau^*,i}^{-1}$$

for $\phi_i$. It follows that

$$[1, \phi_i(t), \phi_i(t)^2, \phi_i(t)^3] = [1, t, t^2, t^3] C_{\tau,i}^{-1} C_{\tau^*,i} .$$

It follows, applying the argument used for the Lagrange polynomials, that $\phi_i = a_i(\tau, \tau^*)t + b_i(\tau, \tau^*)$ for some functions $a_i$ and $b_i$. Lemma 1 may now be applied to get

$$[1, t, t^2, t^3]L_{\phi_i} = [1, t, t^2, t^3]C_{\tau,i}^{-1}C_{\tau^*,i} \text{ for all } t ,$$

where $L_{\phi_i}$ is the matrix

$$L_{\phi_i} = \begin{bmatrix} 1, & b_i, & b_i^2, & b_i^3 \\ 0, & a_i, & 2b_ia_i, & 3b_i^2a_i \\ 0, & 0, & a_i^2, & 3b_ia_i^2 \\ 0, & 0, & 0, & a_i^3 \end{bmatrix} .$$

It follows that $C_{\tau,i}^{-1}C_{\tau^*,i} = L_{\phi_i}$ or

$$C_{\tau^*,i} = C_{\tau,i}L_{\phi_i} .$$

The explicit form of $C_{\tau,i}$ is

$$\begin{bmatrix} 1, & \tau_{i-2}, & \tau_i^2 - 2\tau_i\tau_{i-1} + 2\tau_{i-1}\tau_{i-2}, & \tau_i^3 - 3\tau_{i-1}^2\tau_i + 3\tau_{i-1}^2\tau_{i-2} \\ 1, & \tau_{i-1}, & \tau_{i-1}^2, & \tau_{i-1}^3 \\ 1, & \tau_i, & \tau_i^2, & \tau_i^3 \\ 1, & \tau_{i+1}, & \tau_{i-1}^2 + 2\tau_i\tau_{i+1} - 2\tau_i\tau_{i-1}, & \tau_{i-1}^3 + 3\tau_i^2\tau_{i+1} - 3\tau_i^2\tau_{i-1} \end{bmatrix} ,$$

hence the second column of $C_{\tau^*,i}$ is $(\tau_{i-2}^*, \tau_{i-1}^*, \tau_i^*, \tau_{i+1}^*)^T$, and that of $C_{\tau,i}L_\phi$ is $(a\tau_{i-2} + b, a\tau_{i-1} + b, a\tau_i + b, a\tau_{i+1} + b)^T$ giving

$$\tau_i^* = a_i(\tau, \tau^*)\tau_i + b_i(\tau, \tau^*) ,$$

for $(i - 2) \leq i \leq (i + 1)$ and

$$a_i(\tau, \tau^*) = \frac{\tau_{i-2}^* - \tau_{i-1}^*}{\tau_{i-2} - \tau_{i-1}}$$

$$b_i(\tau, \tau^*) = \frac{\tau_i^*(\tau_{i-2} - \tau_{i-1}) - \tau_i(\tau_{i-2}^* - \tau_{i-1}^*)}{\tau_{i-2} - \tau_{i-1}} .$$

This completes the solution of the local functional equations; it has been shown that $\phi_i$ is affine and the parameters $\tau_{i-2}^*, \tau_{i-1}^*, \tau_i^*, \tau_{i+1}^*$ are affinely related to $\tau_{i-2}, \tau_{i-1}, \tau_i, \tau_{i+1}$ by $\phi_i$. The solutions of the global functional equation $p_S[v, \tau] = p_S[v, \tau^*] \circ \phi$ are a subset of the piecewise defined functions $\phi_{\tau,\tau^*}^{pw}$ on $\tau_0 \leq t \leq \tau_n$ of the form

$$\phi^{pw} = \phi_1 \vee \phi_2 \vee \cdots \vee \phi_n ,$$

constructed from the local solutions $\phi_i$ on $\tau_{i-1} \leq t \leq \tau_i$.

Only globally $C^1$ solutions to the global functional equation are admissible. The function $\phi^{pw}$ is locally affine, i.e., each $\phi_i$ is affine on $\tau_{i-1} \leq t \leq \tau_i$ and the graph of $\phi^{pw}$ has the general shape shown, for $n = 4$, in Fig. 2.

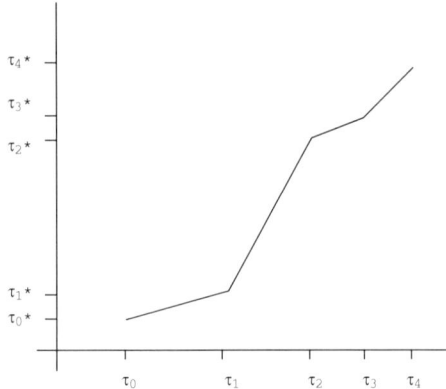

**Fig. 2.** The locally affine nature of the function $\phi^{pw}_{\tau,\tau*}$.

Hence $\phi^{pw}$ is not necessarily differentiable at the points $\tau_i$. It follows that the only admissible solutions to the global functional equation are those for which $\phi^{pw}$ is globally affine, i.e., for some $a(\tau,\tau*)$ and $b(\tau,\tau*)$ we have $a_i = a$ and $b_i = b$ for all $1 \le i \le n$. Equivalently the solutions of the global functional equations must take the form

$$\phi(t) = a(\tau,\tau*)t + b(\tau,\tau*)$$

on $\tau_0 \le t \le \tau_n$, and

$$\tau_i* = a(\tau,\tau*)\tau_i + b(\tau,\tau*)$$

for all $0 \le i \le n$, as required. The proof of Proposition 2 bears many similarities with that for the Lagrange paths. In particular the $L_\phi$ matrix occurs, in $4 \times 4$ form, and plays in the same role in the argument.

### 4.3    Case Study 3: Parametric $C^2$ Cubic Spline Paths

The spans $c_i$ of a parametric $C^2$ cubic spline, $p_c[v,\tau] = c_1 \vee \cdots \vee c_n[v,\tau]$, interpolating $v_0, \ldots, v_n \in V$ with parameter set $\tau \in T$ take the explicit form

$$c_i[v,\tau](t) = [1, t, t^2, t^3]H_i(\tau)[v_{i-1}, v_i, s_{i-1}, s_i]^T$$

where $s_i = c''[v,\tau](\tau_i)$, $h_i = (\tau_i - \tau_{i-1})$ and $H_i(\tau)$ is the matrix

$$H_i(\tau) = \frac{6}{\tau_i - \tau_{i-1}}\begin{bmatrix} 6\tau_i, & -6\tau_{i-1}, & \tau_i^3 - h_i^2\tau_i, & -\tau_{i-1}^3 + h_i^2\tau_{i-1} \\ -6, & 6, & h_i^2 - 3\tau_i^2, & -h_i^2 + 3\tau_{i-1}^2 \\ 0, & 0, & 3\tau_i, & -3\tau_{i-1} \\ 0, & 0, & -1, & 1 \end{bmatrix}.$$

As is well known [3,4], the above produces the $n-1$ linear equations

$$h_i s_{i-1} + 2(h_i + h_{i+1})s_i + h_{i+1}s_{i+1} = 6[\frac{(v_{i+1} - v_i)}{h_{i+1}} - \frac{(v_i - v_{i-1})}{h_i}]; 1 \le i \le n-1$$

in the $n+1$ unknowns $s_0, \ldots, s_n$ and the 'natural' end conditions are $s_0 = 0$ and $s_n = 0$. Under these conditions the matrices $H_1(\tau)$ and $H_n(\tau)$ reduce, respectively, to the following $4 \times 3$ forms:

$$\frac{6}{\tau_1 - \tau_0} \begin{bmatrix} 6\tau_1, & -6\tau_0, & -\tau_0^3 + h_1^2\tau_0 \\ -6, & 6, & ,-h_1^2 + 3\tau_0^2 \\ 0, & 0, & ,-3\tau_0 \\ 0, & 0, & 1 \end{bmatrix} \quad \text{and} \quad \frac{6}{\tau_n - \tau_{n-1}} \begin{bmatrix} 6\tau_n, & -6\tau_{n-1}, & \tau_n^3 - h_n^2\tau_n \\ -6, & 6, & h_n^2 - 3\tau_n^2 \\ 0, & 0, & 3\tau_n \\ 0, & 0, & -1 \end{bmatrix}.$$

The second derivatives $s_1, \ldots, s_{n-1}$ are the solutions of the system of $(n-1)$ linear equations given by

$$J_{n-1}[s_1, s_2, \ldots, s_{n-1}]^T = 6M_n[v_0, v_1, \ldots, v_n]^T$$

where $J_{n-1}$ is the $(n-1) \times (n-1)$ symmetric tri-diagonal matrix

$$J_{n-1} = \begin{bmatrix} 2(h_1 + h_2), & h_2 \\ h_2, & 2(h_2 + h_3), & h_3 \\ & h_3, & 2(h_3 + h_4), & h_4 \\ \\ & & & \ddots \\ \\ & & & & h_{n-2}, & 2(h_{n-2} + h_{n-1}), & h_{n-1} \\ & & & & & h_{n-1}, & 2(h_{n-1} + h_n) \end{bmatrix}$$

and $M_n$ is the $(n-1) \times (n+1)$ matrix defined by

$$M_n = \begin{bmatrix} h_1^{-1}, & -(h_1^{-1} + h_2^{-1}), & h_2^{-1}, 0, & \cdots & & & ,0 \\ 0, & h_2^{-1}, & -(h_2^{-1} + h_3^{-1}), & h_3^{-1}, 0, & \cdots & & ,0 \\ 0, & 0, & h_3^{-1}, & -(h_3^{-1} + h_4^{-1}), & h_4^{-1}, 0, & \cdots & ,0 \\ \vdots & & & & & & \\ \vdots & & & & & & \\ 0, & \cdots & & ,0, h_{n-2}^{-1}, & -(h_{n-2}^{-1} + h_{n-1}^{-1}), & h_{n-1}^{-1}, & 0 \\ 0, & \cdots & & & ,0, h_{n-1}^{-1}, & -(h_{n-1}^{-1} + h_n^{-1}), & h_n^{-1} \end{bmatrix}$$

It can be shown that $det(J_{n-1}) > 0$, hence $J_{n-1}^{-1}$ exists, $s = J_{n-1}^{-1}M_n[v_0, \ldots, v_n]^T$ and the spline $c[v, \tau]$ is determined. The matrix $M_n$ can be written as

$$M_n = \begin{bmatrix} h_1^{-1}, & & ,0 \\ 0, & & ,0 \\ \vdots & M_{n,0} & \vdots \\ 0, & & ,h_n^{-1} \end{bmatrix}$$

where $M_{n,0}$ an $(n-1) \times (n-1)$ symmetric tri-diagonal matrix and the $(n-1) \times (n+1)$ matrix $J_{n-1}^{-1}M_n$ can be expressed as as the two-column extension of the $(n-1) \times (n-1)$ matrix $J_{n-1}^{-1}M_{n,0}$, i.e. we have

$$J_{n-1}^{-1}M_n = \begin{bmatrix} *, & & ,* \\ \vdots & J_{n-1}^{-1}M_{n,0} & \vdots \\ , & & ,* \end{bmatrix}.$$

A number of properties of the matrices $J_{n-1}^{-1}M_n$ and $J_{n-1}^{-1}M_{n,0}$, essential for proof of the symmetry theorem for natural $C^2$ cubic splines, are given in the appendix to the paper. The canonical forms of the spans $c_i$ are now obtained.

*Notation:* Henceforth the elements of the matrix $J_{n-1}^{-1}M_n$ are denoted $\sigma_{i,j}$. We have

$$s_{i-1} = \sigma_{i-1,0}v_0 + \cdots + \sigma_{i-1,n}v_n$$
$$s_i = \sigma_{i,0}v_0 + \cdots + \sigma_{i,n}v_n$$

and, from Corollary 2 of the Appendix, it follows that $\sigma_{i,j}$ and $\sigma_{i,j+2}$ have the same sign for all $\sigma_{i,j}$.

If $e_i \in \mathbb{R}^{n+1}$, for $0 \le i \le n$, is the row vector with a 1 in the $(i+1)th$ position and $0's$ elsewhere, then for $2 \le i \le (n-1)$ we have

$$\begin{bmatrix} v_{i-1} \\ v_i \\ s_{i-1} \\ s_i \end{bmatrix} = \begin{bmatrix} e_{i-1} \\ e_i \\ \sigma_{i-1,0}, \; \sigma_{i-1,1}, \; \ldots, \; \sigma_{i-1,n} \\ \sigma_{i,0}, \quad \sigma_{i,1}, \quad \ldots, \quad \sigma_{i,n} \end{bmatrix} \begin{bmatrix} v_0 \\ v_1 \\ \vdots \\ v_n \end{bmatrix}$$

giving for $2 \le i \le (n-1)$ the canonical form

$$c_i[v,\tau](t) = [1,t,t^2,t^3]H_i(\tau)\Sigma_i(\tau)[v_0,\ldots,v_n]^T$$

for $c_i$ where $\Sigma_i(\tau)$ is the $4 \times (n+1)$ matrix

$$\Sigma_i(\tau) = \begin{bmatrix} e_{i-1} \\ e_i \\ \sigma_{i-1,0}, \; \sigma_{i-1,1}, \; \ldots, \; \sigma_{i-1,n} \\ \sigma_{i,0}, \quad \sigma_{i,1}, \quad \ldots, \quad \sigma_{i,n} \end{bmatrix}.$$

The spans $c_1$ and $c_n$ are special; for $c_1$ we have

$$\begin{bmatrix} v_0 \\ v_1 \\ s_1 \end{bmatrix} = \begin{bmatrix} e_0 \\ e_1 \\ \sigma_{1,0}, \; \sigma_{1,1}, \; \sigma_{1,2}, \; \ldots, \; \sigma_{1,n-1}, \; \sigma_{1,n} \end{bmatrix} \begin{bmatrix} v_0 \\ \vdots \\ v_n \end{bmatrix}$$

and for $c_2$

$$\begin{bmatrix} v_{n-1} \\ v_n \\ s_{n-1} \end{bmatrix} = \begin{bmatrix} e_{n-1} \\ e_n \\ \sigma_{n-1,0}, \; \sigma_{n-1,1}, \; \ldots, \; \sigma_{n-3,n}, \; \sigma_{n-2,n}, \; \sigma_{n-1,n} \end{bmatrix} \begin{bmatrix} v_0 \\ \vdots \\ v_n \end{bmatrix}.$$

From which it follows that canonical forms for $c_1$ and $c_n$ are:

$$c_1[v,\tau](t) = \frac{6[1,t,t^2,t^3]}{\tau_1 - \tau_0} \begin{bmatrix} 6\tau_1, & -6\tau_0, & -\tau_0^3 + h_1^2\tau_0 \\ -6, & 6, & ,-h_1^2 + 3\tau_0^2 \\ 0, & 0, & ,-3\tau_0 \\ 0, & 0, & 1 \end{bmatrix} \begin{bmatrix} e_0 \\ e_1 \\ \sigma_{1,0}, \ldots, \sigma_{1,n} \end{bmatrix} \begin{bmatrix} v_0 \\ \vdots \\ v_n \end{bmatrix}$$

and

$$c_n[v,\tau](t) = \frac{6[1,t,t^2,t^3]}{\tau_n - \tau_{n-1}} \begin{bmatrix} 6\tau_n, & -6\tau_{n-1}, & \tau_n^3 - h_n^2\tau_n \\ -6, & 6, & h_n^2 - 3\tau_n^2 \\ 0, & 0, & 3\tau_n \\ 0, & 0, & -1 \end{bmatrix} \begin{bmatrix} e_{n-1} \\ e_n \\ \sigma_{n-1,0}, \ldots, \sigma_{n-1,n} \end{bmatrix} \begin{bmatrix} v_0 \\ \vdots \\ v_n \end{bmatrix}.$$

The symmetry theorem for $C^2$ cubic splines with natural end conditions is now stated.

**Theorem 1.** *If $\tau, \tau^* \in T$, and $c : V \times T \to \mathcal{P}$ is the $C^2$ parametric natural cubic spline curve function, then the paths $c[v, \tau]$ and $c[v, \tau^*]$ are equivalent for every $v \in V$ if and only if the parameter sets $\tau$ and $\tau^*$ are affine related; i.e., if and only if there are real numbers $a(> 0)$ and $b$, such that $\tau_i = a\tau_i + b$ for all $i$.*

It is necessary to determine the solution space of the functional equation

$$p_c[v, \tau] = p_c[v, \tau^*] \circ \phi_{\tau, \tau^*} \text{ for all } v \in V \ ;$$

where $\phi_{\tau, \tau^*}$ is $C^1$ everywhere and strictly monotone increasing. As for the $C^1$ spline, $\phi_{\tau, \tau^*}$ must satisfy a system of functional equations; i.e.,

$$c_i[v, \tau] = c_i[v, \tau^*] \circ \phi_{\tau, \tau^*} \text{ for all } v \in V \ ,$$

however the proof is more difficult than the $C^1$ case and is presented as a series of lemmas. The 'outer' spans $c_1$ and $c_n$ are special, e.g., the matrices $H_1$ and $H_n$ are not square, and the determination of their symmetries is treated separately to those of $c_2, \ldots, c_{n-1}$. The determination of the symmetries of $c_n$ is very similar to that for $c_1$ and the determination of the symmetries of $c_3, \ldots, c_{n-1}$ is similar to that for $c_2$. Hence details are only given for the spans $c_1$ and $c_2$

**Lemma 2.** *The symmetry condition for $c_1$*

$$[1, t, t^2, t^3] H_1(\tau) \begin{bmatrix} e_0 \\ e_1 \\ \sigma_{1,0}, \ldots, \sigma_{1,n} \end{bmatrix} = [1, \phi, \phi^2, \phi^3] H_1(\tau^*) \begin{bmatrix} e_0 \\ e_1 \\ \sigma_{1,0}^*, \ldots, \sigma_{1,n}^* \end{bmatrix} ;$$

*(i) admits only affine solutions for $\phi$,*

*(ii) has, if $\tau$ and $\tau^*$ are an affinely related pair in $T$, a unique affine solution, $\phi_{\tau, \tau^*}$, for $\phi$*

*(iii) is such that the solution $\phi_{\tau, \tau^*}$ satisfies $\phi_{\tau, \tau^*}(\tau_0) = \tau_0$ and $\phi_{\tau, \tau^*}(\tau_1) = \tau_1$,*

*and*

*(iv) has no other solutions.*

*Proof. Of (i):* If $\phi$ satisfies the symmetry condition for $c_1$ then it is also a solution to the functional equation

$$[1, t, t^2, t^3] H_1(\tau) \begin{bmatrix} e_0 \\ e_1 \\ \sigma_{1,0}, \sigma_{1,1}, \sigma_{1,2} \end{bmatrix} = [1, \phi, \phi^2, \phi^3] H_1(\tau^*) \begin{bmatrix} e_0 \\ e_1 \\ \sigma_{1,0}^*, \sigma_{1,1}^*, \sigma_{1,2}^* \end{bmatrix} .$$

As $\sigma_{i,j}^* \neq 0$ (see Appendix), for all $i, j$, it follows that the matrix

$$\begin{bmatrix} e_0 \\ e_1 \\ \sigma_{1,0}^*, \sigma_{1,1}^*, \sigma_{1,2}^* \end{bmatrix} ,$$

which has determinant $\sigma_{1,2}^*$, is invertible. The function $\phi$ therefore satisfies

$$\begin{bmatrix} 1 \\ t \\ t^2 \\ t^3 \end{bmatrix}^T H_1(\tau) \begin{bmatrix} e_0 \\ e_1 \\ \frac{\sigma_{1,0}\sigma_{1,2}^* - \sigma_{1,0}^*\sigma_{1,2}}{\sigma_{1,2}^*}, & \frac{\sigma_{1,1}\sigma_{1,2}^* - \sigma_{1,1}^*\sigma_{1,2}}{\sigma_{1,2}^*}, & \frac{\sigma_{1,2}}{\sigma_{1,2}^*} \end{bmatrix} = \begin{bmatrix} 1 \\ \phi \\ \phi^2 \\ \phi^3 \end{bmatrix}^T H_1(\tau^*), \quad \text{for all } t \quad (1)$$

which is a set of equations of the form

$$\gamma_0(t) = \frac{1}{\tau_1^* - \tau_0^*}(6\tau_1^* - 6\phi)$$

$$\gamma_1(t) = \frac{1}{\tau_1^* - \tau_0^*}(-6\tau_0^* + \phi)$$

$$\gamma_2(t) = \frac{1}{\tau_1^* - \tau_0^*}[-(\tau_0^*)^3 + (h_1^*)^2\tau_0^* + (-(h_1^*)^2 + 3(\tau_0^*)^2)\phi - 3\tau_0^*\phi^2 + \phi^3]$$

where $\gamma_0$, $\gamma_1$ and $\gamma_2$ are polynomial functions of degree at most 3 in $t$. The first of these equations implies that $\phi$ is a polynomial and degree($\phi$) $\leq 3$. Assuming degree($\phi$) $> 1$ gives degree($\phi^3$) $> 3$ which is incompatible with the last equation for $\phi$; hence degree($\phi$) $\leq 1$ as required for (i) of Lemma 2.

*Proof of (ii)*: from (i) it follows that $\phi$ can be written in the form $\phi(t) = a_1 t + b_1$ for some $a_1, b_1 \in \mathbb{R}$. Hence

$$[1, \phi(t), \phi(t)^2, \phi(t)^3] = [1, t, t^2, t^3]\begin{bmatrix} 1, b_1, & b_1^2, & b_1^3 \\ 0, a_1, 2a_1b_1, & 3b_1^2a_1 \\ 0, 0, & a_1^2, & 3b_1a_1^2 \\ 0, 0, & 0, & a_1^3 \end{bmatrix}$$

and equation 1 becomes

$$H_1(\tau)\begin{bmatrix} 1, & 0, & 0 \\ 0, & 1, & 0 \\ f_0, f_1, f_2 \end{bmatrix} = \begin{bmatrix} 1, b_1, & b_1^2, & b_1^3 \\ 0, a_1, 2a_1b_1, & 3b_1^2a_1 \\ 0, 0, & a_1^2, & 3b_1a_1^2 \\ 0, 0, & 0, & a_1^3 \end{bmatrix} H_1(\tau^*), \quad (2)$$

where

$$f_0 = \frac{\sigma_{1,0}\sigma_{1,2}^* - \sigma_{1,0}^*\sigma_{1,2}}{\sigma_{1,2}^*}, \quad f_1 = \frac{\sigma_{1,1}\sigma_{1,2}^* - \sigma_{1,1}^*\sigma_{1,2}}{\sigma_{1,2}^*}, \quad \text{and} \quad f_2 = \frac{\sigma_{1,2}}{\sigma_{1,2}^*}.$$

It follows from 2 that

$$\frac{1}{\tau_1 - \tau_0}\begin{bmatrix} 6\tau_1 + (-\tau_0^3 + h_1^3\tau_0)f_0, & *, & * \\ , & *, & * \\ , & *, & -3\tau_0 f_2 \\ f_0, & f_1, & f_2 \end{bmatrix} = \frac{1}{\tau_1^* - \tau_0^*}\begin{bmatrix} 6(\tau_1^* - b_1), & *, & * \\ 0, & *, & * \\ 0, & 0, & -3a_1^2(\tau_0^* - b_1) \\ 0, & 0, & a_1^3 \end{bmatrix},$$

where $*$ denotes an element for which the explicit form is not required in the proof. Equating the elements denoted explicitly gives

$$f_0 = 0 \quad (3)$$

$$f_1 = 0 \quad (4)$$

$$\frac{f_2}{\tau_1 - \tau_0} = \frac{a_1^3}{\tau_1^* - \tau_0^*} \quad (5)$$

$$\frac{\tau_1}{\tau_1 - \tau_0} = \frac{\tau_1^* - b_1}{\tau_1^* - \tau_0^*} \quad (6)$$

$$\frac{\tau_0 f_2}{\tau_1 - \tau_0} = \frac{a_1^2(\tau_0^* - b_1)}{\tau_1^* - \tau_0^*}. \quad (7)$$

Equations 5 and 7 give $\tau_0^* = a_1\tau_0 + b_1$, i.e., $\tau_0^* = \phi(\tau_0)$ and using this and equation 6 gives $\tau_1^* = a_1\tau_1 + b_1$, i.e., $\tau_1^* = \phi(\tau_1)$. Hence (ii) of Lemma 2 is proven.

**Lemma 3.** *The symmetry condition for $c_2$*

$$[1, t, t^2, t^3]H_2(\tau)\begin{bmatrix} e_1 \\ e_2 \\ \sigma_{1,0}, \ \sigma_{1,1}, \ \ldots, \ \sigma_{1,n} \\ \sigma_{2,0}, \ \sigma_{2,1}, \ \ldots, \ \sigma_{2,n} \end{bmatrix} = [1, \phi, \phi^2, \phi^3]H_2(\tau^*)\begin{bmatrix} e_1 \\ e_2 \\ \sigma_{1,0}^*, \ \sigma_{1,1}^*, \ \ldots, \ \sigma_{1,n}^* \\ \sigma_{2,0}^*, \ \sigma_{2,1}^*, \ \ldots, \ \sigma_{2,n}^* \end{bmatrix}$$

*admits only affine solutions for $\phi$.*

*Proof.* The condition can be written

$$[1, t, t^2, t^3]H_2(\tau)\Sigma_2(\tau) = [1, \phi, \phi^2, \phi^3]H_2(\tau^*)\Sigma_2(\tau^*)$$

where

$$\Sigma_2(\tau^*) = \begin{bmatrix} 0, & 1, & 0, & 0, \ldots, 0 \\ 0, & 0, & 1, & 0, \ldots, 0 \\ \sigma_{1,0}^*, & \sigma_{1,1}^*, & \sigma_{1,2}^*, & \cdots & , \sigma_{1,n}^* \\ \sigma_{2,0}^*, & \sigma_{2,1}^*, & \sigma_{2,2}^*, & \cdots & , \sigma_{2,n}^* \end{bmatrix}.$$

Columns 2 and 3 of the matrix $\Sigma_2(\tau^*)$ play a key role in the remainder of the proof. It follows from (i) of Corollary 3, of the Appendix, that two main cases can be identified:

**Case A.** there exists a $2 \times 2$ submatrix of

$$\begin{bmatrix} \sigma_{1,0}^*, \ \sigma_{1,1}^*, \ \sigma_{1,2}^*, \ \ldots, \ \sigma_{1,n}^* \\ \sigma_{2,0}^*, \ \sigma_{2,1}^*, \ \sigma_{2,2}^*, \ \ldots, \ \sigma_{2,n}^* \end{bmatrix}$$

having non-zero determinant and not involving either

$$\text{either} \quad \begin{bmatrix} \sigma_{1,1}^* \\ \sigma_{2,1}^* \end{bmatrix} \quad \text{or} \quad \begin{bmatrix} \sigma_{1,2}^* \\ \sigma_{2,2}^* \end{bmatrix}.$$

Hence for Case A, $\phi$ is a solution to

$$[1, t, t^2, t^3]H_2(\tau)\Sigma_2(\tau) = [1, \phi, \phi^2, \phi^3]H_2(\tau^*)\Sigma_2(\tau^*)$$

if it is a solution to the following reduced functional equation for Case A; i.e.,

$$[1, t, t^2, t^3]H_2(\tau)\begin{bmatrix} 0, & 1, & 0, & 0 \\ 0, & 0, & 1, & 0 \\ \alpha, \sigma_{1,1}, \sigma_{1,2}, \beta \\ \gamma, \sigma_{2,1}, \sigma_{2,2}, \delta \end{bmatrix} = [1, \phi, \phi^2, \phi^3]H_2(\tau^*)\begin{bmatrix} 0, & 1, & 0, & 0 \\ 0, & 0, & 1, & 0 \\ \alpha^*, \sigma_{1,1}^*, \sigma_{1,2}^*, \beta^* \\ \gamma^*, \sigma_{2,1}^*, \sigma_{2,2}^*, \delta^* \end{bmatrix}.$$

Here $\alpha^*\delta^* - \beta^*\gamma^* \neq 0$ and

$$\det \begin{bmatrix} 0, & 1, & 0, & 0 \\ 0, & 0, & 1, & 0 \\ \alpha^*, \sigma_{1,1}^*, \sigma_{1,2}^*, \beta^* \\ \gamma^*, \sigma_{2,1}^*, \sigma_{2,2}^*, \delta^* \end{bmatrix} = \det \begin{bmatrix} \alpha^*, \beta^* \\ \gamma^*, \delta^* \end{bmatrix} = \alpha^*\delta^* - \beta^*\gamma^*,$$

hence the inverse of

$$\Sigma_{2,4}(\tau^*) = \begin{bmatrix} 0, & 1, & 0, & 0 \\ 0, & 0, & 1, & 0 \\ \alpha^*, \sigma_{1,1}^*, \sigma_{1,2}^*, \beta^* \\ \gamma^*, \sigma_{2,1}^*, \sigma_{2,2}^*, \delta^* \end{bmatrix}$$

exists and

$$[1, \phi, \phi^2, \phi^3] = [1, t, t^2, t^3] H_2(\tau) \Sigma_{2,4}(\tau) \Sigma_{2,4}^{-1}(\tau^*) H_2^{-1}(\tau^*) \ .$$

It follows, using arguments applied in the previous case studies, that $\phi$ is a polynomial in $t$ with degree($\phi$) $\leq 1$; i.e., $\phi$ is affine and Lemma 3 is proven for Case (A).

Case A is characterized by the existence of an invertible submatrix, $\Sigma_{2,4}(\tau^*)$, of $\Sigma(\tau^*)$ having $(0, 1, 0, 0)$ and $(0, 0, 1, 0)$ as its first two rows.

**Case B.** The case where no invertible $4 \times 4$ submatrix of $\Sigma(\tau^*)$ having $(0, 1, 0, 0)$ and $(0, 0, 1, 0)$ as its first two rows exists; or, equivalently, there does not exist an invertible $2 \times 2$ submatrix of $\Sigma(\tau^*)$ that does not involve

$$\text{either} \quad \begin{bmatrix} \sigma_{1,1}^* \\ \sigma_{2,1}^* \end{bmatrix} \quad \text{or} \quad \begin{bmatrix} \sigma_{1,2}^* \\ \sigma_{2,2}^* \end{bmatrix} \ ,$$

i.e., all $2 \times 2$ submatrices of $\sigma(\tau^*)$ with non-zero determinant involve

$$\begin{bmatrix} \sigma_{1,1}^* \\ \sigma_{2,1}^* \end{bmatrix} \quad \text{and/or} \quad \begin{bmatrix} \sigma_{1,2}^* \\ \sigma_{2,2}^* \end{bmatrix} \ .$$

In this case the matrix $\Sigma(\tau^*)$ takes the form

$$\Sigma(\tau^*) = \begin{bmatrix} 0 & , & 1 & , & 0 & , & 0 & , & 0 & , \ldots, & 0 \\ 0 & , & 0 & , & 1 & , & 0 & , & 0 & , \ldots, & 0 \\ \sigma_{1,0}^* & , & \sigma_{1,1}^* & , & \sigma_{1,2}^* & , & \lambda_3 \sigma_{1,0}^* & , & \lambda_4 \sigma_{1,0}^* & , \ldots, & \lambda_n \sigma_{1,0}^* \\ \sigma_{2,0}^* & , & \sigma_{2,1}^* & , & \sigma_{2,2}^* & , & \lambda_3 \sigma_{2,0}^* & , & \lambda_3 \sigma_{2,0}^* & , \ldots, & \lambda_n \sigma_{2,0}^* \end{bmatrix}$$

for some non-zero $\lambda_i \in \mathbb{R}$ and we know from Corollary 3, of the Appendix, that $\Sigma(\tau^*)$ has at least two non-zero determinants. However columns 4 to $n$ are linearly dependent on the first column

$$\begin{bmatrix} 0 \\ 0 \\ \sigma_{1,0}^* \\ \sigma_{2,0}^* \end{bmatrix}$$

and all give rise to submatrices with zero determinants. Further, these columns add no new information for the solution beyond that provided by columns 1 to 3 and the functional equation for Case B therefore reduces to

$$[1, t, t^2, t^3] H_2(\tau) \begin{bmatrix} 0 & , & 1 & , & 0 \\ 0 & , & 0 & , & 1 \\ \sigma_{1,0} & , & \sigma_{1,1} & , & \sigma_{1,2} \\ \sigma_{2,0} & , & \sigma_{2,1} & , & \sigma_{2,2} \end{bmatrix} = [1, \phi, \phi^2, \phi^3] H_2(\tau^*) \begin{bmatrix} 0 & , & 1 & , & 0 \\ 0 & , & 0 & , & 1 \\ \sigma_{1,0}^* & , & \sigma_{1,1}^* & , & \sigma_{1,2}^* \\ \sigma_{2,0}^* & , & \sigma_{2,1}^* & , & \sigma_{2,2}^* \end{bmatrix} \quad (8)$$

where at least two $2 \times 2$ submatrices of

$$\begin{bmatrix} \sigma_{1,0}^* & , & \sigma_{1,1}^* & , & \sigma_{1,2}^* \\ \sigma_{2,0}^* & , & \sigma_{2,1}^* & , & \sigma_{2,2}^* \end{bmatrix}$$

have non-zero determinant. We write the $2 \times 2$ determinants of the submatrices of the above as

$$\Delta_1 = det \begin{bmatrix} \sigma_{1,0}^* & , & \sigma_{1,1}^* \\ \sigma_{2,0}^* & , & \sigma_{2,1}^* \end{bmatrix} \ , \Delta_2 = det \begin{bmatrix} \sigma_{1,0}^* & , & \sigma_{1,2}^* \\ \sigma_{2,0}^* & , & \sigma_{2,2}^* \end{bmatrix} \ , \Delta_3 = det \begin{bmatrix} \sigma_{1,1}^* & , & \sigma_{1,2}^* \\ \sigma_{2,1}^* & , & \sigma_{2,2}^* \end{bmatrix} \ .$$

The matrix $H_2$ has the form

$$H_2(\tau) = \frac{6}{\tau_2 - \tau_1} \begin{bmatrix} 6\tau_2, & -6\tau_1, & \tau_2^3 - h_2^2\tau_2, & -\tau_1^3 + h_2^2\tau_1 \\ -6, & 6, & h_2^2 - 3\tau_2^2, & -h_2^2 + 3\tau_1^2 \\ 0, & 0, & 3\tau_2, & -3\tau_1 \\ 0, & 0, & -1, & 1 \end{bmatrix}$$

which allows the reduced functional equation for Case B, i.e., equation 8, to be extended in the following way

$$\begin{bmatrix} 1 \\ t \\ t^2 \\ t^3 \end{bmatrix}^T H_2(\tau) \begin{bmatrix} 0 & , & 1 & , & 0 & , & \frac{1}{6(\tau_2-\tau_1)} \\ 0 & , & 0 & , & 1 & , & \frac{1}{6(\tau_2-\tau_1)} \\ \sigma_{1,0} & , & \sigma_{1,1} & , & \sigma_{1,2} & , & 0 \\ \sigma_{2,0} & , & \sigma_{2,1} & , & \sigma_{2,2} & , & 0 \end{bmatrix} = \begin{bmatrix} 1 \\ \phi \\ \phi^2 \\ \phi^3 \end{bmatrix}^T H_2(\tau^*) \begin{bmatrix} 0 & , & 1 & , & 0 & , & \frac{1}{6(\tau_2^*-\tau_1^*)} \\ 0 & , & 0 & , & 1 & , & \frac{1}{6(\tau_2^*-\tau_1^*)} \\ \sigma_{1,0}^* & , & \sigma_{1,1}^* & , & \sigma_{1,2}^* & , & 0 \\ \sigma_{2,0}^* & , & \sigma_{2,1}^* & , & \sigma_{2,2}^* & , & 0 \end{bmatrix}$$ (9)

Writing

$$\Sigma_{2,3}(\tau^*) = \begin{bmatrix} 0 & , & 1 & , & 0 \\ 0 & , & 0 & , & 1 \\ \sigma_{1,0}^* & , & \sigma_{1,1}^* & , & \sigma_{1,2}^* \\ \sigma_{2,0}^* & , & \sigma_{2,1}^* & , & \sigma_{2,2}^* \end{bmatrix}$$

the extended functional equation is equivalent to

$$[1, t, t^2, t^3] \begin{bmatrix} H_2(\tau)\Sigma_{2,3}(\tau) & , & 1 \\ & , & 0 \\ & , & 0 \\ & , & 0 \end{bmatrix} = [1, \phi, \phi^2, \phi^3] \begin{bmatrix} H_2(\tau^*)\Sigma_{2,3}(\tau^*) & , & 1 \\ & , & 0 \\ & , & 0 \\ & , & 0 \end{bmatrix}.$$

It follows that the effect of the extension is to add a fourth column $(1,0,0,0))^T$ to each side of the matrix products $H_2 \Sigma_{2,3}$; which, when multiplied out, simply adds the trivial condition $1 = 1$ to the system of equations for $\phi$. Hence $\phi$ is a solution to the equation for Case B if and only if it is a solution to the extended equation.

We have

$$\det \begin{bmatrix} 0 & , & 1 & , & 0 & , & \frac{1}{6(\tau_2^*-\tau_1^*)} \\ 0 & , & 0 & , & 1 & , & \frac{1}{6(\tau_2^*-\tau_1^*)} \\ \sigma_{1,0}^* & , & \sigma_{1,1}^* & , & \sigma_{1,2}^* & , & 0 \\ \sigma_{2,0}^* & , & \sigma_{2,1}^* & , & \sigma_{2,2}^* & , & 0 \end{bmatrix} = -\frac{1}{6(\tau_2^*-\tau_1^*)}(\Delta_1 + \Delta_2)$$ (10)

and relationship 10 identifies two distinct subcases of Case B; these are identified as B1 and B2 below. It should be noted that the condition:

$$\text{at least two of } \Delta_1, \Delta_2 \text{ and } \Delta_3 \text{ are non-zero}$$ (11)

is applicable to both B1 and B2.

**Case B1.** - as characterized by $\Delta_1 + \Delta_2 \neq 0$ In this case it follows from 10 that only affine solutions, for $\phi$, of the extended functional equation 9 exist.

**Case B2.** - as characterized by $\Delta_1 + \Delta_2 = 0$ The left hand side of the non-extended functional equation for Case B, i.e., equation 8 is a triple of polynomials of degree at most 3 in $t$. Writing the triple as $[\gamma_0, \gamma_1, \gamma_2]$ it follows that the symmetry condition for Case B can be written as

$$[\gamma_0, \gamma_1, \gamma_2] = [1, \phi, \phi^2, \phi^3] \begin{bmatrix} r^*_{1,1} & , & r^*_{1,2} & , & r^*_{1,3} \\ r^*_{2,1} & , & r^*_{2,2} & , & r^*_{2,3} \\ 3(\sigma^*_{1,0}\tau^*_2 - \sigma^*_{2,0}\tau^*_1) & , & 3(\sigma^*_{1,1}\tau^*_2 - \sigma^*_{2,1}\tau^*_1) & , & 3(\sigma^*_{1,2}\tau^*_2 - \sigma^*_{2,2}\tau^*_1) \\ \sigma^*_{2,0} - \sigma^*_{1,0} & , & \sigma^*_{2,1} - \sigma^*_{1,1} & , & \sigma^*_{2,2} - \sigma^*_{1,2} \end{bmatrix}$$

for some $r^*_{i,j}$; from this we obtain

$$\gamma_0 = r^*_{1,1} + \phi r^*_{2,1} + \phi^2 3(\sigma^*_{1,0}\tau^*_2 - \sigma^*_{2,0}\tau^*_1) + \phi^3(\sigma^*_{2,0} - \sigma^*_{1,0})$$
$$\gamma_1 = r^*_{1,2} + \phi r^*_{2,2} + \phi^2 3(\sigma^*_{1,1}\tau^*_2 - \sigma^*_{2,1}\tau^*_1) + \phi^3(\sigma^*_{2,1} - \sigma^*_{1,1})$$
$$\gamma_2 = r^*_{1,3} + \phi r^*_{2,3} + \phi^2 3(\sigma^*_{1,2}\tau^*_2 - \sigma^*_{2,2}\tau^*_1) + \phi^3(\sigma^*_{2,2} - \sigma^*_{1,2}) \ .$$

The non-zero and alternating sign property (see Appendix) of the elements $\sigma^*_{i,j}$ imply that $(\sigma^*_{2,0} - \sigma^*_{1,0}) \neq 0$, hence we have

$$\phi^3 = \frac{1}{\sigma^*_{2,0} - \sigma^*_{1,0}} \left[ \gamma_0 - r^*_{1,1} - r^*_{2,1}\phi - \phi^2 3(\sigma^*_{1,0}\tau^*_2 - \sigma^*_{2,0}\tau^*_1) \right] \qquad (12)$$

which, substituting into the relation for $\gamma_1$ gives

$$\gamma_1 - \frac{\sigma^*_{2,1} - \sigma^*_{1,1}}{\sigma^*_{2,0} - \sigma^*_{1,0}}(\gamma_0 - r^*_{1,1}) - r^*_{1,2} = \phi \left[ \frac{r^*_{2,2}(\sigma^*_{2,0} - \sigma^*_{1,0}) - r^*_{2,1}(\sigma^*_{2,1} - \sigma^*_{1,1})}{\sigma^*_{2,0} - \sigma^*_{1,0}} \right]$$
$$+ \phi^2 \frac{3(\tau^*_2 - \tau^*_1)(-\Delta_1)}{\sigma^*_{2,0} - \sigma^*_{1,0}} \ ,$$

or

$$\gamma^*_1 = \phi \left[ \frac{r^*_{2,2}(\sigma^*_{2,0} - \sigma^*_{1,0}) - r^*_{2,1}(\sigma^*_{2,1} - \sigma^*_{1,1})}{\sigma^*_{2,0} - \sigma^*_{1,0}} \right] + \phi^2 \frac{3(\tau^*_2 - \tau^*_1)(-\Delta_1)}{\sigma^*_{2,0} - \sigma^*_{1,0}}$$

where

$$\gamma^*_1 = \gamma_1 - \frac{\sigma^*_{2,1} - \sigma^*_{1,1}}{\sigma^*_{2,0} - \sigma^*_{1,0}}(\gamma_0 - r^*_{1,1}) - r^*_{1,2} \ .$$

Similarly we obtain

$$\gamma^*_2 = \phi \left[ \frac{r^*_{2,3}(\sigma^*_{2,0} - \sigma^*_{1,0}) - r^*_{2,1}(\sigma^*_{2,2} - \sigma^*_{1,2})}{\sigma^*_{2,0} - \sigma^*_{1,0}} \right] + \phi^2 \frac{3(\tau^*_2 - \tau^*_1)(-\Delta_2)}{\sigma^*_{2,0} - \sigma^*_{1,0}}$$

from the relation for $\gamma_2$. We have

$$\gamma^{**}_1 = \phi \left[ r^*_{2,2}(\sigma^*_{2,0} - \sigma^*_{1,0}) - r^*_{2,1}(\sigma^*_{2,1} - \sigma^*_{1,1}) \right] + \phi^2 3(\tau^*_2 - \tau^*_1)(-\Delta_1) \qquad (13)$$

$$\gamma^{**}_2 = \phi \left[ r^*_{2,3}(\sigma^*_{2,0} - \sigma^*_{1,0}) - r^*_{2,1}(\sigma^*_{2,2} - \sigma^*_{1,2}) \right] + \phi^2 3(\tau^*_2 - \tau^*_1)(-\Delta_2) \qquad (14)$$

where $\gamma^{**}_1$ and $\gamma^{**}_2$ are polynomials in $t$ of degree at most 3. Adding gives, as $\Delta_1 + \Delta_2 = 0$,

$$\gamma^{**}_1 + \gamma^{**}_2 = \phi \left[ (r^*_{2,2} + r^*_{2,3})(\sigma^*_{2,0} - \sigma^*_{1,0}) - r^*_{2,1}(\sigma^*_{2,1} - \sigma^*_{1,1} + \sigma^*_{2,2} - \sigma^*_{1,2}) \right] \ .$$

We note that condition 11 and the defining condition $\Delta_1 + \Delta_2 = 0$ for B2 imply that $\Delta_1 \neq 0$ - for if $\Delta_1 = 0$ then we would also have $\Delta_2 = 0$ and this violates 11. Hence two subcases of B2, characterized by

$$\Delta_1 + \Delta_2 = 0,$$
$$\Delta_1 \neq 0,$$
$$\Delta_2 \neq 0,$$
$$\text{and} \quad \Delta_3 = 0 \quad \text{Case B2 (i)}$$
$$\text{or} \quad \Delta_3 \neq 0 \quad \text{Case B2 (ii)}$$

can be identified.

The B2 condition $\Delta_2 + \Delta_1 = 0$ is equivalent to

$$\sigma_{1,0}^* = \mu(\sigma_{1,1}^* + \sigma_{1,2}^*) \quad \text{and} \quad \sigma_{2,0}^* = \mu(\sigma_{2,1}^* + \sigma_{2,2}^*)$$

where $\mu \in \mathbb{R}$ and, to preserve the alternating sign property of $\sigma_{i,j}^*$, we have

$$\mu > 0 \quad \text{if} \quad \sigma_{1,1}^* + \sigma_{1,2}^* > 0$$
$$\mu < 0 \quad \text{if} \quad \sigma_{1,1}^* + \sigma_{1,2}^* < 0$$

and

$$\mu < 0 \quad \text{if} \quad \sigma_{2,1}^* + \sigma_{2,2}^* > 0$$
$$\mu > 0 \quad \text{if} \quad \sigma_{2,1}^* + \sigma_{2,2}^* < 0 \ .$$

Further, the condition $\sigma_{i,j}^* \neq 0$ for all $i,j$ implies that $\sigma_{1,1}^* + \sigma_{1,2}^* \neq 0$ and $\mu \neq 0$ when $\Delta_2 + \Delta_1 = 0$.

For $\Delta_1 + \Delta_2 = 0$ and $\mu > 0$ it follows that

$$\sigma_{1,1}^* + \sigma_{1,2}^* > 0$$
$$\sigma_{2,1}^* + \sigma_{2,2}^* < 0 \ .$$

Similarly for $\Delta_1 + \Delta_2 = 0$ and $\mu < 0$ we have

$$\sigma_{1,1}^* + \sigma_{1,2}^* < 0$$
$$\sigma_{2,1}^* + \sigma_{2,2}^* > 0 \ .$$

**Case B2.** (i) $\Delta_3 = 0$ ; i.e., $\sigma_{1,1}^* \sigma_{2,2}^* - \sigma_{2,1}^* \sigma_{1,2}^* = 0$ and hence

$$
\begin{bmatrix} 0 & , & 1 & , & 0 \\ 0 & , & 0 & , & 1 \\ \sigma_{1,0}^* & , & \sigma_{1,1}^* & , & \sigma_{1,2}^* \\ \sigma_{2,0}^* & , & \sigma_{2,1}^* & , & \sigma_{2,2}^* \end{bmatrix}
=
\begin{bmatrix} 0 & , & 1 & , & 0 \\ 0 & , & 0 & , & 1 \\ \mu \frac{\sigma_{1,1}^*}{\sigma_{2,1}^*}(\sigma_{2,1}^* + \sigma_{2,2}^*) & , & \sigma_{1,1}^* & , & \frac{\sigma_{1,1}^* \sigma_{2,2}^*}{\sigma_{2,1}^*} \\ \mu(\sigma_{2,1}^* + \sigma_{2,2}^*) & , & \sigma_{2,1}^* & , & \sigma_{2,2}^* \end{bmatrix} .
\tag{15}
$$

Under the substitutions

$$\sigma_{1,0}^* = \mu \frac{\sigma_{1,1}^*}{\sigma_{2,1}^*}(\sigma_{2,1}^* + \sigma_{2,2}^*), \quad \sigma_{2,0}^* = \mu(\sigma_{2,1}^* + \sigma_{2,2}^*), \quad \text{and} \quad \sigma_{1,2}^* = \frac{\sigma_{1,1}^* \sigma_{2,2}^*}{\sigma_{2,1}^*}$$

equations 13 and 14 for $\gamma_1^{**}$ and $\gamma_2^{**}$ become

$$\gamma_1^{**} = \phi \frac{(\sigma_{2,1}^* - \sigma_{1,1}^*)}{\sigma_{2,1}^*} \left[ \mu r_{2,2}^*(\sigma_{2,1}^* + \sigma_{2,2}^*) - r_{2,1}^*\sigma_{2,1}^* \right] + \phi^2 3(\tau_2^* - \tau_1^*)(-\Delta_1) \quad (16)$$

$$\gamma_2^{**} = \phi \frac{(\sigma_{2,1}^* - \sigma_{1,1}^*)}{\sigma_{2,1}^*} \left[ \mu r_{2,3}^*(\sigma_{2,1}^* + \sigma_{2,2}^*) - r_{2,1}^*\sigma_{2,2}^* \right] + \phi^2 3(\tau_2^* - \tau_1^*)(-\Delta_2) \; . \quad (17)$$

The explicit forms of $r_{2,1}^*$, $r_{2,2}^*$ and $r_{2,3}^*$ can be obtained from 8 and 15 as

$$r_{2,1}^* = \mu(\sigma_{2,1}^* + \sigma_{2,2}^*) \left[ \frac{\sigma_{1,1}^*}{\sigma_{2,1}^*}(h_2^2 - 3\tau_2^{*2}) + (-h_2^2 + 3\tau_1^{*2}) \right]$$

$$r_{2,2}^* = -6 + \sigma_{1,1}^*(h_2^2 - 3\tau_2^{*2}) + \sigma_{2,1}^*(-h_2^2 + 3\tau_1^{*2})$$

$$r_{2,3}^* = 6 + \sigma_{2,2}^* \left[ \frac{\sigma_{1,1}^*}{\sigma_{2,1}^*}(h_2^2 - 3\tau_2^{*2}) + (-h_2^2 + 3\tau_1^{*2}) \right] \; .$$

Substituting these expression into equations 16 gives

$$\gamma_1^{**} = \phi \left[ -6\mu(\sigma_{2,1}^* + \sigma_{2,2}^*) \right] + \phi^2 3(\tau_2^* - \tau_1^*)(-\Delta_1) \quad (18)$$

$$\gamma_2^{**} = \phi \left[ 6\mu \frac{(\sigma_{2,1}^* - \sigma_{1,1}^*)(\sigma_{2,1}^* + \sigma_{2,2}^*)}{\sigma_{2,1}^*} \right] + \phi^2 3(\tau_2^* - \tau_1^*)(-\Delta_2) \quad (19)$$

which, on adding, yields

$$\gamma_1^{**} + \gamma_2^{**} = \phi \left[ -6\mu \frac{\sigma_{1,1}^*}{\sigma_{2,1}^*}(\sigma_{2,1}^* + \sigma_{2,2}^*) \right] \; . \quad (20)$$

As $-6\mu \frac{\sigma_{1,1}^*}{\sigma_{2,1}^*}(\sigma_{2,1}^* + \sigma_{2,2}^*) \neq 0$ it follows from 20 that $\phi$ is a polynomial in $t$ of degree $\leq 3$. As $\Delta_1 \neq 0$ and $(\tau_2^* - \tau_1^*) \neq 0$ it now follows from equation 18 that $\phi^2$ is also a polynomial in $t$ of degree $\leq 3$, and then from equation 12 that $\phi^3$ is a polynomial in $t$ of degree $\leq 3$. If degree$(\phi) > 1$ then degree$(\phi^3) > 3$ which contradicts the above. Hence degree$(\phi) \leq 1$ and $\phi$ is affine for Case B2 (i) as required.

**Case B2 (ii).** $\Delta_3 \neq 0$

The proof of this is similar to that for B2 (i).

**Lemma 4.** *If $\tau$ and $\tau^*$ are affinely related elements of $T$ then the symmetry condition for $c_2$*

   *(i) admits a unique affine solution, $\phi_{\tau,\tau^*}$, for $\phi$,*

   *(ii) is such that the solution $\phi_{\tau,\tau^*}$ satisfies $\phi_{\tau,\tau^*}(\tau_1) = \tau_1^*$ and $\phi_{\tau,\tau^*}(\tau_2) = \tau_2^*$.*

*The symmetry condition for $c_2$ has no other solutions.*

*Proof.* **Case A.** Let $\phi(t) = a_2 t + b_2$ to obtain the relationship

$$H_2(\tau)\Sigma_4(\tau)\Sigma_4^{-1}(\tau^*) = \begin{bmatrix} 1, b_2, & b_2{}^2, & b_2{}^3 \\ 0, a_2, 2a_2 b_2, & 3b_2{}^2 a_2 \\ 0, 0, & a_2{}^2, & 3b_2 a_2{}^2 \\ 0, 0, & 0, & a_2{}^3 \end{bmatrix} H_2(\tau^*) .$$

The inverse of the matrix

$$\begin{bmatrix} 0 & 1 & 0 & 0 \\ 0 & 0 & 1 & 0 \\ a & b & c & d \\ e & f & g & h \end{bmatrix}$$

is

$$\frac{1}{ed - ah} \begin{bmatrix} bh - fd, & ch - dg, & -h, & d \\ 1, & 0, & 0, & 0 \\ 0, & 1, & 0, & 0 \\ af - eb, & ag - ec, & e, & -a \end{bmatrix} .$$

It follows that the product $\Sigma_4(\tau)\Sigma_4^{-1}(\tau^*)$ has the generic form

$$\begin{bmatrix} 1 & 0 & 0 & 0 \\ 0 & 1 & 0 & 0 \\ A & B & C & D \\ E & F & G & H \end{bmatrix}$$

for some functions $A, \ldots, H$ of $\tau$ and $\tau^*$. For the purposes of this proof, the explicit form of the functions $A, \ldots, H$ is not required. The relationship between $\tau$ and $\tau^*$ for the affine function $\phi(t) = a_2 t + b_2$ therefore takes the form

$$H_2(\tau) \begin{bmatrix} 1 & 0 & 0 & 0 \\ 0 & 1 & 0 & 0 \\ A & B & C & D \\ E & F & G & H \end{bmatrix} = \begin{bmatrix} 1, b_2, & b_2{}^2, & b_2{}^3 \\ 0, a_2, 2a_2 b_2, & 3b_2{}^2 a_2 \\ 0, 0, & a_2{}^2, & 3b_2 a_2{}^2 \\ 0, 0, & 0, & a_2{}^3 \end{bmatrix} H_2(\tau^*) .$$

The matrix product on the left of this equation evaluates to

$$\frac{6}{\tau_2 - \tau_1} \begin{bmatrix} m_{1,1}, & m_{1,2}, & m_{1,3}, & m_{1,4} \\ m_{2,1}, & m_{2,2}, & m_{2,3}, & m_{2,4} \\ 3(\tau_2 A - \tau_1 E), & 3(\tau_2 B - \tau_1 F), & 3(\tau_2 C - \tau_1 G), & 3(\tau_2 D - \tau_1 H) \\ E - A, & F - B, & G - C, & H - D \end{bmatrix}$$

where the precise form of the $m_{i,j}$ elements is not required in this proof. The right hand product may be computed as

$$\frac{6}{\tau_2^* - \tau_1^*} \begin{bmatrix} 6(\tau_2^* - b_2), & 6(b_2 - \tau_1^*), & n_{1,3}, & n_{1,4} \\ -6a_2, & 6a_2, & n_{2,3}, & n_{2,4} \\ 0, & 0, & n_{3,3}, & n_{3,4} \\ 0, & 0, & -a_2{}^3, & a_2{}^3 \end{bmatrix}$$

where the $n_{i,j}$ elements are not required explicitly. We have

$$\frac{1}{\tau_2 - \tau_1} \begin{bmatrix} m_{1,1}, & m_{1,2}, & m_{1,3}, & m_{1,4} \\ m_{2,1}, & m_{2,2}, & m_{2,3}, & m_{2,4} \\ 3(\tau_2 A - \tau_1 E), & 3(\tau_2 B - \tau_1 F), & 3(\tau_2 C - \tau_1 G), & 3(\tau_2 D - \tau_1 H) \\ E - A, & F - B, & G - C, & H - D \end{bmatrix} =$$

$$\frac{1}{\tau_2^* - \tau_1^*} \begin{bmatrix} 6(\tau_2^* - b_2), & 6(b_2 - \tau_1^*), & n_{1,3}, & n_{1,4} \\ -6a_2, & 6a_2, & n_{2,3}, & n_{2,4} \\ 0, & 0, & n_{3,3}, & n_{3,4} \\ 0, & 0, & -a_2{}^3, & a_2{}^3 \end{bmatrix} .$$

Comparing lower left terms gives $E = A$, the leftmost term on the penultimate row then gives

$$3A(\tau_2 - \tau_1) = 0 \ .$$

As $\tau_2 \neq \tau_1$ it follows that $A = E = 0$; similarly we obtain $B = F = 0$ and, substituting these values in the expressions for $m_{1,1}, m_{1,2}, m_{2,1}$ and $m_{2,2}$, the matrix equation reduces to

$$\frac{1}{\tau_2 - \tau_1} \begin{bmatrix} 6\tau_2, & -6\tau_1, & m_{1,3}, & m_{1,4} \\ -6, & 6, & m_{2,3}, & m_{2,4} \\ 0, & 0, & 3(\tau_2 C - \tau_1 G), & 3(\tau_2 D - \tau_1 H) \\ E - A, & F - B, & G - C, & H - D \end{bmatrix} =$$

$$\frac{1}{\tau_2^* - \tau_1^*} \begin{bmatrix} 6(\tau_2^* - b_2), & 6(b_2 - \tau_1^*), & n_{1,3}, & n_{1,4} \\ -6a_2, & 6a_2, & n_{2,3}, & n_{2,4} \\ 0, & 0, & n_{3,3}, & n_{3,4} \\ 0, & 0, & -a_2^3, & a_2^3 \end{bmatrix} .$$

Comparing terms in the upper left $2 \times 2$ submatrices gives

$$a_2 = \frac{\tau_2^* - \tau_1^*}{\tau_2 - \tau_1}$$

$$b_2 = \frac{\tau_1^*(\tau_2 - \tau_1) - \tau_1(\tau_2^* - \tau_1^*)}{\tau_2 - \tau_1}$$

from which it follows that $\phi(\tau_1) = \tau_1^*$ and $\phi(\tau_2) = \tau_2^*$ - this completes the proof of Lemma 4 for Case A.

**Cases B1, B2 (i) and B2 (ii).** The proof of these is similar to that for Case A

This completes the solution of the local functional equations; the proof may now completed in the same way as that for the $C^1$ spline.

## 5   Summary, Conclusions and Open Problems

Necessary and sufficient conditions for invariant geometry under parametrisation transformations have been established for $C^1$ and natural parametric $C^2$ cubic spline paths in this paper. The results obtained are of theoretical interest and of practical value - for example, they can used as the foundation for embedded tests in modeling or graphics systems to indicate when user-specified transformations of knot values are shape modifying.

The techniques may be applied to other spline paths and parametric paths of the class $p : V \times T \to \mathcal{P}$ and to paths defined on different domains. For example, rational Bézier paths are defined on the domain $V \times \Omega$, where $\Omega = \{(\omega_0, \ldots, \omega_n) : \omega_i > 0\}$. Elements $\omega = (\omega_0, \ldots, \omega_n) \in \Omega$ define the 'weights' associated with the control vector set $v = (v_0, \ldots, v_n) \in V$. The known symmetries [5] of these paths are readily derived using the functional equation techniques described in this paper; i.e., the 'form preserving' rational parameter transformations, and the invariant geometry weight ratios are obtained [6].

Whilst the parameter symmetries discussed in this paper are of both theoretic and of practical value, their determination does not provide a comprehensive picture of shape-invariant transformations of the domain $T$ for path functions of the class $p : V \times T \to \mathcal{P}$. The symmetries obtained are those that that work for all $v \in V$, and these appear to be the only parameter symmetries discussed in the literature. Symmetries that apply at particular $v \in V$ can also be defined and computed. However, the absence of the 'for all $v \in V$' condition means that 'cancellation' of the column of $v$'s - which is the key to obtaining the symmetries in each of the case studies - is not valid, and alternative solution strategies are required. In addition it seems that the techniques presented do not readily extend to the common situation where the parameters $\tau \in V$ are functions of the geometric data set $v \in V$.

# Appendix

This appendix contains a number of properties of the matrices $J_{n-1}^{-1} M_{n,0}$ and $J_{n-1}^{-1} M_n$ required for the proof of the symmetry theorem for $C^2$ natural cubic splines. Recall that the elements of $J_{n-1}^{-1} M_n$ are denoted $\sigma_{i,j}$.

*Property 1.* $|\sigma_{i,j}| \neq 0$ for all $1 \leq i \leq n - 1,\ 0 \leq j \leq n$.

*Property 2.* The elements $\sigma_{i,j}$ have alternating signs and $\sigma_{i,i} < 0$ for all $1 \leq i \leq n - 1$.

*Property 3.* The matrices $J_{n-1}^{-1} M_{n,0}$ and $J_{n-1}^{-1} M_n$ are of rank (n-1).

The following corollaries of properties 1 through 3 are immediate.

**Corollary 1.** *Adjacent rows,*

$$\{(\sigma_{i-1,1}, \cdots, \sigma_{i-1,n-1}), (\sigma_{i,1}, \cdots, \sigma_{i,n-1})\},$$

*of the matrix $J_{n-1}^{-1} M_{n,0}$ are linearly independent.*

**Corollary 2.** *Adjacent rows*

$$\{(\sigma_{i-1,0}, \sigma_{i-1,1}, \cdots, \sigma_{i-1,n-1}, \sigma_{i-1,n}), (\sigma_{i,0}, \sigma_{i,1}, \cdots, \sigma_{i,n-1}, \sigma_{i,n})\},$$

*of $J_{n-1}^{-1} M_n$ are linearly independent in $\mathbb{R}^{n+1}$.*

**Corollary 3.** *The $2 \times (n+1)$ matrix*

$$\Sigma_i = \begin{bmatrix} \sigma_{i-1,0}, \sigma_{i-1,1}, \cdots, \sigma_{i-1,n-1}, \sigma_{i-1,n} \\ \sigma_{i,0}, \quad \sigma_{i,1}, \quad \cdots, \quad \sigma_{i,n-1}, \quad \sigma_{i,n}, \end{bmatrix}$$

*(i) has no non-zero elements,*
*(ii) has either the sign pattern*

$$\begin{bmatrix} +,-,+,-,\dots \\ -,+,-,+,\dots \end{bmatrix}$$

*or the sign pattern*

$$\begin{bmatrix} -,+,-,+,\dots \\ +,-,+,-,\dots \end{bmatrix}$$

*and*

*(iii) has at least two $2 \times 2$ submatrices of the form*

$$\Sigma_{k,l,m} = \begin{bmatrix} \sigma_{k,l}, & \sigma_{k,m} \\ \sigma_{k+1,l}, & \sigma_{k+1,m} \end{bmatrix}$$

*such that*

$$det(\Sigma_{k,l,m}) \neq 0 \ .$$

# References

1. Bez H. E.: Symmetry-a research direction in curve and surface modelling; some results and applications. Proceedings of the IMA Conference on the Mathematics of Surfaces IX (2000) 322-337
2. Bez H. E.: On the relationship between parametrisation and invariance for curve functions. Computer Aided Geometric Design **17** (2000) 793-811
3. de Boor, C.: A practical guide to splines. Springer (1978)
4. Hoschek, J. and Lasser, D.: Fundamentals of computer aided geometric design. A.K.Peters (1993)
5. Patterson R. R.: Projective transformations of the parameter of a Bernstein-Bézier curve. ACM Trans. on Graphics 4 (1985) 276-290
6. Bez H. E.: A functional equation approach to the computation of parameter symmetries for path functions. International Journal of Computer Mathematics (to appear)

# Efficient One-Sided Linearization
of Spline Geometry

Jörg Peters

Computer and Information Science and Engineering
University of Florida
Gainsville
FL 32611-6120, USA
jorg@cise.ufl.edu
http://www.cise.ufl.edu/~jorg/

**Abstract.** This paper surveys a new, computationally efficient technique for linearizing curved spline geometry, bounding such geometry from one side and constructing curved spline geometry that stays to one side of a barrier or inside a given channel. Combined with a narrow error bound, these reapproximations tightly couple linear and nonlinear representations and allow them to be substituted when reasoning about the other. For example, a subdividable linear efficient variety enclosure (sleve, pronounced like Steve) of a composite spline surface is a pair of matched triangulations that sandwich a surface and may be used for interference checks. The average of the sleve components, the *mid-structure*, is a good max-norm linearization and, similar to a control polytope, has a well-defined, associated curved geometry representation. Finally, the ability to fit paths through given channels or keep surfaces near but outside forbidden regions, allows extending many techniques of linear computational geometry to the curved, nonlinear realm.

## 1 Introduction

Compared to piecewise linear, say triangle meshes, higher-order representations, such as b-splines, Bézier patches and subdivision surfaces, offer improved compression (if the target shape is sufficiently smooth) through higher approximation order, arbitrary resolution to any prescribed tolerance, and generally a higher level of abstraction. On the other hand, meshes and triangulations are pervasive, as a result of measuring, for end-uses, such as graphics rendering or driving machine tools, and for the majority of finite element codes. No wonder then that b-spline, Bézier and subdivision control nets look attractive as mediator: on one hand, they approximate a nonlinear shape; on the other hand, they finitely, but completely, define the smooth, nonlinear shape. Armed with the convex hull property and the convergence under subdivision, it is therefore tempting to use the control meshes as end-use or computational meshes. However, control meshes have shortcomings. The control net is far from the best possible geometric approximand for the a given budget of linear pieces. It can cross the curve

M.J. Wilson and R.R. Martin (Eds.): Mathematics of Surfaces 2003, LNCS 2768, pp. 297–319, 2003.
© Springer-Verlag Berlin Heidelberg 2003

or surface and therefore does not provide a one-sided approximation. Finally, until recently, there was no theory giving easy access to the error (not just the rate of decay under subdivision) of the distance of the control net to the actual curved object. Consequently, despite their geometrically indicative control structure, objects in b-spline, Bézier or generalized subdivision representation pose numerical and implementation challenges, say when measuring distance between objects, re-approximating for format conversion, meshing with tolerance, or detecting the silhouette. It should be emphasized that naive linearization, such as triangulation by sampling, reapproximates without known error and not safely from one side.

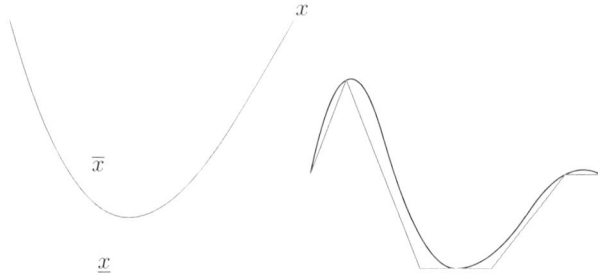

**Fig. 1.** (*left*) A cubic Bézier segment $x$ and its optimized, grey 3-piece slefe bounded by $\overline{x}$ and $\underline{x}$. (*right*) Spline curve computed to be above but close to a given piecewise linear barrier.

Subdividable linear efficient function enclosures [7,8,9], short slefes (pronounced like sleve), by contrast, are a low-cost technique for constructing piecewise linear bounds that sandwich nonlinear functions. The slefe of a function $x$ consists of 2 one-sided bounds $\overline{x}, \underline{x}$ so that $\underline{x} \leq x \leq \overline{x}$ over the domain of interest. Here $\overline{x}$ and $\underline{x}$ are, for example, piecewise linear functions bounding the grey region in Figure 1, *left*. In practice, slefe bounds are observed to be very tight. Analytically, at present, only the case of cubic functions in one variable is fully understood: the slefe width of a convex $x$ is within 6% of the optimal, narrowest; at an inflection, the error ratio is at most 3:5. The extension of slefes to free-form surfaces requires some care – but the resulting algorithm is still rather simple and sleves inherit many of the properties of slefes such as near-optimality in the $L^\infty$ norm and refinability for adaptive multiresolution.

The average $\overline{\underline{x}} := (\overline{x} + \underline{x})/2$ is called *mid-structure*. It is well-defined also for vector-valued functions $\mathbf{x}$ and its construction does not require the construction of the sleve of $\mathbf{x}$. By making the mid-structure along a boundary depend only on the boundary, e.g. a space curve for a patch in $\mathbb{R}^3$ or the endpoint for a curve, mid-structures join continuously if their patches join continuously. In contrast to approximation theory, which strives to establish optimality and uniqueness over all sufficiently smooth functions, mid-structures are a concrete, efficiently computable approximation with a small, quantifiable $L^\infty$ error. (We recall that

**Fig. 2.** Enclosures based on control points: cubic curve with Bézier control polygon, axis-aligned box, bounding circle, Filip et al. bound (scaled by 1/2), bounding ellipse convex hull, oriented bounding box, 'fat arc', 3-piece sleve.

Chebyshev economization applies to degree reduction by one polynomial degree and only to a single polynomial segment. The problem at hand is to determine a best continuous, piecewise linear, max-norm approximation of a nonlinear curve.)

Since the construction of slefes is simple, inverse problem of one-sided linearization can also be efficiently addressed: to find a spline (from a fixed spline space) that stays close to but to one side of a given piecewise linear curve (Figure 1, *right*). In the related CHANNEL problem, a channel is given and a smooth spline is sought that stays inside that channel. Inverse problems address underconstrained design problems where the emphasis is not on optimality or uniqueness of the solution but on feasibility or on pinpointing the cause of infeasibility (which might trigger a refinement of the spline space). Such tools should improve design and shorten the design cycle and extend techniques of computational geometry to the domain of curved smooth paths and surfaces. For example, solutions to the inverse problem yield a postprocessing scheme that smoothes edges to one side while preserving the essential quality of the series of straight edges generated as: upper hulls, edges of a binary space partition, parts of triangulations, road maps or visibility graphs.

**Overview:** Section 1.1 below contrasts sleves with other commonly used bounding constructs and points to prior work. Section 2, page 301, reviews the slefe construction, for functions in one variable and for tensored multivariate functions. Section 3, p. 307, discusses midstructures in more detail. Curve and surface sleves are explained in Section 4.1, p.309 and a solution strategy for inverse problems, in Section 5, p.312.

## 1.1   Commonly Used Bounding Constructs

The enclosure of a geometric object is a *bounding construct*, consisting typically of two sheets (or more, say in the case of a space curve), such that each sheet is guaranteed not to intersect the object, i.e. each sheet lies to 'one side' of the object. For example, a surface without boundary can be enclosed by an inner and an outer triangulation.

We distinguish between elementary bounding constructs and hierarchical structures that employ these elementary bounding constructs as their oracles. enclosures fall into the category of elementary bounding constructs. A gallery of elementary bounding constructs is shown in Figure 2. That is, enclosures add to the arsenal of axis-aligned bounding boxes (AABB), oriented bounding boxes (OBB), quantized bounding boxes also called 'k-dops' or discrete orientation polytopes (convex polytopes whose facets are determined by halfspaces whose outward normals come from a small fixed set of $k$ orientations) [5,11,12], fat arcs [28], convex hulls, bounding spheres and minimal enclosing ellipsoids [30] and the Filip et al. bound [9]. The Filip et al. bound is based on the observation that, on [0..1], the difference between a function $f$ and its linear interpolant at 0 and 1 is bounded by $\|f''\|_{[0..1]}/8$. For a polynomial of degree $n$ the latter is bounded by $n(n-1)$ times the maximal second difference of the Bézier control points. Note that all bounds can be improved by subdividing the curve segment as part of a recursive process. Publications [10] and [13] give a good overview of how elementary bounding constructs are used in the context of hierarchical interference detection (for space partitioning methods see e.g. [2]): simpler constructs like AABBs and spheres provide fast rejection tests in sparse arrangements, while more expensive $k$-dops and OBBs perform better on complex objects in close proximity; sleves with adaptive resolution fall in the more expensive comparison category and promise to outperform other bounding constructs for curved, non-polyhedral objects in close proximity, due to their basis-specific pre-optimization that is done off-line (and tabulated once and for all) and local refinability.

The theory of slefes has its roots in bounds on the distance of piecewise polynomials to their Bézier or B-spline control net [18,25]. Compared to these constructions, slefes yield dramatically tighter bounds for the underlying functions since they need not enclose the control polygon. Approximation theory has long recognized the problems of one-sided approximation and two-sided approximation [3]. Algorithmically, though, according to the seminal monograph [24], page 181, the convergence of the proposed Remez-type algorithms is already in one variable 'generally very slow'. The only termination guarantee is that a subsequence must exist that converges. By contrast, the slefes provide a solution with an explicit error very fast and with a guarantee of error reduction under refinement.

Surface simplification for triangulated surfaces has been modified to generate (locally) inner and outer hulls [4,26]. This requires solving a sequence of linear programs at runtime and applies to already triangulated surfaces. The object oriented bounding boxes for subdivision curves or surfaces in [14] are based on a min–max criterion and require the evaluation of several points and normals on the curve or surface. Thus the dependence on the coefficients is not linear. Linearity of the slefe construction allows us to solve inverse problems, like the CHANNEL problem mentioned earlier. Farin [8] shows that for rational Bézier–

curves, the convex hull property can be tightened to the convex hull of the first and the last control point and so-called *weight* points.

Starting with Farouki and Sederberg [27] the use of *interval spline representation* for tolerancing, error maintenance and data fitting has been promoted in a series of papers, collected in a recent book by Patrikalakis and Maekawa [19] (see also [20]). The key ingredient is the solution of nonlinear polynomial systems in the Bernstein basis through rounded interval arithmetic. This, in turn, relies on AABBs based on the positivity and partition of unity property of spline representations. sleves complement this work: while interval spline representations focus on uncertainty of the control points, sleves offer tight two-sided bounds for nonlinear (bounding) curves or surfaces.

## 2  Subdividable Linear Efficient Function Enclosures

### 2.1  The Basic Idea

The subdividable linear efficient function enclosure, or slefe of a function $x$ with respect to a domain $U$ is a piecewise linear pair, $\overline{x}, \underline{x}$, of upper and lower bounds that sandwich the function on $U$: $\overline{x} \geq x \geq \underline{x}$. The goal is to minimize the width,

$$w(x, U) := \overline{x} - \underline{x},$$

in the *recursively applied* $L^\infty$ *norm*: the width is as small as possible where it is maximal – and, having fixed the breakpoint values where the maximal width is taken on (zeroth and first breakpoint in Fig. 3), the width at the remaining breakpoints is recursively minimized subject to matching the already fixed break point values.

Slefes are based on the two general lemmas [16,17] (Section 2.2 gives concrete examples), and the once-and-for-all tabulation of best recursive $L^\infty$ enclosures of a small set of functions, $\mathbf{a}_i$, $i = 1, \ldots, s$ collected into a vector $\mathbf{a}$ below. With $\mathbf{b}$ is a vector of (basis) functions and $\mathbf{x}$ a vector of coefficients, we use the following notation below:

$$x = \mathbf{b} \cdot \mathbf{x} := \sum \mathbf{b}_i \mathbf{x}_i.$$

**Lemma 1 (change of basis).** *Given two finite-dimensional vector spaces of functions, $\mathcal{B} \neq \mathcal{H}$, $s := \dim \mathcal{B} - \dim(\mathcal{B} \cap \mathcal{H})$, $(\mathbf{b}_i)_{i=1,\ldots,\dim \mathcal{B}}$ a basis of $\mathcal{B}$, $(\mathbf{a}_i)_{i=1,\ldots,s}$ functions in $\mathcal{B}$, and linear maps*

$$L : \mathcal{B} \to \mathcal{H}, \ \Delta : \mathcal{B} \to \mathbb{R}^s,$$

*such that (i) $(\Delta_j \mathbf{a}_i)_{i,j}$ is the identity in $\mathbb{R}^{s \times s}$ and (ii) $\ker \Delta = \ker(E - L)$ (where E is the embedding identity) then for any $x := \mathbf{b} \cdot \mathbf{x} \in \mathcal{B}$,*

$$(\mathbf{b} - L\mathbf{b}) \cdot \mathbf{x} = (\mathbf{a} - L\mathbf{a}) \cdot (\Delta x).$$

*Proof.* By (i) $\Delta(I - \mathbf{a}\Delta)\mathbf{x} = 0$ and hence, by (ii), $(E - L)(I - \mathbf{a}\Delta)\mathbf{x} = 0$.

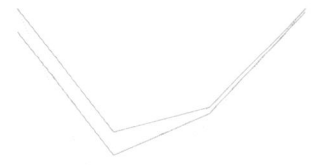

**Fig. 3.** $a := \mathbf{a}_1^3 := -(2\mathbf{b}_1^3 + \mathbf{b}_2^3)/3$ with control polygon and slefe.

*Remarks:* We can extend the lemma, say to the bi-infinite spline setting, by defining $s = \infty$ if $\dim \mathcal{B} = \infty$; however, for practical computation, $(\mathbf{a} - L\mathbf{a}) \cdot (\Delta x)$ has to have finitely many terms. In (ii), $\ker \Delta \subset \ker(E - L)$ is needed since for any $x \in \ker \Delta \setminus \ker(E - L)$, $(\mathbf{a} - L\mathbf{a}) \cdot (\Delta x)$ is zero, but not $(\mathbf{b} - L\mathbf{b}) \cdot \mathbf{x}$. Since the width of the enclosure changes under addition of any element in $\ker(E - L) \setminus \ker \Delta$, we also want $\ker(E - L) \subset \ker \Delta$.

**Lemma 2 (bounds).** *If, with the definitions of Lemma 1, additionally the maps* $x \mapsto \underline{x} : \mathcal{B}^s \to \mathcal{H}^s$ *and* $x \mapsto \overline{x} : \mathcal{B}^s \to \mathcal{H}^s$ *satisfy* $\underline{\mathbf{a} - L\mathbf{a}} \leq \mathbf{a} - L\mathbf{a} \leq \overline{\mathbf{a} - L\mathbf{a}}$ *componentwise on every point of a domain* $U$, *and* $(\Delta x)_+(i) := \max\{0, \Delta x(i)\}$, *and* $(\Delta x)_-(i) := \min\{0, \Delta x(i)\}$ *then*

$$\underline{x} := Lx + \underline{\mathbf{a} - L\mathbf{a}} \cdot (\Delta x)_+ + \overline{\mathbf{a} - L\mathbf{a}} \cdot (\Delta x)_-,$$

$$\overline{x} := Lx + \underline{\mathbf{a} - L\mathbf{a}} \cdot (\Delta x)_- + \overline{\mathbf{a} - L\mathbf{a}} \cdot (\Delta x)_+$$

*sandwich* $x$ *on* $U$: $\quad \underline{x} \leq x \leq \overline{x}.$

The general **slefe construction** is as follows where (1),(2),(3),(4) are precomputed, off-line and (5) is easy to compute.

1. Choose $U$, the domain of interest, and the space $\mathcal{H}$ of enclosing functions.
2. Choose a difference operator $\Delta : \mathcal{B} \mapsto \mathbb{R}^s$, with $\ker \Delta = \mathcal{B} \cap \mathcal{H}$.
3. Compute $\mathbf{a} : \mathbb{R}^s \mapsto \mathcal{B}$ so that $\Delta \mathbf{a}$ is the identity on $\mathbb{R}^s$ and each $\mathbf{a}_i$ matches the same $\dim(\mathcal{B} \cap \mathcal{H})$ additional independent constraints.
4. Compute $\underline{\mathbf{a} - L\mathbf{a}}$ and $\overline{\mathbf{a} - L\mathbf{a}} \in \mathcal{H}$.
5. Compute $(\Delta x)_+$ and $(\Delta x)_-$ and assemble $\underline{x}$ and $\overline{x}$ according to Lemma 2.

## 2.2 Example in One Variable

For a concrete example of the general framework in the previous section, let

- $\mathcal{B}$ be the space of univariate polynomials of degree $d$, in Bézier form

$$x(u) := \sum_{i=0}^{d} x_i \mathbf{b}_i^d(u), \quad \mathbf{b}_i^d(u) := \frac{d!}{(d-i)! i!}(1 - u)^{d-i} u^i.$$

Specifically, we choose $d = 3$.

- $\mathcal{H}$ the space of piecewise linear functions $\mathbf{h}_j$ with break points at $j/m$, $j \in \{0, \ldots, m\}$. Specifically, we choose $m = 3$ segments.
- $\Delta x$ the $d - 1$ second differences of the Bézier coefficients

$$\Delta x := \begin{bmatrix} \Delta_1 x \\ \Delta_2 x \end{bmatrix} := \begin{bmatrix} x_0 - 2x_1 + x_2 \\ x_1 - 2x_2 + x_3 \end{bmatrix}$$

- $Lx(u) := x_0(1 - u) + x_d u$, and therefore $\mathbf{a}_i^d - L\mathbf{a}_i^d = \mathbf{a}_i^d$.
- $U = [0..1]$.

This yields

$$\mathbf{a}_1^d := -(2\mathbf{b}_1^d + \mathbf{b}_2^d)/d, \quad \mathbf{a}_2^d := -(\mathbf{b}_1^d + 2\mathbf{b}_2^d)/d,$$

$$x - Lx = \mathbf{a}_1^d \Delta_1 x + \mathbf{a}_2^d \Delta_2 x.$$

By symmetry, it is sufficient to compute the optimal enclosures for $a := \mathbf{a}_1^3$. Due to the convexity of $a$ (see Fig. 3), the piecewise linear interpolant at $j/m$ is an upper bound. We write $\bar{a} := \bar{a}^m$ (where $m$ indicates the number of linear segments of the upper bound) as the vector of its breakpoint values, e.g. the value of $\bar{a}$ at $1/3$ is $-10/27$:

$$27\bar{a} \simeq [0, -10, -8, 0].$$

The lower bound is computed by recursive minimization. The first segment is the dominant segment in the sense that its tightest bound has the largest width among the three segments (see Figure 3 – the genral case is covered in Lemma 5 of [21]). Therefore, we calculate the values of $\underline{a}$ at 0 and $1/3$ by shifting down the first segment of the upper bound until it is tangent to $a$. The other two break point values are computed by calculating the tangent line to $a$, keeping one end fixed. This procedure yields the $m + 1$ break point values of the lower bound

$$27\underline{a} \simeq \left[30, 20, 25 + \frac{\beta_1 - 9}{2}\beta_2, \beta_3\right] - \frac{38\beta_1}{9},$$

where

$$\beta_1 := \sqrt{57}, \quad \beta_2 := \sqrt{-10 + 2\beta_1},$$

$$\beta_3 := \frac{261}{8} + \frac{\beta_1 - 9}{4}\beta_2 + \frac{3\beta_2 - \beta_1}{8}\sqrt{11 - 12\beta_2 - 2\beta_1 + 2\beta_1\beta_2}.$$

An approximation of the values is $\underline{a}_m \approx [-.0695, -.4399, -.3154, -.0087]$. Evidently, the width

$$w_{\text{slefe}}(x; U) := \max_U \bar{x} - \underline{x} = \max_U \sum_{i=1}^{d-1} (\overline{\mathbf{a}_i - L\mathbf{a}_i}^m - \underline{\mathbf{a}_i - L\mathbf{a}_i}_m) |\Delta_i x|$$

is invariant under addition of constant and linear terms to $x$ and one (DeCasteljau) subdivision step at the midpoint, $t = 1/2$ cuts the width to roughly *a quarter* (see Figure 4) since $(\overline{\mathbf{a}_i - L\mathbf{a}_i}^m - \underline{\mathbf{a}_i - L\mathbf{a}_i}_m)$ stays fixed and the maximal $\Delta_i x$ shrinks to $1/4$ its size.

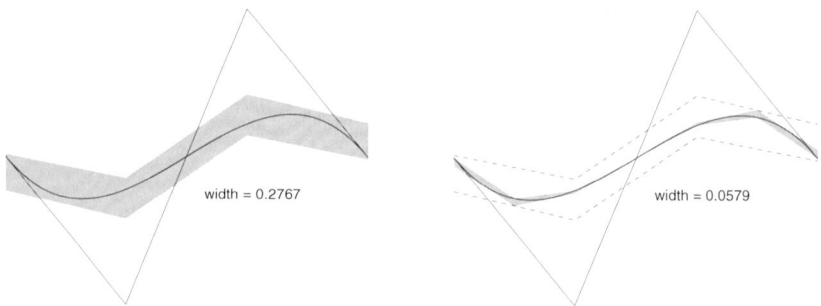

**Fig. 4.** (*left*) A cubic Bézier segment with coefficients $0, -1, 1, 0$. The control polygon exaggerates the curve far more than the *grey* 3-piece slefe. (*right*) After one subdivision at the midpoint, the width of the slefe (*grey*) is roughly 1/4th of the width of the unsubdivided slefe (*dashed*).

## 2.3    How Good Are slefes?

slefe-based bounds are observed to be very tight. Yet, being linear, the slefe construction cannot be expected to provide the best two-sided max-norm approximation, a difficult nonlinear problem. Therefore, it is of interest to see how close to optimal the slefe construction actually is by deriving and comparing it with the narrowest possible enclosure with the same breakpoints. In this section, we determine, for a class of functions, the optimal enclosure width, $w_{\text{opt}}$, and compare it with $w_{\text{slefe}}$. The simplest nontrivial case is when the function $x$ is a univariate quadratic polynomial; however, in this case, the slefe construction is optimal, because the vector of functions $\mathbf{a} - L\mathbf{a}$ is a singleton and slefes are based on the optimal enclosures of $\overline{\mathbf{a} - L\mathbf{a}}, \underline{\mathbf{a} - L\mathbf{a}}$. Since explicit determination of the least-width, piecewise linear enclosure is a challenge, we consider polynomials $x$ of degree $d = 3$ in Bézier representation on the interval $U = [0..1]$. Generalization of the results to $m \neq 3$ pieces is not difficult; generalization of exact bounds to degree $d > 3$ appears to be non-trivial. Without loss of generality, we assume $\Delta_1 x \geq |\Delta_2 x|$ in the following.

**Computing $w_{\text{slefe}}$.** Let $w_i := (\overline{a}^m - \underline{a}_m)(i/m)$, $i = 0, 1, \ldots, m$. Then, due to the symmetry $\mathbf{a}_1^3(t) = \mathbf{a}_2^3(1 - t)$,

$$w_{\text{slefe}}(x; U) = \max_{i \in \{0,1,2,3\}} \{|\Delta_1 x|w_i + |\Delta_2 x|w_{m-i}\}.$$

Since $w_0 = w_1 > w_2 > w_3$, the term with $i = 1$ is the maximal term, and

$$w_{\text{slefe}}(x; U) = |\Delta_1 x|w_1 + |\Delta_2 x|w_2 \approx 0.0695|\Delta_1 x| + 0.0191|\Delta_2 x|.$$

If we set $|\Delta_1 x| := 1 + \epsilon$, $\epsilon \in [0..\infty]$, and $|\Delta_2 x| := 1$ then

$$w_{\text{slefe}}(x; U) = -\frac{1}{243}\left(270\epsilon + 567 + \frac{9\beta_2}{2}(\beta_1 - 9) - 38\beta_1(\epsilon + 2)\right).$$

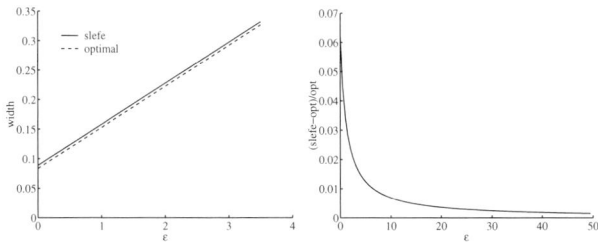

**Fig. 5.** Convex cubic $x$. (*left*) Value of $w_{\text{slefe}}(x;U)$ vs $w^{\cup}_{\text{opt}}(x;U)$. (*right*) The ratio $\frac{w_{\text{slefe}} - w^{\cup}_{\text{opt}}}{w^{\cup}_{\text{opt}}}(x;U)$.

**Computing $w_{\text{opt}}$.** To determine the width of the *narrowest* possible piecewise linear enclosure for $x$, $w_{\text{opt}}(x, [0..1])$, with breakpoints at $i/3$, elementary considerations show that it is sufficient to compare the width of functions with first and last coefficient equal zero, that then an increase of the second derivative of $x$ increases $w_{\text{opt}}$; and finally, since $\Delta_1 x > |\Delta_2 x|$, $w_{\text{opt}}(x, [0..1/3]) > w_{\text{opt}}(x, [0..1])$, i.e. the first segment determines $w_{\text{opt}}$.

**Comparison of $w_{\text{opt}}$ and $w_{\text{slefe}}$.** We first consider the case of *no inflection*. Without loss of generality, $\Delta_1 x := 1 + \epsilon, \epsilon \in [0, \infty]$, $\Delta_2 x := 1$ and with $A := \sqrt{57\epsilon^2 + 135\epsilon + 81}$,

$$w^{\cup}_{\text{opt}}(x;U) := -\frac{1}{243}\frac{(9 + 9\epsilon - A)(-3A(1+\epsilon) + 11\epsilon^2 + 36\epsilon + 27)}{\epsilon^2}.$$

Figure 5, *left* plots $w_{\text{slefe}}$ against $w^{\cup}_{\text{opt}}$. The gap between $w_{\text{slefe}}$ and $w^{\cup}_{\text{opt}}$ increases with $\epsilon$ but is finite at infinity:

$$w_{\text{slefe}}(x, \infty) - w^{\cup}_{\text{opt}}(x, \infty) \approx .0053353794.$$

The relative difference has a maximum of ca. 6% when $\epsilon = 0$ (c.f. Fig. 5,*right*), i.e. when $x$ is of degree 2.

If $x$ has an inflection point, we may assume that $\Delta_1 x := 1 + \epsilon$, and $\Delta_2 x := -1$ and get

$$w_{\text{slefe}}(x, \infty) - w^{\sim}_{\text{opt}}(x, \infty) \approx .032775216.$$

The worst ratio $\frac{w_{\text{slefe}} - w^{\sim}_{\text{opt}}}{w^{\sim}_{\text{opt}}}(x;U)$ occurs when $x$ is of the type depicted in Figure 4: if $\Delta_1 x = -\Delta_2 x = 1$ then $w_{\text{opt}}(x, U) = .05593616039$ and $w_{\text{slefe}}(x, U) = .08857673214$. Although the ratio is almost 3:5, the slefe is considerably tighter than the convex hull of the control polygon (c.f. Figure 6, *left*).

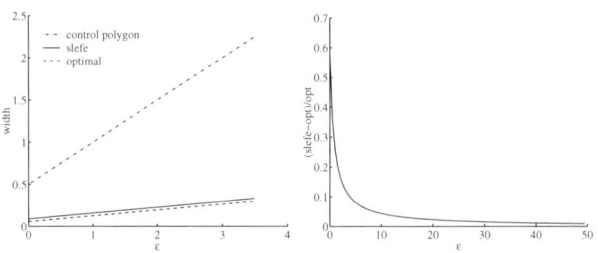

**Fig. 6.** Cubic $x$ with inflection: (*left*) Value of $w_{\mathrm{slefe}}(x;U)$, $w_{\widetilde{\mathrm{opt}}}(x;U)$ and the width based on the convex hull of the control polygon. (*right*) The ratio $\frac{w_{\mathrm{slefe}} - w_{\widetilde{\mathrm{opt}}}}{w_{\widetilde{\mathrm{opt}}}}(x;U)$.

## 2.4   Tensoring slefes

We can bootstrap univariate slefes by tensoring. A tensor-product polynomial $x(s,t)$ of degree $d_1, d_2$ is in Bézier form if

$$x(s,t) = \sum_{i=0}^{d_1}\sum_{j=0}^{d_2} x_{ij}\mathbf{b}_j^{d_2}(t)\mathbf{b}_i^{d_1}(s).$$

For example, a bi-cubic has 16 coefficients $x_{ij}$. We enclose $x(s,t)$ by a linear combination of $n \cdot m$-piecewise bilinear hat functions $\mathbf{h}_j^m(t)\mathbf{h}_i^n(s) \in \mathcal{H}^n \times \mathcal{H}^m$, for $(s,t) \in U := [0..1]^2$. Let $x_i(s)$ be the univariate Bézier polynomial with coefficients $x_{i0}, x_{i1}, \ldots, x_{id_2}$. We compute

$$\overline{x}_i(t) := \sum_{j=0}^m \mathsf{u}_{ij}^t\mathbf{h}_j^m(t) := b_{i0}(1-t) + b_{id_2}t$$

$$+ \sum_{j=1}^{d_2-1} \underline{\mathbf{a}_j^{d_2}}_m \min\{\Delta_j x_i, 0\} + \overline{\mathbf{a}_j^{d_2}}^m \max\{\Delta_j x_i, 0\},$$

and, with $\mathbf{u}_j(s) := \sum_{i=0}^{d_1} \mathsf{u}_{ij}^t \mathbf{b}_i^{d_1}(s)$,

$$\sum_{i=0}^n \mathsf{u}_{ij}\mathbf{h}_i^n(s) := \mathsf{u}_{0j}(1-s) + \mathsf{u}_{d_1 j}s$$

$$+ \sum_{i=1}^{d_1-1} \underline{\mathbf{a}_i^{d_1}}_n \min\{\Delta_i\mathbf{u}_j, 0\} + \overline{\mathbf{a}_i^{d_1}}^n \max\{\Delta_i\mathbf{u}_j, 0\},$$

Similarly, bounding $x(s,t)$ from below, we get

$$\underline{x}(s,t) := \sum_{j=0}^{m} \sum_{i=0}^{n} 1_{ij}\mathbf{h}_i^n(s)\mathbf{h}_j^m(t)$$

$$\leq \sum_{j=0}^{m}\left(\sum_{i=0}^{d_1} 1_{ij}^t\mathbf{b}_i^{d_1}(s)\right)\mathbf{h}_j^m(t) = \sum_{i=0}^{d_1}\sum_{j=0}^{m} 1_{ij}^t\mathbf{h}_j^m(t)\mathbf{b}_i^{d_1}(s)$$

$$\leq x(s,t) = \sum_{i=0}^{d_1}\sum_{j=0}^{d_2} x_{ij}\mathbf{b}_j^{d_2}(t)\mathbf{b}_i^{d_1}(s)$$

$$\leq \sum_{i=0}^{d_1}\sum_{j=0}^{m} \mathbf{u}_{ij}^t\mathbf{h}_j^m(t)\mathbf{b}_i^{d_1}(s) = \sum_{j=0}^{m}\left(\sum_{i=0}^{d_1} \mathbf{u}_{ij}^t\mathbf{b}_i^{d_1}(s)\right)\mathbf{h}_j^m(t)$$

$$\leq \overline{x}(s,t) := \sum_{j=0}^{m}\sum_{i=0}^{n} \mathbf{u}_{ij}\mathbf{h}_i^n(s)\mathbf{h}_j^m(t).$$

It is not difficult, although the generation of optimal approximation tables for $\underline{a}$ and $\overline{a}$ requires care, to extend the slefe construction to box-splines and to rational splines.

## 3    Mid-structures

The mid-structure

$$\overline{\underline{x}} := (\overline{x} + \underline{x})/2$$

is well-defined for a vector-valued curve or surface $\mathbf{x}$. By making the mid-structure along a boundary depend only on the boundary, e.g. a space curve for a patch in $\mathbb{R}^3$ and the endpoint for a curve, mid-structures join continuously if their patches join continuously. Mid-structures are good $L^\infty$ approximands and may be used, for example, to render more accurately than based on sampling (Figure 7). Certain mid-structures are invertible, e.g. in one variable, if $m = d$, then we can obtain $\mathbf{x}$ from $\overline{\underline{x}}$.

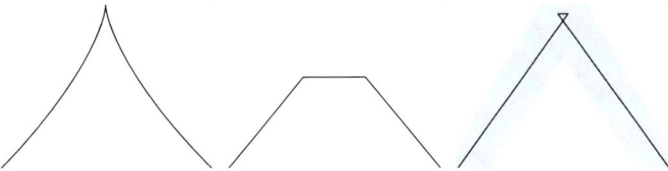

**Fig. 7.** A cubic Bézier segment (*left*) finely evaluated, (*middle*) approximated by 4 sample values, (*right*) approximated by a 3-piece mid-path.

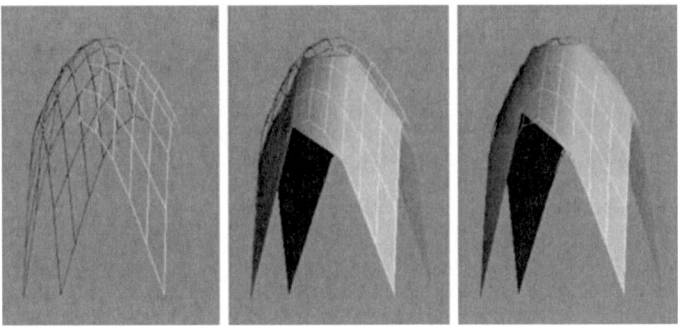

**Fig. 8.** A bi-quadratic Bézier patch (*left*) finely evaluated, (*middle*) approximated by sampling, (*right*) approximated by a mid-patch.

## 3.1   Univariate and Bivariate Mid-structures

We define the *mid-path*, $\underline{\overline{x}}$, of $x$ as the $m$-piece linear function in $\mathcal{H}^n$ with values

$$\underline{\overline{x}}\left(\frac{i}{m}\right) := \begin{cases} \frac{1}{2}(\overline{x}^m + \underline{x}_m)\left(\frac{i}{m}\right) & \text{if } 0 < i < n, \\ x_i & \text{if } i = 0 \text{ or } i = m. \end{cases}$$

The choice for $i = 0$ and $i = m$ guarantees that mid-paths of continuously joined Bézier pieces match up at their endpoints. The distance between the polynomial $x$ and $\overline{x}$ on the interval $[\frac{i}{m}..\frac{i+1}{m}]$ is bounded by the linear average of the distances at the endpoints; and these distances are evidently bounded by

$$|\mathbf{x} - \underline{\overline{x}}|\left(\frac{i}{m}\right) \le \frac{\epsilon_i}{2}(\overline{x}^m - \underline{x}_m)\left(\frac{i}{m}\right)$$

where $\epsilon_i = 2$ for $i = 0$ or $i = m$ and $\epsilon_i = 1$ otherwise. We can bound the derivative $x'$ of $x$ in the same manner and with the same number of $m = 3$ linear pieces:

$$(x')^+ := L(x') + \sum_{i=1}^{d-1} \underline{\mathbf{a}_i^{d-1}}_m \min\{\Delta_i x', 0\} + \overline{\mathbf{a}_i^{d-1}}^m \max\{\Delta_i x', 0\}.$$

For example, if $x'$ is of degree $d = 2$, there is only one function $\mathbf{a}_1^2$ to bound and all we need are the numbers

$$\begin{bmatrix} \overline{\mathbf{a}_1^2}^3 \\ \underline{\mathbf{a}_1^2}_3 \end{bmatrix} \simeq \begin{bmatrix} 0.0 & -0.2\overline{2} & -0.2\overline{2} & 0.0 \\ -0.02777780 & -0.25 & -0.25 & -0.02777780 \end{bmatrix}$$

to assemble $\underline{\overline{x'}}$. This allows associating a midnormal with every breakpoint of the midpath.

Analogous to midpaths in one variable, the *mid-patch* $\underline{\overline{\mathbf{x}}}$ of $\mathbf{x}$ as the $n\cdot m$-piece bilinear function in $\mathcal{H}^n \times \mathcal{H}^m$. We associate the average of a bivariate enclosure

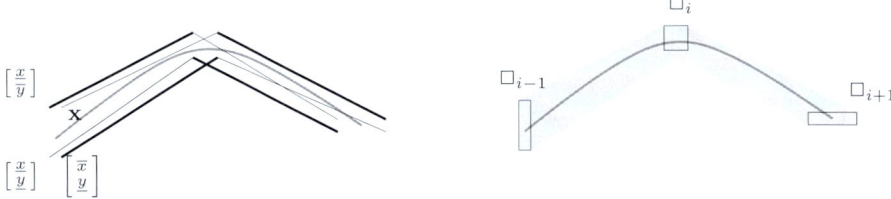

**Fig. 9.** (*left:*) The curve **x** is bounded by a 'union' of four component enclosures. The extreme, outermost components that stay to one side of the curve, are emphasized as fat line segments. Note the gap and the intersection between consecutive extreme segments. (*right:*) The bounding region is equivalently generated as the piecewise linear combination of point enclosures (axis-aligned rectangles) $\square_i$.

with the interior, the average of a univariate enclosure with the boundaries, and a constant with the vertices of $U$, so that adjacent midpatches match up continuously along boundaries. Let $x_j(t)$ be the polynomial with coefficients $x_{0j}$ and $x_i(s)$ the polynomial with coefficients $x_{i0}$ and $U = [0..1] \times [0..1]$. Then

$$\overline{x}(\frac{i}{n},\frac{j}{m}) := \begin{cases} \frac{1}{2}(\overline{x}+x)(\frac{i}{n},\frac{j}{m}) & \text{if } i \notin \{0,n\} \text{ and } j \notin \{0,m\}, \\ \frac{1}{2}(\overline{x}_i+x_i)(\frac{j}{m}) & \text{if } i \in \{0,n\} \text{ and } j \notin \{0,m\}, \\ \frac{1}{2}(\overline{x}_j+x_j)(\frac{i}{n}) & \text{if } i \notin \{0,n\} \text{ and } j \in \{0,m\}, \\ x_{ij} & \text{if } i \in \{0,n\} \text{ and } j \in \{0,m\}. \end{cases}$$

## 4   Extension to Curves and Surfaces

Slefes can be leveraged to generate Subdividable Linear Efficient Variety Enclosures, short **sleves**, i.e. enclosures of *varieties* such as curves and surfaces in parametric or implicit representation [22].

### 4.1   Planar Curve Enclosures

Since both the $x$ and the $y$ component of a planar curve **x** provide an upper and a lower bound, we obtain four segments

$$\begin{bmatrix} \overline{x} \\ y \end{bmatrix}, \begin{bmatrix} \overline{x} \\ \overline{y} \end{bmatrix}, \begin{bmatrix} x \\ \overline{y} \end{bmatrix}, \begin{bmatrix} x \\ y \end{bmatrix}$$

for each interval between breakpoints (see Figure 9, *left*). A certain 'union' of these bounds appears to enclose the curve. A simple way to give some structure to this 'soup' of line segments, is to observe that, due to linearity, each piece of the enclosure is a convex combination of consecutive point enclosures $\square_i$, $\square_{i+1}$, where $\square_j$ has the four vertices

$$\square_j \sim \begin{bmatrix} \overline{x} \\ y \end{bmatrix}(\frac{j}{n}), \begin{bmatrix} \overline{x} \\ \overline{y} \end{bmatrix}(\frac{j}{n}), \begin{bmatrix} x \\ \overline{y} \end{bmatrix}(\frac{j}{n}), \begin{bmatrix} x \\ y \end{bmatrix}(\frac{j}{n}).$$

That is, a *point enclosure* $\square_j$ is an axis-parallel rectangle or box (Figure 9 right). Differently put, the function enclosures directly yield a *piecewise linear interval enclosure.* Here, linearity is crucial, since the outer curves of general interval Bézier curves (see Farouki and Sederberg [27]) are non-trivial to represent and to work with: only the case where all $\square_j$ are of equal size has to date a short representation [29], p 50. Even in the linear case, deciding which combinations of linear function bounds are outer bounds for the curve, is not immediate. While the case where all $\square_j$ are of equal size is again straightforward, the general characterization requires several (simple) tests since it depends on the relative size and distance of $\square_{i-1}$ and $\square_i$.

Moreover, even linear interval enclosures have two shortcomings: multiplicity, and intersections or gaps. By keeping information on all four components, interference checking between two interval objects would require 16 intersection tests. Moreover, the piecewise linear outer bounds have more pieces or need to be trimmed due to the intersections and gaps between adjacent pieces (fat lines in Figure 9, *left*).

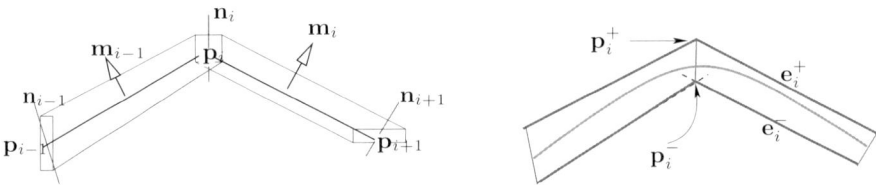

**Fig. 10.** *left:* Anchor points $\mathbf{p}_j$, normals $\mathbf{n}_j$ and directions $\mathbf{m}_j$ for identifying extreme segments of the enclosure. *right:* Antipodal points $\mathbf{p}_i$ and enclosure pieces $\mathbf{e}_i$.

**Gaps and Intersections.** To address the problem of gaps and intersections, we associate, with each local parameter $i/n$, a point $\mathbf{p}_i$ that lies in *all* point enclosures associated with that location (there may be two point enclosures of differing size if $i \in \{0, n\}$, say the end of one curve segment and the start of another). We call the point, *anchor point*, because we have in mind to attach a line segment to it with direction $\mathbf{n}_i$, roughly normal to the curve (c.f. Figure 10 left). The two endpoints $\mathbf{p}_i^+$ and $\mathbf{p}_i^-$ will serve as the vertices of the two sheets $\mathbf{e}_i^+$ and $\mathbf{e}_i^-$ of the curve enclosure (c.f. Figure 10 right). If the Bézier pieces join with tangent continuity,

$$\mathbf{p}_i := \begin{bmatrix} \overline{x} \\ \overline{y} \end{bmatrix} (\frac{i}{n}), \text{ and } \mathbf{n}_i := \begin{bmatrix} \overline{y'} \\ -\overline{x'} \end{bmatrix} (\frac{i}{n}) \qquad (3)$$

fit the bill and we can process each piece independent of its neighbor. If the curves meet just with continuity of position then (the normalized) $\mathbf{n}_n$ of the first and (the normalized) $\mathbf{n}_0$ of the second segment need to be averaged.

**Multiplicity.** To address the problem of multiplicity, we observe that, due to linearity, there is always a pair of linear function enclosures, $\begin{bmatrix} \overline{x} \\ y \end{bmatrix}$ and $\begin{bmatrix} x \\ y \end{bmatrix}$ for the

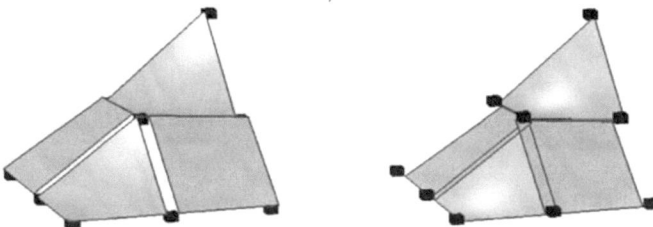

**Fig. 11.** Four pieces of an upper piecewise bilinear enclosure $\mathbf{x}^+$ and four pieces of a lower bilinear facets $\mathbf{x}^-$. The nine cubes represent point enclosures. Note the gaps and overlaps.

left segment in Figure 9, whose linear extensions or trims enclose the other two enclosures over the region of interest. The computationally efficient selection of this *extreme pair* of line segments from the four possible choices, as well as the full algorithm is presented in [23] (see also [22]). Given a (per segment or global) tolerance, the algorithm refines (subdivides) the representation locally until a sleve is obtained whose width is below the prescribed tolerance.

## 4.2   Interval Patch Enclosures

For $x, y, z$ we each have an upper and a lower bound yielding eight candidates for enclosures (Figure 12) for $u, v$ in the domain square $[i..i+1]/m_1 \times [j..j+1]/m_2$:

$$\begin{bmatrix}\overline{x}\\\overline{y}\\\overline{z}\end{bmatrix},\begin{bmatrix}\overline{x}\\\overline{y}\\\underline{z}\end{bmatrix},\begin{bmatrix}\overline{x}\\\underline{y}\\\overline{z}\end{bmatrix},\begin{bmatrix}\overline{x}\\\underline{y}\\\underline{z}\end{bmatrix},\begin{bmatrix}\underline{x}\\\overline{y}\\\overline{z}\end{bmatrix},\begin{bmatrix}\underline{x}\\\overline{y}\\\underline{z}\end{bmatrix},\begin{bmatrix}\underline{x}\\\underline{y}\\\overline{z}\end{bmatrix},\begin{bmatrix}\underline{x}\\\underline{y}\\\underline{z}\end{bmatrix}.$$

All combinations with positive weights summing to 1 of the eight enclosures form a shell that is a 3D enclosure of the surface piece (see the surfaces with parameter grid in Figure 12). The union of the shells of all patches form a sleve.

Since the pieces are bilinear, we can also view the shell as a bilinear combination of the four point enclosures $\square_{i+\alpha,j+\beta}, \alpha, \beta \in \{0,1\}$ of the corner points $[x(i+\alpha, j+\beta), y(i+\alpha, j+\beta), z(i+\alpha, j+\beta)]^T$. A *point enclosure* $\square_{i,j}$ is now an axis-parallel box whose vertices are the eight combinations of the corner points of the component slefes (the boxes displayed in Figures 11, 12 and 12) and slefes directly yield a *bilinear interval Bézier enclosure*.

The bilinear interval enclosures just defined have three shortcomings for efficient use: nonlinearity, multiplicity and gaps or intersections. The bilinearity of the facets implies that intersections between enclosures result in algebraic curves of degree 4 and force iterative techniques for intersections with rays as opposed to short explicit formulas for triangles. Slivers arise when computing the exact union of the shells which entails intersection of bilinear facets and trimming bilinear patches. Multiplicity, i.e. the choice from eight possible bilinear function enclosures implies up to 64 nonlinear intersection tests when intersecting two patch enclosures. The algorithm in [23,22] remedies gaps, intersections and slivers just as in the case of one variable using points $\mathbf{p}_{ij}$ and directions $\mathbf{n}_{ij}$ (Figure

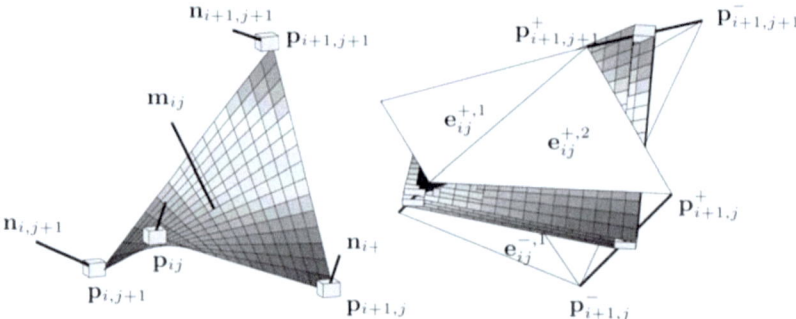

**Fig. 12.** (*left*) A single bilinear facet of the midpatch with direction $\mathbf{m}_{ij}$, anchor points $\mathbf{p}_{ij}$ and normal direction $\mathbf{n}_{ij}$. (*right*) Antipodal pairs $\mathbf{p}_{ij}^{+}$ and $\mathbf{p}_{ij}^{-}$ as interpolation points of the two sheets $\mathbf{e}_{ij}^{+,1}$, $\mathbf{e}_{ij}^{+,2}$ and $\mathbf{e}_{ij}^{-,1}$, $\mathbf{e}_{ij}^{-,2}$ of the surface enclosure of eight bilinear facets (with parameter grid) whose extreme pair is $\mathbf{h}_{ij}^{+}$ and $\mathbf{h}_{ij}^{-}$. (The black spot is due to Matlab's depth sorting algorithm in the presence of many overlapping surfaces).

**Fig. 13.** Teapot inside a sleve.

12) to construct serve as the vertices points $\mathbf{p}_{ij}^{+}$ and $\mathbf{p}_{ij}^{-}$ that support two triangle pairs $\mathbf{e}_{ij}^{+,1}, \mathbf{e}_{ij}^{+,2}$ and $\mathbf{e}_{ij}^{-,1}, \mathbf{e}_{ij}^{-,2}$. The surface enclosure is thus a tent-like construction with support beams in the direction $\mathbf{n}_{ij}$ as shown in Figure 12. Multiplicity is addressed by picking an approximate normal $\mathbf{m}_{ij}$ analoguous to the univariate case and finding an extreme pair of bilinear patches.

Finally, bilinear slefes are replaced by two pairs of triangles per original facet, a subtle operation, since extrapolation of a triangle interpolating three vertices of a bilinear facet may intersect the extrapolated bilinear facet and it cease to be a one-sided approximation.

## 5    Inverse Problems

The simplicity and linearity of the slefe construction allows solving problems like the

CHANNEL Problem: Given a channel or tube, construct a smooth spline that stays within that channel.

(see Figure 14). Solutions can be used to thread pipes past a set of obstacles or determine robot motion paths. The CHANNEL problem is a two-sided version of the

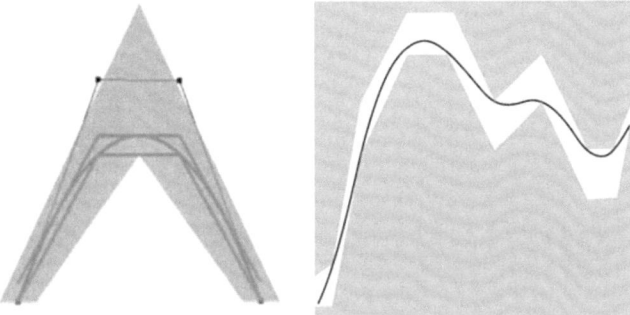

**Fig. 14.** Simple 2D examples. (*left*) the curve segment is fit into the *shaded* channel by fitting its **sleve** (piecewise linear) into the channel. Characteristically, the control polygon (*square control points*) does not stay within the channel. (*right*) A more complex, but still function-based channel scenario.

SUPPORT Problem (see Figure 1): given an input polygon, find a spline *(black)* that stays above but close to the polygon.

## 5.1    The Channel Problem for Space Curves

The emphasis in this problem is neither on the optimality nor the uniqueness of the solution, although we will see that we can easily augment the feasibility problem with a linear (or quadratic) optimization function. Our approach to solving the CHANNEL problem is to construct a 'sleeve' around the candidate space curve; then it is sufficient to constrain this sleeve – rather than the original nonlinear curve – to stay within the channel. This approximation reduces the original, complexity-wise intractable, *continuous feasibility problem* to a simple *linear feasibility problem* that is solvable by any Simplex or Linear Program solver!

Three properties are crucial to make this approach work.

(i)   The enclosure must depend linearly on the coefficients of the spline representation.

(ii)  The enclosure should be near-optimal in the max-norm, i.e. as narrow as possible.

(iii) The enclosure should be refinable to adaptively adjust to where the channel is particularly narrow or tricky to navigate.

Requirement (i) rules out oriented bounding box and convex hull-based approaches (see Section 1.1, page 299) since the coefficients of the curve will be variable as well as [14]. Requirement (ii) rules out the use of looser bounding constructs such as bounding spheres and axis-aligned bounding boxes (c.f. Figure 16). The best match of linearity, tightness and refinability for the CHANNEL problem are therefore **slefe**-based constructions. For *functions* in one variable, the CHANNEL problem was first formulated in [15]. A closely related set of problems appears in graph layout [7] where, however, the emphasis is on a large number of piecewise linear curves with few pieces. By tightly linking discrete and non-linear techniques, our new approach may be viewed as bridging a gap between established techniques of *computational geometry and geometric design*.

**Channel Definition.** We define a channel as a sequence of *cross-sections*. Each cross-section has $n_{c\_sides}$ edges. For example, in Figure 15, each cross-section is quadrilateral with the four vertices not necessarily co-planar. Adjacent cross-section pairs constitute a channel segment, or *c-segment*. Two points of one cross-section and the corresponding pair of points from its neighboring cross-section form a face of the channel. A face is not necessarily planar and is split into two triangles. The normals $\mathbf{n}_j^c$ are consistently oriented outwards. We combine

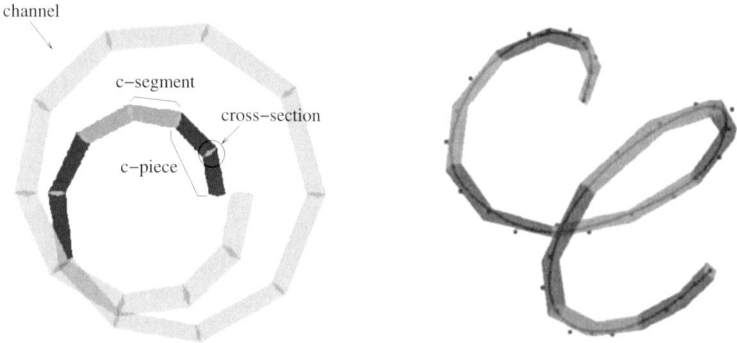

**Fig. 15.** Composition of a CHANNEL. (*right:*) A solution curve (control points indicated as black squares) fit into the given, transparently rendered channel whose consecutive c-pieces alternate between yellow and grey (transparent).

$n_{c\_segs}$ c-segments to form a *c-piece* (this is analogous to line segments connected to form a control polygon). In Figure 15, each c-piece consists of two adjacent c-segments. The union of $n_{pcs}$ c-pieces forms a channel. In the following, one curve piece in Bézier form is fitted through each c-piece, and $C^1$ continuity constraints are enforced between adjacent polynomial pieces. The sleves are calculated for each $c \in (1, \dots, n_{pcs})$, $i = 1 \dots, d-1$ and $v \in \{x, y, z\}$ as

- $\triangle_{i,v}^c := x_{i-1,v}^c - 2x_{i,v}^c + x_{i+1,v}^c$
- $\triangle_{i,v}^c{}^+ := \max(\triangle_{i,v}^c, 0)$
- $\triangle_{i,v}^c{}^- := \min(\triangle_{i,v}^c, 0)$
- $\overline{e}_v^c := L(\mathbf{x}^c) + \overline{\mathbf{a}^d} \cdot \triangle_v^{c-} + \underline{\mathbf{a}^d} \cdot \triangle_v^{c+}$
- $\underline{e}_v^c := L(\mathbf{x}^c) + \underline{\mathbf{a}^d} \cdot \triangle_v^{c+} + \overline{\mathbf{a}^d} \cdot \triangle_v^{c-}$

**The Constraint System.** We can now formulate the CHANNEL problem as a feasibility problem for fitting a degree $d$ polynomial piece through each of $n_{pcs}$

c-pieces. For each $c \in (1, \ldots, n_{pcs})$, $i = 1 \ldots, d-1$ and $v \in \{x, y, z\}$, we have following constraints:

1. $\triangle_{i,v}^{c\,+} \geq \triangle_{i,v}^{c}$ and $\triangle_{i,v}^{c\,+} \geq 0$
2. $\triangle_{i,v}^{c\,-} \leq \triangle_{i,v}^{c}$ and $\triangle_{i,v}^{c\,-} \leq 0$
3. $\triangle_{i,v}^{c\,+} + \triangle_{i,v}^{c\,-} = \triangle_{i,v}^{c}$
4. Set $x_{d,v}^{c} = x_{0,v}^{c+1}$ equal (e.g. to the center of a cross-section) to ensure $C^0$ continuity.
5. Set $x_{d,v}^{c} = \frac{1}{2}\left(x_{d-1,v}^{c} + x_{0,v}^{c+1}\right)$ to ensure $C^1$ continuity.

The remaining constraints combine the $x$, $y$, and $z$ components to ensure that the curve stays within the channel. Here the linearity of the sleve construction pays off.

1. At each sleve breakpoint, $e_i^c := (e_{i,x}^c, e_{i,y}^c, e_{i,v}^c)$, where $e_{i,v}^c \in \{\underline{e}_{i,v}^c, \overline{e}_{i,v}^c\}$, needs to be within the channel. Hence, for every such $e_i^c$ and each point $tript_j^c$ in the corresponding c-segment, we enforce

$$\mathbf{n}_j^c \cdot (e_i^c - tript_j^c) \leq 0,$$

where $\mathbf{n}_j^c$ is the outward-pointing normal:

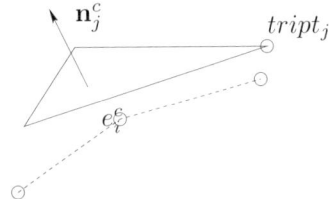

2. At each concave (inward-pointing) channel breakpoint $tript_j^c$, and the normals of the triangles incident to $tript_j^c$, constrain any point on the corresponding enclosure segment, $e'^c := u e_{i-1}^c + v e_i^c$, to satisfy

$$\mathbf{n}_j^c \cdot (e'^c - tript_j^c) \leq 0$$

with $u + v = 1$ and $u > 0$ and $v > 0$.

**The Objective Function.** A typical linear program has the form

$$\min_b f(b), \qquad \text{subject to linear equality and inequality constraints.}$$

The previous section listed the constraints and we may add an objective function $f$. (If there is no objective function, then the problem is called a linear feasibility problem.) A possible choice for $f$ is to minimize the sum of the absolute second-differences. Since $\triangle_{i,v}^{c\,+} \geq 0$, $\triangle_{i,v}^{c\,-} \leq 0$, and $\triangle_{i,v}^{c\,+} + \triangle_{i,v}^{c\,-} = \triangle_{i,v}^{c}$, it follows that $|\triangle_{i,v}^c| = \triangle_{i,v}^{c\,+} - \triangle_{i,v}^{c\,-}$. Hence, to minimize kinks, the objective function is:

$$\sum_{c=1}^{n_{pcs}} \left( \sum_{i=1}^{d-1} (\triangle_{i,x}^{c\,+} - \triangle_{i,x}^{c\,-} + \triangle_{i,y}^{c\,+} - \triangle_{i,y}^{c\,-} + \triangle_{i,z}^{c\,+} - \triangle_{i,z}^{c\,-}) \right)$$

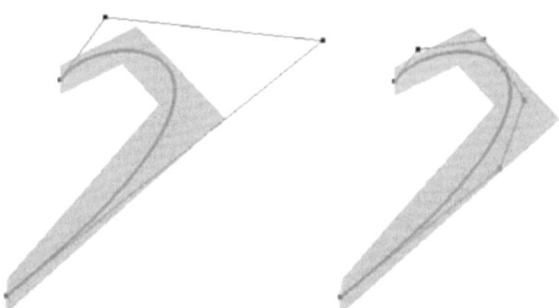

**Fig. 16.** A sharp 2D channel. (*left*) The control polygon of the solution lies well outside the channel. (*right*) The control polygon of the solution after subdivision still violates the channel boundaries illustrating the need for the tight slefe bounds.

**Fig. 17.** Bilinear barrier surface (*left*) and a solution to the SUPPORT problem (*right*).The surface covers the barrier from the top. Its control points, as well as the control points of the bilinear barrier are indicated as small cubes. The line on the left of each figure indicates the height axis.

**Examples.** Using the open-source *PCx* [6], the channel in Figure 15 to be traversed by a twenty segment, degree 4 spline $n_{c\_segs} = 2$ and one polynomial piece for each pair of c-segments, results in 3560 constraints (567 equations, and 720 variables, generated by a scripting language) and is solved in 5.45 seconds on a generic PC. The example in Figure 16 highlights the crucial role played by the near-optimal width of the bounding construct.

## 5.2   The Support Problem for Surfaces

The support problem is a simpler optimization problem, using only one-sided constraints. Remarkably, if we were to attempt to solve SUPPORT over all nonsmooth functions with arbitrary break points, rather than for splines with fixed break points, the problem would be NP hard [1].

# 6    Open Problems, Future Work

Much of the usefulness of slefes and sleves relies on the narrowness of slefes. To date, this near-optimality has only been *proven and quantified* for cubic functions in one variable (Section 2.3). Further investigation of polynomials of degree $d$ leads, after removal of symmetries, to $2^{d-2}$ distinct cases of minimization problems in $d-2$ variables. An essential difficulty is to explicitly determine the narrowest enclosure for a class of functions – if that were easy, we would not need sleves. Numerical analysis for small $d$ leads to the conjecture that the ratio (optimal max-norm enclosure width : the width of the slefe construction) is minimal when all second differences are equal. If true, this would settle the question for univariate polynomials.

The proper use of the observed near-optimality of sleves must be established by looking in detail at applications. Here the balance between the cost of generating the enclosure and the cost of using the enclosure has to be evaluated. For example, while sleves have clear advantages in applications that value tightness and linearity of the bounds, such as the inverse problems sketched in Section 5 and certain applications in marine cartography, the relative merits of computing sleves for intersection testing and root finding *vis a vis* established techniques, say as summarized in [19], has to be characterized for specific classes of problems.

## Acknowledgments

This survey has drawn from joint work and work in progress with Xiaobin Wu, David Lutterkort and Ashish Myles. Thanks also to the referees for asking for more details on [9] and pointing out further references on interval spline computations.

## References

1. Pankaj K. Agarwal and Subhash Suri. Surface approximation and geometric partitions. *SIAM Journal on Computing*, 27(4):1016–1035, August 1998.
2. Julien Basch. *Kinetic Data Structures*. PhD thesis, Stanford University, 1999.
3. R. C. Buck. Applications of duality in approximation theory. In *Approximation of Functions (Proc. Sympos. General Motors Res. Lab., 1964)*, pages 27–42. Elsevier Publ. Co., Amsterdam, 1965.
4. Jonathan Cohen, Amitabh Varshney, Dinesh Manocha, Greg Turk, Hans Weber, Pankaj Agarwal, Frederick P. Brooks, Jr., and William Wright. Simplification envelopes. In Holly Rushmeier, editor, *SIGGRAPH 96 Conference Proceedings*, Annual Conference Series, pages 119–128. ACM SIGGRAPH, Addison Wesley, August 1996. held in New Orleans, Louisiana, 04-09 August 1996.
5. A. Crosnier and J. R. Rossignac. Technical section — tribox bounds for three-dimensional objects. *Computers and Graphics*, 23(3):429–437, June 1999.
6. Joe Czyzyk, Sanjay Mehrotra, Michael Wagner, and Stephen Wright. http://www-fp.mcs.anl.gov/otc/Tools/PCx.

7. D. Dobkin, E.R. Gansner, E. Koutsofios, and S.C. North. A path router for graph drawing. In *Proceedings of the 14th annual symposium on Computational Geometry*, pages 415–416. ACM Press, New York, 1998. June 1-10 1998, Minneapolis, MN.

8. Gerald Farin. Tighter convex hulls for rational Bézier curves. *Comput. Aided Geom. Design*, 10:123–125, 1993.

9. Daniel Filip, Robert Magedson, and Robert Markot. Surface algorithms using bounds on derivatives. *CAGD*, 3(4):295–311, 1986.

10. S. Gottschalk, M. C. Lin, and D. Manocha. OBBTree: A hierarchical structure for rapid interference detection. *Computer Graphics*, 30(Annual Conference Series):171–180, 1996.

11. Timothy L. Kay and James T. Kajiya. Ray tracing complex scenes. In David C. Evans and Russell J. Athay, editors, *Computer Graphics (SIGGRAPH '86 Proceedings)*, volume 20, pages 269–278, August 1986.

12. James T. Klosowski, Joseph S. B. Mitchell, Henry Sowizral, and Karel Zikan. Efficient Collision Detection Using Bounding Volume Hierarchies of k-DOPs. *IEEE Transactions on Visualization and Computer Graphics*, 4(1):21–36, January 1998.

13. James Thomas Klosowski. *Efficient collision detection for interactive 3d graphics and virtual environments.* PhD thesis, State Univ. of New York at Stony Brook, May 1998.

14. Leif P. Kobbelt, Katja Daubert, and Hans-Peter Seidel. Ray tracing of subdivision surfaces. In *Rendering Techniques '98 (Proceedings of the Eurographics Workshop)*, pages 69–80, New York, June 1998. Springer-Verlag.

15. D. Lutterkort and J. Peters. Smooth paths in a polygonal channel. In *Proceedings of the 15th annual symposium on Computational Geometry*, pages 316–321, 1999.

16. D. Lutterkort and J. Peters. Optimized refinable enclosures of multivariate polynomial pieces. *Comput. Aided Geom. Design*, 18(9):851–863, 2001.

17. David Lutterkort. *Envelopes for Nonlinear Geometry.* PhD thesis, Purdue University, May 2000.

18. D. Nairn, J. Peters, and D. Lutterkort. Sharp, quantitative bounds on the distance between a polynomial piece and its Bézier control polygon. *Computer Aided Geometric Design*, 16(7):613–633, Aug 1999.

19. N. M. Patrikalakis and T. Maekawa. *Shape Interrogation for Computer Aided Design and Manufacturing.* Springer-Verlag, Berlin, 2002.

20. N.M. Patrikalakis and T. Maekawa. Intersection problems. In *Handbook of Computer Aided Geometric Design.* Springer, 2001.

21. J. Peters and X. Wu. On the optimality of piecewise linear max-norm enclosures based on slefes. In *Proceedings of the 2002 St Malo conference on Curves and Surfaces*, 2003.

22. Jörg Peters and Xiaobin Wu. Optimized refinable surface enclosures. Technical report, University of Florida, 2000.

23. Jörg Peters and Xiaobin Wu. Sanwiching spline surfaces. *Computer-Aided Geometric Design*, 200x.

24. Allan M. Pinkus. *On $L^1$-approximation.* Cambridge University Press, Cambridge, 1989.

25. U. Reif. Best bounds on the approximation of polynomials and splines by their control structure. *Comput. Aided Geom. Design*, 17(6):579–589, 2000.

26. P.V. Sander, Xianfeng Gu, S.J. Gortler, H. Hoppe, and J. Snyder. Silhouette clipping. *Computer Graphics*, 34(Annual Conference Series):327–334, 2000.

27. T. W. Sederberg and R. T. Farouki. Approximation by interval bezier curves. *IEEE Computer Graphics and Applications*, 12(5):87–95, September 1992.

28. Thomas W. Sederberg, Scott C. White, and Alan K. Zundel. Fat arcs: A bounding region with cubic convergence. *Comput. Aided Geom. Design*, 6:205–218, 1989.
29. G. Shen and N.M. Patrikalakis. Numerical and geometric properties of interval B-splines. *International Journal of Shape Modeling*, 4(1,2):35–62, 1998.
30. E. Welzl. Smallest enclosing disks (balls and ellipsoids). In Hermann Maurer, editor, *Proceedings of New Results and New Trends in Computer Science*, volume 555 of *LNCS*, pages 359–370, Berlin, Germany, June 1991. Springer.

# Procedural Modelling, Generative Geometry, and the International Standard ISO 10303 (STEP)

Michael J. Pratt

LMR Systems, 21 High Street, Carlton, Bedford, MK43 7LA, UK
mike@lmr.clara.co.uk

**Abstract.** The paper describes current work towards the development of an international standard for the exchange of procedurally defined shape models between CAD systems. Such models are specified by the sequence of operations used to construct them rather than by the explicit elements they contain. They have the advantage of being easy to manipulate in a receiving system, by contrast with the explicit models that can be exchanged using current standards. The transfer of comparatively simple procedural shape models has already been demonstrated. Some of the problems faced are briefly outlined, and comparisons are made between the approach taken in this work and a formal method for procedural shape representation recently advocated by Leyton.

## 1 Introduction

Papers on the exchange of geometric data between CAD systems have been presented at earlier conferences in this series [4,14]. The first of the cited papers, originally presented in 1990, described a new international standard for this purpose that was still under development at the time. The first release of that standard, ISO 10303 or STEP (STandard for the Exchange of Product modelling data) was published by ISO in 1994, after ten years of development [9]. STEP is a large and complex standard, consisting of a set of parts, of which some 50 have so far been published [16,17]. The scope of the standard covers many different types of engineered products, and a range of different life-cycle stages including design, finite element analysis and manufacturing. At the core of the standard is a set of Integrated Generic Resource parts; of primary interest in the context of this conference is ISO 10303-42 (referred to as 'Part 42' of the standard), *Geometric and topological representation* [8]. This formed part of the initial release of the standard in 1994; a second, expanded, edition appeared in 2000, and a third edition is due to be published shortly.

ISO 10303 specifies a 'neutral' (system-independent) representation for the information transferred. Its use requires the existence, for each CAD system, of two translators, a *preprocessor* to translate the native format of the sending system into the neutral format, and a *postprocessor* to translate from the neutral format into the native format of the receiving system. This has the obvious

M.J. Wilson and R.R. Martin (Eds.): Mathematics of Surfaces 2003, LNCS 2768, pp. 320–337, 2003.

advantage of only requiring $2n$ translators for transfers between all possible pairs of $n$ CAD systems, rather than $n(n-1)$ if direct translators are used between all possible pairs. The most widely used form of model exchange employs file transfer, but it is also possible to create a dynamic model built from STEP data structures, and to access this via an interface defined in the standard.

Part 42 is one of the most fundamental resources of STEP; it forms the basis of several Application Protocols (APs), which are the implementable parts of the standard. The application protocol that has been most frequently implemented with CAD systems, and which is most widely used in industry, is AP203 (ISO 10303-203: 'Configuration controlled design'). This is intended essentially for the exchange of geometric models, though some supplementary administrative information is also tranferred. Part 42 provides AP203 with the capability for capturing various types of CAD model including surface models and boundary representation solid models. The spectrum of curve and surface geometry available includes the following:

**Curves:**

- lines (unbounded),
- conics,
- B-splines, polynomial and rational (with Bézier curves as special cases),
- trimmed curves,
- composite curves,
- various forms of curves defined on surfaces,
- offset curves.

**Surfaces:**

- planes (unbounded),
- 'natural quadrics',
- torus and Dupin cyclide,
- B-spline surfaces, polynomial and rational; Bézier surfaces,
- trimmed surfaces,
- composite surfaces,
- various forms of 'swept' surfaces (extrusion, rotation, lofted),
- offset surfaces.

These are all familiar forms of geometry widely implemented in almost every CAD system.

Following the initial launch of the standard there was a period when the percentage of successful translations of geometric models between different systems was low. The primary reasons for this are listed below. The percentages of translations suffering from the various causes are taken from a 1997 survey.

*CAD System Accuracy Mismatch (16%):* Different CAD systems work to different internal numerical tolerances. Typically, the range of distances within which two points are judged to be identical ranges between $10^{-4}$ and $10^{-7}$ units. Thus a boundary representation model in which two edges meet at a point in one system may be deemed to have disconnected edges in another system, and to lack topological integrity. This problem has been largely overcome by the provision of greater flexibility in CAD systems, and a certain level of 'intelligence' in interpreting imported models.

*Non-conformance to the Standard (10%):* Errors of this kind arose from misunderstandings of the standard by translator writers. In fact, it was found impossible to remove all ambiguity from the standard itself, and a range of 'implementer agreements' have been arrived at to overcome some commonly occurring problems of interpretation.

*Excessive File Size Growth (10%):* The difficulty here arises because most CAD systems have certain proprietary algorithms which are not shared with other systems. Blend surfaces provide an example — there are many different ways of modelling these, some of them unique to a single system. In such a case it is necessary to approximate the 'proprietary' surface representation for purposes of model exchange by some other surface form in common use, for example NURBS. In the early days such approximations often generated excessive amounts of data, but more concise approximation methods have now been developed.

*Translator Bugs (2%):* Little needs to be said on this topic, except that most such bugs were removed in the mid-1990s.

*Unknown (11%):* Although the origin of these problems was a mystery in 1997, eventually it became clear that most of them were attributable to

*'Model Quality' Deficiencies:* These arise, typically, through the use of what is called 'variational' modelling, in which the values of dimensional parameters may be changed to vary the shape of a model subject to imposed constraints [6]. This can sometimes give rise to models containing edges whose lengths are very close to zero, or faces with areas very close to zero. The effect is that the geometric accuracy problems mentioned above are exacerbated. In some cases the originating system itself can become so confused that it loses consistency between the geometric and topological aspects of the modelled shape. Another manifestation of model quality deficiencies arises (unfortunately) because of poor practice by system users, who may adopt the view that if a model looks satisfactory on the CAD system screen then it is suitable for its purpose. High levels of magnification often show that this is far from the truth. One of the worst practices is the 'stitching together' of several surface models to create what purports to be a 'solid' model. Software tools have become available in recent years for the checking and enhancement of model quality, and their use before the invocation of a STEP translator significantly improves the success rate of model exchanges using the standard.

## 2   Editability of Transferred Models

Since the first release of STEP in 1994 much has been learned about the exchange
of solid models in particular. The experience gained has enabled the problems
described above to be largely overcome. However, the level of success achieved
has now highlighted a new problem. This is, essentially, that the geometric model
as received after a transmission does not behave in the receiving system as it did
in the originating system when an attempt is made to modify it. What is received
has been described as a 'dumb' model — it has the correct appearance on the
CAD system screen, and further details can be added to it, but the geometry of
the received model itself cannot (at least in most CAD systems) be edited. This
is a severe limitation. To see why, consider the following scenarios where CAD
data exchange may be important:

- A single company may use different CAD systems for different product life-
  cycle stages (e.g., design, manufacturing). This happens often in practice,
  because different CAD systems have their greatest strengths in different
  application areas.
- Companies with contractor/subcontractor relationships may use different
  CAD systems but need to collaborate on designing the geometry of interface areas (e.g., the attachment of engines made by one manufacturer to an
  airframe and piping systems made by another).
- Companies collaborating in what are known as 'virtual enterprises' may form
  a grouping and wish to collaborate on a design activity. In such a case it
  is often necessary to transfer the design between the systems used by the
  different partners in such a way that any of them can modify it, subject to
  suitable approval by the others.

It is evident from these examples that a transferred model that cannot be
edited (e.g., to optimize it against a set of relevant criteria in the receiving
system) poses a severe problem. To see how to overcome that problem it is
necessary to look more deeply into the nature of geometric modelling in CAD.

Early work in 3D shape modelling for CAD revealed several possible approaches, most notably surface modelling, boundary representation (B-rep) solid
modelling and constructive solid geometry (CSG) [5]. Pure CSG proved insuf-
ficiently flexible for CAD purposes, and no mainstream commercial systems of
that type have existed for ten years or more. However, CSG left an important
legacy, as will be discussed below. Surface modelling and B-rep solid modelling
are closely related. Surface modelling is appropriate for certain specialized applications, but solid modelling provides a wider range of geometric capabilities
and supplements them with an underlying topological structure of faces, edges
and vertices, each topological element having associated geometric information.
Solid modelling systems use this structure to provide a certain level of integrity
checking that is lacking from pure surface modelling systems. In the mid-1980s
it seemed that most major commercial CAD systems were gravitating towards
the use of B-rep technology.

However, it was found that B-rep models, once created, were not easy to work with. Like the 'dumb models' referred to earlier, they could be added to but not otherwise modified. This is because a boundary representation model merely records the outcome of a modelling session — it does not contain any details of the method of construction. In the late 1980s CAD systems emerged on the market in which B-rep modelling was supplemented by an additional type of shape representation, at a higher level of abstraction. Essentially, this form of representation describes a shape, not in terms of its constituent elements but in terms of the sequence of operations used to build it. In systems of this kind the boundary representation is ephemeral. It is used to generate the picture on the CAD system screen, as the basis for certain kinds of user interaction (e.g., the picking of an element of the model to be modified), and for geometric computations, which require the availability of explicit geometric information. But if the user makes a modification, the current B-rep model is discarded and replaced by an updated model reflecting the change. For that reason, the problem of maintaining consistency between the two modelling levels is minimal.

It was earlier remarked that CSG modelling, although no longer mainstream, has left a major legacy in today's CAD systems. Recall that constructive solid geometry defines shapes in terms of *Boolean operations* (union, intersection or difference)[1] performed on volumes or half-spaces regarded as point sets. It is therefore a procedural method. In the 1980s a rigorous theory was developed for this mode of model creation [19], but the method was found to be inadequate for the CAD modelling of general shapes. However, current procedural modelling systems may be regarded as descendents of the early CSG systems in which the ranges of both operations and operands have been greatly extended — Boolean operations between volumes are in fact still provided by most systems, but the operands are now B-rep volumes rather than volumetric point sets. The original theory relating to CSG unfortunately does not cover the whole spectrum of operations in modern CAD systems, which mostly combine a collection of capabilities in a rather *ad hoc* manner without any pretence of a unified theoretical foundation for them.

The high-level models mentioned above are referred to as *procedural* or *construction history* models. Apart from a history of the model's construction, this kind of model typically also contains information concerning the following:

**features** — high-level geometric constructs used during the design process to create shape configurations in the model that are usually related to the intended functionality of the designed artefact [20];

**parameters** — values of quantities in the model (typically, dimensions) that may be regarded as variable;

**constraints** — relationships between geometric elements in the model (e.g., parallelism, tangency) that are required to be maintained if the model is edited.

---

[1] Strictly, the Boolean operations are required to be of a special kind called *regularized*.

The presence of parameters and constraints in a model constitutes at least part of what is known as *design intent*. This permits the model to be edited easily, by variation of parameter values. If this is done, the model should automatically readjust in such a way that the constraints imposed on it remain satisfied. It may be wondered how geometric constraints can be asserted between elements of a model described entirely in terms of constructional operations, and therefore containing no explicit geometry. The answer is that certain explicit constructs are typically embedded in the construction history. For example, a 2D *profile* or *sketch*, represented explicitly in terms of points and curve segments, may be used as the basis of a class of constructional operations that create *swept surfaces* or *swept volumes*. The most common forms are *extrusions*, where the profile is swept normal to its plane, and rotational sweeps, where the profile is rotated about an axis lying in the same plane but not intersecting it. If the profile is regarded as a curve it sweeps out a surface, but if it is closed and regarded as enclosing an area then it sweeps out a solid. Another important application of constraints is for the relative positioning and orientation of components in an assembly model.

To summarize, most modern CAD systems generate what has been called a *dual model*, in which the procedural component gives the master description of the shape and the B-rep component provides visual feedback and a basis for geometric computations. This has been the case for the past ten years or more. However, the development of the STEP standard commenced in 1984 when the B-rep was regarded as the master model. The nature of the slow and meticulous international standardization process made it impossible to add major new capabilities to the standard before its first publication in 1994. Consequently, what STEP currently transfers in the B-rep model, which has now come to be regarded as a secondary model, and which is difficult to edit, as earlier stated. Discussions about adding a procedural capability to STEP started soon after the first publication of the standard, but it took considerable time to identify an approach that would be upwardly compatible with those parts of the standard already published. This is a major consideration with the CAD vendors; they have to implement STEP translators for their systems, and want to reuse as much as possible of the code they have already written. Meanwhile, manufacturing industry has become increasingly frustrated by the fact that, despite an increasingly high level of superficially successful STEP model exchanges it is exteremely difficult to work effectively with a model after it has been transferred into another system.

The situation at the time of writing is that an approach to procedural modelling within STEP has been identified, and internal conflicts with other STEP resources have been resolved. New parts of the standard are now under development for the capture and transfer of procedural models, and the remainder of this paper is mainly concerned with descriptions of these parts and their underlying rationales. Before concluding, a comparison is made with the concept of *generative geometry* as defined in a recent book by Michael Leyton [15].

# 3    Status of Procedural Modelling in STEP

The following subsections describe various aspects of the development work, which is being undertaken by the Parametrics Group in Working Group 12 of ISO TC184/SC4. The areas of responsibility of the parent Technical Committee (TC) and its relevant Subcommittee (SC) are *Industrial data* and *Industrial automation systems and integration*, respectively. It should be noted that ISO 10303 is composed of computer-interpretable *schemas*, each defining entities involved in a model exchange in terms of an information modelling language called EXPRESS, which forms part of the standard itself [7].

## 3.1    The Capture and Exchange of Operation Sequences

The standardized representation of operation sequences is a fundamental requirement for the exchange and sharing of procedural CAD models. Early progress on this topic was slow, because

1. The modelling methodology differs significantly from anything previously undertaken in STEP;
2. Nevertheless, the new resource must be consistent with the previously published parts of STEP without requiring them to be modified.

The required new fundamental resource is ISO 10303-55 [10]. At the time of writing, this has just passed its international Committee Draft ballot, which allows progression to Draft International Standard status.

Previously published parts of the STEP standard have provided representations for explicit models, which may be regarded as snapshots of product specifications at some specific point in time. A B-rep model, for example, is essentially a collection of static constituents built into a structure in which geometric elements and topological relationships are represented. Provided the set of constituents transferred is complete, their sequencing in an exchange file is irrelevant; a complete model will be reconstructed in the receiving system regardless of the order in which the constituents are added to it. In a procedural model, by contrast, the correct ordering of operations is vital, because the same set of constructional operations performed in a different order will in general lead to a different shape model. An ISO 10303-55 entity **procedural_representation_sequence** has therefore been defined as the basic component of a procedural model.

The most appropriate representation for an individual constructional operation was only resolved after a good deal of discussion, and eventually the views of the CAD vendor community prevailed. The EXPRESS language defines a syntax for the representation of functions and procedures, and the original proposal was to use this. However, the approach finally adopted relies on a statement in the EXPRESS manual that the definition of a static entity may also be regarded as a creation procedure for an instance of that type of entity. In existing STEP usage, an instance of **plane** in an exchange file indicates that a plane existed in the model in the originating system and that it is being transferred explicitly into the receiving system. However, the interpretation of the plane instance in a **procedural_representation_sequence** is quite different; here it invokes the procedure in the receiving system that actually creates a new plane with the specified characteristics. In the first case the action is copying, and in the second it is re-creation.

The use of entity definitions as creation procedures in procedural model exchange was not envisaged when the EXPRESS language was developed. The intended application lay in rules for validity checking during the translation phase in the exchange of explicit models. However, the proposed usage for parametric model exchange, while not originally foreseen, does not contravene any guidelines of the standard, and this is the mechanism which has been adopted by the Parametrics Group. It has the major advantage of not requiring separate procedures to be written for the creation of instances of any entities for which EXPRESS specifications already exist. Nevertheless, many new entities will need to be defined to take into account the different constructional methods provided by CAD systems for elements that are fundamentally of the same type. For example, the **plane** entity specified in Part 42 of STEP defines an unbounded plane containing a given point and having a specified normal direction. No new entity therefore needs to be written for the creation of a plane instance in this canonical form. On the other hand, CAD systems may provide many other ways of constructing planes, for example

1. a plane through three points;
2. a plane tangent to three spheres;
3. a plane containing a given point and tangent to a cylinder.

In some cases (e.g., the first example above) there is also a possible distinction between bounded and unbounded planes. All the commonly provided plane generation operations need to be made available for use in an **operation_sequence**. At present only the canonical specification exists.

While the **procedural_representation_sequence** is the basis of the proposed procedural model representation, other capabilities provided in the new resource are as follows:

– The import of explicit elements as the basis for procedural operations, e.g. a 2D profile or sketch that is subsequently extruded or rotated to sweep out a volume;
– The capture of explicit elements picked from the screen display of the originating system, also used as the basis for subsequent modelling operations, e.g. an edge which is selected for the application of a rounding operation.

The distinction between these is that the first type of element does not exist in the receiving system until imported, while the second already exists there. It is an element that has been generated by partial evaluation of a **procedural_representation_sequence**, and which must be identified with a corresponding element in the originating system (see the example in Section 3.5). A model that is basically procedural but also contains imported explicit elements is described as a *hybrid model*.

It is intended that a procedural or hybrid model will always be transmitted together with an explicit model representing the intended result of its evaluation. This will enable possible ambiguities in interpretation of the procedural model to be resolved by reference to the explicit model, which will also provide a check that the result of the evaluation is acceptably close to the original explicit model. Some slight variation will of course arise due to differences in the computational environments of the originating and receiving systems.

Another capability provided in this resource is that for capturing 'design rationale', i.e. the logic underlying the constructional procedure used. This is felt necessary as an aid to subsequent human understanding of the transferred procedural model. Unfortunately there is no known means at present of capturing such information automatically during the design process, and so provision has been made for its representation as human-interpretable text strings.

Finally, although its primary intended application is the procedural representation of CAD shape models, the document under development is proposed as a fundamental resource for the whole of the STEP standard. It is therefore written in general terms so that it can serve for the procedural modelling of any physical or non-physical objects, including products, plans, processes or organizations.

## 3.2    Representation of Parameterized Design Features

As already mentioned, existing STEP shape modelling entities can be used as constructional operations in procedural models. These are mostly defined at a low level, representing such things as individual points, curves, surfaces, vertices, edges, faces and so on. Most CAD systems, on the other hand, provide feature-based modelling capabilities that shield the user from having to work at such a low level. Features are high-level entities, sometimes composed of many of the low-level entities mentioned above [20].

The representation of features in STEP is a controversial topic. At one time a generic feature modelling resource was proposed, but for historical reasons this was not proceeded with. Instead, feature representations were separately provided in several of the STEP Application Protocols (APs). These APs are concerned with different phases of the product life-cycle (manufacture by machining, in particular), and there has been some harmonization of the feature modelling methodology used for each of these applications. However, at present no AP defines features for design, as needed for procedural shape modelling, and it has been found inconvenient to model design features in a manner consistent with the features in existing APs. In particular, that would lead to a significant increase in the size of transfer files with no compensating benefit. Initially, the proposal that design features should be modelled differently led to opposition from other groups in ISO TC184/SC4, but the need for a new approach has recently been agreed.

Concerning the actual range of design features to be provided, the Parametrics Group has written a prioritized list of the more commonly implemented feature types, and is working on writing representations of them. Some are very complex; e.g., blend or fillet features encompass many special cases that all have to be taken into account. It is intended that a future edition of AP203 ('Configuration controlled design') will include representations for a range of design features that have been identified as common to all major CAD systems.

## 3.3   Representation of Constraints

Two types of constraints can occur in procedural models, implicit and explicit. Implicit constraints are inherent in the operation of many constructional operations. For example, the operation 'create rectangular block' will lead to the creation of a volume having three pairs of parallel opposite faces, and in which all adjacent faces are perpendicular to each other. The parallelism and perpendicularity constraints are not explicitly represented, but are automatically imposed every time the procedure is invoked because they are characteristic of a rectangular block. If the dimensional parameters are edited, the resulting new block will of course be subject to the same constraints.

Explicit constraints, by contrast, are present as entities in their own right in the model data structure. Consider a boundary representation model of a planar-faced hexahedron. With no constraints imposed, this may be edited by changing the definition of any of the planes containing its faces; the resulting changes in face, edge and vertex geometry will need to be computed by the system. If three parallelism constraints are defined, each referring to a pair of opposite faces, the shape will be constrained to be a parallelepiped. Two further constraints must be applied to enforce perpendicularity of adjacent sides before the shape is that of a rectangular parallelepiped, i.e. a rectangular block.

The most usual applications of explicit constraints are for constraining (i) geometric elements of 2D profiles or sketches, and (ii) positions and orientations of parts in assembly models. As mentioned earlier, sketches are often used as the basis for constructional operations that generate surfaces or volumes. Sketches are typically composed of 2D curves, and their elements may be subject to constraints. An example is provided by a rectangle with rounded corners. The explicit elements making up this shape are four line segments and four circular arcs. Opposite lines will be subject to parallelism constraints, and tangency constraints will apply where the arcs join the line segments. Extrusion of such a sketch normal to itself will generate planar and cylindrical surfaces.

A STEP resource, ISO 10303-108 [12] has been written to provide representations of explicit constraints. It also allows parameters to be defined and associated with attributes of elements in a model, e.g. the radius attribute of a circle or cylinder. A wide range of commonly implemented geometric constraints is specified, including parallelism, perpendicularity, tangency, incidence and symmetry. Additionally, provision is made for defining mathematical relationships between parameters. The document includes specialized methods for modelling 2D profiles with or without constraints.

It is worth noting that parameterization and explicit constraint data associated with elements of an explicit model are redundant at the time of model exchange - they simply reaffirm relationships that are already present in the model. It is not until the model is edited following a transfer that this information comes into play. Then it determines the nature of the permissible changes that may be made to the model, and prevents impermissible ones.

ISO 10303-108 is the most highly developed of the new parametric capabilities in STEP. At the time of writing it has recently been submitted for acceptance as a Draft International Standard.

## 3.4   Parametric Assembly Modelling

Currently, a STEP assembly model is a collection of positioned, oriented parts, suitable for bill-of-materials and parts-list applications. The future use of parameterization, combined with an existing kinematics resource (ISO 10303-105), will allow the exchange of representations of dynamic mechanisms, for example. A Committee Draft of ISO 10303-109, a new STEP resource for this purpose [11] has just passed its first ISO international ballot at the time of writing. It provides representations for

- Logical relationships between parts, expressed as assembly feature relations;
- Different types of contact between parts (continuous, intermittent);
- Kinematic degrees of freedom between parts;
- Parameterized assemblies, with inter-part constraints.

In this context, a geometric constraint may be specified between two surfaces belonging to the shape representations of different parts in the assembly. These surfaces may also participate in the representation of assembly features, and in this case a constraint relationship may have the additional semantics of a mating feature relationship.

## 3.5   Proof-of-Concept Demonstrations

Several demonstrations have been performed to validate the concepts developed in the Parametrics Group for the exchange and sharing of parametric models. Some of these demonstrations are briefly described below.

**PDES Inc. Demonstrations.** PDES Inc. is a US-based international industry/government consortium dedicated to the development and testing of STEP-related methodologies and software.

Some years ago PDES Inc. undertook a project called ENGEN ('Enabling Next GENeration design'), mainly concerned with the exchange of constrained 2D profile data. The documentary output of this project (the 'ENGEN Data Model') provided valuable input for the development of ISO 10303-108 mentioned earlier. The constrained profile of an automotive connecting-rod was successfully exchanged between three different CAD systems using this model. An account of this project was given by Anderson and Ansaldi [1].

More recently the exchange of a procedurally defined CAD model was demonstrated by the British company Theorem Solutions, a PDES Inc. member. The exchange representation used was the one under development by the Parametrics Group. The CAD model exchanged was constructed as follows in the sending system:

1. An explicit 2D profile was defined, made up of line segments enclosing an L-shaped area;
2. The profile was extruded normal to its plane to sweep out a block with a step feature;
3. The concave edge of the step feature was selected by a pick from the screen display;
4. A blend operation was performed to fillet the selected edge;
5. A cylindrical volume and a conical volume were created, representing the shaft and bottom of a drilled blind hole;
6. The cylinder and cone were united by a Boolean union operation;
7. The resulting volume was subtracted from the L-block using a Boolean difference operation to give the final model, an L-block with a blind hole (see Figure 1 below).

Two-way exchange was demonstrated between the Dassault Systems CATIA and PLM Solutions Unigraphics CAD systems. However, the construction history exchanged had no constraints defined upon the extruded profile, and the ISO 10303-108 capabilities were not used. Work is currently under way to include these capabilities.

**NIST Demonstration.** NIST is the US National Institute of Standards and Technology in Gaithersburg, MD, a government agency. Here the L-block model has also been used as the test part.

At NIST, two-way transfer was achieved between Parametric Technology Corporation's ProEngineer and the SolidWorks CAD system. Initially the exchange was based on an EXPRESS schema influenced by ProEngineer data structures, but more recently versions of the 'official' STEP schema have been used. Parameterization and constraint capabilities were included, and parametric changes to the model, subject to the imposed constraints, were found to be possible following the transfer. A paper describing this work is in preparation [13].

**KAIST Demonstration.** KAIST is the Korean Advanced Institute of Science and Technology in Seoul, Korea. Here a team of graduate students is working with five CAD systems (CATIA, I-DEAS, ProEngineer, SolidWorks and Unigraphics). The L-block model has again been used, in this case as one of a range of test models of increasing complexity. Parametric model exchange was demonstrated by the KAIST group at a meeting in Fukuoka, Japan, in October 2001.

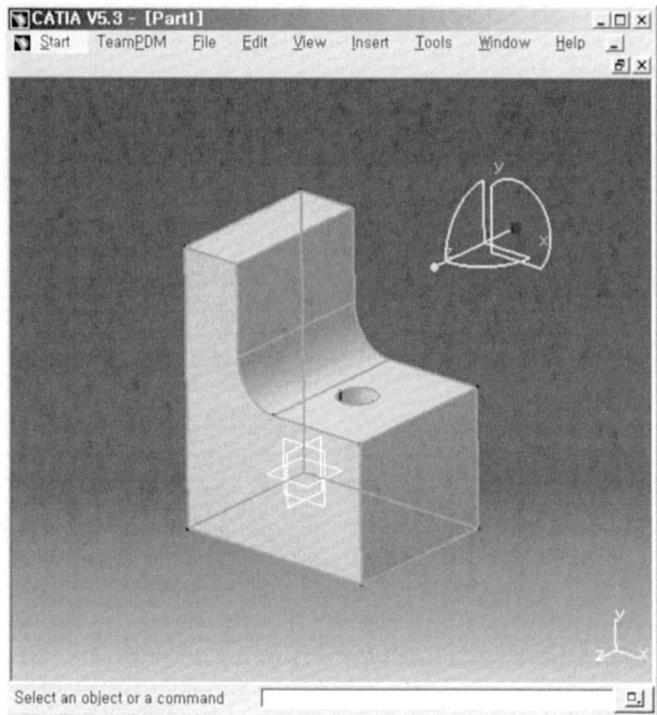

**Fig. 1.** L-block model as depicted by Dassault CATIA v.5 (courtesy of KAIST)

Initially a non-STEP-related exchange method was used, but the KAIST team is now collaborating in the context of the Parametrics Group in the development of EXPRESS-based capabilities. The earliest experiments by the KAIST team demonstrated the exchange of shape models between CATIA and Solid-Works. These systems both write native construction history files in Visual Basic, and the initial method used was direct translation between these two native formats [3]. The other systems employed require access to the model via system procedural interfaces rather than file input/output, and the use of a neutral format model in the STEP manner is proving more convenient in these cases.

The KAIST team has identified a core set of some 160 constructional operations for CAD models, common (with minor variations) to the major CAD systems. This is providing valuable input for the STEP parametric modelling work.

## 4    Leyton's Generative Theory of Shape

Michael Leyton has recently published a book entitled 'A Generative Theory of Shape' [15]. Leyton expresses many interesting viewpoints, one of them being that the standard approach to the study of geometry, expressed by Klein in his Erlangen program as the study of invariants under transformations, is not appropriate for many practical applications. Leyton is also a psychologist, with an interest in the human perception of shape. His experiments have led him to the view that shape is perceived dynamically rather than statically. In the case of a square, for example, the eye follows one edge, and then effectively perceives that the subsequent edges are the results of applying the members of a transformation group sucessively to the initial one. Leyton is concerned with the maximization of two abstract quantities:

**transfer** — the reuse of previously defined geometric constructs in the creation of more complex configurations (following the manner of human perception of such configurations), and

**recoverability** — the possibility of inferring from the representation of a shape the method by which that shape was generated.

Transfer is illustrated, in the case of the 'square' example given above, by the reuse of the method of generating the first edge for all subsequent edges, modified only by the application of transformations. Leyton achieves recoverability by expressing the operations used in generating a shape as sequences of wreath products between symmetry groups. His primary objection to the formulation of geometry in terms of invariants is that geometric invariants are 'memoryless', i.e. that all information about how they were generated has been deliberately erased.

There is a clear parallel here with the use of procedural or construction history representations in CAD. It has already been noted that the explicit availability of the constructional procedure gives a clear advantage in subsequent modification of the resulting shape: if it is known how that shape was initially generated, appropriate modifications can be made to the constructional sequence and the modified shape created by following the revised sequence. In CAD, the modifications are often simple changes of values of dimensional parameters in constructional operations.

However, the two representations proposed for generative geometry are very different. Leytons's group-theoretic representation defines every detail at a very low level, and is not intuitively understandable by a non-mathematician. The proposed STEP approach, on the other hand, allows the use of high-level constructional operations, and the ASCII format of the ISO 10303 transfer file, containing explicit entity names, allows a moderate level of comprehensibility for the non-mathematician. Nevertheless, it should be emphasized that the two forms of representation appear to be equivalent. Leyton claims that CAD operations are all expressible in terms of wreath products between symmetry groups. A major part of his book is devoted to this application; in his research for the book its author actually studied in depth the range of operations provided by several major CAD systems and formulated group-theoretic representations for them. It is possible that some future standardization effort could capture the geometric semantics of CAD operations precisely using this type of approach – at present these semantics are only defined descriptively.

The work of Leyton is interesting in highlighting the virtues of the procedural approach to shape representation. In particular, it strongly confirms the advantages of 'recoverability', or knowledge of how the shape was created, and may therefore be regarded as providing some theoretical underpinning for the work described earlier in the paper.

## 5    Other Geometric Developments in STEP

Before concluding, it is appropriate to mention that the ISO 10303 geometric coverage is currently being extended to cover algebraic surface representation. The initial enhancement will handle general quadric surfaces and appropriate means for controlling them, but more general algebraic surfaces may be added later. The initial intended application is the use of these surfaces to define half-spaces. These will be combined using generalizations of the standard regularized Boolean operations, primarily to create models for visualization and animation.

## 6    Summary and Conclusions

This paper has described a major conceptual extension of the STEP standard (ISO 10303) for the electronic exchange of product data. The emphasis has been on the standardized representation of shape, and in particular on a procedural shape representation expressed in terms of constructional operations rather than static structures. Leyton [15] has shown that a formal mathematical basis exists for representations of this kind, but political and commercial reality dictates that in the short and medium term the standard for data exchange must be founded on a pragmatic selection of constructional procedures defined at a much higher level than Leyton's elementary operations. A tacit assumption will be made that each chosen operation can be formally expressed in Leyton's terms, or in some equivalent terms that are equally well-founded mathematically. This assumption can be tested as time permits, and some rigorously justified formulation could form the basis of a future standard for the exchange of shape information.

As regards the experimental transfers of procedural models referred to in Section 3.5, these have mostly been successful. The primary problems encountered have been concerned with model elements selected from the screen of the sending system as the basis of future modelling operations (e.g., the selection of an edge to be rounded). The question is, how can such elements be identified in the receiving system when the model being transferred is procedural and contains no explicit elements? The approach adopted involves a specialized form of hybrid model; the selected elements are transferred explicitly from the sending system and compared with the elements of an explicit model under reconstruction in the receiving system by evaluation of the transferred procedural history. The efficacy of this solution depends upon the nature of the access allowed by the CAD system vendors to their systems' internal data structures — some systems are much more 'open' in this respect than others, and these are the ones for which the best progress has been demonstrated. Nevertheless, only comparatively simple models have so far been transferred in the reported demonstrations. More complex models, particularly assembly models containing multiple instances of the same part positioned and oriented by the use of transformations, are anticipated to present further challenges. It is possible that recourse must be made to some form of standardized 'persistent naming' mechanism [2,18] to handle such cases. This is a potential source of friction with the CAD vendor companies, because their systems' internal persistent naming strategies are closely guarded commercial secrets. Thus they may not even wish to commit themselves to an agreement that their particular approach to the naming of model elements will map onto any proposed standard approach.

To conclude, CAD systems in general use a procedural or generative representation as their primary means for capturing shape information. CAD technology has evolved towards this situation for good reasons; in the early days, explicit geometric representations were the norm, but they were found hard for designers and application engineers to work with effectively. In his book [15] Leyton has pointed out the virtues of procedural shape representation for a variety of other scientific and engineering purposes. The primary topic of this paper has been the development of a standard means for exchanging procedural shape models between CAD systems. The methodology used is not directly based on Leyton's formulation, but is equivalent to it. There seems to be ample scope for further work on generative approaches to geometry, which could possibly lead in the future to a standard representation for shape that is based on firm mathematical foundations.

## Acknowledgments

The author is grateful for financial support from the US National Institute for Standards and Technology under Contract SB134102C0014.

As leader of the ISO TC184/SC4/WG12 Parametrics Group, the author also gratefully acknowledges major contributions to the work of the group by Bill Anderson (Advanced Technology Institute, USA), Noel Christensen (Honeywell, USA), Ray Goult (LMR Systems, UK), Soonhung Han (KAIST, Korea), Pascal Huau (GOSET, France), Tom Kramer (NIST, USA), Akihiko Ohtaka (Nihon Unisys, Japan), Ullrich Pfeifer-Silberbach (TU Darmstadt, Germany), Guy Pierra (LISI/ENSMA, France), Chia Hui Shih (Pacific STEP, USA), Vijay Srinivasan (IBM/Columbia University, USA) Tony Ranger (Theorem Solutions, UK), and Nobuhiro Sugimura (Osaka Prefectural University, Japan). He would also like to thank the many other people who have assisted the work of the group in various ways.

## References

1. W. Anderson and S. Ansaldi. ENGEN data model: A neutral model to capture design intent. In G. Jacucci, G. J. Olling, K. Preiss, and M. J. Wozny, editors, *CDROM Procs. of IFIP PROLAMAT'98 conference, Trento, Italy, Sept. 1998.* Kluwer Academic Publishers, 1998.

2. V. Capoyleas, X. Chen, and C. M. Hoffmann. Generic naming in generative constraint-based design. *Computer Aided Design*, **28**, 1, 17 – 26, 1996.

3. G.-H. Choi, D.-W. Mun, and S.-H. Han. Exchange of CAD part models based on the macro-parametric approach. *International J. of CADCAM*, **2**, 2, 23 – 31, 2002. (Online at `http://www.ijcc.org`).

4. R. J. Goult. Standards for curve and surface data exchange. In A.D. Bowyer, editor, *Computer-aided Surface Geometry and Design*. Oxford University Press, 1994. (Proc. 4th IMA Conf. on the Mathematics of Surfaces, Bath, England, Sept. 1990).

5. C. M. Hoffmann. *Geometric and Solid Modeling*. Morgan Kaufmann, San Mateo, CA, 1989.

6. J. Hoschek and W. Dankwort, editors. *Parametric and Variational Design*. Teubner-Verlag, Stuttgart, 1994.

7. International Organization for Standardization, Geneva, Switzerland. *ISO 10303-11:1994 – Industrial Automation Systems and Integration – Product Data Representation and Exchange, Part 11 – Description methods: The* EXPRESS *language reference manual*, 1994.

8. International Organization for Standardization, Geneva, Switzerland. *ISO 10303-42:1994 – Industrial Automation Systems and Integration – Product Data Representation and Exchange, Part 42 – Integrated Generic Resources: Geometrical and Topological Representation*, 1994.

9. International Organization for Standardization, Geneva, Switzerland. *ISO 10303:1994 – Industrial Automation Systems and Integration – Product Data Representation and Exchange*, 1994. (The ISO catalogue is at http://www.iso.ch/cate/cat.html; search on 10303 for a listing of parts of the standard).

10. International Organization for Standardization, Geneva, Switzerland. *Industrial Automation Systems and Integration – Product Data Representation and Exchange, Part 55 – Integrated generic resource: Procedural and hybrid representation*, 2002. (Committee Draft of International Standard ISO 10303-55).

11. International Organization for Standardization, Geneva, Switzerland. *Industrial Automation Systems and Integration – Product Data Representation and Exchange, Part 109 – Integrated application resource: Component relationships and assembly constraints for assembly models of products*, 2002. (Committee Draft of International Standard ISO 10303-109).

12. International Organization for Standardization, Geneva, Switzerland. *Industrial Automation Systems and Integration – Product Data Representation and Exchange, Part 108 – Integrated application resource: Parameterization and constraints for explicit geometric product models*, 2003. (Draft International Standard ISO 10303-108).

13. H. Jiang and M. J. Pratt. Exchange of parametric CAD models using ISO 10303 (STEP). In preparation, 2003.

14. M. A. Lachance. Data proliferation in the exchange of surfaces. In R. R. Martin, editor, *The Mathematics of Surfaces II*. Oxford University Press, 1987. (Proc. 2nd IMA Conf. on the Mathematics of Surfaces, Cardiff, Wales, Sept. 1986).

15. M. Leyton. *A Generative Theory of Shape*. Springer-Verlag, Berlin, 2001. (Lecture Notes in Computer Science, Volume LNCS 2145).

16. J. Owen. *STEP: An Introduction*. Information Geometers, Winchester, UK, 2nd edition, 1997.

17. M. J. Pratt. Introduction to ISO 10303 — the STEP standard for product data exchange. *ASME J. Computing & Information Science in Engineering*, **1**, 1, 102 – 103, 2001.
18. S. Raghothama and V. Shapiro. Boundary representation deformation in parametric solid modeling. *ACM Trans. on Computer Graphics*, **17**, 4, 259 – 286, 1998.
19. A. A. G. Requicha. Representations of rigid solids — Theory, methods and systems. *ACM Computing Surveys*, **12**, 437 – 464, 1980.
20. J. J. Shah and M. Mäntylä. *Parametric and Feature-based CAD/CAM*. Wiley, 1995.

# Variable-Free Representation of Manifolds via Transfinite Blending with a Functional Language

Alberto Paoluzzi

Dipartimento di Informatica e Automazione, Università Roma Tre
Via della Vasca Navale 79, 00146 Roma, Italy
paoluzzi@dia.uniroma3.it

**Abstract.** In this paper a variable-free parametric representation of manifolds is discussed, using transfinite interpolation or approximation, i.e. function blending in some functional space. This is a powerful approach to generation of curves, surfaces and solids (and even higher dimensional manifolds) by blending lower dimensional vector-valued functions. Transfinite blending, e.g. used in Gordon-Coons patches, is well known to mathematicians and CAD people. It is presented here in a very simple conceptual and computational framework, which leads such a powerful modeling to be easily handled even by the non mathematically sophisticated user of graphics techniques. In particular, transfinite blending is discussed in this paper by making use of a very powerful and simple functional language for geometric design.

## 1 Introduction

Parametric curves and surfaces, as well as splines, are usually defined [2] as vector-valued functions generated from some vector space of polynomials or rationals (i.e. ratio of polynomials) over the field of real numbers. In this paper it is conversely presented an unified view of curves, surfaces, and multi-variate manifolds as vector-valued functions generated from the same vector spaces, but over the field of polynomial (or rational) functions itself. This choice implies that the *coefficients* of the linear combination which uniquely represents a curved mapping in a certain basis are not real numbers, as usually, but vector-valued functions.

This approach is a strong generalization, which contains the previous ones as very special cases. For example, the well-known approach of Hermite cubic interpolation of curves, where two extreme points and tangents are interpolated, can so be applied to surfaces, where two extreme curves of points are interpolated with assigned derivative curves, or even to volume interpolation of two assigned surfaces with assigned normals. Such an approach is not new, and is quite frequently used in CAD applications, mainly to ship and airplane design, since from the times that Gordon-Coons patches were formulated [3,5]. It is sometime called *function blending* [5,8], or *transfinite interpolation* [6,4]. Transfinite methods have been recently applied to interpolating implicitly defined sets of different cardinality [11].

M.J. Wilson and R.R. Martin (Eds.): Mathematics of Surfaces 2003, LNCS 2768, pp. 338–354, 2003.

Transfinite interpolation, that the author prefers to call *transfinite blending*, because it can also be used for approximation, is quite hard to handle by using standard imperative languages. In particular, it is quite difficult to be abstracted, and too often *ad hoc* code must be developed to handle the specific application class or case. A strong mathematical background is also needed to both implement and handle such kind of software. This fact discouraged the diffusion of such a powerful modeling technique outside the close neighbourhood of automobile, ship and airplane shell design.

The contribution of this paper is both in introducing a general algebraic setting which simplifies the description of transfinite blending by the use of functions without variables, and in embedding such an approach into a modern functional computing environment [10], where functions can be easily multiplied and added exactly as numbers. This results in an amazing descriptive power when dealing with parametric geometry. Several examples of this power are given in the paper. Consider, e.g., that multi-variate transfinite Bezier blending of any degree with both domain and range spaces of any dimension is implemented with few lines of quite readable source code.

Last but not least, in the paper we limited our exposition to Bézier and Hermite cases for sake of space. Actually, the same approach can be applied to any kind of parametric representation of geometry, including splines. Notice in particular that different kinds of curves surfaces and splines can be freely blended, so giving a considerable amount of design freedom.

## 2 Syntax

We use in our discussion the *design language* PLaSM, that is a geometric extension of the functional language FL [1]. We cannot recap the PLaSM syntax here; it is described in [10,9] [1]. In order to understand the examples, it may suffice to remember that the FL programming style is strongly based upon functional *composition*. In particular, the composition operator is denoted by "$\sim$", whereas function *application* is denoted by "$:$". Therefore, the PLaSM translation of the mathematical expression $(f \circ g)(x)$ is:

$$(\texttt{f} \sim \texttt{g}):\texttt{x}$$

Let us just remember that $< \texttt{x}_1, \ldots, \texttt{x}_n >$, where $\texttt{x}_1, \ldots, \texttt{x}_n$ are arbitrary expressions, is a *sequence*, and that language *operators* (i.e., functions) are normally *prefix* to their argument sequence.

A function with $n$ parameter lists is called a function of $n$-th *order*. Such a function, when applied to actual arguments for the first parameter list, returns a function of order $(n-1)$. This one, when further applied to actual arguments for the second parameter list, returns a function of order $(n-2)$, and so on. Finally, when all the parameter lists are *bound* to actual arguments, the function returns

---

[1] The language IDE for *Windows*, *Linux* and *Mac OS X* platforms can be downloaded from http://www.plasm.net/download/.

the value generated by the evaluation of its *body*. Functions of order higher than one are called *higher-order* functions. The functions returned from the application of higher-order functions to some (ordered) *subset* of their parameter lists are called *partial* functions. A higher-order function

$$h : A \to B \to C \equiv h : A \to (B \to C)$$

can be defined in PLaSM, using formal parameters, as

```
DEF h (a::isA)(b::isB) = body_expr
```

where isA and isB are predicates used to run-time test the set-membership of arguments. The function h is applied to actual arguments as h:a:b ≡ (h:a):b, returning[2] a value c ∈ C. Finally, notice that the value generated by evaluating the expression h:a is a *partial* function $h_{\mathtt{a}} : B \to C$.

## 3   Variable-Free Representation

We may define a curve $\boldsymbol{c} : \mathbb{R} \to \mathbb{E}^d$, when a Cartesian system $(\boldsymbol{o}, (\boldsymbol{e}_i))$ is given, as the point-valued mapping $\boldsymbol{c} = \boldsymbol{o} + \boldsymbol{\alpha}$, with $\boldsymbol{\alpha} = (\alpha_i)$, where $\alpha_i : \mathbb{R} \to \mathbb{R}$, for $1 \le i \le d$. Therefore, accordingly with the functional character of our language, we may denote a vector-valued function of one real parameter by using the variable-free notation:

$$\boldsymbol{\alpha} = (\alpha_i)^T$$

with $\alpha_i = \boldsymbol{\alpha} \cdot \boldsymbol{e}_i$, with $1 \le i \le d$. It should be clearly understood that each $\alpha_i$ is here a map $\mathbb{R} \to \mathbb{R}$ and that each $\boldsymbol{e}_i$ has the constant maps $\underline{0} : \mathbb{R} \to 0$ and $\underline{1} : \mathbb{R} \to 1$ as components.

The variable-free notation [7] discussed in this section, where functions are directly added and multiplied, exactly like numbers, is very useful for easily implementing curves and surfaces in our language, that allows for a very direct translation of such a notation.

*Combinators.* Some combinators are used [7] to perform such variable-free calculus with functions. The model of computation supported by combinatory logic, as the study of combinators is known, is reduction or re-writing, where certain rules are used to re-write an expression over and over again, simplifying or reducing it each time, until to get the answer. The same type of reduction using combinators in used by *Mathematica* [12]. The combinators given below have a direct translation in FL and in its geometry-oriented extension PLaSM.

1. $\mathrm{id} : \mathbb{R} \to \mathbb{R} : x \mapsto x$                                              (identity)
2. $\underline{c} : \mathbb{R} \to \mathbb{R} : x \mapsto c$                                            (constant)
3. $\sigma : \{1, \ldots, d\} \times \mathbb{R}^d \to \mathbb{R} : (i, (x_1, \ldots, x_d)) \mapsto x_i$       (selection)

---

[2] Since application is left-associative.

A computer scientist would probably prefer the following specification, just to point out that $\sigma$ is often used as a *partial* function, i.e. a function which may be applied to a subset of its arguments:

3. $\sigma : \{1, \ldots, d\} \to (\mathbb{R}^d \to \mathbb{R}); i \mapsto ((x_1, \ldots, x_d) \mapsto x_i)$      (selection)

Actually, the FL primitives ID, K and SEL used by the PLaSM language for the functions above, respectively, have no domain restrictions, and can be applied to arbitrary types of data objects.

*Algebraic Operations.* We also need to recall how to perform algebraic operations in the algebra of maps $\mathbb{R} \to \mathbb{R}$. In particular, for each map $\alpha : \mathbb{R} \to \mathbb{R} : u \mapsto \alpha(u)$, $\beta : \mathbb{R} \to \mathbb{R} : u \mapsto \beta(u)$ and each scalar $a \in \mathbb{R}$ we have

$$\alpha + \beta : u \mapsto \alpha(u) + \beta(u), \quad \alpha\beta : u \mapsto \alpha(u)\beta(u), \quad a\beta : u \mapsto a\,\beta(u).$$

Consequently, we have that

$$\alpha - \beta : u \mapsto \alpha(u) + (-1)\beta(u), \quad \alpha/\beta : u \mapsto \alpha(u)/\beta(u).$$

*Coordinate Representation.* Finally, remember that the coordinate functions of a curve $\alpha = (\alpha_i)$ are maps $\mathbb{R} \to \mathbb{R}$. The variable-free vector notation stands for the linear combination of coordinate functions with the basis vectors of the target space:

$$(\alpha_1, \ldots, \alpha_d)^T : \mathbb{R} \to \mathbb{R}^d; \; u \mapsto \sum_{i=1}^{d} \alpha_i e_i.$$

*Example 1 (Circular arc).* Some different representations of a circle arc are given here. They have the same image in $\mathbb{E}^2$ but different coordinate representation in the space of functions $\mathbb{R} \to \mathbb{R}$. All such curves generate a circular arc of unit radius centered at the origin.

$$\alpha(u) = \left( \cos\left(\frac{\pi}{2}u\right), \sin\left(\frac{\pi}{2}u\right) \right)^T \quad u \in [0, 1] \tag{1}$$

$$\beta(u) = \left( \frac{1 - u^2}{1 + u^2}, \frac{2u}{1 + u^2} \right)^T \quad u \in [0, 1] \tag{2}$$

$$\gamma(u) = \left( u, \sqrt{1 - u^2} \right)^T \quad u \in [0, 1] \tag{3}$$

Below we give a *variable-free representation* of the three maps on the $[0, 1]$ interval shown in Example 1, that is exactly the representation we need for a PLaSM implementation of such maps, provided in Script 1:

$$\alpha = \left( \cos \circ \left(\frac{\pi}{2}\mathrm{id}\right), \; \sin \circ \left(\frac{\pi}{2}\mathrm{id}\right) \right)^T \tag{4}$$

$$\beta = \left( \frac{1 - \mathrm{id}^2}{1 + \mathrm{id}^2}, \; \frac{2\,\mathrm{id}}{1 + \mathrm{id}^2} \right)^T \tag{5}$$

$$\gamma = \left( \mathrm{id}, \; \mathrm{id}^{\frac{1}{2}} \circ (1 - \mathrm{id}^2) \right)^T \tag{6}$$

## 3.1   Implementation

The circle segment representations of Example 1 are directly used in the PLaSM implementation of curves in Script 1. To understand the implementation, notice that we generate a polyhedral complex by mapping the vector-valued function (either $\alpha$, $\beta$ or $\gamma$) on a cell decomposition of the $[0, 1]$ domain.

According to the semantics of the MAP operator, the curve mapping is applied to all vertices of a simplicial decomposition of a polyhedral domain. But all vertices are represented as *sequences* of coordinates, say <u> for a curve, so that in order to act on $u$ the mapping must necessarily *select* it from the sequence. Hence we might substitute each *id* function instance with the PLaSM denotation S1 for the $\sigma(1)$ selector function. Exactly the same result is obtained by using either $\alpha \circ \sigma(1)$, $\beta \circ \sigma(1)$ or $\gamma \circ \sigma(1)$, as done in the following code.

---

**Script 1 (Circular arc maps)**
```
DEF SQR = ID * ID;
DEF SQRT  = ID ** K:(1/2);

DEF alpha = < cos ~ (K:(PI/2) * ID), sin ~ (K:(PI/2) * ID) >;
DEF beta  = < (K:1 - SQR)/(K:1 + SQR), (K:2 * ID)/(K:1 + SQR) >;
DEF gamma = < ID, SQRT ~ (K:1 - SQR) >;

MAP:(CONS:alpha ~ S1):(interval:<0,1>:10);
MAP:(CONS:beta  ~ S1):(interval:<0,1>:10);
MAP:(CONS:gamma ~ S1):(interval:<0,1>:10);
```

---

*Remarks.* Let us note that, e.g., alpha is a *sequence* of coordinate functions. Conversely, CONS:alpha is the correct implementation of the *vector-valued* function $\alpha$, which only can be *composed* with other functions, say S1.

Notice also that SQR (square), given in Script 1, is the PLaSM implementation of the $id^2$ function and that the language explicitly requires the operator * to denote the product of functions.

Finally, we remark that SQRT (square root), which is actually primitive in PLaSM, can be also defined easily using standard algebraic rules, where ** is the predefined power operator.

*Toolbox.* Some predicates and functions needed by the operators in this chapter are given in Script 2. In particular, the interval operator provides a simplicial decomposition with n elements of the real interval [a, b], whereas the interval2D operator returns a decomposition with n1 × n2 subintervals of the domain [a1, b1] × [a2, b2] $\subset \mathbb{R}^2$.

Few other functions of general utility are used in the remainder. In particular, the predicates IsVect and IsPoint are used to test if a data object is a vector or a point, respectively. Some vector operations are also given above, where AA stands for *apply-to-all* a function to a sequence of arguments, producing the sequence of applications, and TRANS and DISTL respectively stand for *transpose* a sequence of sequences and for *distribute left* a value over a sequence, returning a sequence of pairs, with the value distributed as the left element of each pair.

---

**Script 2 (Toolbox)**

```
DEF interval (a,b::IsReal)(n::IsIntPos) =
    (T:1:a ~ QUOTE ~ #:n):((b-a)/n);
DEF interval2D (a1,a2,b1,b2::IsReal)(n1,n2::IsIntPos) =
    interval:<a1,b1>:n1 * interval:<a2,b2>:n2;

DEF vectsum = AA:+ ~ TRANS;
DEF scalarvectprod = AA:* ~ DISTL;
```

---

*Coordinate Maps.* It should be remembered that curves, as $\alpha$, $\beta$ and $\gamma$ in the previous example, are vector-valued functions. In order to obtain their coordinate maps, say $\alpha_i : \mathbb{R} \to \mathbb{R}$, a composition with the appropriate selector function is needed:

$$\alpha_i = \sigma(i) \circ \alpha$$

The conversion from a 3D vector-valued function   `curve := CONS:alpha` to the sequence of its coordinate functions may be obtained in PLaSM as:

```
< S1 ~ curve, S2 ~ curve, S3 ~ curve >;
```

Such an approach is quite expensive and inefficient, because the curve function is repeatedly evaluated to get its three component functions. So, for the sake of efficiency, we suggest maintaining a coordinate representation as a *sequence* of scalar-valued functions, and then `CONS` it into a single *vector-valued* function only when it is strictly necessary.

## 3.2    Reparametrization

A *smooth curve* is defined as a curve whose coordinate functions are smooth. If $c : I \to \mathbb{E}^d$, with $I \subset \mathbb{R}$, is a smooth curve and $\rho : I \to I$ is a smooth invertible function, i.e. a *diffeomorphism*, then also

$$c_\rho = c \circ \rho$$

is a smooth curve. It is called a *reparametrization* of $c$. A very simple reparametrization is the *change of origin*. For example $c_\rho$ is called a change of origin when

$$\rho = \mathrm{id} + \underline{c}.$$

A reparametrization $c_\tau$ by an affine function

$$\tau = \underline{a}\,\mathrm{id} + \underline{c},$$

with $a \neq 0$, is called an *affine reparametrization*.

*Example 2 (Circle reparametrization).*

The circle with the center in the origin and unit radius may be parametrized on different intervals:

$$c_1(u) = \big(\cos u \; \sin u\big), \qquad\qquad u \in [0, 2\pi]$$
$$c_2(u) = \big(\cos 2\pi u \; \sin 2\pi u\big), \qquad\qquad u \in [0, 1]$$

or with a different starting point:

$$c_3(u) = \big(\cos(2\pi u + \tfrac{\pi}{2}) \; \sin(2\pi u + \tfrac{\pi}{2})\big), \qquad u \in [0, 1]$$

The reparametrization becomes evident if we use a variable-free representation:

$$c_1 = \big(\cos \; \sin\big),$$
$$c_2 = \big(\cos \; \sin\big) \circ (\underline{2\,\pi}\,\mathrm{id}),$$
$$c_3 = \big(\cos \; \sin\big) \circ (\underline{2\,\pi}\,\mathrm{id} + \underline{\pi/2}).$$

*Orientation.* Two curves with the same image can be distinguished by the sense in which the image is traversed for increasing values of the parameter. Two curves which are traversed in the same way are said to have the same *orientation*. Actually, an orientation is an equivalence class of parametrizations.

A *reversed orientation* of a curve $c : \mathbb{R} \to \mathbb{E}^d$, with image $c[a, b]$, is given by the affine reparametrization $c_\lambda = c \circ \lambda$, where $\lambda : \mathbb{R} \to \mathbb{R}$ such that $x \mapsto -x + (a + b)$. This map can be written as:

$$\lambda = \underline{a + b} - \mathrm{id}.$$

*Example 3 (Reversing orientation).*

It is useful to have a PLaSM function REV, which reverses the orientation of any curve parametrized on the interval $[a, b] \subset \mathbb{R}$. The REV function will therefore be abstracted with respect to the bounds a and b, which are real numbers.

In Script 3 we also give a polyhedral approximation of the boundary of the unit circle centered at the origin, as seen from the angle interval $[0, \frac{\pi}{2}]$. The curve is a map from $[0, 1]$ with reversed orientation.

---

**Script 3**

```
DEF REV (a,b::IsReal) = K:(a+b) - ID;
DEF alpha = [ COS, SIN ] ~ (K:(PI/2) * ID);

MAP:(alpha ~ REV:<0,1> ~ S1):(interval:<0,1>:20);
```

---

## 4   Transfinite Methods

*Transfinite blending* stands for interpolation or approximation in *functional spaces*. In this case a bi-variate mapping is generated by blending some uni-variate maps with a suitable basis of polynomials. Analogously, a three-variate mapping is generated by blending some bi-variate maps with a polynomial basis, and so on. To implement transfinite blending with PLaSM consists mainly in using functions without variables that can be combined, e.g. multiplied and added, exactly as numbers.

*Definition.* A *d-variate parametric mapping* is a point-valued polynomial function $\boldsymbol{\Phi} : U \subset X \to Y$ with degree $k$, domain $U$, support $X = \mathbb{R}^d$ and range $Y = \mathbb{E}^n$.

As commonly used in Computer Graphics and CAD, such point-valued polynomials $\boldsymbol{\Phi} = (\Phi_j)_{j=1,\dots,n}$ belong component-wise to the vector space $\mathbb{P}_k[\mathbb{R}]$ of polynomial functions of bounded integer degree $k$ over the $\mathbb{R}$ field.

*Coordinates.* Since the set $\mathbb{P}_k$ of polynomials is also a vector space $\mathbb{P}_k[\mathbb{P}_k]$ over $\mathbb{P}_k$ itself *as a field*, then each mapping component $\Phi_j$, $1 \leq j \leq n$, can be expressed uniquely as a linear combination of $k+1$ basis elements $\phi_i \in \mathbb{P}_k$ with *polynomial coordinates* $\xi_j^i \in \mathbb{P}_k$, so that

$$\Phi_j = \xi_j^0 \phi_0 + \cdots + \xi_j^k \phi_k, \qquad 1 \leq j \leq n.$$

Hence a unique coordinate representation

$$\Phi_j = \left( \xi_j^0, \dots, \xi_j^k \right)^T, \qquad 1 \leq j \leq n$$

of the mapping is given, after a basis $\{\phi_0, \dots, \phi_k\} \subset \mathbb{P}_k$ has been chosen.

*Choice of a Basis.* If the basis elements are collected into a vector $\phi = (\phi_i)$, then it is possible to write:

$$\boldsymbol{\Phi} = \boldsymbol{\Xi} \, \phi, \qquad \phi \in \mathbb{P}_k^{k+1}, \ \boldsymbol{\Phi} \in \mathbb{P}_k^n.$$

where

$$\boldsymbol{\Xi} = \left( \xi_j^i \right), \qquad 1 \leq i \leq n, \ 0 \leq j \leq k.$$

is the coordinate representation of a linear operator in $Lin[n \times (k+1), \mathbb{P}_k]$ that maps the $k+1$ basis polynomials of $k$ degree, into the $n$ polynomials which transform the vectors in the domain $U \subset \mathbb{R}^d$ into the $\mathbb{E}^n$ points of the manifold. A quite standard choice in computer-aided geometric modeling is $U = [0, 1]^d$. The *power* basis, the *cardinal* (or *Lagrange*) basis, the *Hermite* basis, the *Bernstein/Bézier* basis and the *B-spline* basis are the most common and useful choices for the $\phi = (\phi_i)$ basis. This approach can be readily extended to the space $\mathcal{Z}_k$ of rational functions of degree $k$.

*Blending Operator.* The *blending operator* $\boldsymbol{\Xi}$ specializes the maps generated by a certain basis, to fit and/or to approximate a given set of specific data (points, tangent vectors, boundary curves or surfaces, boundary derivatives, and so on). Its coordinate functions $\xi_j^i$ may be easily generated, as will be shown in the following, by either

1. transforming the *geometric handles* of the mapping into vectors of constant functions, in the standard (non-transfinite) case. These are usually points or vectors $\boldsymbol{x}_j = (x_j^i) \in \mathbb{E}^n$ to be interpolated or approximated by the set $\boldsymbol{\Phi}(U)$;
2. assuming directly as $\xi_j^i$ the components of the curve (surface, etc.) maps to be fit or approximated by $\boldsymbol{\Phi}$, in the transfinite case.

*Notation.* For the sake of readability, only greek letters, either capitals or lower-case, are used here to denote functions. Latin letters are used for numbers and vectors of numbers. As usually in this paper, bold letters denote vectors, points or tensors. Please remember that $B$ and $H$ are also Greek capitals for $\beta$ and $\eta$, respectively.

## 4.1   Uni-variate Case

Let consider the uni-variate case $\boldsymbol{\Phi} : U \subset X \to Y$, where the dimension $d$ of support space $X$ is one. To generate the coordinate functions $\xi_j^i$ it is sufficient to transform each data point $\boldsymbol{x}_i = (x_j^i) \in Y$ into a vector of constant functions, so that

$$\xi_j^i = \kappa(x_j^i), \quad \text{where} \quad \kappa(x_j^i) : U \to Y : u \mapsto x_j^i.$$

We remember that, using the functional notation with explicit variables, the *constant* function $\kappa : \mathbb{R} \to (\mathbb{R} \to \mathbb{R})$ is defined such that

$$\kappa(x_j^i)(u) = x_j^i$$

for each parameter value $u \in U$. The PLaSM implementation clearly uses the constant functional K at this purpose.

*Example 4 (Cubic Bézier curve in the plane).*
   The cubic Bézier plane curve depends on four points p0,p1,p2,p3 $\in E^2$, which are given in Script 4 as formal parameters of the function Bezier3, that generates the $\Phi$ mapping by linearly combining the basis functions with the coordinate functions. The local functions b0,b1,b2,b3 implement the cubic Bernstein/Bézier basis functions $\beta_i^3 : \mathbb{R} \to \mathbb{R}$ such that $u \mapsto \binom{3}{i} u^i (1-u)^{3-i}$, $0 \le i \le 3$.

---

**Script 4**

```
DEF Bezier3 (p0,p1,p2,p3::IsSeqOf:isReal) =
    [ (x:p0 * b0) + (x:p1 * b1) + (x:p2 * b2) + (x:p3 * b3),
      (y:p0 * b0) + (y:p1 * b1) + (y:p2 * b2) + (y:p3 * b3) ]
    WHERE
        b0 = u1 * u1 * u1,
        b1 = K:3 * u1 * u1 * u,
        b2 = K:3 * u1 * u * u,
        b3 = u * u * u,
        x = K ~ S1, y = K ~ S2, u1 = K:1 - u, u = S1
    END;
```

---

The x and y functions, defined as composition of a selector with the constant functional K, respectively select the first (second) component of their argument sequence and transform such a number in a constant function. The reader should notice that + and ∗ are used as operators *between functions*.

## 4.2   Multi-variate Case

When the dimension $d$ of the support space $X$ is greater than one, two main approaches can be used to construct a parametric mapping $\Phi$. The first approach is the well-known *tensor-product* method. The second approach, discussed below, is called here *transfinite blending*.

*Transfinite Blending.* Let consider a polynomial mapping $\Phi : U \to Y$ of degree $k_d$, where $U$ is $d$-dimensional and $Y$ is $n$-dimensional. Since $\Phi$ depends on $d$ parameters, in the following will be denoted as $^d\Phi = (^d\Phi_j)$, $1 \le j \le n$. In this case $^d\Phi$ is computed component-wise by linear combination of coordinate maps, depending on $d-1$ parameters, with the *uni-variate* polynomial basis $\phi = (\phi_i)$ of degree $k_d$. Formally we can write:

$$^d\Phi_j = {}^{d-1}\Phi_j^0 \, \phi_0 + \cdots {}^{d-1}\Phi_j^{k_d} \, \phi_{k_d}, \qquad 1 \le j \le n.$$

The coordinate representation of $^d\Phi_j$ with respect to the basis $(\phi_0, \ldots, \phi_{k_d})$ is so given by $k_d + 1$ maps depending on $d-1$ parameters:

$$^d\Phi_j = \left( {}^{d-1}\Phi_j^0, \ldots, {}^{d-1}\Phi_j^{k_d} \right).$$

In matrix notation, after a polynomial basis $\phi$ has been chosen, it is

$$^d\Phi = \Xi \, \phi, \quad \text{where} \quad \Xi = ({}^{d-1}\Phi_j^i), \quad \phi = (\phi_i), \quad 0 \le i \le k_d, \ 1 \le j \le n.$$

*Example.* As an example of transfinite blending consider the generation of a bicubic Bézier surface mapping $B(u_1, u_2)$ as a combination of four Bézier cubic curve maps $B_i(u_1)$, with $0 \le i \le 3$, where some curve maps may possibly reduce to a constant point map:

$$B(u_1, u_2) = \sum_{i=0}^{3} B_i(u_1) \, \beta_i^3(u_2)$$

where

$$\beta_i^3(u) = \binom{3}{i} u^i (1-u)^{3-i}, \qquad 0 \le i \le 3,$$

is the Bernstein/Bézier cubic basis. Analogously, a three-variate Bézier solid body mapping $B(u_1, u_2, u_3)$, of degree $k_3$ on the last parameter, may be generated by uni-variate Bézier blending of surface maps $B_i(u_1, u_2)$, some of which possibly reduced to a curve map or even to a constant point map:

$$B(u_1, u_2, u_3) = \sum_{i=0}^{k_3} B_i(u_1, u_2) \, \beta_i^{k_3}(u_3)$$

*Note.* The more interesting aspects of transfinite blending are *flexibility* and *simplicity*. Conversely than in tensor-product method, there is no need that all component geometries have the same degree, and even neither that were all generated using the same function basis. For example, a quintic Bézier surface map may be generated by blending both Bézier curve maps of lower (even zero) degree together with Hermite and Lagrange curve maps. Furthermore, it is much simpler to combine lower dimensional geometries (i.e. maps) than to meaningfully assembly the multi-index tensor of control data (i.e. points and vectors) to generate multi-variate manifolds with tensor-product method.

## 4.3   Transfinite Bézier

The full power of the PLaSM functional approach to geometric programming is used in this section, where dimension-independent transfinite Bézier blending of any degree is implemented in few lines of code, by easily combining coordinate maps which may depend on an arbitrary number of parameters.

We remark that the `Bezier` : $[0,1]^d \to \mathbb{E}^n$ mapping given here can be used:

1. to blend points to give curve maps;
2. to blend curve maps to give surface maps;
3. to blend surface maps to give solid maps;
4. and so on ...

Notice also that the given implementation is independent on the dimensions $d$ and $n$ of support and range spaces.

*Implementation.* At this purpose, first a small toolbox of related functions is needed, to compute the factorial function, the binomial coefficients, the $\beta^k = (\beta_i^k)$ Bernstein basis of degree $k$, and the $\beta_i^k$ Bernstein/Bézier polynomials.

---

**Script 5 (Transfinite Bézier toolbox)**
```
DEF Pred = ID - K:1;
DEF Fact (n::IsNat) = IF:< C:EQ:0, K:1, * ~ INTSTO >;
DEF Choose (n,k::IsNat) = IF:<
    OR ~ [C:EQ:0 ~ S2, EQ], K:1, Choose ~ AA:Pred * / >:< n,k >;
DEF Bernstein (u::IsFun)(n::IsInt)(i::IsInt) =
    ~ [K:(Choose:<n,i>),** ~ [ID,K:i],
      ** ~ [- ~ [K:1,ID],K:(n-i)]] ~ u;
DEF BernsteinBasis (u::IsFun)(n::IsInt) = AA:(Bernstein:u:n):(0..n);
```

---

Then the `Bezier:u` function is given, to be applied on the sequence of `Data`, which may contain either control points $\boldsymbol{x}_i = (x_j^i)$ or control maps $^{d-1}\boldsymbol{\Phi}_i = (^{d-1}\Phi_j^i)$, with $0 \leq i \leq k$, $1 \leq j \leq n$. In the former case each component $x_j^i$ of each control point is firstly transformed into a constant function.

The body of the `Bezier:u` function just linearly combines component-wise the sequence $(\xi_j^i)$ of coordinate maps generated by the expression

```
(TRANS~fun):Data
```

with the basis sequence $(\beta_i^k)$ generated by `BernsteinBasis:u:degree`, where `degree` equates the number of geometric handles minus one.

---

**Script 6 (Dimension-independent transfinite Bézier mapping)**

```
    DEF Bezier (u::IsFun) (Data::IsSeq) = (AA:InnerProd ~ DISTR):
        < (fun ~ TRANS):Data, BernsteinBasis:u:degree >
    WHERE
        degree = LEN:Data - 1,
        fun = (AA ~ AA):(IF:< IsFun, ID, K >)
    END;
```

---

It is much harder to explain in few words what actual argument to pass (and why) for the formal parameter u of the `Bezier` function. As a rule of thumb let pass either the selector S1 if the function must return a uni-variate (curve) map, or S2 to return a bi-variate (surface) map, or S3 to return a three-variate (solid) map, and so on.

*Example 5 (Bézier curves and surface).*

Four Bézier $[0, 1] \to \mathbb{E}^3$ maps C0, C1, C2, and C3, respectively of degrees $1, 2, 3$ and 2 are defined in Script 7.

It may be useful to notice that the control points have three coordinates, so that the generated maps C0, C1, C2 and C3 will have three co-ordinate functions. Such maps can be blended with the Bernstein/Bézier basis to produce a cubic transfinite bi-variate (i.e. surface) mapping:

$$B(u_1, u_2) = \text{C0}(u_1)\beta_0^3(u_2) + \text{C1}(u_1)\beta_1^3(u_2) + \text{C2}(u_1)\beta_2^3(u_2) + \text{C3}(u_1)\beta_3^3(u_2).$$

Such a linear combination of coordinate functions with the Bézier basis (here working on the second coordinate of points in $[0, 1]^2$) is performed by the PLaSM function `Surf1`, defined by using again the `Bezier` function.

A simplicial approximation (with triangles) of the surface $B[0, 1]^2 \subset \mathbb{E}^3$ is finally generated by evaluating the last expression of Script 7.

---

**Script 7**

```
    DEF C0 = Bezier:S1:<<0,0,0>,<10,0,0>>;
    DEF C1 = Bezier:S1:<<0,2,0>,<8,3,0>,<9,2,0>>;
    DEF C2 = Bezier:S1:<<0,4,1>,<7,5,-1>,<8,5,1>,<12,4,0>>;
    DEF C3 = Bezier:S1:<<0,6,0>,<9,6,3>,<10,6,-1>>;

    DEF Surf1 = Bezier:S2:<C0,C1,C2,C3>;

    MAP:Surf1:(Intervals:1:20 * Intervals:1:20);
```

---

According to the semantics of the MAP operator, `Surf1` is applied to all vertices of the automatically generated simplicial decomposition $\Sigma$ of the 2D product (`Intervals` : 1 : 20 * `Intervals` : 1 : 20) $\subset \mathbb{R}^2$. A simplicial approximation

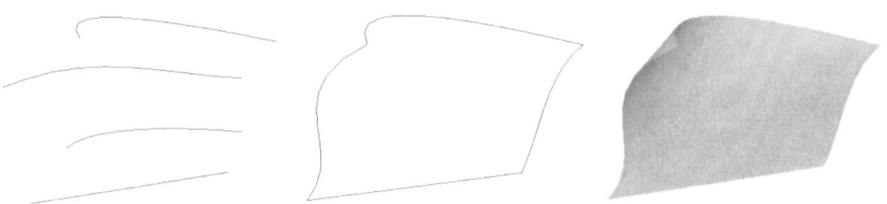

**Fig. 1.** (a) Graphs of four Bézier curve maps c0, c1, c2 and c3; (b) graphs of c0 and c3 together with graphs of bicubic maps b0 and b1 generated by extreme control points; (c) graph of Surf1 surface

$B(\Sigma)$ of the surface $B([0,1]^2) \subset \mathbb{E}^3$ is finally produced and displayed in Figure 1c.

The four generating curves and the generated cubic blended surface are displayed in Figures 1a. It is possible to see (Figure 1b) that such surface interpolates the four boundary curves defined by the extreme control points, exactly as in the case of tensor-product method, but obviously with much greater generality, since any defining curve may be of any degree (and actually of any different type).

## 4.4   Transfinite Hermite

The cubic Hermite uni-variate map is the unique cubic polynomial $\boldsymbol{H} : [0,1] \to \mathbb{E}^n$ which matches two given points $p_0, p_1 \in \mathbb{E}^n$ and derivative vectors $t_0, t_1 \in \mathbb{R}^n$ for $u = 0, 1$ respectively. Let denote as $\eta^3 = \left(\eta_0^3, \eta_1^3, \eta_2^3, \eta_3^3\right)$ the cubic Hermite function basis, with

$$\eta_i^3 : [0,1] \to \mathbb{R}, \qquad 0 \le i \le 3,$$

and such that

$$\eta_0^3(u) = 2u^3 - 3u^2 + 1, \quad \eta_1^3(u) = 3u^2 - 2u^3, \quad \eta_2^3(u) = u^3 - 2u^2 + u, \quad \eta_3^3(u) = u^3 - u^2.$$

Then the mapping $\boldsymbol{H}$ can be written, in vector notation, as

$$\begin{aligned}
\boldsymbol{H} &= \boldsymbol{\xi}_0\, \eta_0^3 + \boldsymbol{\xi}_1\, \eta_1^3 + \boldsymbol{\xi}_2\, \eta_2^3 + \boldsymbol{\xi}_3\, \eta_3^3 \\
&= \kappa(\boldsymbol{p}_0)\, \eta_0^3 + \kappa(\boldsymbol{p}_1)\, \eta_1^3 + \kappa(\boldsymbol{t}_0)\, \eta_2^3 + \kappa(\boldsymbol{t}_1)\, \eta_3^3.
\end{aligned}$$

It is easy to verify, for the uni-variate case, that:

$$\begin{aligned}
\boldsymbol{H}(0) &= \kappa(\boldsymbol{p}_0)(0) = \boldsymbol{p}_0, & \boldsymbol{H}(1) &= \kappa(\boldsymbol{p}_1)(1) = \boldsymbol{p}_1, \\
\boldsymbol{H}'(0) &= \kappa(\boldsymbol{t}_0)(0) = \boldsymbol{t}_0, & \boldsymbol{H}'(1) &= \kappa(\boldsymbol{t}_1)(1) = \boldsymbol{t}_1,
\end{aligned}$$

and that the image set $\boldsymbol{H}[0,1]$ is the desired curve in $\mathbb{E}^n$.

A multi-variate transfinite Hermite map $\boldsymbol{H}^n$ can be easily defined by allowing the blending operator $\boldsymbol{\Xi} = (\boldsymbol{\xi}_j) = (\xi_j^i)$ to contain maps depending at most on $d-1$ parameters.

*Implementation.* A transfinite `CubicHermite` mapping is implemented here, with four data objects given as formal parameters. Such data objects may be either points/vectors, i.e. sequences of numbers, or $1/2/3/d$-variate maps, i.e. sequences of (curve/surface/solid/etc) component maps, or even mixed sequences, as will be shown in the following examples.

---

**Script 8 (Dimension-independent transfinite cubic Hermite)**

```
DEF fun = (AA ~ AA):(IF:<IsFun,ID,K>);

DEF HermiteBasis (u::IsFun) = < h0,h1,h2,h3 >
WHERE
    h0 = k:2 * u3 - k:3 * u2 + k:1,
    h1 = k:3 * u2 - k:2 * u3,
    h2 = u3 - k:2 * u2 + u,
    h3 = u3 - u2, u3 = u*u*u, u2 = u*u
END;

DEF CubicHermite (u::IsFun) (p1,p2,t1,t2::IsSeq) =
    (AA:InnerProd ~ DISTL):
        < HermiteBasis:u, (TRANS ~ fun):<p1,p2,t1,t2> >;
```

---

## 4.5   Connection Surfaces

The creation of surfaces smoothly connecting two given curves with assigned derivative fields by cubic transfinite blending is discussed in this section. The first applications concern the generation of surfaces in 3D space, the last ones concern the generation of planar grids as 1-skeletons of 2D surfaces, according to the dimension-independent character of the given PLaSM implementation of transfinite methods.

The curve maps $c1(u)$ and $c2(u)$ of Example 8 are here interpolated in 3D by a `Surf2` mapping using the cubic Hermite basis $\eta^3 = (\eta_j^3)$, $0 \leq j \leq 3$, with the further constraints that the tangent vector field $\mathtt{Surf2}^v(u,0)$ along the first curve are constant and parallel to $(0,0,1)$, whereas $\mathtt{Surf2}^v(u,v)$ along the second curve is also constant and parallel to $(0,0,-1)$. The resulting map

$$\mathtt{Surf2} : [0,1]^2 \to I\!E^3$$

has unique vector representation in $I\!P_3^3[I\!P_3]$ as

$$\mathtt{Surf2} = \mathtt{c1}\ \eta_0^3 + \mathtt{c2}\ \eta_1^3 + (\kappa(0),\kappa(0),\kappa(1))\ \eta_2^3 + (\kappa(0),\kappa(0),\kappa(-1))\ \eta_3^3.$$

*Example 6 (Surface interpolation of curves).*

Such a map is very easily implemented by the following PLaSM definition. A simplicial approximation $\mathtt{Surf2}(\Sigma)$ of the point-set $\mathtt{Surf2}([0,1]^2)$ is generated by the MAP expression in Script 9, and is shown in Figure 2.

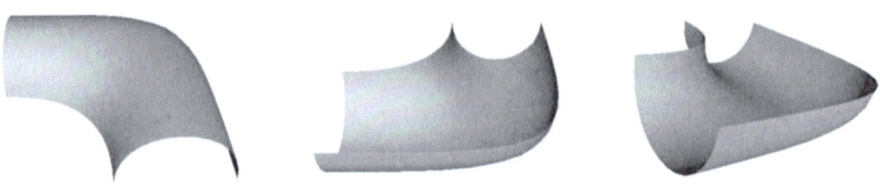

**Fig. 2.** Some pictures of the surface interpolating two plane Hermite curves with constant vertical tangent vectors along the curves

**Fig. 3.** Some pictures of a new surface interpolating the same Hermite curves with constant oblique tangent vectors

---

**Script 9 (Surface by Hermite's interpolation (1))**
```
DEF Surf2 = CubicHermite:S2:< c1,c2,<0,0,1>,<0,0,-1> >;
MAP: Surf2: (Domain:14 * Domain:14);
```

---

*Example 7 (Surface interpolation of curves).*
A different surface interpolation of the two plane curves c1 and c2 is given in Script 10, where the boundary tangent vectors are constrained to be constant and parallel to $(1,1,1)$ and $(-1,-1,-1)$, respectively. Some pictures of the resulting surface are given in Figure 3.

*Example 8 (Grid generation).* Two planar Hermite curve maps c1 and c2 are defined, so that the curve images c1$[0,1]$ and c2$[0,1]$.

Some different grids are easily generated from the plane surface which interpolates the curves c1 and c2. At this purpose it is sufficient to apply the CubicHermite:S2 function to different tangent curves.

The grids generated by maps grid1, grid2 and grid3 are shown in Figure 4. The tangent map d is simply obtained as component-wise difference of the curve maps c2 and c1.

It is interesting to notice that the map grid1 can be also generated as linear (transfinite) Bézier interpolation of the two curves, as given below. Clearly the solution as cubic Hermite is more flexible, as it is shown by Figures 4b and 4c.

```
DEF grid1 = Bezier:S2:<c1,c2>;
MAP:(CONS:grid1):(Domain:8 * Domain:8);
```

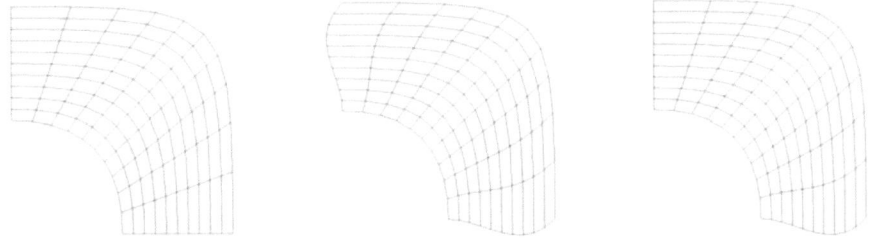

**Fig. 4.** The simplicial complexes generated by the `MAP` operator on the `grid1`, `grid2` and `grid3` maps given in Example 8

---

**Script 10 (Surface by Hermite's interpolation (2))**
```
DEF Surf3 = CubicHermite:S2:<c1,c2,<1,1,1>,<-1,-1,-1>>;
MAP: Surf3: (Domain:14 * Domain:14);
```

---

**Script 11 (Examples of planar grids)**
```
DEF c1 = CubicHermite:S1:<<1,0>,<0,1>,<0,3>,<-3,0>>;
DEF c2 = CubicHermite:S1:<<0.5,0>,<0,0.5>,<0,1>,<-1,0>>;
DEF d = (AA:- ~ TRANS):<c2,c1>;

DEF grid1 = CubicHermite:S2:<c1,c2,d,d>;
DEF grid2 = CubicHermite:S2:<c1,c2,<-0.5,-0.5>,d>;
DEF grid3 = CubicHermite:S2:<c1,c2,<S1:d,-0.5>,d>;
```

---

# References

1. BACKUS, J., WILLIAMS, J., AND WIMMERS, E. An introduction to the programming language FL. In *Research Topics In Functional Programming*, D. Turner, Ed. Addison-Wesley, Reading, Massachusetts, 1990, ch. 9, pp. 219–247.

2. BARTELS, R., BEATTY, J., AND BARSKY, B. *An Introduction to Splines for Use in Computer Graphics & Geometric Modeling*. Morgan Kaufmann, Los Altos, CA, 1987.

3. COONS, S. Surfaces for computer-aided design of space forms. Tech. Rep. MAC-TR-41, MIT, Cambridge, MA, 1967.

4. GOLDMAN, R. The role of surfaces in solid modeling. In *Geometric Modeling: Algorithms and New Trends*, G. Farin, Ed. SIAM Publications, Philadelphia, Pennsylvania, 1987.

5. GORDON, W. Blending function methods of bivariate and multivariate interpolation and approximation. Tech. Rep. GMR-834, General Motors, Warren, Michigan, 1968.

6. GORDON, W. Spline-blended surface interpolation through curve networks. *Journal of Mathematical Mechanics 18* (1969), 931–952.

7. JONES, A., GRAY, A., AND HUTTON, R. *Manifolds and Mechanics*. No. 2 in Australian Mathematical Society Lecture Series. Cambridge University Press, Cambridge, UK, 1987.

8. LANCASTER, P., AND SALKAUSKAS, K. *Curve and Surface Fitting. An Introduction*. Academic Press, London, UK, 1986.

9. PAOLUZZI, A. *Geometric programming for computer aided design*. J. Wiley & Sons, Chichester, UK, 2003.

10. PAOLUZZI, A., PASCUCCI, V., AND VICENTINO, M. Geometric programming: a programming approach to geometric design. *ACM Transactions on Graphics 14*, 3 (July 1995), 266–306.

11. RVACHEV, V. L., SHEIKO, T. I., SHAPIRO, V., AND TSUKANOV, I. Transfinite interpolation over implicitly defined sets. *Computer Aided Geometric Design 18* (2001), 195–220.

12. WOLFRAM, S. *A new kind of science*. Wolfram Media, Inc., 2002.

# Modified Affine Arithmetic Is More Accurate than Centered Interval Arithmetic or Affine Arithmetic

Huahao Shou[1,2], Hongwei Lin[1], Ralph Martin[2], and Guojin Wang[1]

[1] State Key Laboratory of CAD & CG, Zhejiang University, Hangzhou, China
[2] Department of Applied Mathematics, Zhejiang University of Technology
Hangzhou, China
[3] Department of Computer Science, Cardiff University, Cardiff, UK
Ralph.Martin@cs.cf.ac.uk

**Abstract.** In this paper we give mathematical proofs of two new results relevant to evaluating algebraic functions over a box-shaped region: (i) using interval arithmetic in centered form is always more accurate than standard affine arithmetic, and (ii) modified affine arithmetic is always more accurate than interval arithmetic in centered form. Test results show that modified affine arithmetic is not only more accurate but also much faster than standard affine arithmetic. We thus suggest that modified affine arithmetic is the method of choice for evaluating algebraic functions, such as implicit surfaces, over a box.

## 1 Introduction

Affine arithmetic (AA) was first introduced by Comba and Stolfi in 1993 [3] as an improvement to interval arithmetic (IA). When used for finding the range of a multivariate polynomial function over a box, AA can be more resistant to over-conservatism due to its ability to keep track of correlations between various quantities, doing so using a linear series of "noise terms".

AA has been successfully applied as a replacement for IA in many geometric and computer graphics applications such as surface intersection [5], adaptive enumeration of implicit surfaces [6], ray tracing procedural displacement shaders [7], sampling procedural shaders [8], ray casting implicit surfaces [4], linear interval estimations for parametric objects [2] and in a CSG geometric modeller [1].

However, standard AA still has an over-conservatism problem because it uses an imprecise approximation in the multiplication of two affine forms, and we have shown how it can be further improved to give so-called modified affine arithmetic (MAA). MAA uses a matrix form for bivariate polynomial evaluation which keeps all noise terms without any unprecise approximation [9,11]. Of course, this more precise MAA involves more complex formulas.

In practical applications such as curve drawing, typically recursive methods are used to locate the curve. The extra accuracy provided by MAA means that fewer recursive calls are likely to be needed—some regions of the plane can be

M.J. Wilson and R.R. Martin (Eds.): Mathematics of Surfaces 2003, LNCS 2768, pp. 355–365, 2003.

rejected as not containing the curve when using MAA which would need further subdivision when using ordinary AA. However, the amount of work to be done for each recursive call is greater for MAA than AA, and so it is still not clear whether MAA's advantage in accuracy is worth the extra cost and complexity in terms of overall algorithm performance. We thus give an empirical comparison between standard AA and MAA used in a curve drawing algorithm, as well as a theoretical proof that over a given interval, MAA is more accurate. Our test results in Section 3 show that MAA is not only more accurate but also much faster than standard AA in a curve drawing application.

We also demonstrate that MAA is actually the same as interval arithmetic on the centered form (IAC) [10] together with a consideration of the properties of even or odd powers of polynomial terms. In detail, we show in Section 2 that IAC is always more accurate than standard AA for polynomial evaluation, and that the MAA method is always somewhat more accurate than the IAC method. Overall, we conclude that the MAA method is better than the IAC method, and the IAC method is better than the standard AA method. These results hold in one, two and three dimensions.

The subdivision quadtree based implicit curve plotting algorithm described in [9] can be easily generalized to a subdivision octree based implicit surface plotting algorithm. The MAA in matrix form method proposed in [11] for bivariate polynomial evaluation and algebraic curve plotting problem can also be readily generalized to an MAA in tensor form method for trivariate polynomial evaluation and algebraic surface plotting. Thus, a 3D subdivision based algebraic surface plotting method and the 2D subdivision based algebraic curve plotting method have many similarities. Although further experiments are needed, due to these similarities we believe that the experimental conclusions we draw from 2D algebraic curve plotting problem are also applicable to the 3D algebraic surface plotting problem.

This paper is organized as follows. In Section 2 we theoretically prove two new results: that IAC is more accurate than AA, and MAA is more accurate than IAC, for evaluation of multivariate polynomials over a box-shaped region. In Section 3 we use some examples to test whether the MAA method is more efficient than the AA method when used in a practical curve drawing application. Finally in Section 4 we give some conclusions.

## 2   Why MAA Is More Accurate than AA

In this Section we prove theoretically two new results: that IAC is more accurate than AA, and MAA is more accurate than IAC. For definitions and explanations of how to evaluate functions using IAC, AA and MAA, see [9].

**Theorem 1:** IAC is more accurate than AA for bounding the range of a polynomial.

**Proof:** We only prove the theorem here the one dimensional case to avoid much more complex formulae needed in the 2D and 3D cases. However, the same basic idea works in all dimensions. Suppose we wish to find bounds on the range of $f(x)$ over some interval $x \in [\underline{x}, \overline{x}]$.

Let

$$f(x) = \sum_{i=0}^{n} a_i x^i.$$

Let $\hat{x} = x_0 + x_1 \varepsilon_1$ be the affine form of the interval $[\underline{x}, \overline{x}]$, where $\varepsilon_1$ is the noise symbol whose value is unknown but is assumed to be in the range $[-1, 1]$, $x_0 = (\underline{x} + \overline{x})/2$, and $x_1 = (\overline{x} - \underline{x})/2 > 0$.

Using AA we may write:

$$f(\hat{x}) = \sum_{i=0}^{n} a_i x_0^i + \sum_{i=1}^{n} i a_i x_0^{i-1} x_1 \varepsilon_1 + \sum_{k=2}^{n} (\sum_{i=k}^{n} a_i x_0^{i-k}) x_1 [(|x_0| + x_1)^{k-1} - |x_0|^{k-1}] \varepsilon_k,$$

$$(1)$$

where $\varepsilon_k, k = 2, 3, \cdots, n$ are also noise symbols whose values are assumed to be in the range $[-1, 1]$.

Therefore the upper bound of $f(\hat{x})$ computed using AA is:

$$\overline{x}_{AA} = \sum_{i=0}^{n} a_i x_0^i + |\sum_{i=1}^{n} i a_i x_0^{i-1}| x_1 + \sum_{k=2}^{n} |\sum_{i=k}^{n} a_i x_0^{i-k}| x_1 [(|x_0| + x_1)^{k-1} - |x_0|^{k-1}]$$

$$= \sum_{i=0}^{n} a_i x_0^i + |\sum_{i=1}^{n} i a_i x_0^{i-1}| x_1 + \sum_{l=2}^{n} (\sum_{k=l}^{n} C_{k-1}^{l-1} |x_0|^{k-l} |\sum_{i=k}^{n} a_i x_0^{i-k}|) x_1^l$$

Using IAC we may write:

$$f(\hat{x}) = \sum_{i=0}^{n} a_i x_0^i + \sum_{i=1}^{n} i a_i x_0^{i-1} x_1 \varepsilon_1 + \sum_{k=2}^{n} (\sum_{i=k}^{n} a_i C_i^k x_0^{i-k}) x_1^k \varepsilon_1^k \quad (2)$$

Therefore the upper bound of $f(\hat{x})$ computed using IAC is:

$$\overline{x}_{IAC} = \sum_{i=0}^{n} a_i x_0^i + |\sum_{i=1}^{n} i a_i x_0^{i-1}| x_1 + \sum_{k=2}^{n} |\sum_{i=k}^{n} a_i C_i^k x_0^{i-k}| x_1^k$$

$$= \sum_{i=0}^{n} a_i x_0^i + |\sum_{i=1}^{n} i a_i x_0^{i-1}| x_1 + \sum_{l=2}^{n} |\sum_{k=l}^{n} C_{k-1}^{l-1} x_0^{k-l} (\sum_{i=k}^{n} a_i x_0^{i-k})| x_1^l$$

Since the first term and the second term of $\overline{x}_{AA}$ and $\overline{x}_{IAC}$ are the same, while for the third term it always holds that

$$|\sum_{k=l}^{n} C_{k-1}^{l-1} x_0^{k-l} (\sum_{i=k}^{n} a_i x_0^{i-k})| \leq \sum_{k=l}^{n} C_{k-1}^{l-1} |x_0|^{k-l} |\sum_{i=k}^{n} a_i x_0^{i-k}|,$$

we thus obtain that $\overline{x}_{IAC} \leq \overline{x}_{AA}$.

In a similar way we can prove that the lower bounds of AA and IAC satisfy $\underline{x}_{IAC} \geq \underline{x}_{AA}$.

Therefore we have proved that the bounds provided by IAC are more accurate than those provided by AA when evaluating a univariate polynomial over a range.

In addition, we can clearly see from equations (1) and (2) that the expression which must be evaluated in AA is actually more complicated and contains more arithmetic operations than the corresponding expression in IAC. We therefore conclude that IAC is always to be preferred to AA for polynomial evaluation.

**Theorem 2:** MAA is more accurate than IAC for bounding the range of a polynomial.

**Proof:** We only prove the theorem here in the 2D case. The proof is similar in the 1D and 3D cases. Let

$$f(x,y) = \sum_{i=0}^{n} \sum_{j=0}^{m} a_{ij} x^i y^j, \quad (x,y) \in [\underline{x}, \overline{x}] \times [\underline{y}, \overline{y}].$$

Let $\hat{x} = x_0 + x_1 \varepsilon_x$, $\hat{y} = y_0 + y_1 \varepsilon_y$ be the affine forms of the intervals $[\underline{x}, \overline{x}]$ and $[\underline{y}, \overline{y}]$ respectively, where $\varepsilon_x, \varepsilon_y$ are noise symbols whose values are unknown but are assumed to be in the range $[-1, 1]$, $x_0 = (\underline{x} + \overline{x})/2$, $x_1 = (\overline{x} - \underline{x})/2 > 0$, and $y_0 = (\underline{y} + \overline{y})/2$, $y_1 = (\overline{y} - \underline{y})/2 > 0$. Let

$$f(\hat{x}, \hat{y}) = \sum_{i=0}^{n} \sum_{j=0}^{m} d_{ij} \varepsilon_x^i \varepsilon_y^j$$

be the centered form of the polynomial.

Using MAA, the upper bound for the range of the function over this region is

$$\overline{x}_{MAA} = d_{00} + \sum_{j=1}^{m} \left\{ \begin{array}{ll} \max(0, d_{0j}), & \text{if } j \text{ is even} \\ |d_{0j}|, & \text{otherwise} \end{array} \right\}$$
$$+ \sum_{i=1}^{n} \sum_{j=0}^{m} \left\{ \begin{array}{ll} \max(0, d_{ij}), & \text{if } i, j \text{ are both even} \\ |d_{ij}|, & \text{otherwise} \end{array} \right\}.$$

Using IAC, the upper bound for the range of the function over this region is

$$\overline{x}_{IAC} = d_{00} + \sum_{j=1}^{m} |d_{0j}| + \sum_{i=1}^{n} \sum_{j=0}^{m} |d_{ij}|.$$

Since it always holds that $\max(0, d_{0j}) \le |d_{0j}|$ and $\max(0, d_{ij}) \le |d_{ij}|$ we get that $\overline{x}_{MAA} \le \overline{x}_{IAC}$.

in a similar way, we can prove that the lower bounds obtained using MAA and IAC satisfy $\underline{x}_{MAA} \ge \underline{x}_{IAC}$.

Therefore we have proved that MAA provides more accurate bounds on the range of a bivariate polynomial over a rectangular region than does IAC.

The weak point of standard AA is that it uses a new noise symbol with a conservative coefficient to replace the quadratic term generated when multiplying two affine forms. This error due to conservativism is accumulated and magnified during long chains of multiplication operations, resulting in an "error explosion" problem, well known to also arise in standard IA. Thus, while standard AA is aimed at reducing this tendency of IA, and does so to some extent as shown in the examples in Section 3, it is possible to better with MAA.

As proved above, the MAA method provides more accurate bounds on a polynomial function over a range than the standard AA method. Whether the more accurate MAA method is also faster than the standard AA in such algorithms as one for recursive curve plotting over a region is not so obvious—while less subdivision is needed as some parts of the plane can be discarded sooner due to the higher accuracy of MAA, the amount of computation needed for each range evaluation is greater using MAA. Whether the advantages outwiegh the disadvantages must be determined by experiment. In the following section we give some examples to see what happens when AA and MAA are applied to the same bivariate polynomial evaluation and subdivision based algebraic curve plotting problem. Also see [9] for further results of this kind.

## 3    Experimental Comparison of AA and MAA for Curve Plotting

In this section we use ten carefully chosen example curves to compare the relative speed of AA and MAA, and to confirm the theoretical results concerning relative accuracy. Each example consists of plotting an implicit curve $f(x, y) = 0$ using the algorithm given in [9] on a grid of $256 \times 256$ pixels. We recursively compute whether a region can contain the curve by computing a bound on the range of the function over the region. If the range does not contain zero, the curve cannot be present in the region, and is discarded. If the range does contain zero, the region is subdivided in $x$ and $y$, and retested. We continue down to $1 \times 1$ regions, which are plotted in black if they still potentially contain the curve. We used Visual C++ 6.0 running on Windows 2000 on a Pentium IV 2.00GHz computer with 512MB RAM for all the tests.

Overall, these examples were chosen to illustrate curves of varying polynomial degree, with differing numbers of both open and closed components, and include cusps, self-intersections and tangencies as special cases. Obviously, no finite set of test cases can establish universal truths, but we have aimed to capture a range of curve behaviour with these test cases, to at least give some hope that any conclusions we draw are relevant to many practical cases of interest.

We not only show the generated graphical output for these examples, but also present in tabular form an analysis of accuracy and computational load for each example. When comparing the performance and efficiency of AA and MAA methods, a number of quantities were measured:

- The number of pixels plotted, the fewer the better: plotted pixels may or may not contain the curve in practice.
- The CPU time used, the less the better.
- The number of subdivisions involved, the fewer the better.

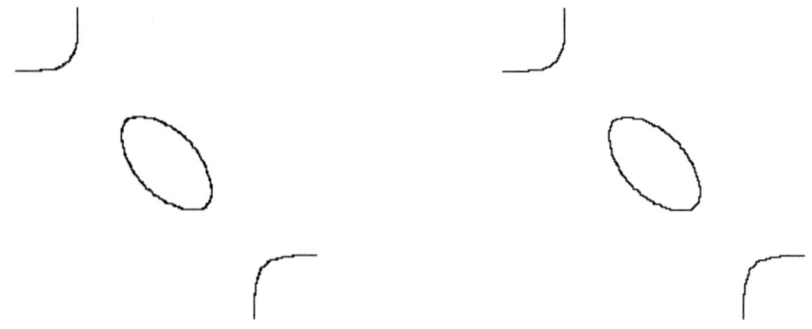

**Fig. 1.** Example 1. $\frac{15}{4} + 8x - 16x^2 + 8y - 112xy + 128x^2y - 16y^2 + 128xy^2 - 128x^2y^2 = 0$, plotted using AA.

**Fig. 2.** Example 1. $\frac{15}{4} + 8x - 16x^2 + 8y - 112xy + 128x^2y - 16y^2 + 128xy^2 - 128x^2y^2 = 0$, plotted using MAA.

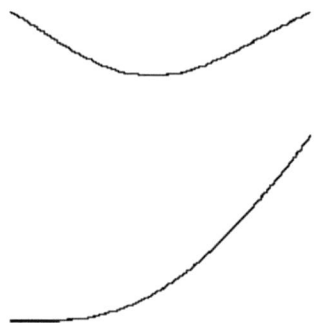

**Fig. 3.** Example 2. $20160x^5 - 30176x^4 + 14156x^3 - 2344x^2 + 151x + 237 - 480y = 0$, plotted using AA.

**Fig. 4.** Example 2. $20160x^5 - 30176x^4 + 14156x^3 - 2344x^2 + 151x + 237 - 480y = 0$, plotted using MAA.

The ten algebraic curve examples we used here for comparison of AA and MAA are all chosen from [9]. The graphical outputs for all these ten curve examples using AA and MAA methods respectively are shown in Figure 1 to Figure 20. The tabulated results are presented in Table 1.

From Figures 1–20 and Table 1 we can see that in general the MAA method is much more accurate *and* quicker than AA method. The AA method is particularly bad in Examples 5, 6, 9, and 10. In Example 5, AA completely fails to reveal the shape of the curve while MAA successfully reveals it. In Example 6 the curve generated by AA is much thicker than the one generated by MAA. In Example 9 AA is unable to distinguish two concentric circles of very similar radii, while MAA can do this. In Example 10 AA has an overconservativism problem near the tangency point of two circles which MAA does not. MAA is slightly more accurate than IAC, and MAA takes almost the same CPU time as IAC. Overall, the performance of MAA is slightly better than IAC.

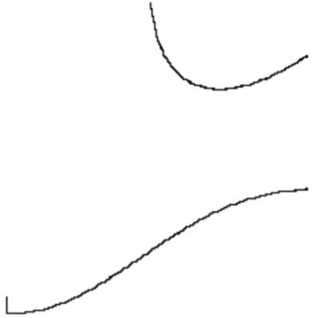

**Fig. 5.** Example 3. $0.945xy - 9.43214x^2y^3 + 7.4554x^3y^2 + y^4 - x^3 = 0$, plotted using AA.

**Fig. 6.** Example 3. $0.945xy - 9.43214x^2y^3 + 7.4554x^3y^2 + y^4 - x^3 = 0$, plotted using MAA.

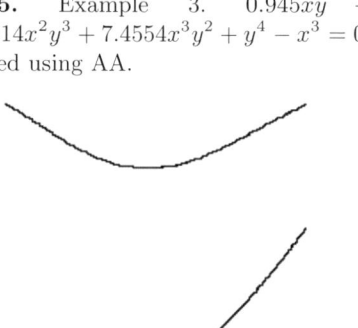

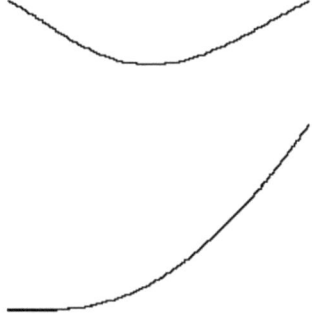

**Fig. 7.** Example 4. $x^9 - x^7y + 3x^2y^6 - y^3 + y^5 + y^4x - 4y^4x^3 = 0$, plotted using AA.

**Fig. 8.** Example 4. $x^9 - x^7y + 3x^2y^6 - y^3 + y^5 + y^4x - 4y^4x^3 = 0$, plotted using MAA.

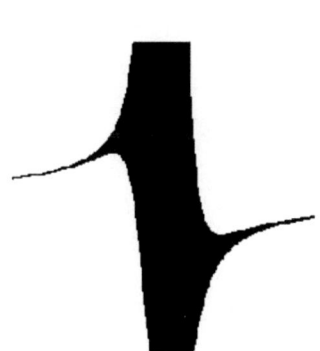

**Fig. 9.** Example 5. $-\frac{1801}{50} + 280x - 816x^2 + 1056x^3 - 512x^4 + \frac{1601}{25}y - 512xy + 1536x^2y - 2048x^3y + 1024x^4y = 0$, plotted using AA.

**Fig. 10.** Example 5. $-\frac{1801}{50} + 280x - 816x^2 + 1056x^3 - 512x^4 + \frac{1601}{25}y - 512xy + 1536x^2y - 2048x^3y + 1024x^4y = 0$, plotted using MAA.

**Fig. 11.** Example 6. $\frac{601}{9} - \frac{872}{3}x + 544x^2 - 512x^3 + 256x^4 - \frac{2728}{9}y + \frac{2384}{3}xy - 768x^2y + \frac{5104}{9}y^2 - \frac{2432}{3}xy^2 + 768x^2y^2 - 512y^3 + 256y^4 = 0$, plotted using AA.

**Fig. 12.** Example 6. $\frac{601}{9} - \frac{872}{3}x + 544x^2 - 512x^3 + 256x^4 - \frac{2728}{9}y + \frac{2384}{3}xy - 768x^2y + \frac{5104}{9}y^2 - \frac{2432}{3}xy^2 + 768x^2y^2 - 512y^3 + 256y^4 = 0$, plotted using MAA.

**Fig. 13.** Example 7. $-13 + 32x - 288x^2 + 512x^3 - 256x^4 + 64y - 112y^2 + 256xy^2 - 256x^2y^2 = 0$, plotted using AA.

**Fig. 14.** Example 7. $-13 + 32x - 288x^2 + 512x^3 - 256x^4 + 64y - 112y^2 + 256xy^2 - 256x^2y^2 = 0$, plotted using MAA.

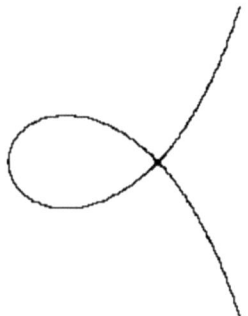

**Fig. 15.** Example 8. $-\frac{169}{64} + \frac{51}{8}x - 11x^2 + 8x^3 + 9y - 8xy - 9y^2 + 8xy^2 = 0$, plotted using AA.

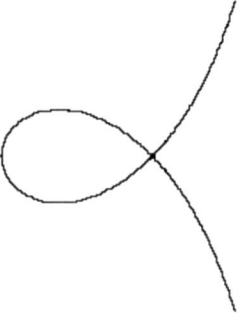

**Fig. 16.** Example 8. $-\frac{169}{64} + \frac{51}{8}x - 11x^2 + 8x^3 + 9y - 8xy - 9y^2 + 8xy^2 = 0$, plotted using MAA.

**Fig. 17.** Example 9. $47.6 - 220.8x + 476.8x^2 - 512x^3 + 256x^4 - 220.8y + 512xy - 512x^2y + 476.8y^2 - 512xy^2 + 512x^2y^2 - 512y^3 + 256y^4 = 0$, plotted using AA.

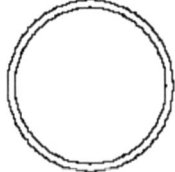

**Fig. 18.** Example 9. $47.6 - 220.8x + 476.8x^2 - 512x^3 + 256x^4 - 220.8y + 512xy - 512x^2y + 476.8y^2 - 512xy^2 + 512x^2y^2 - 512y^3 + 256y^4 = 0$, plotted using MAA.

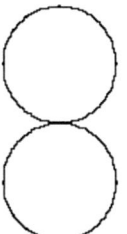

**Fig. 19.** Example 10. $\frac{55}{256} - x + 2x^2 - 2x^3 + x^4 - \frac{55}{64}y + 2xy - 2x^2y + \frac{119}{64}y^2 - 2xy^2 + 2x^2y^2 - 2y^3 + y^4 = 0$, plotted using AA.

**Fig. 20.** Example 10. $\frac{55}{256} - x + 2x^2 - 2x^3 + x^4 - \frac{55}{64}y + 2xy - 2x^2y + \frac{119}{64}y^2 - 2xy^2 + 2x^2y^2 - 2y^3 + y^4 = 0$, plotted using MAA.

**Table 1.** Comparison of AA, MAA and IAC methods by examples.

| Example | Method | Pixels plotted | Subdivisions involved | CPU time used |
|---|---|---|---|---|
| 1 | AA | 604 | 900 | 1.047 sec |
| 1 | IAC | 530 | 587 | 0.047 sec |
| 1 | MAA | 526 | 563 | 0.047 sec |
| 2 | AA | 513 | 815 | 1.219 sec |
| 2 | IAC | 435 | 471 | 0.063 sec |
| 2 | MAA | 433 | 459 | 0.063 sec |
| 3 | AA | 625 | 715 | 1.187 sec |
| 3 | IAC | 609 | 638 | 0.094 sec |
| 3 | MAA | 608 | 634 | 0.094 sec |
| 4 | AA | 832 | 934 | 4.969 sec |
| 4 | IAC | 819 | 880 | 0.547 sec |
| 4 | MAA | 816 | 857 | 0.562 sec |
| 5 | AA | 15407 | 9027 | 49.468 sec |
| 5 | IAC | 470 | 659 | 0.063 sec |
| 5 | MAA | 464 | 611 | 0.062 sec |
| 6 | AA | 1287 | 2877 | 10.266 sec |
| 6 | IAC | 466 | 596 | 0.109 sec |
| 6 | MAA | 460 | 560 | 0.110 sec |
| 7 | AA | 933 | 1409 | 1.766 sec |
| 7 | IAC | 532 | 675 | 0.078 sec |
| 7 | MAA | 512 | 627 | 0.078 sec |
| 8 | AA | 891 | 989 | 0.938 sec |
| 8 | IAC | 838 | 853 | 0.078 sec |
| 8 | MAA | 818 | 827 | 0.078 sec |
| 9 | AA | 5270 | 4314 | 13.75 sec |
| 9 | IAC | 1208 | 1373 | 0.250 sec |
| 9 | MAA | 1144 | 1269 | 0.250 sec |
| 10 | AA | 2071 | 2796 | 8.625 sec |
| 10 | IAC | 812 | 905 | 0.172 sec |
| 10 | MAA | 784 | 845 | 0.171 sec |

The reasons why MAA is much faster than AA are as follows. Firstly, the most crucial reason which we can clearly see from Section 2 is that the expression used in AA is actually more complicated than that used in IAC or MAA. Therefore AA in fact involves more arithmetic operations than IAC or MAA. Secondly AA is less accurate than MAA, and therefore AA needs more subdivisions. Furthermore, many more incorrect pixels cannot be discarded, which increases the computation load of the AA method. Thirdly, although MAA looks more complicated, actually MAA only contains matrix manipulations which are easy to implement using loops, while AA, although looking simple, requires dynamic lists to represent affine forms with varying numbers of noise symbols. AA operations $(+, -, *)$ must be performed through insertion and deletion of elements of the lists, which are not as efficient the simper arithmetic operations in MAA.

# 4   Conclusions

From the above theoretical proofs and the experimental test results we conclude that the MAA method for estimating bounds on a polynomial over a range is not only more accurate but also much faster than the standard AA method. We also have demonstrated that the MAA method is very similar to the IAC method but also takes into consideration the special properties of even and odd powers of polynomial terms. Therefore the MAA method is always at least as or slightly more accurate than the IAC method. We have also shown that the IAC method is more accurate than the standard AA method. In conclusion we strongly recommend that the MAA method is used instead of standard AA method in geometric computations on implicit curves and surfaces.

# Acknowledgements

This work was supported jointly by the National Natural Science Foundation of China (Grant 60173034), National Natural Science Foundation for Innovative Research Groups (Grant 60021201) and the Foundation of State Key Basic Research 973 Item (Grant 2002CB312101).

# References

1. Bowyer, A., Martin, R., Shou, H., Voiculescu, I.: Affine intervals in a CSG geometric modeller. In: Winkler J., Niranjan, M. (eds.): Uncertainty in Geometric Computations. Kluwer Academic Publisher (2002) 1-14
2. Bühler, K.: Linear interval estimations for parametric objects theory and application. Computer Graphics Forum 20(3) (2001) 522–531
3. Comba, J.L.D., Stolfi, J.: Affine arithmetic and its applications to computer graphics. Anais do VII SIBGRAPI (1993) 9–18
   Available at http://www.dcc.unicamp.br/~stolfi/

4. De Cusatis, A., Jr., De Figueiredo, L.H., Gattass, M.: Interval Methods for Ray Casting Implicit Surfaces with Affine Arithmetic. XII Brazilian Symposium on Computer Graphics and Image Processing (1999) 65–71
5. De Figueiredo, L.H.: Surface intersection using affine arithmetic. Proceedings of Graphics Interface (1996) 168–175
6. De Figueiredo, L.H., Stolfi, J.: Adaptive enumeration of implicit surfaces with affine arithmetic. Computer Graphics Forum 15(5) (1996) 287–296
7. Heidrich, W., Seidel, H.P.: Ray tracing procedural displacement shaders. Proceedings of Graphics Interface (1998) 8–16
8. Heidrich, W., Slusallek, P., Seidel, H.P.: Sampling of procedural shaders using affine arithmetic. ACM Trans. on Graphics 17(3) (1998) 158–176
9. Martin, R., Shou, H., Voiculescu, I., Bowyer, A., Wang, G.: Comparison of interval methods for plotting algebraic curves. Computer Aided Geometric Design 19(7) (2002) 553–587
10. Ratschek, H., Rokne, J.: Computer Methods for the Range of Functions. Ellis Horwood (1984)
11. Shou, H., Martin, R., Voiculescu, I., Bowyer, A., Wang, G.: Affine arithmetic in matrix form for polynomial evaluation and algebraic curve drawing. Progress in Natural Science 12(1) (2002) 77–80

# On the Spine of a PDE Surface

Hassan Ugail

Department of Electronic Imaging and Media Communications
School of Informatics, University of Bradford
Bradford BD7 1DP, UK
h.ugail@bradford.ac.uk

**Abstract.** The spine of an object is an entity that can characterise the object's topology and describes the object by a lower dimension. It has an intuitive appeal for supporting geometric modelling operations.
The aim of this paper is to show how a spine for a PDE surface can be generated. For the purpose of the work presented here an analytic solution form for the chosen PDE is utilised. It is shown that the spine of the PDE surface is then computed as a by-product of this analytic solution.
This paper also discusses how the of a PDE surface can be used to manipulate the shape. The solution technique adopted here caters for periodic surfaces with general boundary conditions allowing the possibility of the spine based shape manipulation for a wide variety of free-form PDE surface shapes.

## 1 Introduction

Generally speaking the spine of an object is the trace of the center of all spheres (disks in the case of two dimensions) that are maximally inscribed in the object [6]. The spine of an object has a very close geometric resemblance to the more widely known shape entity called the medial axis or the skeleton [9]. Bearing in mind the general definition for a spine, one could therefore imagine that the spine of a shape brings out the symmetries in that shape. It can also be noted that the spine in general has far richer topologies than the shape it is derived from. Other important properties of the spine of a shape include its use in the intermediate representation of the object and its canonical general form that can be used to represent the object by a lower dimensional description.

Apart from the rich geometric properties the spine posses, many have also noted its intuitive appeal in applications in geometric manipulations. For example Blum [6] suggested the spine or the skeleton as a powerful mechanism for representing the shape of two dimensional objects at a level higher than cell-enumeration. He proposed a technique that can uniquely decompose a shape into a collection of sub-objects that can be readily identified with a set of basic primitive shapes. Many others have affirmed the flexibility of the spine and its ability to naturally capture important shape characteristics of an object [12,11,7].

Despite its intuitive appeal the spine is rarely used in CAD systems for supporting geometric modelling operations. Among the reasons for this include the

M.J. Wilson and R.R. Martin (Eds.): Mathematics of Surfaces 2003, LNCS 2768, pp. 366–376, 2003.

lack of robust implementations of spine generating procedures for existing CAD techniques, and the inability to demonstrate the wide range of shape manipulations that can be potentially performed using the spine of a shape.

In spite of this a number of methods for constructing the spine of polyhedral models as well as free-form shapes have been proposed. These include topological thinning [13], Euclidian distance transform [1] and the use of deformable snakes [10]. The majority of the existing techniques use numerical schemes to scan the domain of the whole object in order to generate its spine. Thus, these algorithms not only consume excessive CPU time to perform their operations but also are prone to errors.

The focus of this paper is on the spine of the PDE surfaces. PDE surfaces are generated as solutions to elliptic Partial Differential Equations (PDEs) where the problem of surface generation is treated as a boundary-value problem with boundary conditions imposed around the edge of the surface patch [2,3,16]. PDE surfaces have emerged as a powerful shape modelling technique [8,14,15]. It has been demonstrated how a designer sitting in front of a workstation is able to create and manipulate complex geometry interactively in real time [14]. Furthermore, it has been shown that complex geometry can be efficiently parameterised both for intuitive shape manipulation [15] and for design optimisation [5].

The aim of this paper is to show how the spine of a PDE surface can be created and utilised in order to characterise PDE surfaces as well as to enable the development of further intuitive techniques for powerful shape manipulations. By exploiting the structural form of a closed form solution for the chosen PDE, it is shown how the spine of a PDE surface can be generated as a by-product of this solution. Furthermore, it is shown that the spine of the PDE surface patch is represented as a cubic polynomial that can be used as a shape manipulation tool to deform the shape in an intuitive fashion. It is also shown that, by exploiting a general form of an analytic solution method, the spine for PDE surfaces with general boundary conditions can equally be represented as a by-product of the solution that generates the surface shape. To demonstrate the ideas presented here, practical examples of shapes involving PDE surfaces are discussed throughout the paper.

## 2 PDE Surfaces

A PDE surface is a parametric surface patch $\underline{X}(u, v)$, defined as a function of two parameters $u$ and $v$ on a finite domain $\Omega \subset R^2$, by specifying boundary data around the edge region of $\partial\Omega$. Typically the boundary data are specified in the form of $\underline{X}(u, v)$ and a number of its derivatives on $\partial\Omega$. Moreover, this approach regards the coordinates of point $u$ and $v$ as a mapping from that point in $\Omega$ to a point in the physical space. To satisfy these requirements the surface $\underline{X}(u, v)$ is regarded as a solution of a PDE of the form,

$$D_{u,v}^m \underline{X}(u, v) = \underline{F}(u, v), \tag{1}$$

where $D_{u,v}^m \underline{X}(u, v)$ is a partial differential operator of order $m$ in the independent variables $u$ and $v$, while $\underline{F}(u, v)$ is a vector valued function of $u$ and $v$. Since boundary-value problems are of concern here, it is natural to choose $D^m_{u,v}\underline{X}(u,v)$ to be elliptic.

Various elliptic PDEs could be used, although the most widely used PDE is based on the biharmonic equation namely,

$$\left(\frac{\partial^2}{\partial u^2} + a^2 \frac{\partial^2}{\partial v^2}\right)^2 \underline{X}(u,v) = 0. \tag{2}$$

Here the boundary conditions on the function $\underline{X}(u,v)$ and its normal derivatives $\frac{\partial X}{\partial n}$ are imposed at the edges of the surface patch.

With this formulations one can see that the elliptic partial differential operator in Equation (2) represents a smoothing process in which the value of the function at any point on the surface is, in some sense, a weighted average of the surrounding values. In this way a surface is obtained as a smooth transition between the chosen set of boundary conditions. The parameter $a$ is a special design parameter which controls the relative smoothing of the surface in the $u$ and $v$ directions [3].

## 2.1   Solution of the PDE

There exist many methods to determine the solution of Equation (2). In some cases, where the boundary conditions can be expressed as relatively simple analytic functions of $u$ and $v$, it is possible to find a closed form solution. However, for a general set of boundary conditions a numerical method often need to be employed.

For the work on the spine to be described here, restricting to periodic boundary conditions the closed form analytic solution of Equation (2) is utilised. Choosing the parametric region to be $0 \leq u \leq 1$ and $0 \leq v \leq 2\pi$, the periodic boundary conditions can be expressed as,

$$\underline{X}(0,v) = \underline{p}_0(v), \tag{3}$$

$$\underline{X}(1,v) = \underline{p}_1(v), \tag{4}$$

$$\underline{X}_u(0,v) = \underline{d}_0(v), \tag{5}$$

$$\underline{X}_u(1,v) = \underline{d}_1(v). \tag{6}$$

Note that the boundary conditions $\underline{p}_0(v)$ and $\underline{p}_1(v)$ define the edges of the surface patch at $u = 0$ and $u = 1$ respectively. Using the method of separation of variables, the analytic solution of Equation (2) can be written as,

$$\underline{X}(u,v) = \underline{A}_0(u) + \sum_{n=1}^{\infty}[\underline{A}_n(u)\cos(nv) + \underline{B}_n(u)\sin(nv)], \tag{7}$$

where

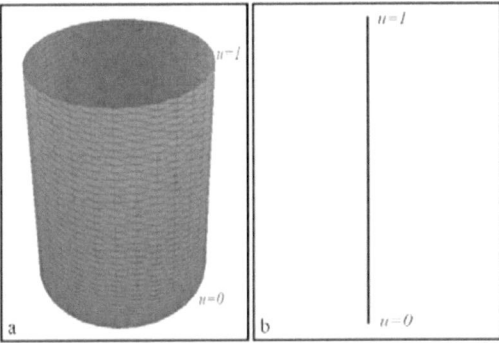

**Fig. 1.** Description of the spine of a PDE surface. (**a**) A 'cylindrical' PDE surface. (**b**) The spine described by the $\underline{A}_0$ term.

$$\underline{A}_0 = \underline{a}_{00} + \underline{a}_{01}u + \underline{a}_{02}u^2 + \underline{a}_{03}u^3, \tag{8}$$

$$\underline{A}_n = \underline{a}_{n1}e^{anu} + \underline{a}_{n2}e^{anu} + \underline{a}_{n3}e^{-anu} + \underline{a}_{n4}e^{-anu}, \tag{9}$$

$$\underline{B}_n = \underline{b}_{n1}e^{anu} + \underline{b}_{n2}e^{anu} + \underline{b}_{n3}e^{-anu} + \underline{b}_{n4}e^{-anu}, \tag{10}$$

where $\underline{a}_{00}, \underline{a}_{01}, \underline{a}_{02}, \underline{a}_{03}$ $\underline{a}_{n1}, \underline{a}_{n2}, \underline{a}_{n3}, \underline{a}_{n4}, \underline{b}_{n1}$ $\underline{b}_{n2}, \underline{b}_{n3}$ and $\underline{b}_{n4}$ are vector-valued constants, whose values are determined by the imposed boundary conditions at $u = 0$ and $u = 1$.

## 3  The Spine of a PDE Surface

Taking the form of Equation (7) one could observe the following properties of the analytic solution that allows us to extract the spine of a PDE surface as a by-product of the solution. Firstly the term $\underline{A}_0$ in Equation (7) is a cubic polynomial of the parameter $u$. Secondly it can be seen that for each point $\underline{X}(u, v)$ on the surfaces the term $\sum_{n=1}^{\infty}[\underline{A}_n(u)\cos(nv)+\underline{B}_n(u)\sin(nv)]$ in Equation (7) describes the 'radial' position of the point $\underline{X}(u, v)$ relative to a point at $\underline{A}_0$.

Thus, the term $\underline{A}_0$ which is a cubic polynomial of the parameter $u$ and lies within the periodic surface patch. Therefore, using the solution technique described in Equation (7) a surface point $\underline{X}(u, v)$ may be regarded as being composed of sum of a vector $\underline{A}_0$ giving the position on the spine of the surface and a radius vector defined by the term $\sum_{n=1}^{\infty}[\underline{A}_n(u)\cos(nv) + \underline{B}_n(u)\sin(nv)]$ providing the position of $\underline{X}(u, v)$ relative to the spine. What follows in the rest of this section describes some examples of PDE surfaces and the corresponding spines relating to the $\underline{A}_0$ term given in Equation (7).

Fig. 1(**a**) shows a typical PDE surface where the boundary conditions are taken as,

$$\underline{p}_0(v) = (0.5\cos v, 0.5\sin v, 0), \tag{11}$$

$$\underline{p}_1(v) = (0.5\cos v, 0.5\sin v, 0.5), \tag{12}$$

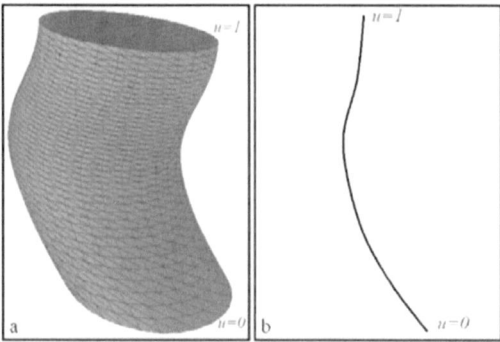

**Fig. 2.** Description of the spine of a PDE surface. (**a**) A deformed 'cylindrical' PDE surface. (**b**) The spine described by the $\underline{A}_0$ term.

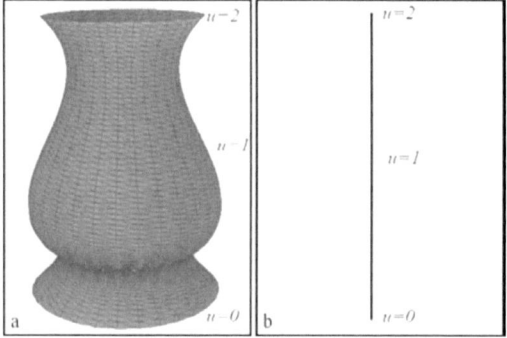

**Fig. 3.** Description of the spine of a composite PDE surface. (**a**) A composite PDE surface describing a vase shape. (**b**) The spine of the vase shape.

$$\underline{d}_0(v) = (0.5\cos v, 0.5\sin v, 1), \tag{13}$$

$$\underline{d}_1(v) = (0.5\cos v, 0.5\sin v, 1). \tag{14}$$

Fig. 1(**b**) shows the image of the cubic polynomial described by the $\underline{A}_0$ term corresponding to the spine for this surface patch.

Fig. 2(**a**) shows another example of a single PDE patch where the boundary conditions were taken to be that described by the previous example with the exception that the circle defining $\underline{d}_1(v)$ was translated by an amount of $0.2$ units along the negative x-axis. The resulting spine for this surface patch is shown in Fig. 2(**b**). As can be noted in both these examples the spine closely describes the midline or the skeleton of the surfaces patch.

Fig. 3(**a**) shows a composite shape that looks like the shape of a vase. This shape is created by means of two surface patches with a common boundary at $u = 1$. Again the boundary conditions for these surface patches are circular and similar to those used to create Fig. 1(**a**). Furthermore, for the two surface patches the derivative conditions at $u = 1$ were taken in such a way to ensure

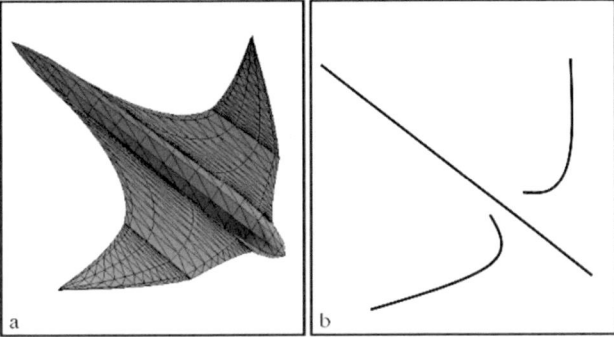

**Fig. 4.** Aircraft created using a composite of 5 PDE surface patches. (**a**) The aircraft shape. (**b**) The corresponding composite spine.

that tangent plane continuity between them is maintained. The corresponding spine for the vase shape is shown is Fig. 3(**b**).

Fig. 4(**a**) shows a composite shape that looks like the shape of an aircraft. This shape is created by means of four surface patches. The corresponding composite spine for the aircraft shape is shown is Fig. 4(**b**).

## 4   Shape Manipulation Using the Spine

One of the many attractive features of the PDE surfaces is the ability to be able to create and manipulate complex shapes with ease. Previous work on interactive design has demonstrated that the user having little or no knowledge about solving PDEs and how the boundary conditions effect the solutions of the PDEs is able to use the method to create complex geometry with ease [14,15].

The aim of this section is to show that the spine of a PDE surface can be utilised to create design tools for further efficient shape manipulation. As shown in the previous section the spine of a PDE surface comes as a by-product of the analytic solution used. By virtue of the very definition of the spine it can be seen as a powerful and intuitive mechanism to manipulate the shape of surface once it is defined. There are many ways by which one could utilise the spine to manipulate a PDE surface. One such possibility is described here.

We can express the cubic polynomial described by $\underline{A}_0$ in Equation (7) to be a Hermite curve of the form,

$$\underline{H}(u) = \underline{B}_1(u)\underline{p}_0 + \underline{B}_2(u)\underline{p}_1 + \underline{B}_3(u)\underline{v}_0 + \underline{B}_4(u)\underline{v}_1, \tag{15}$$

where the $\underline{B}_i$ are the Hermite basis functions, $\underline{p}_0, \underline{p}_1$ and $\underline{v}_0, \underline{v}_1$ define the position and the speed of the Hermite curve at $u = 0$ and $u = 1$ respectively. By comparing the Hermite curve given in Equation (15) with the cubic for the spine given by the $\underline{A}_0$ term in Equation (7), the terms $\underline{a}_{00}, \underline{a}_{01}, \underline{a}_{02}$ and $\underline{a}_{03}$ described in Equation (8) can be related to the position vectors and its derivatives at the end points of the spine as,

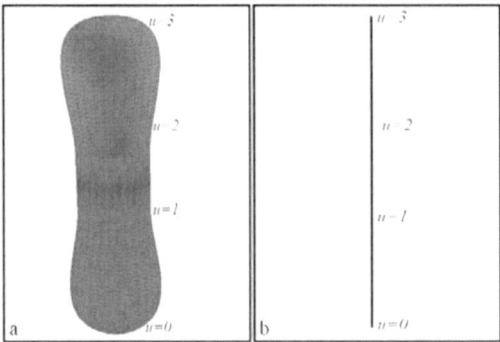

**Fig. 5.** PDE vesicle shape. (**a**) The vesicle shape created using three surface patches. (**b**) The spine of the vesicle shape.

**Table 1.** Vector values for the position vectors and its derivatives at the end of points of the spine for the vesicle shapes shown in Figs. 5 and 6.

| vector | Fig. 5 $(x, y, z)$ | Fig. 6 $(x, y, z)$ |
|---|---|---|
| $\underline{p}_0$ | $(0.0, -1.0, 0.0)$ | $(0.1, -1.0, 0.0)$ |
| $\underline{p}_1$ | $(0.0, 1.0, 0.0)$ | $(0.4, 1.0, 0.0)$ |
| $\underline{v}_0$ | $(0.0, -0.2, 0.0)$ | $(-0.1, 0.1, 0.0)$ |
| $\underline{v}_1$ | $(0.0, 0.2, 0.0)$ | $(-0.3, -0.2, 0.0)$ |

$$a_{00} = \underline{p}_0, \tag{16}$$

$$a_{01} = 3\underline{p}_1 - \underline{v}_1 - 3\underline{v}_0, \tag{17}$$

$$a_{02} = \underline{v}_1 + 2\underline{v}_0 - 2\underline{p}_1, \tag{18}$$

$$a_{02} = \underline{v}_0. \tag{19}$$

Since the $\underline{A}_0$ term in Equation (7) is an integral part of the solution that generates the surface shape, any change in the shape of the spine will of course results in a change in the shape of the surface. A useful mechanism to change the shape of the spine would be to manipulate its position and the derivative at the two end points. Therefore, the position vectors and its derivatives at the end of points of the spine can be used as shape parameter to manipulate the shape.

To demonstrate this idea consider the vesicle shape, similar to that of a human red blood cell, shown in Fig. 5(**a**) where the corresponding spine is also shown in 5(**b**). The vesicle shape is created using three surface patches with common boundaries between the adjacent patches at $u = 1$ and $u = 2$. The boundary conditions for this problem are circles similar to those used in the example shown in Fig. 1, with the positional boundary conditions at $u = 0$ and $u = 3$ taken to be points in 3-space.

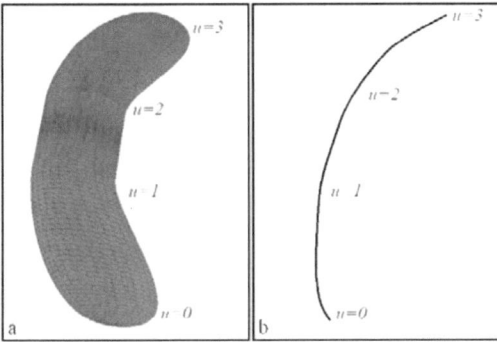

**Fig. 6.** New PDE vesicle shape. (**a**) The vesicle shape created by manipulating the spine. (**b**) The manipulated spine of the new vesicle shape.

The vesicle can be then manipulated by making changes to the shape of the spine via the vectors $\underline{p}_0, \underline{p}_1$, $\underline{v}_0$ and $\underline{v}_1$. Table 1 shows the value for the $x$, $y$ and $z$ components of the vectors $\underline{p}_0, \underline{p}_1$, $\underline{v}_0$ and $\underline{v}_1$ for the vesicle shown in Fig. 5(**a**) and those for the manipulated shape shown in Fig. 6(**a**). Fig. 6(**b**) shows the manipulated spine. Note that throughout this shape manipulation process the derivative conditions at $u = 1$ and $u = 2$ were taken in a manner to ensure tangent plane continuity between the adjacent surface patches is maintained. This also ensures that the corresponding spine has tangent continuity at $u = 1$ and $u = 2$.

In previous work discussed in [5], involving the problem of determining the shapes of the stable structures occurring in human red blood cells, a similar vesicle structure as shown in Fig. 5 was used as a starting shape of an optimisation processes. These shapes were parameterised using the appropriate Fourier coefficients, where some of the intermediate shapes resulted during the optimisation process resembled the shape of the vesicle shown in Fig. 6. An interesting and rather intuitive way to parameterise the vesicle geometry would be the spine approach outlined here.

## 5   General PDE Boundary Conditions and the Spine

One could note that the examples described above are somewhat simple where the shapes are generated using simple boundary conditions possessing analytic forms. However, to cater for a wide range of possible free-form shape manipulations the spine of shapes with general boundary conditions need to be addressed. For this purpose the approach adopted here is to use a previously developed solution technique which can handle general periodic boundary conditions [4]. The method is based on a spectral approximation providing an approximate analytic solution form of the chosen PDE. The basic idea behind this solution method is presented here with details on how the solution affects the spine. For detailed discussions of this solution method the interested reader is referred to [4].

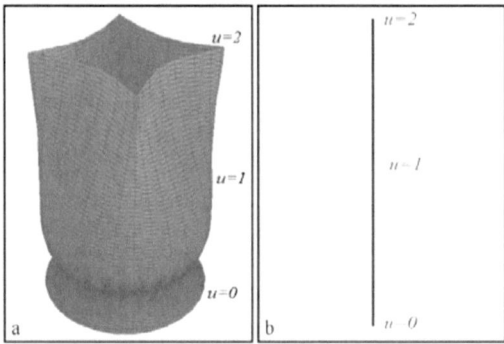

**Fig. 7.** Description of the spine of a PDE surface with general boundary conditions. (**a**) A vase shape with general boundary conditions at $u = 2$ defined as cubic B-spline curves. (**b**) The spine of the vase shape.

For a general set of boundary conditions, in order to define the various constants in the solution, it is necessary to Fourier analyse the boundary conditions and identify the various Fourier coefficients. Where the boundary conditions can be expressed exactly in terms of a finite Fourier series, the solution given in Equation (7) will also be finite. However, this is often not possible, in which case the solution will be the infinite series given Equation (7).

The technique for finding an approximation to $\underline{X}(u, v)$ is based on the sum of the first few Fourier modes and a 'remainder term', i.e.,

$$\underline{X}(u, v) \simeq \underline{A}_0(u) + \sum_{n=1}^{N} [\underline{A}_n(u) \cos(nv) + \underline{B}_n(u) \sin(nv)] + \underline{R}(u, v), \qquad (20)$$

where usually $N \leq 6$ and $\underline{R}(u, v)$ is a remainder function defined as,

$$\underline{R}(u, v) = \underline{r}_1(v)e^{wu} + \underline{r}_2(v)e^{wu} + \underline{r}_3(v)e^{-wu} + \underline{r}_4(v)e^{-wu}, \qquad (21)$$

where $\underline{r}_1$, $\underline{r}_2$, $\underline{r}_3$, $\underline{r}_4$ and $w$ are obtained by considering the difference between the original boundary conditions and the boundary conditions satisfied by the function,

$$\underline{F}(u, v) = \underline{A}_0(u) + \sum_{n=1}^{N} [\underline{A}_n(u) \cos(nv) + \underline{B}_n(u) \sin(nv)]. \qquad (22)$$

An important point to note here is that although the solution is approximate this new solution technique guarantees that the chosen boundary conditions are exactly satisfied since the remainder function $\underline{R}(u, v)$ is calculated by means of the difference between the original boundary conditions and the boundary conditions satisfied by the function $\underline{F}(u, v)$.

It is noteworthy that the introduction of the $R(u, v)$ term in the new solution described in Equation (20) has virtually no effect in the interior shape of the surfaces. This is because, for large enough $n$, the Fourier modes make negligible contributions to the interior of the patch. Therefore, by taking a reasonable truncation of the Fourier series at some finite $N$, (say $N = 6$) of the boundary conditions an approximate PDE surface can be quickly generated satisfying the boundary conditions exactly. Furthermore, as far as the spine is concerned since the spine does not represent the detailed geometry of the shape the $\underline{A}_0(u)$ term is left unchanged in the approximate solution and hence the spine of the shape is left unchanged. Fig. 7 exemplifies this.

Fig. 7(**a**) shows a vase shape similar to that shown in Fig. 3(**a**), where the position and derivative boundary conditions at $u = 2$ were taken as periodic cubic B-spline curves with cusps. The curves along with the rest of the boundary conditions which are in analytic form were used in the solution outlined to create the shape shown in Fig. 7(**a**). The corresponding spine for the new vase shape is shown in Fig. 7(**b**).

## 6    Conclusions

This paper describes how the spine of a PDE surfaces can be generated. Due to the analytic form of the solution used to generate the surface shape the spine is computed as a by-product of the solution. This outlines the advantage of using PDE surfaces for modelling since in the case of most other techniques for shape generation the spine has to be computed separately.

Due to the canonical and intuitive nature of the spine it can be used to manipulate the shape once the shape is defined. It has been demonstrated how simple shape manipulation can be carried out using the spine of a PDE surface. The solution technique adopted here caters periodic surfaces with general boundary conditions with the spine derived as a by-product of the solution. This allows the possibility of the spine based shape manipulation for a wide variety of free-form surface shapes.

As shown here the shape manipulation using the spine can be seen as an added bonus to the existing intuitive tools available for efficient shape manipulation of PDE surfaces. An interesting future direction of study would be to parameterise the shapes based on the spine described here. This would allow one to create geometry that can handle not only complex shapes but also shapes with changing topology. Such a parameterisation scheme then can be applied to design optimisation problems where a wide variety of geometry with changing topology would be available to the optimisation scheme.

## References

1. Arcelli, C., and Sanniti di Baja, G.: Ridge Points in Euclidean Distance Maps. Pattern Recognition Letters, **13**(4) (1992) 237-243
2. Bloor, M. I. G., and Wilson, M. J.: Generating Blend Surfaces Using Partial Differential Equations, Computer-Aided Design. **21** (1989) 165–171
3. Bloor, M. I. G., and Wilson, M. J.: Using Partial Differential Equations to Generate Freeform Surfaces. Computer Aided Design, **22** (1990) 202–212

4. Bloor, M.I.G., and Wilson, M.J.: Spectral Approximations to PDE Surfaces. Computer-Aided Design, **28** (1996) 145–152

5. Bloor, M. I. G. and Wilson, M. J.: Method for Efficient Shape Parametrization of Fluid Membranes and Vesicles. Physical Review E, **61**(4) (2000) 4218–4229

6. Blum, H.: A transformation for Extracting New Descriptors of Shape. In: Wathen-Dunn, W. (ed.): Models for Perception of Speech and Visual Form, MIT Press (1976) 362-381

7. Boissonnat, J. D.: Geometric Surfaces for 3-Dimensional Shape Representation. ACM Transactions on Graphics, **3**(4) (1984) 244-265

8. Du, H., and Qin, H.; Direct Manipulation and Interactive Sculpting of PDE Surfaces. Computer Graphics Forum (Proceedings of Eurographics 2000), **19**(3) (2000) 261-270

9. Dutta, D and Hoffmann, C.M.: On the Skeleton of Simple CSG Objecs. ASME Journal of Mechanical Design, **115**(1) (1992) 87-94

10. Leymarie, F., and Levine, M.D.: Skeleton from Snakes, Progress in Image Analysis and Processing, World Scientific Singapore (1990)

11. Nackman, L.R., and Pizer, S.M.: Three-Dimensional Shape Description using Symmetric Axis Transform. IEEE Transactions on Pattern Analysis and Machine Intelligence, **7**(2) (1985) 187-202

12. Patrikalakis, N.M., and Gursoy, H.N.: Shape Interrogation by Medial Axis Transform. In: Ravani, B. (ed.): Advances in Design Automation: Computer-Aided Computational Design, ASME **1** (1990) 77-88

13. Tsao, Y.F., and Fu, K.S.: A Parallel Thinning Algorithm for 3-D Pictures. Computer Graphics Image Process, **17** (1981) 315-331

14. Ugail, H., Bloor, M.I.G., and Wilson, M.J.: Techniques for Interactive Design Using the PDE Method. ACM Transactions on Graphics, **18**(2) (1999) 195–212

15. Ugail, H., Bloor, M.I.G., and Wilson, M.J.: Manipulations of PDE Surfaces Using an Interactively Defined Parameterisation. Computers and Graphics, **24**(3) (1999) 525–534

16. Vida, J., Martin, R.R., and Varady, T.: A Survey of Blending Methods that use Parametric Surfaces. Computer-Aided Design, **26**(5) (1994) 341–365

# Application of PDE Methods
# to Visualization of Heart Data

Ognyan Kounchev[1] and Michael J. Wilson[2]

[1] Institue of Mathematics and Informatics - Bulgarian Academy of Sciences
Acad. G. Bonchev 8, 1113 Sofia, Bulgaria
kounchev@math.bas.bg
[2] University of Leeds, School of Mathematics
Leeds LS2 9JT, UK
mike@maths.leeds.ac.uk

**Abstract.** We apply new methods based on Partial Differential Equations techniques (polysplines) to the visualization of the heart surface.

## 1 Introduction

### 1.1 The Medical Perspective

There is considerable effort in the area of medical imaging to capture and analyse the motion of the heart, using a variety of imaging techniques, e.g. *X-ray Computed Tomography* (CT) or *Magnetic Image Resonancing* (MRI). Because of the complicated motion of the heart and the fact that its surface is deformable, interpreting and analysing image data in order to deduce the underlying motion is not straight-forward and most approaches to the problem are model-based. For example simple shapes such as spheres, ellipsoids, or cylinders are sometimes used to approximate the shape of the *Left Ventricle* (LV) [2,4,15]. Recently techniques based on the usage of deformable models for reconstructing the 3D surface shape and motion of the *LV* from *CT* or *MRI* data have been developed (e.g. [1,7,13,17,18,23,20]). They use a variety of physics-based or geometrical techniques to model the rigid and non-rigid motion of the ventricles (usually the left).

There are two common approaches to modelling the shape of the ventricles. The first aims to construct a generic parametric model to describe the main features of the heart's shape during the cardiac cycle, whilst the second uses *MRI* scan data to construct a more accurate, and hence more complex, patient-specific model.

The latter approach has been used by Taylor and Yamaguchi [37], Park et al. [21] and Haber et al. [12] amongst others. While these techniques can provide good approximations to the actual geometry in an individual case, a general investigation into the effects of modifications to the shape can be more difficult to perform, due to the large amounts of data typically involved in describing the surface.

M.J. Wilson and R.R. Martin (Eds.): Mathematics of Surfaces 2003, LNCS 2768, pp. 377–391, 2003.
© Springer-Verlag Berlin Heidelberg 2003

With regard to the former approach, that is to create a generic representation of the heart, the work of Peskin and McQueen [24] is some of the most advanced to date. Their model encompasses both ventricles and atria and also the major arteries connected to the heart. They build up the heart surface by specifying the position of muscle fibres in the heart walls which are connected to the fluid flow using their Immersed Boundary Method. Yoganathan et al. [39] have also adapted Peskin's method to study a thin-walled *LV* during early systole. The computational time required to perform these calculations, however, makes it difficult to conduct general investigations into different aspects of the motion so other work has used greatly simplified generic ventricle geometry to look at the effects of disease upon the fluid flow in the heart. For example Schoephoerster [28] uses a spherical *LV* to examine the effects of abnormal wall motion on the flow dynamics.

Other work of note that attempts to combine a geometric model of the heart and its structure with a biomechanical model of its functionality, is that of Ayache and coworkers, who create a volumetric mesh of the ventricles and couple this with electrical and biomechanical models of its functionality, e.g. [29,30,31,3,32].

The aim of the work described *here* is to create a parametric model of the ventricles of the heart which lies in between the above extremes. Parametric in this sense means that the geometry is defined by a set of 'shape' or 'design' parameters and can be altered by varying these numbers in a controlled way. The work uses a new flexible method for *Computer Aided Geometric Design* (CAGD) which is based on application of Partial Differential Equations techinques.

Let us stress at this early point of our exposition that the surfaces which we use for representing the heart are interpolating the scanned data and are obtained by a variational principle – by minimization of a *curvature functional* (to be more precise, the functional is very close to the curvature functional). As such they are in a certain sense *surfaces of minimal curvature*.

## 1.2    The Geometric Design Perspective

In order to better explain the apparatus which we use for modelling the heart geometry, let us make a brief account of its history.

The application of Partial Differential Equations to analyzing and visualizing data has a rather long history. In the late fifties–early sixties, harmonic functions have been used for interpolation of data (for modelling the shape of the cars, aircrafts, etc.) – the basis for using harmonic functions was the existence of analog devices which were able to quickly compute the values of physical quantities which are expressed as harmonic functions[1], in fact they were much faster than digital computations. However, there was not enough smoothness and flexibility of the harmonic functions as it was clear that only problems of Dirichlet type are in fact soluble in a 'stable way'. Later on with the development of the digital computers there was no need to apply analog devices, and conventional

---

[1] See more on that in [10, p. 5 top and p. 9 top].

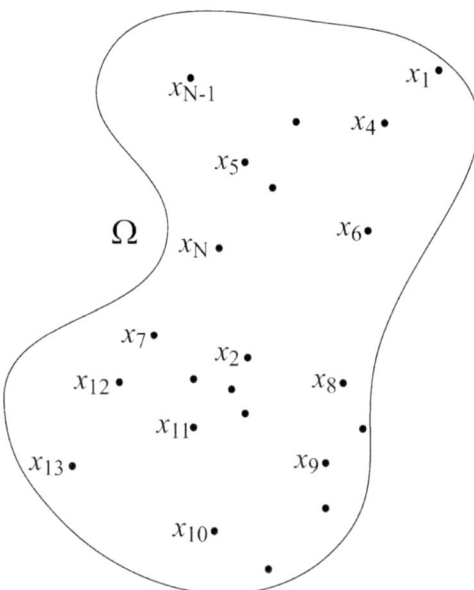

**Fig. 1.** Scattered data in a plane domain.

splines (piecewise polynomial) have taken over in popularity as tools for data interpolation and approximation.

A drawback of such methods is that piecewise polynomial splines (and every kind of simpler objects) need, in general, the data to interpolated to be arranged in a special geometry. On the other hand, data from real applications such as Geophysics, often involve huge amounts of data scattered in an unstructured way. For this reason several methods appeared over the years for handling such data. Some of them are closely related to Partial Differential Equations. One of the first, and nowadays very popular method, is the so–called Minimum curvature method, see Briggs [6], which is a solution of a minimization problem. Namely, suppose that we are given some measurements $c_j$ at data points $(x_j, y_j)$ for $j = 1, 2, ..., N$, in a plane domain $D$, see Figure 1.

Then the problem of finding the surface of minimum curvature may be formulated as follows: Find a surface $f(x, y)$ which satisfies

$$f(x_j, y_j) = c_j \qquad \text{for } j = 1, 2, ..., N,$$

and which has minimum curvature in the differential–geometric sense of the word. Since the exact expression for the curvature is rather complicated, one takes the expression $\Delta f(x, y)$ to be a relatively good approximation to it. Or written more precisely, we have the following extremal problem,

$$\begin{cases} \inf_f \int_D [\Delta f(x, y)]^2 \, dx dy \\ f(x_j, y_j) = c_j \qquad \text{for } j = 1, 2, ..., N, \\ \text{and some boundary condition for } f \text{ on } \partial D. \end{cases} \tag{1}$$

There are very efficient algorithms for solving such problems and the results obtained are rather satisfactory for the purposes of Geophysics; see [6] as an initial reference, and for more recent account of these methods see [11], [34]. Let us mention that another method called "Kriging" has appeared about the same time, see the package "Surfer" for a software implementation of these and other algorithms, [35].

In the mid–seventies, the ultimate generalization of the above approach was obtained in the works of Duchon [8] and Meinguet [19]. They laid the mathematical foundation of a new area called nowadays Radial Basis Functions. This direction has developed theoretically as well as practically, see the survey papers [25] and [9]. Let us note that the above methods have some drawbacks in analyzing huge amounts of data which have points of concentration (in the mathematical sense of this word), in particular if the data are densely located on some curves. Their computation is based on solving large linear systems, and with increasing the number of data points the condition number becomes very small. Their major drawback for the purposes of CAGD is that the surfaces which they create show some artificial oscillations (called "pockmarks" by the geophysicists) occurring mainly near the data points.

Parallel to the above development, during the last decade new interest has appeared towards methods based on Partial Differential Equations in Approximation Theory, Spline Theory and especially in CAGD, see [5]. In these new methods solutions of PDEs are used for interpolation and approximation of the measured data. As an alternative to the "classical" spline theory which relies upon polynomials a new method has appeared which relies upon solutions of PDEs, see [16]. It is based on the *minimum curvature functional* but has a different concept of the data. The **polysplines** are a solution of a problem similar to the one in (1) but the data (when we consider the two–dimensional case) are lying on a set of curves, see e.g. the geometrical configuration on Figure 2.

We will assume throughout the present paper that the exterior–most curve $\Gamma_N$ is the boundary of the domain $D$, i.e. $\partial D = \Gamma_N$. Let us note that the domain $D$ may be unbounded and the curve $\Gamma_N$ may consist of several disconnected pieces.

In the above setting, the polysplines are a solution to the following problem

$$\begin{cases} \inf_f \int_D \left[ \Delta f\left(x,y\right) \right]^2 dx dy \\ f\left(x,y\right) = g_j \qquad \text{on } \Gamma_j \text{ for } j = 1, 2, ..., N, \\ \text{and some boundary condition for } f \text{ on } \partial D. \end{cases} \tag{2}$$

Here the "data functions" $g_j$ are prescribed on the curves $\Gamma_j$.

The theory of polysplines has been extensively studied in the recent monograph [16]. The polysplines have proved to be efficient for smoothing data in Magnetism, in Geophysics, as well as for application to a number of CAGD data, in particular in cases where the data curves have rather irregular form, see Chapter 6 of [16].

The present paper is devoted to the application of the polysplines to heart data. These data are measured by scanning the surface of the heart at different

bdry($\Omega$)=$\Gamma_N$

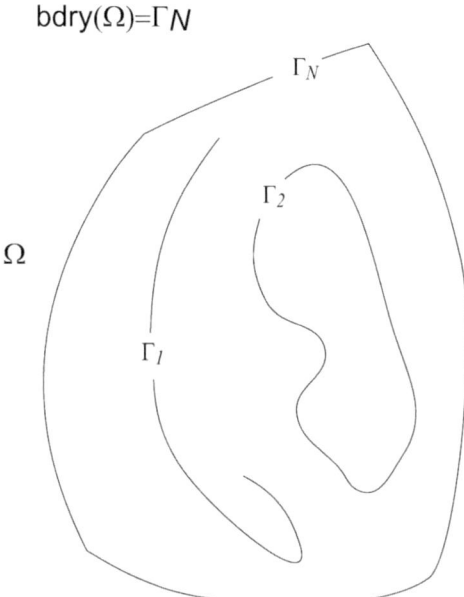

**Fig. 2.** The data are lying on the curves $\Gamma_j$. They are considered to be "scattered".

levels, i.e. for different $z$ in the three dimensional coordinates $(x, y, z)$. In the data available we have measurements at 7 different levels, i.e. for $z = Z^1, Z^2, ..., Z^7$. The main point is to extend these scanned data and to create a visualization of the heart (the left ventricle) as well as to observe the dynamics of the heart activity by creating its form at different phases. The data used in this paper has been extracted from MRI images of horizontal sections through a human heart taken at various stages in the cardiac cycle[2].

The method which we introduce below is applicable to more general situations when the data measured need not to be at the same level in the $z$ direction.

## 2   Introduction of Appropriate Coordinates for the Heart Surface

In the present Section we discuss the introduction of appropriate coordinates in which the surface of the heart is conveniently represented[3]. We will see that in these new coordinates the part of the heart surface of interest ($LV$) becomes a function on a cylinder. The scanned data will be respectively the sections at

---

[2] The data used supplied by the Department of Medical Physics at Leeds General Infirmary.

[3] Speaking mathematically precisely, we define a diffeomorphism on the heart surface which maps it to a cylinder surface. We provide a description of such diffeomorphism in a case somewhat more general than the concrete data require.

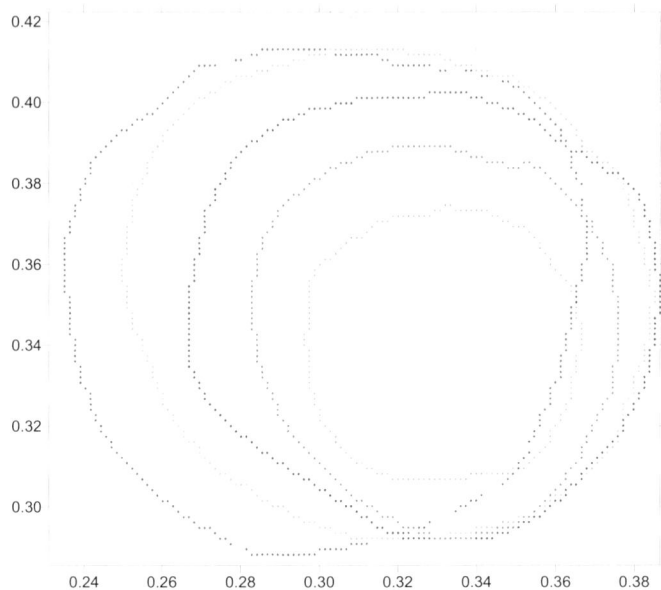

**Fig. 3.** This is the projection of the section curves $G_j$ on the $(x, y)$ plane for the Phase 2 of the heart dynamics.

some levels of this surface. Thus we will be able to apply to this "heart surface function" the ready–made device called "polysplines on cylinder"[4].

We assume that all sectional curves $G_j$ (which are closed curves) are made in the $z$ direction at the level $z = Z^j$ for $j = 1, 2, ... N$, where

$$Z^1 < Z^2 < ... < Z^N.$$

In all available data we have in particular

$$Z^1 = 0.00, \quad Z^2 = 0.01, \quad Z^3 = 0.02,$$
$$Z^4 = 0.03, \quad Z^5 = 0.04, \quad Z^6 = 0.05,$$
$$Z^7 = 0.07.$$

On the next Figure 3 we have a picture showing the projections of the 7 curves on the $(x, y)$ plane for *Phase 2* of the heart dynamics (needless to say these curves and their projections vary with the phases).

Further we make the **assumption** that all curves $G_j$ are star–shaped, i.e. for every $G_j$ there exists a point $(X^j, Y^j, Z^j)$ such that for every point $P \in G_j$ the line interval connecting $P$ and $(X^j, Y^j, Z^j)$ lies entirely inside the curve $G_j$.

---

[4] In [16] see Sections 5.2 and 5.3, p. 60, for a detailed definition of "polysplines on cylinder"; see Section 22.2, p. 448 for the proof of the existence of "interpolation polysplines on cylinder".

*Remark 1.* All practical data which we use further in our study satisfy this assumption – the curves $G_j$ are **star–shaped**. Even more, we may observe that they are star–shaped with respect to their **center of gravity** which provides an algorithm for finding the center of star–shaped-ness.

After we have done such simplifying assumption our algorithm runs as follows: We choose the new coordinate system by first finding the points $(X^j, Y^j, Z^j)$ for all sections $G_j$. Then we join them in the space by an interpolation spline in the $z$ direction, for example, by interpolation cubic spline. In fact we have to take two such spline functions of the variable $z$, namely $g_1(z)$ and $g_2(z)$. They will give the curve $(g_1(z), g_2(z), z)$ described by the parameter $z$. The interpolation condition means

$$g_1(Z^j) = X^j, \quad g_2(Z^j) = Y^j \qquad \text{for } j = 1, 2, ..., N.$$

Clearly, the simplest situation is when we are able to choose all centers $(X^j, Y^j, Z^j)$ coinciding, i.e.

$$(X^j, Y^j, Z^j) = (X^0, Y^0, Z^j) \qquad \text{for all } j = 1, 2, ..., N.$$

In such a case we have a trivial step in the algorithm since the "joining spline" is a constant, namely

$$g_1(z) = X^0, \quad g_2(z) = Y^0.$$

For example, for our particular data we may choose

$$X^0 = 0.33, \quad Y^0 = 0.34;$$

these coordinates are obviously located on Figure 3.

Then we will introduce new variables by using cylindric coordinates. At every level $z$ these new parameters have the form

$$\theta = \arctan \frac{y - Y^j}{x - X^j} \qquad \text{for } (x, y, Z^j) \in G_j \tag{3}$$

$$r_j = r_j(\theta) = r(Z^j, \theta) = \sqrt{(x - X^j)^2 + (y - Y^j)^2}.$$

In the above notations we have assumed implicitly that the part of the heart surface of interest $(LV)$ is described by the function $r(z, \theta)$ in the cylindric coordinates $(z, \theta, r)$. Thus the given data are in the form of $N$ periodic functions $r_j(\theta)$ for $j = 1, 2, ..., N$ and $0 \leq \theta \leq 2\pi$, corresponding to the curves $G_j$. Finally, we obtain the image of the curves $G_j$ (which we will denote by $\widetilde{G_j}$) in the new coordinates $(z, \theta, r)$, namely, for every $j$ the curve $G_j$ from the heart surface is mapped into the graph of the function $r_j$, i.e. into the curve

$$\widetilde{G_j} := \left\{ (Z^j, \theta, r_j(\theta)) : 0 \leq \theta \leq 2\pi \right\}.$$

See Figure 4 where we have the images of $\widetilde{G_j}$ on the cylinder projected on the plane $(z, \theta)$ – these are $N = 7$ parallel lines.

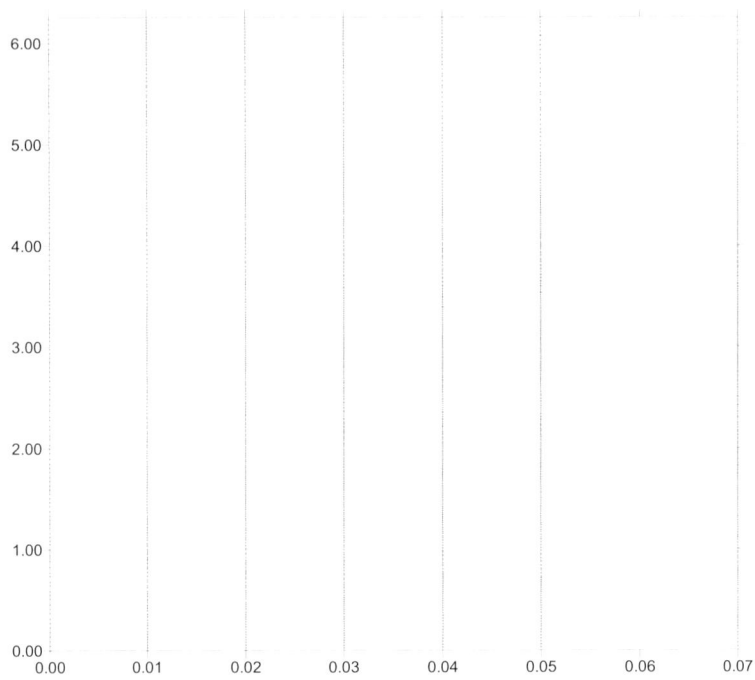

**Fig. 4.** The projections of the seven curves $\widetilde{G}_j$ are shown in the plane $(z, \theta)$.

With this we finish the introduction of the new coordinates.

The visualization in the new coordinates $(z, \theta, r)$ is relatively simple, and to that end we are able to apply directly the polysplines. The result will be a periodic polyspline, namely, a function $r(z, \theta)$ defined for

$$Z^1 \leq z \leq Z^N,$$
$$0 \leq \theta \leq 2\pi,$$

which is $2\pi$-periodic with respect to the variable $\theta$. Recall that in our particular data we have $N = 7$ with $Z^1 = 0$ and $Z^7 = 0.07$.

Now we need the **inverse map** in order to come back to the original variables. By using the original variables $(x, y, z)$ we obtain the following representation of the heart surface in the variables $(z, \theta, r)$,

$$Z^1 \leq z \leq Z^N, \qquad 0 \leq \theta \leq 2\pi$$
$$\theta = \arctan \frac{y - g_2(z)}{x - g_1(z)}$$
$$r(z, \theta) = \sqrt{(x - g_1(z))^2 + (y - g_2(z))^2}.$$

Solving with respect to $x$ and $y$ we obtain the surface

$$H = \begin{cases} x = g_1\left(z\right) + r\left(z, \theta\right)\cos\theta \\ y = g_2\left(z\right) + r\left(z, \theta\right)\sin\theta \\ Z^1 \leq z \leq Z^N, \qquad 0 \leq \theta \leq 2\pi. \end{cases} \qquad (4)$$

## 3   Polysplines on a Cylinder

Let us say some words about the polysplines on a cylinder.

As we saw above the part of the heart surface of interest $(LV)$ is mapped to a periodic function on the rectangle $[0, 0.07] \times [0, 2\pi]$ with coordinates $(z, \theta)$. The measured data lie on the lines $\gamma_j$ defined by $z = Z^j$, so we have

$$\gamma_j := \left\{ (z, \theta) : z = Z^j, \ 0 \leq \theta \leq 2\pi \right\}$$

or shortly

$$\gamma_j = \left\{ z = Z^j \right\}.$$

Recall that $\gamma_j$ is the projection of the curve $\widetilde{G}_j$. We have

$$\gamma_1 = \{z = 0\}, \quad \gamma_2 = \{z = 0.01\}, \quad \gamma_3 = \{z = 0.02\},$$
$$\gamma_4 = \{z = 0.03\}, \quad \gamma_5 = \{z = 0.04\}, \quad \gamma_6 = \{z = 0.05\}, \quad \gamma_7 = \{z = 0.07\}$$

To that configuration we apply "polysplines on a cylinder", as we already referred to [16, Sections 5.2 and 5.3, p. 60]. To give a short summary of the main result: the polyspline in our case will be a function $u\left(z, \theta\right)$ consisting of 6 pieces $u_j\left(z, \theta\right)$ where $u_j\left(z, \theta\right)$ is a function defined between the lines $\gamma_j$ and $\gamma_{j+1}$ for $j = 1, 2, ..., 6$; and $u_j$ satisfies there

– the **biharmonic** equation

$$\Delta^2 u_j\left(z, \theta\right) = \left( \frac{\partial^2}{\partial z^2} + \frac{\partial^2}{\partial \theta^2} \right)^2 u_j\left(z, \theta\right) = 0 \qquad \text{between } \gamma_j \text{ and } \gamma_{j+1};$$

– the **interpolation** to the experimental data

$$u_j\left(Z^{j+1}, \theta\right) = u_{j+1}\left(Z^{j+1}, \theta\right) = r_j\left(Z^{j+1}, \theta\right) \qquad \text{for all } 0 \leq \theta \leq 2\pi,$$

– and the **smoothness** condition

$$u \in C^2,$$

i.e.

$$u_j\left(Z^{j+1}, \theta\right) = u_{j+1}\left(Z^{j+1}, \theta\right) \qquad \text{for all } 0 \leq \theta \leq 2\pi,$$

$$\frac{\partial}{\partial z} u_j\left(Z^{j+1}, \theta\right) = \frac{\partial}{\partial z} u_{j+1}\left(Z^{j+1}, \theta\right) \qquad \text{for all } 0 \leq \theta \leq 2\pi,$$

$$\frac{\partial^2}{\partial z^2} u_j\left(Z^{j+1}, \theta\right) = \frac{\partial^2}{\partial z^2} u_{j+1}\left(Z^{j+1}, \theta\right) \qquad \text{for all } 0 \leq \theta \leq 2\pi,$$

The main result of [16] is the existence of interpolation polyspline: If we are given the scanned data $r\left(Z^j, \theta\right)$ then we may find the polyspline with the above properties. Let us remark that the functions $u_j$ are real analytic, i.e. the polysplines enjoy infinite smoothness away from the "interfaces" $\gamma_j$ and on $\gamma_j$ they are $C^2$.

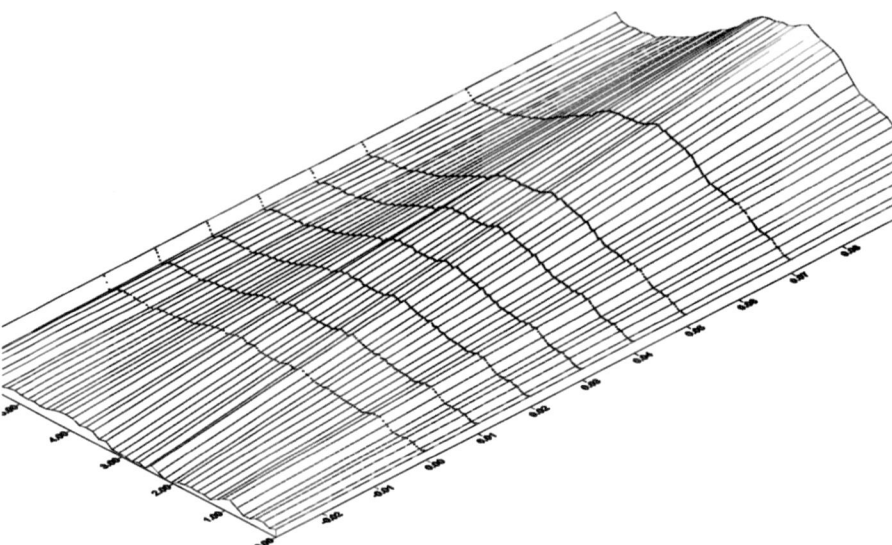

**Fig. 5.** This is the smoothing with Kriging in the $(z, \theta, r)$ coordinates. The seven curves $\widetilde{G}_j$ with the measured data are displayed on the surface.

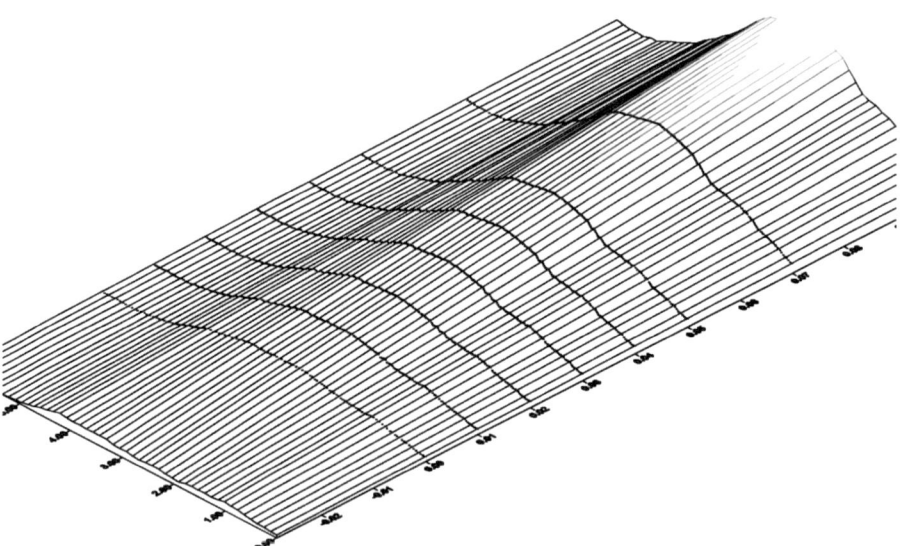

**Fig. 6.** This is the smoothing with the polyspline in the $(z, \theta, r)$ coordinates. The seven curves $\widetilde{G}_j$ with the measured data are displayed on the surface.

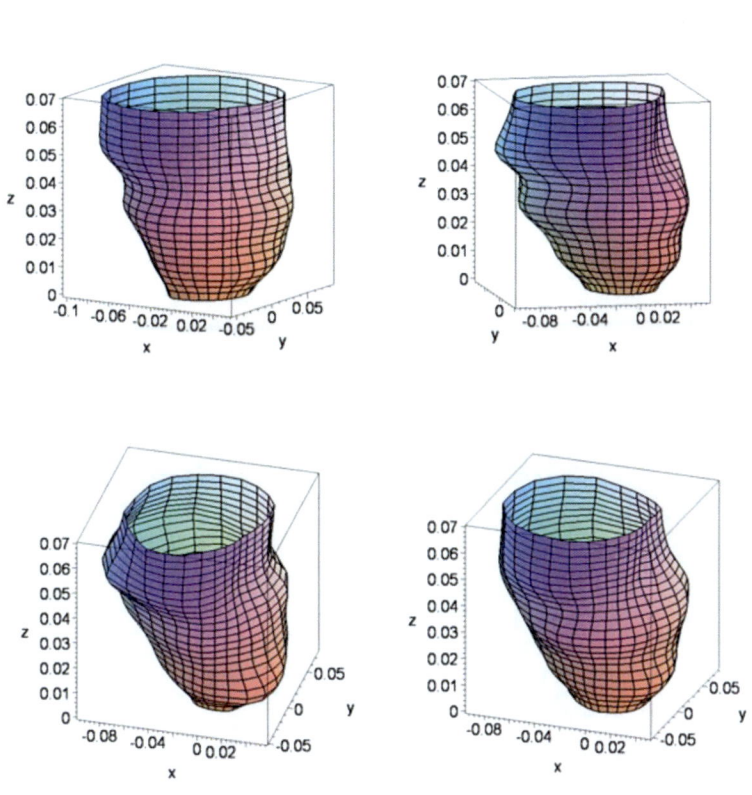

**Fig. 7.** This is the heart surface of *Phases 1,3,5,7* resulting from the application of the polysplines.

## 4   The Results

In the $(z, \theta, r)$ coordinates the result of smoothing of the given data $r\left(Z^j, \theta\right)$ with the Kriging method as implemented in [35] is shown on Figure 5. What we have on this Figure is in fact an approximation to *Phase 1* of the heart surface in the $(z, \theta, r)$ coordinates.

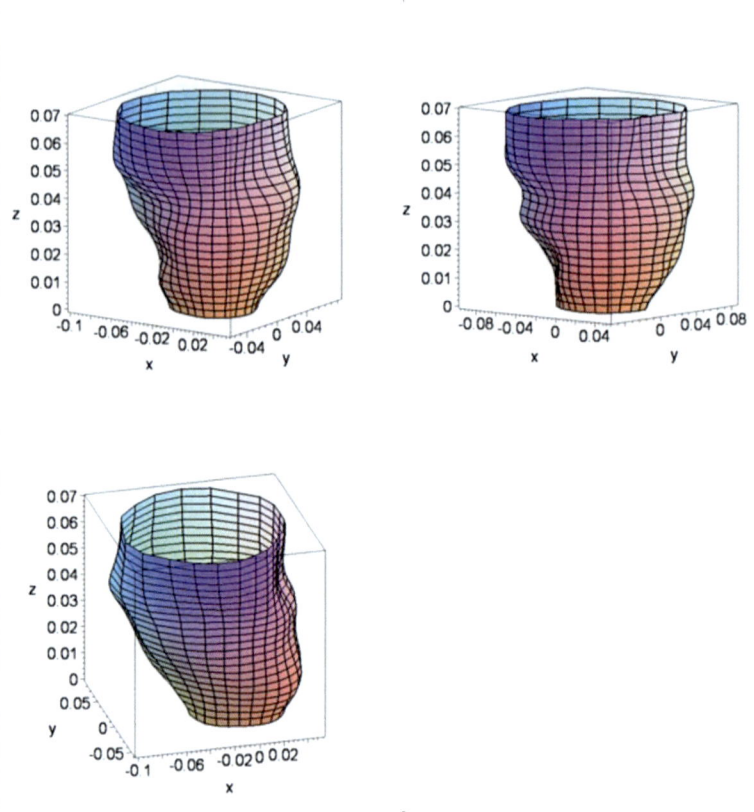

**Fig. 8.** This is the heart surface of *Phases 9,11,13* resulting from the application of the polysplines.

The result of smoothing the same data with interpolation polysplines is shown on Figure 6.

One has to compare the two Figures 5 and 6 and to see that the surface obtained by the polyspline method is visually smoother than the one obtained with the Kriging method.

Next we provide the visualization result using the original $(x, y, z)$ coordinates by means of formula (4). We provide the results in a series of phases of the heart dynamics viewed from different perspectives.

Let us note that we take the same "center of star–shaped-ness" $(X^0, Y^0)$ in all cases, namely

$$X^0 = 0.33, \qquad Y^0 = 0.34.$$

## 4.1   Visualization of the Left Ventrical at Different Phases

On Figure 7 *Phases 1,3,5,7* are visualized.
  On Figure 8 we have *Phases 9,11,13* visualized.

**Conclusion 1** *We have shown that by means of appropriate coordinate change the Left Ventricle may be conveniently represented in cylindrical coordinates where the axis variable is z. In this setting if scanning has been performed at different levels in z then we have seen that a ready-made tool called "polyspline on a cylinder" is available. This polyspline is a surface which interpolates the measured data and provides as seen from Figures 7 and 8 visually very satisfactory approximation to the heart surface. The above scheme is applicable to a lot more general setting for modelling surfaces which may be transformed by a diffeomorphism to become "functions on a cylinder".*

## Acknowledgement

The current research of the first author has been sponsored by the Nuffield foundation during his stay at the University of Leeds, and by the Alexander von Humboldt Foundation during his stay at the University of Duisburg. The authors would like to acknowledge the help of Chris Evans (Applied Mathematics, University of Leeds), A. Radjenovic , and J.P. Ridgeway of the Department of Medical Physics of the University of Leeds, in providing the data for these studies.

## References

1. Amini, A., Duncan, J., "Pointwise tracking of Left-Ventricular Motion in 3D", *Proc. IEEE Workshop on Visual Motion*, Princeton, NJ, 1991, pp. 294-298.
2. Arts, T., Hunter, W. C., Douglas, A., Muijtjens, M. M., Reneman, R.S., "Description of the deformation of the left ventricle by a kinematic model", *J. Biomechanics*, Vol. 25, No. 10, 1992, pp. 1119-1127.
3. N. Ayache, D. Chapelle, F. Clément, Y. Coudière, H. Delingette, J.A. Désidéri, M. Sermesant, M. Sorine, and J. Urquiza, "Towards Model-Based Estimation of the Cardiac Electro-Mechanical Activity from ECG Signals and Ultrasound Images", In T. Katila, I.E. Magnin, P. Clarysse, J. Montagnat, and Nenonen J., editors, Functional Imaging and Modeling of the Heart (FIMH'01), Helsinki, Finland, VOl 2230 Lecture Notes in Computer Science, pp. 120-127, 2001. Springer.
4. Beyar, R., Sideman, S., "Effect of the twisting motion on the non-uniformities of transmural fiber mechanics and energy demands—A theoretical Study", *IEEE Trans. Biomed. Eng.*, Vol, 32, 1985, pp. 764-769.
5. Bloor, M.I.G.; Wilson, M.J. "The PDE method in geometric and functional design". Goodman, Tim (ed.) et al., The mathematics of surfaces. VII. Proceedings of the 7th conference, Dundee, Great Britain, September 1996. Winchester: Information Geometers, Limited. 281-307 (1997).
6. Briggs, J.C. "Machine contouring using minimum curvature", Geophysics, 39 (1974), 39-48.

7. Cohen, L.D. and Cohen, I., "A Finite Element Method Applied to New Active Contour Models and 3d Reconstruction from Cross-sections", *Proc. 3rd Int. Conf. on Computer Vision*, Osaka, Japan, 1990, pp. 587-591.

8. Duchon, J. "Interpolation des fonctions de deux variables suivant le principe de la flexion des plaques minces", R.A.I.R.O. Analyse numerique, vol. 10, no. 12 (1976), pp. 5-12.

9. Dyn, N. "Interpolation and approximation by radial and related functions", In: Approximation Theory VI, C. K. Chui, L. L. Schumaker and J. D. Ward (eds.), Academic Press, New York, 1989, pp. 211-234.

10. Farin, G. E. "Curves and Surfaces for Computer Aided Geometric Design: a Practical Guide". 4th ed, Boston, MA: Academic Press, 1997.

11. Gonzalez-Casanova, P.; R. Alvarez, "Splines in geophysics", Geophysics, 50, No. 12 (1985), 2831-2848.

12. Haber E., Metaxas D.N. and Axel L., "Motion Analysis of the Right Ventricle from MRI Images", Lecture Notes in Computer Science, 1998, Vol. 1496, pp 177-188.

13. Huang, W.C. and Goldgof, D., "Adaptive-Size Meshes for Rigid and Nonrigid Shape Analysis and Synthesis", *IEEE Transactions on Pattern Analysis*, Vol. 15, 1993, pp. 611-616.

14. Kim W.Y et al., "Left-ventricular blood-flow patterns in normal subjects", Journal of the American College of Cardiology, 1995, Vol. 26, 1, pp 224-37.

15. Kim, H. C., Min, B. G., Lee, M. M., Seo, J. D., Lee, Y. W., Han, M. C., "Estimation of local cardiac wall deformation and regional stress from biplane coronary cineangiograms", *IEEE Trans. Biomed. Eng.*, Vol. 32, 1985, pp. 503-511.

16. Kounchev, O. "Multivariate Polysplines. Applications to Numerical and Wavelet Analysis", Academic Press, San Diego–London, 2001.

17. McInerney, T. and Terzopoulos, D., "A Finite Element Model for 3D Shape Reconstruction and Nonrigid Motion Tracking", *Proc. 4th Internation Conference on Computer Vision*, Berlin, Germany, 1993, pp. 518-523.

18. McInerney, T. and Terzopoulos, D. "A Dynamic Finite-Element Surface Model for segmentation and tracking in Multi-dimensional Medical Images with application to Cardiac 4D Image-Analysis", *Computerized Medical Imaging and Graphics*, Vol 19, 1995, pp. 69-83.

19. Meinguet, J. "Multivariate interpolation at arbitrary points made simple", Z. Angew. Math. Phys. **30** (1979), pp. 292–304.

20. Park, J., Metaxas, D., Young, A. A., Axel, L., "Deformable models with parameter functions for cardiac motion analysis from tagged MRI data", *IEEE Trans. Medical Imaging*, Vol. 15, 1996, pp. 278-289.

21. Park J., Metaxas D., Young A.A and Axel L, "Deformable Models with Parameter Functions for Cardiac Motion Analysis from Tagged MRI Data", IEEE Transactions on Medical Imaging, 1996, Vol. 15, 3.

22. Pedley T.J., "The Fluid Dynamics of Large Blood Vessels", Cambridge University Press, 1980.

23. Pentland, A., Horowitz, B., Sclaroff, S., "Recovery of Nonrigid Motion and Structure from Contour", *Proc. IEEE Workshop on Visual Motion*, Princeton, NJ, 1991, pp. 288-293.

24. Peskin C.S. and McQueen D.M., "Cardiac Fluid Dynamics", Critical Reviews in Biomedical Engineering, 1992, Vol. 20, 5-6, pp 451 et seq.

25. Powell, M. J. D. "The theory of radial basis function approximation in 1990". In: Advances in numerical analysis. Vol. 2: Wavelets, subdivision algorithms, and radial basis functions, Proc. 4th Summer Sch., Lancaster/UK 1990, pp. 105-210 (1992).

26. Roma A.M, Peskin C.S. and Berger M.J., "An Adaptive Version of the Immersed Boundary Method", Journal of Computational Physics, 1999, Vol. 147, 2, pp 509-534.

27. J. Saghri and J. Freeman, "Analysis of the Precision of Generalized Chain Codes for the Representation of Planar Curves", *IEEE Transactions on Pattern Analysis and Machine Intelligence*, 1981, PAMI-3, pp. 533-539

28. Schoephoerster R.T., Silva C.L. and Ray G., "Evaluation of Left-Ventricular Function based on Simulated Systolic Flow Dynamics Computed from Regional Wall-Motion", Journal of Biomechanics, 1994, Vol. 27, 2, pp 125-136.

29. M. Sermesant, Y. Coudière, H. Delingette, N. Ayache, J. Sainte-Marie, D. Chapelle, F. Clément, and M. Sorine, "Progress Towards Model-Based Estimation of the Cardiac Electromechanical Activity from ECG Signals and 4D Images", *In* Marc Thiriet, editor, Modelling and Simulation for Computer-aided Medicine and Surgery (MS4CMS'02), Vol 12 of ESAIM: PROC, pp. 153-162, 2002. European Series in Applied and Industrial Mathematics

30. M. Sermesant, Y. Coudière, H. Delingette, and N. Ayache, "Progress towards an Electro-Mechanical Model of the Heart for Cardiac Image Analysis", *In* IEEE International Symposium on Biomedical Imaging (ISBI'02), pp. 10-14, 2002.

31. M. Sermesant, C. Forest, X. Pennec, H. Delingette, and N. Ayache, "Biomechanical Model Construction from Different Modalities: Application to Cardiac Images", *In* Takeyoshi Dohi and Ron Kikinis, editors, Medical Image Computing and Computer-Assisted Intervention (MICCAI'02), Vol 2488 Lecture Notes in Computer Science, Tokyo, pp. 714-721, September 2002. Springer.

32. M. Sermesant, Y. Coudière, H. Delingette, N. Ayache, and J.A. Désidéri, "An Electro-Mechanical Model of the Heart for Cardiac Image Analysis". In W.J. Niessen and M.A. Viergever, editors, 4th Int. Conf. on Medical Image Computing and Computer-Assisted Intervention (MICCAI'01), Vol 2208 of Lecture Notes in Computer Science, Utrecht, The Netherlands, pp. 224-231, October 2001.

33. Smith J.J. and Kampine J.P., "Circulatory Physiology : the essentials", Baltimore : Williams and Wilkins, 1990, 3rd ed.

34. Smith, M. H. F.; P. Wessel, "Gridding with continuous curvature splines in tension", Geophysics, 55 (1990), No. 3, 293-305.

35. "SURFER – Golden Software", Golden, Colorado, Release of 1997.

36. Taylor D.E.M. and Wade J.D., "The pattern of flow around the atrioventricular valves during diastolic ventricular filling", Journal of Physiology, 1970, Vol. 207, pp 71-2.

37. Taylor T.W. and Yamaguchi T., "Flow Patterns in 3-dimensional Left-ventricular Systolic and Diastolic flows determined from Computational Fluid-Dynamics", Biorheology, 1995, Vol. 32, 1, pp 61-71.

38. Vierendeels J.A., Riemslagh K., Dick E. and Verdonck P.R. "Computer Simulation of Intraventricular Flow and Pressure Gradients During Diastole", Journal of Biomechanical Engineering-Transactions of the ASME, 2001, Vol. 122, 6, pp 667-674.

39. Yoganathan A.P., Lemmon J.D., Kim Y.H., Walker P.G., Levine R.A., Vesier C.C, "A Computational Study of a Thin-Walled 3-Dimensional Left Ventricle during Early Systole", Journal of Biomechanical Engineering-Transactions of the ASME, 1994, Vol. 116, 3, pp 307-314.

# Author Index

# Lecture Notes in Computer Science

For information about Vols. 1–2721
please contact your bookseller or Springer-Verlag